CLASS NOTES

FOR CAMPBELL'S
BIOLOGY
THIRD EDITION

Nina Caris, Ph.D.
Texas A&M University

Harold T. Underwood, Ph.D.
Texas A&M University

The Benjamin/Cummings Publishing Company, Inc.

Redwood City, California ■ Menlo Park, California
Reading, Massachusetts ■ New York ■ Don Mills, Ontario
Wokingham, U.K. ■ Amsterdam ■ Bonn ■ Sydney
Singapore ■ Tokyo ■ Madrid ■ San Juan

Sponsoring Editor: Edith Beard Brady
Developmental Editor: Valerie Kuletz
Production Supervisor: Larry Olsen
Cover Design: Yvo Riezebos

Cover: "Aspens, Northern New Mexico, 1958."
Photograph by Ansel Adams. Copyright © 1992
by the Trustees of the Ansel Adams Publishing Rights Trust.

Copyright © 1993 by The Benjamin/Cummings Publishing Company, Inc.

All rights reserved. No part of this publication may be reproduced, stored in a database or retrieval system, or transmitted in any form or by any means, electronic, mechanical, photocopying, recording, or otherwise, without the prior written permission of the publisher. Printed in the United States of America. Published simultaneously in Canada.

ISBN 0-8053-1887-9

2 3 4 5 6 7 8 9 10--KE--98 97 96 95 94

The Benjamin/Cummings Publishing Company, Inc.
390 Bridge Parkway
Redwood City, California 94065

TABLE OF CONTENTS

To the Student — v

Chapter 1   Introduction: Themes in the Study of Life   1

UNIT ONE   **The Chemistry of Life**

Chapter 2	Atoms, Molecules, and Chemical Bonds	12
Chapter 3	Water and the Fitness of the Environment	22
Chapter 4	Carbon and Molecular Diversity	35
Chapter 5	The Structure and Function of Macromolecules	35
Chapter 6	An Introduction to Metabolism	52

UNIT TWO   **The Cell**

Chapter 7	A Tour of the Cell	63
Chapter 8	Membrane Structure and Function	84
Chapter 9	Cellular Respiration: Harvesting Chemical Energy	98
Chapter 10	Photosynthesis	121
Chapter 11	The Reproduction of Cells	140

UNIT THREE   **The Gene**

Chapter 12	Meiosis and Sexual Life Cycles	153
Chapter 13	Mendel and the Gene Idea	164
Chapter 14	The Chromosomal Basis of Inheritance	185
Chapter 15	The Molecular Basis of Inheritance	200
Chapter 16	From Gene To Protein	212
Chapter 17	Microbial Models: The Genetics of Viruses and Bacteria	233
Chapter 18	Genome Organization and Expression in Eukaryotes	261
Chapter 19	DNA Technology	276

UNIT FOUR   **Mechanisms of Evolution**

Chapter 20	Descent with Modification: A Darwinian View of Life	298
Chapter 21	How Populations Evolve	314
Chapter 22	The Origin of the Species	335
Chapter 23	Tracing Phylogeny: Macroevolution, the Fossil Record, and Systematics	351

UNIT FIVE   **The Evolutionary History of Biological Diversity**

Chapter 24	Early Earth and the Origin of Life	371
Chapter 25	Prokaryotes and the Origins of Metabolic Diversity	376
Chapter 26	Protists and the Origin of Eukaryotes	388
Chapter 27	Plants and the Colonization of Land	400
Chapter 28	Fungi	415
Chapter 29	Invertebrates and the Origin of Animal Diversity	424
Chapter 30	The Vertebrate Genealogy	450

UNIT SIX   Plants: Form and Function

Chapter 31	Plant Structure and Growth	476
Chapter 32	Transport in Plants	489
Chapter 33	Plant Nutrition	500
Chapter 34	Plant Reproduction and Development	509
Chapter 35	Control Systems in Plants	523

UNIT SEVEN   Animals: Form and Function

Chapter 36	An Introduction to Animal Structure and Function	534
Chapter 37	Animal Nutrition	540
Chapter 38	Circulation and Gas Exchange	549
Chapter 39	The Body's Defenses	564
Chapter 40	Controlling the Internal Environment	583
Chapter 41	Chemical Signals in Animals	601
Chapter 42	Animal Reproduction	612
Chapter 43	Animal Development	626
Chapter 44	Nervous Systems	640
Chapter 45	Sensory and Motor Mechanisms	656

UNIT EIGHT   Ecology

Chapter 46	Ecology: Distribution and Adaptations of Organisms	668
Chapter 47	Population Ecology	688
Chapter 48	Community Ecology	700
Chapter 49	Ecosystems	713
Chapter 50	Behavior	727

TO THE STUDENT

Class Notes really began as lecture notes for instructors who teach introductory biology courses. Several faculty using the *Instructor's Guide* wanted to share these notes with their students, so we have transformed the original teacher-centered lecture notes into student-centered class notes.

The original lecture notes included course objectives for faculty to help focus lecture material on important concepts and to inform students of their expectations. Through our experience with thousands of introductory biology students, we have found another effective way to use these objectives. We asked our students to bring course objectives to class and, just before lecture, to change the objectives from declarative statements to questions. During the lecture, they were to listen for the answers. To our amazement, students who used this process not only raised their course grades but also found that it increased their interest level and attention span in class and ultimately enhanced learning. We urge you to try this simple technique and learn for yourself the powerful effect of questioning.

How To Use These Notes

Examine the Course Syllabus At the beginning of the semester, carefully examine the course syllabus to determine the scope and sequence of topics your instructor will cover. In this way you can anticipate what the teacher will discuss during each class period, and you can properly prepare for class by reading the textbook.

Read the Textbook We want to emphasize how important it is to read the material before going to lecture. Also, these class notes are intended just to be a learning aid to use during class. They are not a substitute for reading the text, using the *Student Study Guide*, and attending class. The class notes should be used as a link between the text and lecture.

Take the Notes to Class For these notes to be most helpful, take them to class to maximize your learning time in the classroom.

1. *Read the study questions just before lecture.* About five to ten minutes before class, read through the study questions in anticipation of the answers to be revealed during lecture.

2. *Actively participate in class.* You will learn much more as an active participant in class than as a passive observer. Listen carefully and take notes directly in the margins and at the bottom of the page. You will be more efficient at learning and remembering if you summarize key points *in your own words* rather than copying your instructor's words.

    Listen for the answers to the study questions. If a question goes unanswered, ask the instructor or go back to the textbook to find the answer. Page references to Campbell's *BIOLOGY* have been included in the *Class Notes* by major heading to help you quickly find the material. Note that some questions will be easier to answer than others. Some questions require you to synthesize information learned earlier in the lecture or course.

3. *Ask your own questions.* There are so many benefits to formulating and asking questions during or after class. The first benefit is that you help your own learning by enhancing concentration and attention during class. By asking questions in class, you also are making a positive contribution to the classroom community. It helps other students who have the same question but are too timid to ask; and, though you may not realize it, it also helps the teacher. Your questions allow the instructor to check students' understanding and to respond better to student needs. Don't be shy. Most teachers want your feedback—in fact, they need it to be effective. One caveat here is to respect time limitations. If the instructor wants you to hold questions, ask your questions after class; it is a good opportunity to interact directly with your teacher. Also, ask only relevant and timely questions so as not to break the teacher's rhythm or train of thought. Be constructive, supportive, and respectful.

4. *Review questions and answers shortly after class.* Finally, to enhance retention and understanding, review your notes and answers to questions as soon after class as possible. This will allow you to make any corrections or further investigations while the material is still fresh in your mind.

Closing the Circle

Class Notes began as lecture notes for faculty teaching introductory biology. As all experienced teachers will tell you, the way to learn material in a meaningful way is to teach it. The intellectual process of reorganizing and restructuring material for presentation challenges and clarifies thinking. Consider forming study groups. Take turns playing the role of instructor and use these notes as lecture notes. Perhaps you will eventually close the circle and become a teacher yourself.

Acknowledgments

The authors wish to acknowledge the faculty colleagues who suggested a student version of *Class Notes* and the many introductory biology students who have inspired us to write it. We are especially grateful to Kenneth Brobst, the compositor for this project. His judgment and expertise were integral to every aspect of production. Kenneth decided on the format for this manuscript, typed the text, made suggestions for clarity, and designed the tables and graphics. The final version of *Class Notes* is a reflection of his professionalism, good-natured support, and willingness to suffer through the anxiety of meeting deadlines.

It is a pleasure to acknowledge the staff from The Benjamin/Cummings Publishing Company, especially Korinna Sodic, Assistant Editor, and Valerie Kuletz, Developmental Editor. Their amiable support and professionalism are greatly appreciated. We thank them for their patience and help through all phases of this project. We also thank Don O'Neal, Acquisitions Editor, who successfully brought us to the finish line. We offer a special note of gratitude to Edith Beard Brady, Sponsoring Editor for Campbell's *BIOLOGY*. Completion of this project is a direct reflection of her unwavering support and vision. Finally, we thank Neil Campbell for writing such a wonderful book.

1   INTRODUCTION: THEMES IN THE STUDY OF LIFE

CHAPTER OUTLINE

Biology, the study of life, is a human endeavor resulting from an innate attraction to life in its diverse forms (E.O. Wilson's *biophilia*).

The science of biology is enormous in scope.
- ☞ It reaches across size scales from submicroscopic molecules to the global distribution of biological communities.
- ☞ It encompasses life over huge spans of time from contemporary organisms to ancestral life forms stretching back to nearly four billion years.

As a science, biology is an ongoing process.
- ☞ As a result of new research methods developed over the past few decades, there has been an information explosion.
- ☞ Technological advances yield new information that may change the conceptual framework accepted by the majority of biologists.

What are some of the unifying themes that pervade the science of biology?

With rapid information flow and new discoveries, biology is in a continuous state of flux. There are, however, enduring unifying themes that pervade the science of biology:
- ☞ A hierarchy of organization.
- ☞ Emergent properties.
- ☞ The cellular basis of life.
- ☞ Heritable information.
- ☞ A feeling for organisms.
- ☞ The correlation between structure and function.
- ☞ The interaction of organisms with their environment.
- ☞ Unity in diversity.
- ☞ Evolution: the core theme.
- ☞ Scientific process: the hypothetico-deductive method.

I.   A HIERARCHY OF ORGANIZATION (p. 3–4)

How is the hierarchy of structural levels in biology organized?

A characteristic of life is a high degree of order. Biological organization is based on a hierarchy of structural levels, with each level building on the levels below it.

: Themes in the Study of Life

```
                        Atoms
                          ↓
              Complex biological molecules.
                          ↓
                 Subcellular Organelles
                          ↓  are ordered into
                        Cells
                          ↓  in multicellular organisms similar cells are
                             organized into
                       Tissues
                          ↓
                       Organs
                          ↓
                   Organ systems
                          ↓
                  Complex organism
```

There are levels of organization beyond the individual organism:

Population = Localized group of organisms belonging to the same species.

Community = Populations of species living in the same area.

Ecosystem = An energy-processing system of community interactions that include abiotic environmental factors such as soil and water.

Biomes = Large scale communities classified by predominant vegetation type and distinctive combination of plants and animals.

Biosphere = The sum of all the planet's ecosystems.

II. EMERGENT PROPERTIES (p. 4-7)

How do the properties of life emerge from complex organization?

Emergent property = Property that emerges as a result of interactions between components.

Introduction: Themes in the Study of Life **3**

- With each step upward in the biological hierarchy, new properties emerge that were not present at the simpler organizational levels.
- Life is difficult to define because it is associated with numerous emergent properties that reflect a hierarchy of structural organization.

What are some emergent properties and processes associated with life?

Some of the emergent properties and processes associated with life are:

1. *Order*. Organisms are highly ordered, and other characteristics of life emerge from this complex organization.
2. *Reproduction*. Organisms reproduce; life comes only from life (*biogenesis*). — *Biogenesis*
3. *Growth and Development*. Heritable programs stored in DNA direct the species-specific pattern of growth and development.
4. *Energy Utilization*. Organisms take in and transform energy to do work, including the maintenance of their ordered state.
5. *Response to Environment*. Organisms respond to stimuli from their environment.
6. *Homeostasis*. Organisms regulate their internal environment to maintain a steady-state, even in the face of a fluctuating external environment.
7. *Evolutionary Adaptation*. Life evolves in response to interactions between organisms and their environment.

What is the difference between holism and reductionism? How does the study of biology involve both strategies?

Because properties of life emerge from complex organization, it is impossible to fully explain a higher level of order by breaking it into its parts.

<u>Holism</u> = The principle that a higher level of order cannot be meaningfully explained by examining component parts in isolation. — *Holism*
- An organism is a living whole greater than the sum of its parts.
- For example, a cell dismantled to its chemical ingredients is no longer a cell.

It is also difficult to analyze a complex process without taking it apart.

<u>Reductionism</u> = The principle that a complex system can be understood by studying its component parts. — *Reductionism*
- Has been a powerful strategy in biology.
- For example, Watson and Crick deduced the role of DNA in inheritance by studying its molecular structure.

The study of biology balances the reductionist strategy with the goal of understanding how the parts of cells, organisms and populations are functionally integrated.

III. THE CELLULAR BASIS OF LIFE (*p. 7–8*)

The cell is an organism's basic unit of structure and function.
- Lowest level of structure capable of performing all activities of life.

4 Introduction: Themes in the Study of Life

☞ All organisms are composed of cells.
☞ May exist singly as unicellular organisms or as subunits of multicellular organisms.

How have technological breakthroughs contributed to the formulation of the cell theory and our current knowledge about the cell?

The invention of the microscope led to the discovery of the cell and the formulation of the cell theory.
☞ Robert Hooke (1665) reported a description of his microscopic examination of cork. Hooke described tiny boxes which he called "cells" (really cell walls). The significance of this discovery was not recognized until 150 years later.
☞ Antonie van Leeuwenhok (1600's) used the microscope to observe living organisms such as microorganisms in pond water, blood cells and animal sperm cells.
☞ Matthias Schleiden and Theodor Schwann (1839) reasoned from their own microscopic studies and those of others, that all living things are made of cells. This formed the basis for the *cell theory*.
☞ The cell theory has since been modified to include the idea that all cells come from preexisting cells.

Over the past 40 years, use of the electron microscope has revealed the complex ultrastructure of cells:
☞ Cells are bounded by *plasma membranes* that regulate passage of materials between the cell and its surroundings.
☞ All cells, at some stage, contain DNA.

Based on structural organization, there are two major kinds of cells: *prokaryotic* and *eukaryotic*.

How do prokaryotic and eukaryotic cells differ? How are they similar?

<u>Prokaryotic cell</u> = Cell lacking membrane-bound organelles and a membrane-enclosed nucleus.
☞ Found only in the Kingdom Monera, the bacteria and blue-green algae.
☞ Generally much smaller than eukaryotic cells.
☞ Contains DNA that is <u>not</u> separated from the rest of the cell, as there is no membrane-bound nucleus.
☞ Lacks membrane-bound organelles.
☞ Almost all have tough external walls.

<u>Eukaryotic cell</u> = Cell with a membrane-enclosed nucleus and membrane-enclosed organelles.
☞ Found in protists, plants, fungi and animals.
☞ Subdivided by internal membranes into different functional compartments called *organelles*.
☞ Contains DNA that is segregated from the rest of the cell. DNA is organized with proteins into *chromosomes* that are located within the *nucleus*, the largest organelle of most cells.

<u>Cell theory</u>

<u>Prokaryotic cell</u>

<u>Eukaryotic cell</u>

- ☞ *Cytoplasm* surrounds the nucleus and contains various *organelles* of different functions.
- ☞ Some cells have a tough *cell wall* outside the plasma membrane (e.g. plant cells). Animal cells lack cell walls.

Though structurally different, eukaryotic and prokaryotic cells have many similarities, especially in their chemical processes.

IV. HERITABLE INFORMATION (*p. 8*)

Biological instructions for an organism's complex structure and function are encoded in DNA.
- ☞ Each DNA molecule is made of four types of chemical building blocks called *nucleotides*.
- ☞ The linear sequence of these four *nucleotides* encode the precise information in a *gene*, the unit of inheritance from parent to offspring.
- ☞ An organism's complex structural organization is specified by an enormous amount of coded information.

Inheritance itself is based on:
- ☞ A complex mechanism for copying DNA.
- ☞ Passing the information encoded in DNA from parent to offspring.

All forms of life use essentially the same genetic code.
- ☞ A particular nucleotide sequence provides the same information to one organism as it does to another.
- ☞ Differences among organisms reflect differences in nucleotide sequence.

V. A FEELING FOR ORGANISMS (*p. 8-9*)

We may study life at structural levels below or above the organism, but it is organisms that first attract us. A feeling for the organisms we seek to understand enriches our study of life at all levels.

Barbara McClintock is a good example of a biologist whose feeling for the organism contributed to an important scientific discovery.
- ☞ McClintock investigated inheritance of kernel color in Indian corn and uncovered a type of genetic change not yet recognized by the scientific community in the 1940's and 50's.
- ☞ In McClintock's biography, *A Feeling for the Organism* by Evelyn Keller, the author highlights that McClintock's success was due to her familiarity with corn. McClintock knew the individual corn plants in her experiments and enjoyed these attachments.
- ☞ Because she was familiar with the corn plants in her experiments, McClintock was able to recognize unexpected results from corn breeding experiments and deduce the existence of transposable genetic elements.

6 Introduction: Themes in the Study of Life

☞ The significance of McClintock's discovery was not recognized until researchers in the 1970's found a similar phenomenon in bacteria. She later received due recognition with the Nobel Prize in 1983.

VI. THE CORRELATION BETWEEN STRUCTURE AND FUNCTION (p. 9–10)

What is meant by the expression, "form fits function"?

There is a relationship between an organism's structure and how it works. Form fits function.
- ☞ Biological structure gives clues about what it does and how it works.
- ☞ Knowing a structure's function gives insights about its construction.
- ☞ This correlation is apparent at many levels of biological organization. (See Campbell, Figure 1.6)

VII. THE INTERACTION OF ORGANISMS WITH THEIR ENVIRONMENT (p. 10)

Organisms interact with their environment, which includes other organisms as well as abiotic factors.
- ☞ Both organism and environment are affected by the interaction between them.
- ☞ Ecosystem dynamics include two major processes:
   1. Nutrient cycling.
   2. Energy flow.

VIII. UNITY IN DIVERSITY (p. 10–11)

Biological diversity is enormous.
- ☞ Estimates of total diversity range from 5 million to over 30 million species.
- ☞ About 1.5 million species have been identified and named, including approximately 260,000 plants, 50,000 vertebrates and 750,000 insects.

To make this diversity more comprehensible, biologists classify species into categories.

<u>Taxonomy</u>

<u>Taxonomy</u> = Branch of biology concerned with naming and classifying organisms.
- ☞ Taxonomic groups are ranked into a hierarchy from the most to least inclusive category: *kingdom, phylum, class, order, family, genus, species.*

What are the kingdoms of life in the five-kingdom system? How do you distinguish among them?

- ☞ The kingdoms of life recognized in the five-kingdom system are: Monera, Protista, Plantae, Fungi and Animalia. (See Campbell, Figure 1.9)

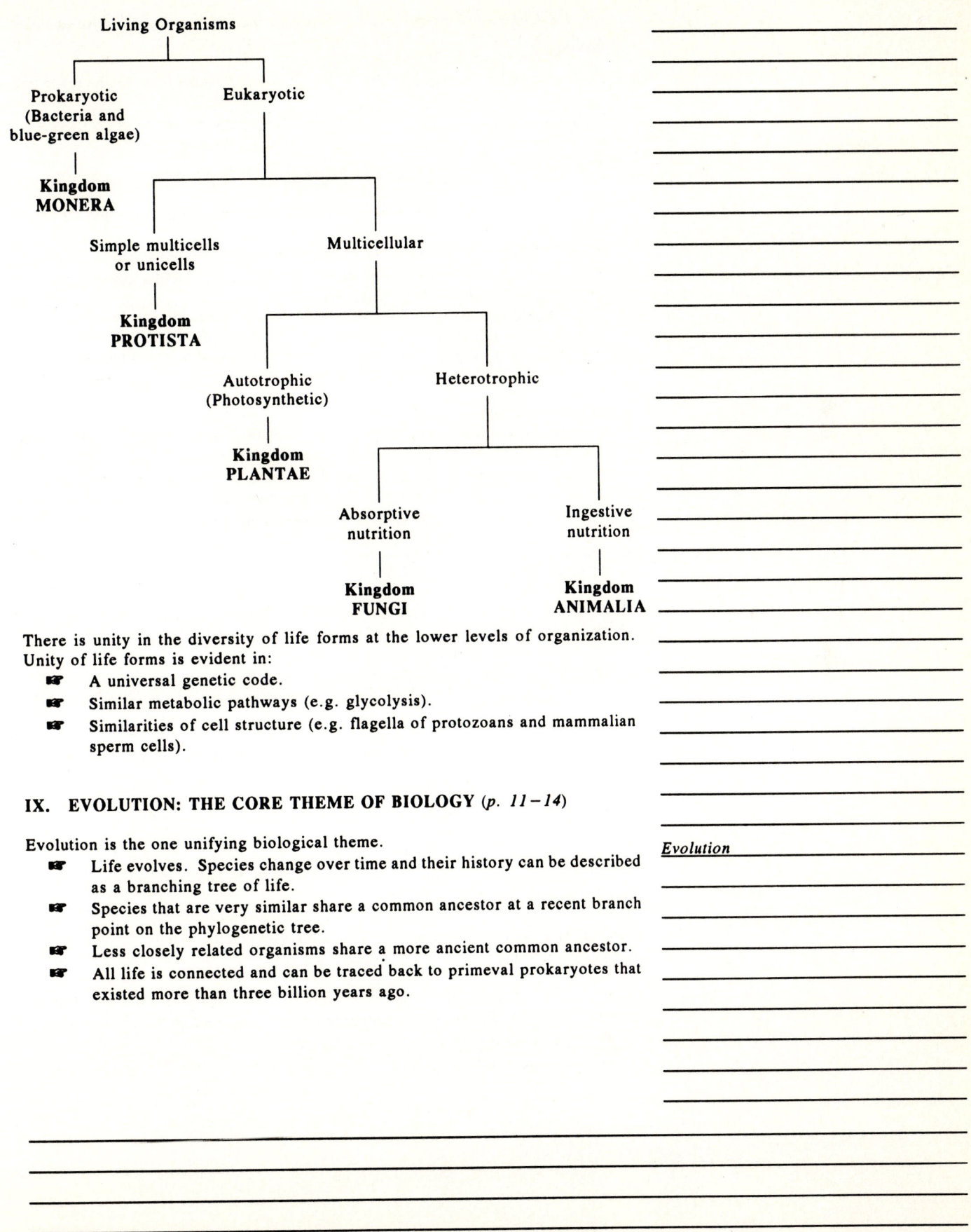

There is unity in the diversity of life forms at the lower levels of organization. Unity of life forms is evident in:
- A universal genetic code.
- Similar metabolic pathways (e.g. glycolysis).
- Similarities of cell structure (e.g. flagella of protozoans and mammalian sperm cells).

IX. EVOLUTION: THE CORE THEME OF BIOLOGY (p. 11–14)

Evolution is the one unifying biological theme.
- Life evolves. Species change over time and their history can be described as a branching tree of life.
- Species that are very similar share a common ancestor at a recent branch point on the phylogenetic tree.
- Less closely related organisms share a more ancient common ancestor.
- All life is connected and can be traced back to primeval prokaryotes that existed more than three billion years ago.

Evolution

How did Charles Darwin's ideas contribute to the conceptual framework of biology?

In 1859, Charles Darwin published *On the Origin of Species* in which he made two major points:

1. Species change, and contemporary species arose from a succession of ancestors through a process of "descent with modification."
2. A mechanism of evolutionary change is *natural selection*.

Darwin synthesized the concept of natural selection based upon the following observations:

- Individuals in a population of any species vary in many inheritable traits.
- Populations have the potential to produce more offspring than will survive or than the environment can support.
- Individuals with traits best suited to the environment leave a larger number of offspring, which increases the proportion of inheritable variations in the next generation. This differential reproductive success is what Darwin called *natural selection*.

Organisms' adaptations to their environments are the products of natural selection.

- Natural selection does <u>not create</u> adaptations; it merely increases the frequency of inherited variants that arise by chance.
- Adaptations are the result of the editing process of natural selection. When exposed to specific environmental pressures, certain inheritable variations favor the reproductive success of some individuals over others.

Darwin proposed that cumulative changes in a population over long time spans could produce a new species from an ancestral one.

Descent with modification accounts for both the unity and diversity of life:

- Similarities between two species may be a reflection of their descent from a common ancestor.
- Differences between species may be the result of natural selection modifying the ancestral equipment in different environmental contexts.

X. SCIENTIFIC PROCESS: THE HYPOTHETICO-DEDUCTIVE METHOD
(p. 15–19)

As the science of life, biology has the characteristics associated with science in general.

What is science? What method is the key ingredient of the scientific process?

Science is a way of knowing. It is a human endeavor that emerges from our curiosity about ourselves, the world and the universe. Good scientists are people who:

- Ask questions about nature and believe those questions are answerable.
- Are curious, observant and passionate in their quest for discovery.
- Are creative, imaginative and intuitive.
- Are generally skeptics.

Scientific method = Process which outlines a series of steps used to answer questions.
- Is not a rigid procedure.
- Based on the conviction that natural phenomena have natural causes.
- Requires evidence to logically solve problems.

The key ingredient of the scientific process is the *hypothetico-deductive method*, which is an approach to problem-solving that involves:
1. asking a question and formulating a tentative answer or *hypothesis* by *inductive reasoning*.
2. using *deductive reasoning* to make predictions from the hypothesis and then testing the validity of those predictions.

Hypothesis = Educated guess proposed as a tentative answer to a specific question or problem.

What is the difference between inductive and deductive reasoning?

Inductive reasoning = Making an inference from a set of specific observations to reach a general conclusion.

Deductive reasoning = Making an inference from general premises to specific consequences, which logically follow if the premises are true.
- Usually takes the form of *If...then* logic.
- In science, deductive reasoning usually involves predicting experimental results that are expected <u>if</u> the hypothesis is true.

Useful hypotheses have the following characteristics:
- *Hypotheses are possible causes.* Generalizations formed by induction are not necessarily hypotheses. Hypotheses should also be tentative explanations for observations or solutions to problems.
- *Hypotheses reflect past experience with similar questions.* Hypotheses are not just blind propositions, but are <u>educated</u> guesses based upon available evidence.
- *Multiple hypotheses should be proposed whenever possible.* The disadvantage of operating under only one hypothesis is that it might restrict the search for evidence in support of this hypothesis; scientists might bias their search, as well as neglect to consider other possible solutions.
- *Hypotheses must be testable via the hypothetico-deductive method.* Predictions made from hypotheses must be testable by making observations or performing experiments. This limits the scope of questions that science can answer.
- *Hypotheses can be eliminated, but not confirmed with absolute certainty.* If repeated experiments consistently disprove the predictions, then we can assume that the hypothesis is false. However, if repeated experimentation supports the deductions, we can only assume that the hypothesis <u>may</u> be true; accurate predictions can be made from false hypotheses. The more deductions that are tested and supported, the more confident we can be that the hypothesis is true.

Another feature of the scientific process is the *controlled experiment* which includes *control* and *experimental groups*.

Control group

Variable

Experimental group

Scientific theories

Control group = In a controlled experiment, the group in which all *variables* are held constant.
- ☞ Controls are a necessary basis for comparison with the experimental group, which has been exposed to a single treatment variable.
- ☞ Allows conclusions to be made about the effect of experimental manipulation.
- ☞ Setting up the best controls is a key element of good experimental design.

Variable = Condition of an experiment that is subject to change and that may influence an experiment's outcome.

Experimental group = In a controlled experiment, the group in which one factor or treatment is varied.

```
              ┌─→ Hypothesis ←─┐
              │       ↓        │
              │  Test = experiment
              │   or observations
              │    ↓       ↓
          Support        Reject
         hypothesis     hypothesis
         (may be true)      ↓
    Reject ↑                Modify or
           │                abandon hypothesis
           └── Test repeatedly
                    ↓
                 Accept
                    ↓
                 Verify
                    ↓
            Hypothesis becomes
                 theory
```

Science is an ongoing process that is a self-correcting way of knowing. Scientists:
- ☞ Build on prior scientific knowledge.
- ☞ Try to replicate the observations and experiments of others to check on their conclusions.
- ☞ Share information through publications, seminars, meetings and personal communication.

What really advances science is not just an accumulation of facts, but a new concept that collectively explains observations that previously seemed to be unrelated.
- ☞ Newton, Darwin and Einstein stand out in the history of science because they synthesized ideas with great explanatory power.
- ☞ *Scientific theories* are comprehensive conceptual frameworks which are well supported by evidence and are widely accepted by the scientific community.

Biology is a multi-disciplinary science that requires knowledge of chemistry, physics and mathematics.

XI. SCIENCE, TECHNOLOGY, AND SOCIETY (p. 19)

How are science and technology interdependent?

Science and technology are interdependent in the following ways.
- Technology extends our ability to observe and measure, which enables scientists to work on new questions that were previously unapproachable.
- Science, in turn, generates new information that makes technological inventions possible.
- For example, Watson and Crick's scientific discovery of DNA structure led to further investigation that enhanced our understanding of DNA, the genetic code, and how to transplant foreign genes into microorganisms. The biotechnology industry has capitalized on this knowledge to produce valuable pharmaceutical products such as human insulin.

We have a love-hate relationship with technology.
- Technology has improved our standard of living.
- The consequence of using technology also includes the creation of new problems such as increased population growth, acid rain, deforestation, global warming, nuclear accidents, ozone holes, toxic wastes and endangered species.
- Solutions to these problems have as much to do with politics, economics, culture and values as with science and technology.

A better understanding of nature must remain the goal of science. Scientists should:
- Try to influence how technology is used.
- Help educate the public about the benefits and hazards of specific technologies.

2

ATOMS, MOLECULES, AND CHEMICAL BONDS

CHAPTER OUTLINE

I. MATTER: ELEMENTS AND COMPOUNDS (p. 24–26)

Chemistry, the study of *matter*, is fundamental to an understanding of life.

Matter = Anything that takes up space and has mass.

Mass = A measure of the amount of matter an object contains.

A. Elements Essential to Life (p. 26)

Element = A substance that cannot be broken down into other substances by chemical reactions.
- All matter made of elements.
- 92 naturally occurring.
- Designated by a symbol of one or two letters.

What elements are essential to life? Which four of these elements make up most of living matter?

About 25 of the 92 naturally occurring elements are essential to life. Biologically important elements include: (See Campbell, Table 2.1)
- C (carbon), O (oxygen), H (hydrogen), and N (nitrogen) make up 96% of living matter.
- Ca (calcium), P (phosphorus), K (potassium), S (sulfur), Na (sodium), Cl (chlorine), Mg (magnesium), and *trace elements* make up the remaining 4% of an organism's weight.

Trace element = Element required by an organism in extremely minute quantities.
- Though in small quantity, are indispensable for life.
- For example: B, Cr, Co, Cu, F, I, Fe, Mn, Mo, Se, Si, Sn, V and Zn.

B. Compounds (p. 26)

Compound = A molecule composed of two or more elements combined in a fixed ratio.
- For example: NaCl (sodium chloride).
- Has unique emergent properties beyond those of its combined elements. (Na and Cl have very different properties from NaCl.)

Atoms, Molecules, and Chemical Bonds 13

II. THE STRUCTURE AND BEHAVIOR OF ATOMS (p. 27–32)

Atom = Smallest possible unit that retains the physical and chemical properties of its element.
- ☞ Atoms of the same element share similar chemical properties.
- ☞ Made up of *subatomic particles*.

What are the three most stable subatomic particles? How do they contribute to the structure of an atom?

A. Subatomic Particles (p. 27)

The three <u>most</u> stable subatomic particles are:
1. *Neutrons* [no charge (neutral)].
2. *Protons* [+1 electrostatic charge].
3. *Electrons* [−1 electrostatic charge].

NEUTRON	PROTON	ELECTRON
no charge	+1 charge	−1 charge
found together in a dense core called the *nucleus* (positively charged because of protons)		orbit around nucleus (held by electrostatic attraction to positively charged nucleus)
1.009 dalton	1.007 dalton	1/2000 dalton
masses of both are about the same (about 1 dalton)		mass is so small, usually not used to calculate atomic mass

NOTE: The *dalton* is a unit used to express mass at the atomic level. One dalton (d) is equal to 1.67×10^{-24} g.

If an atom is electrically neutral, the number of protons equals the number of electrons, which yields an electrostatically balanced charge.

B. Atomic Number and Atomic Weight (p. 28)

What information about an atom can be derived from the atomic number and mass number?

Atomic number = Number of protons in an atom of a particular element.
- ☞ All atoms of an element have the same atomic number.
- ☞ Written as a subscript to the left of the element's symbol (e.g. $_{11}Na$).
- ☞ In a neutral atom, # protons = # electrons.

Mass number = Number of protons and neutrons in an atom.
- ☞ Written as a superscript to left of an element's symbol (e.g. ^{23}Na).
- ☞ Is approximate mass of the whole atom, since the mass of protons and neutrons is close to 1 dalton.
- ☞ Number of neutrons can vary in an element, but number of protons is constant.

14 *Atoms, Molecules, and Chemical Bonds*

☞ Is not the same as an element's *atomic weight*, which is the weighted mean of the masses of an element's constituent isotopes.

Given the atomic number and mass number of an atom, how can you determine the number of neutrons?

☞ Can deduce the number of neutrons by subtracting *atomic number* from *mass number*.

Examples:

(Mass #) $^{23}_{11}$Na # of electrons ____ $^{12}_{6}$C # of electrons ____
(Atomic #)   # of protons ____   # of protons ____
                    # of neutrons ____   # of neutrons ____

C. Isotopes (p. 28–30)

Isotopes = Atoms of an element that have the same atomic number but different mass number.
☞ Have the same number of protons, but different number of neutrons.
☞ Different isotopes of the same element react chemically in same way.
☞ Some isotopes are radioactive.

Radioactive isotope = Unstable isotope in which the nucleus spontaneously decays emitting sub-atomic particles and/or energy as radioactivity.
☞ Loss of nuclear particles may transform one element to another (e.g. $^{14}_{6}$C → $^{14}_{7}$N).
☞ Has a fixed *half life*.

Half life = Time for 50% of radioactive atoms in a sample to decay.

Why are radioisotopes important to biologists?

Useful applications of radioactive isotopes in biology:

1. Dating of geological strata and fossils (See Campbell, Methods Box, Chapter 23)
   ☞ Radioactive decay is at a fixed rate.
   ☞ By comparing the ratio of radioactive and stable isotopes in a fossil with the ratio of isotopes in living organisms, one can estimate the age of a fossil.
   ☞ The ratio of ^{14}C to ^{12}C is frequently used to date fossils less than 50,000 years old.

2. Radioactive tracers
   ☞ Chemicals labelled with radioactive isotopes are used to trace the steps of a biochemical reaction or to determine the location of a particular substance within an organism.
   ☞ Isotopes of P, N and H were used to determine DNA structure.
   ☞ Used to diagnose disease (e.g. PET scanner).

3. Treatment of cancer
   ☞ e.g. radioactive cobalt.

D. Energy Levels (p. 30-31)

<u>Electrons</u> = Light negatively charged particles that orbit around nucleus.
- ☞ Equal in mass and charge.
- ☞ Are directly involved in the chemical behavior of an atom by interacting with electrons of other atoms.
- ☞ Have *potential energy* because of their position relative to the positively charged nucleus.

<u>Energy</u> = Ability to do work.

<u>Potential energy</u> = Energy that matter stores because of its position or location.
- ☞ There is a natural tendency for matter to move to the lowest state of potential energy.
- ☞ Potential energy of electrons only exists in discrete steps called *energy levels* or *electron shells*.
- ☞ Electrons with lowest potential energy are in the electron shell closest to the nucleus.
- ☞ Electrons with greater energy are in electron shells further from nucleus.

<u>Energy Levels of Electrons</u>:
Electrons may move from one shell to another. In the process, they gain or lose energy equal to the difference in potential energy between the old and new electron shell.

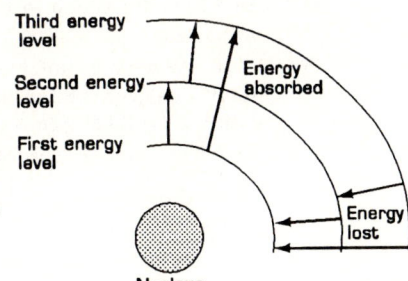

E. Electron Orbitals (p. 31)

<u>Orbital</u> = Three-dimensional space where an electron will most likely be found 90% of the time.
- ☞ Viewed as a probability cloud — a statistical concept.
- ☞ No more than two electrons can occupy same orbital.

<u>Electron Orbitals</u>: (See Campbell, Figure 2.9)

First electron shell:
- ☞ Has one spherical *s* orbital (1*s* orbital).
- ☞ Holds a maximum of two electrons.

Second electron shell:
- ☞ Holds a maximum of 8 electrons.
- ☞ One spherical *s* orbital (2*s* orbital).
- ☞ 3 dumbbell-shaped *p* orbitals each oriented at right angles to the other two ($2p_x$, $2p_y$, $2p_z$ orbitals).

Higher electron shells:
- ☞ May hold additional electrons in orbitals with more complex shapes.

16 Atoms, Molecules, and Chemical Bonds

F. Electron Configuration and Chemical Properties (p. 31–32)

How does electron configuration influence the chemical behavior of an atom?

Chemical behavior of an atom is determined by the *electron configuration* of the outermost electron shell (*valence shell*).

<u>Electron configuration</u> = Distribution of electrons in an atom's electron shells.

<u>Electron Configuration of the First 18 Elements</u>: (See Campbell, Figure 2.10) In this abbreviated periodic table, the first 18 elements are arranged sequentially by atomic number into 3 rows (periods). In reference to these <u>representative</u> elements, note the following:
- ☞ Outermost shell of these atoms never have more than 4 orbitals (one *s* and three *p*) or 8 electrons.
- ☞ Electrons must first occupy lower electron shells before the higher shells can be occupied. (A reflection of the natural tendency for matter to move to the lowest possible state of potential energy — the most stable state.)
- ☞ Electrons are added to each of the *p* orbitals singly, before they can be paired.
- ☞ If an atom does not have enough electrons to fill all shells, the outer shell will be the only one partially filled. For example: O_2 with a total of 8 electrons:

OXYGEN

Two electrons have the 1*s* orbital of the first electron shell.

First two electrons in the second shell are both in the 2*s* orbital.

Next three electrons each have a *p* orbital ($2p_x$, $2p_y$, $2p_z$).

Eighth electron is paired in the $2p_x$ orbital.

1*s*	2*s*	$2p_x$	$2p_y$	$2p_z$
2	2	2	1	1

Chemical properties of an atom depend upon the number of *valence electrons*.

<u>Valence electrons</u> = Electrons in the outermost energy shell (valence shell).

What is the octet rule? How can it be used to predict the number of bonds an atom will form?

Why are the noble gases so unreactive?

<u>Octet rule</u> = Rule that a valence shell is complete when it contains 8 electrons (except H and He).
- ☞ An atom with a complete valence shell is unreactive or *inert*.
- ☞ Noble elements (e.g. helium, argon and neon) have filled outer shells in their elemental state and are thus inert.

- ☞ An atom with an incomplete valence shell is chemically reactive (tends to form chemical bonds until it has 8 electrons to fill the valence shell).
- ☞ Atoms with the same number of valence electrons show similar chemical behavior.

NOTE: The consequence of this unifying chemical principle is that the valence electrons are responsible for the atom's bonding capacity. This rule applies to most of the representative elements, but <u>not all</u>.

III. CHEMICAL BONDS AND MOLECULES (p. 32–36)

Atoms with incomplete valence shells tend to fill those shells by interacting with other atoms. These interactions of electrons among atoms may allow atoms to form *chemical bonds*.

<u>Chemical bonds</u> = Attractions that hold molecules together.

<u>Molecules</u> = Two or more atoms held together by chemical bonds.

A. Covalent Bonds (p. 33–34)

<u>Covalent bond</u> = Chemical bond between atoms formed by <u>sharing</u> a pair of valence electrons.
- ☞ Strong chemical bond.
- ☞ For example: molecular hydrogen (H_2). When 2 hydrogen atoms come close enough for their $1s$ orbitals to overlap, they <u>share</u> electrons — thus completing the valence shell of each atom.

<u>Structural formula</u> = Formula which represents the atoms and bonding within a molecule (e.g. H–H). The line represents a shared pair of electrons.

<u>Molecular formula</u> = Formula which indicates the number and type of atoms (e.g. H_2).

<u>Single covalent bond</u> = Bond between atoms formed by sharing a single pair of valence electrons.
- ☞ Atoms may freely rotate around the axis of the bond.

<u>Double covalent bond</u> = Bond formed when atoms share <u>two</u> pairs of valence electrons (e.g. O_2).

<u>Triple covalent bond</u> = Bond formed when atoms share three pairs of valence electrons (e.g. N_2 or N≡N).

NOTE: Double and triple covalent bonds are rigid and do not allow rotation.

18 Atoms, Molecules, and Chemical Bonds

Valence

Compound

Electronegativity

How can you use the octet rule to predict the bonding capacity of an atom?

Valence = Bonding capacity of an atom which is the number of covalent bonds that must be formed to complete the outer electron shell.
- ☞ Valences of some common elements: hydrogen=1, oxygen=2, nitrogen=3, carbon=4, phosphorus=3, sulfur=2.

Compound = Substance whose molecules contain more than one element.
- ☞ For example: water (H_2O), methane (CH_4).
- ☞ Note that two hydrogens are necessary to complete the valence shell of oxygen in water, and four hydrogens are necessary for carbon to complete the valence shell in methane.

Biological Importance of Molecular Shapes: (p. 34)

The function of many molecules depends upon their shape.
- ☞ Molecules with only two atoms are linear.
- ☞ Molecules with more than two atoms have more complex shapes.

When an atom forms covalent bonds, orbitals in the valence shell rearrange into the most stable configuration. To illustrate, consider atoms with valence electrons in the *s* and three *p* orbitals:
- ☞ The *s* and three *p* orbitals hybridize into four new orbitals.
- ☞ The new orbitals are teardrop shaped, extend from the nucleus and spread out as far apart as possible.
- ☞ For example, if outer tips of orbitals in methane (CH_4) are connected by imaginary lines, the new molecule has a tetrahedral shape with C at the center. (See Campbell, Figure 2.13)

Nonpolar and Polar Covalent Bonds: (p. 34)

What is electronegativity?

Electronegativity = Atom's ability to attract and hold electrons.
- ☞ The more electronegative an atom, the more strongly it attracts shared electrons.
- ☞ Scale determined by Linus Pauling:

   O = 3.5   S and C = 2.5
   N = 3.0   P and H = 2.1

What are the differences among nonpolar covalent, polar covalent and ionic bonds?

How does electronegativity influence the formation of chemical bonds?

Nonpolar covalent bond = Covalent bond formed by an equal sharing of electrons between atoms.
- Occurs when electronegativity of both atoms is about the same (e.g. CH_4).
- Molecules made of one element usually have nonpolar covalent bonds (e.g. H_2, O_2, Cl_2, N_2).

Polar covalent bond = Covalent bond formed by an unequal sharing of electrons between the atoms.
- Occurs when the atoms involved have different electronegativities.
- Shared electrons spend more time around the more electronegative atom.
- In H_2O, for example, the oxygen is strongly electronegative, so negatively charged electrons spend more time around the oxygen than the hydrogens. This causes the oxygen atom to have a slight negative charge and the hydrogens to have a slight positive charge.

B. Ionic Bonds (*p. 34–35*)

Ion = Charged atom or molecule.

Anion = An atom that has gained one or more electrons from another atom and has become negatively charged. A negatively charged ion.

Cation = An atom that has lost one or more electrons and has become positively charged. A positively charged ion.

Ionic bond = Bond formed by the electrostatic attraction after the complete transfer of an electron from a donor atom to an acceptor.
- The acceptor atom attracts the electrons because it is much more electronegative than the donor atom.
- Are strong bonds in crystals, but are fragile bonds in water; salt crystals will readily dissolve in water and dissociate into ions.
- Ionic compounds are called salts (e.g. NaCl or table salt).

NOTE: The difference in electronegativity between interacting atoms determines if electrons are shared equally (nonpolar covalent), shared unequally (polar covalent), gained or lost (ionic bond). Nonpolar covalent bonds and ionic bonds are two extremes of a continuum from interacting atoms with similar electronegativities to interacting atoms with very different electronegativities.

Atoms, Molecules, and Chemical Bonds

C. Hydrogen Bonds (p. 36)

What is a hydrogen bond and how does it differ from a covalent or ionic bond?

<u>Hydrogen bond</u> = Bond formed by the charge attraction when a hydrogen atom covalently bonded to one electronegative atom is attracted to another electronegative atom.
- Weak attractive force that is about 20 times easier to break than a covalent bond.
- Is a charge attraction between oppositely charged portions of polar molecules.
- Can occur between a hydrogen that has a slight positive charge when covalently bonded to an atom with high electronegativity (usually O and N).
- For example: NH_3 in H_2O.

D. The Biological Importance of Weak Bonds (p. 36)

Why are weak bonds important to living organisms?

Biologically important weak bonds:
- Include hydrogen bonds, ionic bonds in aqueous solutions and other weak forces such as Van der Waals and hydrophobic interactions.
- Make chemical signaling possible in living organisms, because they are only temporary associations. Signal molecules can briefly and reversibly bind to receptor molecules on a cell, causing a short-lived response.
- Can form between molecules or between different parts of a single large molecule.
- Help stabilize the three-dimensional shape of large molecules (e.g. DNA and proteins).

IV. CHEMICAL REACTIONS (p. 36–37)

What are reactants and products in a chemical reaction? What effect does their relative concentration have on a chemical reaction?

<u>Chemical reactions</u> = process of making and breaking chemical bonds leading to changes in the composition of matter.
- Process where *reactants* undergo changes into *products*.
- Matter is conserved, so all reactant atoms are only rearranged to form products.
- Some reactions go to completion (all reactants converted to products), but most reactions are *reversible*. For example:

$$3H_2 + N_2 \rightleftharpoons 2NH_3$$

- The relative concentration of reactants and products affects reaction rate (the higher the concentration, the greater probability of reaction).

Margin terms: Hydrogen bond; Weak bonds; Chemical reactions; Reactants, Products

Chemical equilibrium = Equilibrium established when the rate of forward reaction equals the rate of the reverse reaction.
- Is a <u>dynamic</u> equilibrium with reactions continuing in both directions.
- Relative concentrations of reactants and products stay the same.

V. CHEMICAL CONDITIONS ON THE EARLY EARTH: SETTING THE STAGE FOR THE ORIGIN AND EVOLUTION OF LIFE (p. 37–38)

How were the chemical conditions on early Earth different from today?

Matter in our solar system formed from the debris of supernovae that exploded over 5 billion years ago.
- Most of the resulting swirling matter condensed in the center as the sun.
- The Earth formed about 4.5 billion years ago from peripheral material spinning around the early sun.

The early Earth was initially cold, but melted from heat generated by compaction, radioactive decay and meteorite impact. The resulting molten material sorted into layers of varying density.
- Most nickel and iron sank to the center.
- Less dense material concentrated into a mantle.
- Least dense material settled on the surface solidifying into a thin crust.
- Present continents are attached to plates of crust that float on the semisolid, flexible mantle.

Chemical conditions on the early earth were probably quite different from today. Evidence indicates that:
- The first early atmosphere was mostly hot hydrogen gas made of small molecules that escaped the earth's gravitational pull.
- The second early atmosphere consisted mostly of water vapor, carbon monoxide, carbon dioxide, nitrogen, methane and ammonia released by volcanoes and vents through the Earth's crust.
- As the earth cooled, water vapor condensed and rainfall formed the first seas.
- Lightning, volcanic activity and ultraviolet radiation were more intense than at present.

Under these conditions of early Earth, chemical evolution may have given rise to the first cells.

3 WATER AND THE FITNESS OF THE ENVIRONMENT

CHAPTER OUTLINE

How does water contribute to the fitness of the environment to support life? (Try to answer this question as the lecture progresses and as you proceed through Chapter 3 in the text.)

Water contributes to the fitness of the environment to support life.
- ☞ Life on earth probably evolved in water.
- ☞ Living cells are 70%–95% H_2O.
- ☞ Water covers about 3/4 of the earth.
- ☞ In nature, water naturally exists in all three physical states of matter — solid, liquid and gas.

Water's extraordinary properties are emergent properties resulting from water's structure and molecular interactions.

I. WATER MOLECULES AND HYDROGEN BONDING (p. 40–41)

What is the structure and geometry of the water molecule? What properties emerge as a result of this structure?

Water is a *polar* molecule. Its polar bonds and asymmetrical shape give water molecules opposite charges on opposite sides.
- ☞ Four valence orbitals of O point to corners of a tetrahedron.
- ☞ 2 corners are orbitals with unshared pairs of electrons and weak negative charge.
- ☞ 2 corners are occupied by H atoms which are in polar covalent bonds with O. Oxygen is so electronegative, that shared electrons spend more time around the O causing a weak positive charge near H's.

What is the relationship between the polar nature of water and its ability to form hydrogen bonds?

Water and the Fitness of the Environment

The polar molecules of water are held together by hydrogen bonds.

- ☞ Positively charged H of one molecule is attracted to the negatively charged O of another water molecule.
- ☞ Each water molecule can form a maximum of four hydrogen bonds with neighboring water molecules.
- ☞ Hydrogen bonding orders water into a higher level of structural organization.

II. SOME EXTRAORDINARY PROPERTIES OF WATER (p. 41–46)

What are five characteristics of water that are emergent properties resulting from hydrogen bonding?

A. Liquid Water is Cohesive (p. 41)

What is the biological significance of the cohesiveness of water?

<u>Cohesion</u> = Phenomenon of a substance being held together by hydrogen bonds.
- ☞ Though hydrogen bonds are transient, enough water molecules are hydrogen bonded at any given time to give water more structure than other liquids.
- ☞ Contributes to upward water transport in plants by holding the water column together.

<u>Surface tension</u> = Measure of how difficult it is to stretch or break the surface of a liquid.
- ☞ Water has a greater surface tension than most liquids.
- ☞ Function of the fact that at the air/H$_2$O interface, surface water molecules are hydrogen bonded to each other and to the water molecules below.
- ☞ Causes H$_2$O to bead (shape with smallest area to volume ratio and allows maximum hydrogen bonding).

<u>Adhesion</u> = Clinging of water to *hydrophilic* substances (e.g. glass).

B. Water Has a High Specific Heat (p. 42–43)

1. Heat and Temperature (p. 42)

What is the difference between heat and temperature?

<u>Kinetic energy</u> = The energy of motion.

<u>Heat</u> = Total kinetic energy due to molecular motion in a body of matter.

<u>Calorie (cal)</u> = Amount of heat it takes to raise the temperature of one gram of water by one degree Celsius.

24 Water and the Fitness of the Environment

Kilocalorie

Temperature

Kilocalorie (kcal or Cal) = Amount of heat required to raise the temperature of one kilogram of water by one degree Celsius (1000 cal).

Temperature = Measure of heat intensity due to the <u>average</u> kinetic energy of molecules in a body of matter.

The Celsius Scale at Sea Level:

Celsius Scale	Scale Conversion
100°C (212°F) = water boils	°C = $\frac{5(°F - 32)}{9}$
37°C (98.6°F) = human body temperature	°F = $\frac{9°C}{5}$ + 32
23°C (72°F) = room temperature	°K = °C + 273
0°C (32°F) = water freezes	

Specific heat

Specific heat = Amount of heat that must be absorbed or lost for one gram of a substance to change its temperature by one degree Celsius.

Specific heat of water = One calorie per gram per degree Celsius (1 cal/g/°C).

2. How Water Stabilizes Temperature (*p. 43*)

Water has a high specific heat, which means that it resists temperature changes when it absorbs or releases heat.
- ☞ As a result of hydrogen bonding among water molecules, it takes a relatively large heat loss or gain for each 1°C change in temperature.
- ☞ Hydrogen bonds must absorb heat to break, and they release heat when they form.
- ☞ Much absorbed heat energy is used to disrupt hydrogen bonds before water molecules can move faster (increase temperature).

How do water's high specific heat, high heat of vaporization and expansion upon freezing affect both aquatic and terrestrial ecosystems?

A large body of water can act as a heat sink — absorbing heat from sunlight during the day and summer and releasing heat during the night and winter as the water gradually cools. As a result:
- ☞ Water, which covers three-fourths of the planet, keeps temperature fluctuations within a range suitable for life.
- ☞ Coastal areas have milder climates than inland.
- ☞ The marine environment has a relatively stable temperature.

C. Water Has a High Heat of Vaporization (p. 43–44)

<u>Vaporization</u> (evaporation) = Transformation from liquid to a gas.
- Molecules with enough kinetic energy to overcome the mutual attraction of molecules in a liquid, can escape into the air.

<u>Heat of vaporization</u> = Quantity of heat a liquid must absorb for 1 g to be converted to the gaseous state.
- For water molecules to evaporate, hydrogen bonds must be broken which requires heat energy.
- Water has a relatively high heat of vaporization (540 cal/g).

<u>Evaporative cooling</u> = Cooling of a liquid's surface when a liquid evaporates.
- The surface molecules with the highest kinetic energy are most likely to escape into gaseous form; the average kinetic energy of the remaining surface molecules is thus lower.

Water's high heat of vaporization:

1. Moderates the earth's climate.
   - Solar heat absorbed by tropical seas dissipates when surface water evaporates (*evaporative cooling*).
   - As moist tropical air moves poleward, water vapor releases heat as it condenses into rain.
2. Stabilizes temperature in aquatic ecosystems (*evaporative cooling*).
3. Helps organisms from overheating by *evaporative cooling*.

D. Water Expands When It Freezes (p. 44–45)

Water is densest at 4°C.
- Water contracts as it cools to 4°C.
- As water cools from 4°C to freezing (0°C), it expands and becomes <u>less dense</u> than liquid water (ice floats).
- When water begins to freeze, the molecules do not have enough kinetic energy to break hydrogen bonds.
- As the crystalline lattice forms, each water molecule forms a maximum of 4 hydrogen bonds, which keeps water molecules further apart than they would be in the liquid state.

Expansion of water contributes to the fitness of the environment for life:
- Prevents deep bodies of water from freezing solid from the bottom up.
- Since ice is less dense, it forms on the surface first. As water freezes it releases heat to the water below and insulates it.
- Makes the transitions between seasons less abrupt. As water freezes, hydrogen bonds form releasing heat. As ice melts, hydrogen bonds break absorbing heat.

E. Water is a Versatile Solvent (p. 45–46)

<u>Solution</u> = A liquid that is a homogenous mixture of two or more substances.

<u>Solvent</u> = Dissolving agent of a solution.

Solute

Aqueous solution

Solute = Substance dissolved in a solution.

Aqueous solution = Solution in which water is the solvent.

How does the polarity of the water molecule make it a versatile solvent?

Water is a versatile solvent owing to the polarity of the water molecule.

Hydrophilic
1. Ionic compounds dissolve in water.
   - Charged regions of polar water molecule have an electrical attraction to charged ions.
   - Water surrounds individual ions, separating and shielding them from one another.
2. Polar compounds in general, are water-soluble.
   - Charged regions of polar H_2O molecules have an affinity for oppositely charged regions of other polar molecules.

Hydrophobic
3. Nonpolar compounds (which have symmetric distribution in charge) are NOT water-soluble.

Ionic and polar substances are *hydrophilic*, but nonpolar compounds are *hydrophobic*.

Hydrophilic

Hydrophilic = (Hydro=water; philo=loving) Property of having an affinity for water.

Hydrophobic

Hydrophobic = (Hydro=water; phobos=fearing) Property of not having an affinity for water, and thus not being water-soluble.

III. AQUEOUS SOLUTIONS (p. 46–49)

Most biochemical reactions involve solutes dissolved in water. There are two important properties of aqueous solutions: solute concentration and pH.

A. Solute Concentration (p. 46–47)

Molecular weight

Molecular weight = Sum of the weight of all atoms in a molecule (expressed in daltons).

Mole

Mole = Amount of a substance that has a mass in grams numerically equivalent to its molecular weight in daltons.

For example: To determine a mole of sucrose ($C_{12}H_{22}O_{11}$),
- Calculate molecular weight:
  - C = 12 dal   12 dal × 12 =   144 dal
  - H = 1 dal   1 dal × 22 =   22 dal
  - O = 16 dal   16 dal × 11 =   176 dal
  -   342 dal
- Express it in grams (342 g).

What is molarity? What are some advantages to measuring substances in moles?

<u>Molarity</u> = Number of moles of solute per liter of solution.

For example: To obtain 1M sucrose solution, weigh out 342 g of sucrose and add water up to 1L.

Advantage of measuring in moles:
- Rescales weighing of single molecules in daltons to grams, which is more practical for laboratory use.
- A mole of one substance has the <u>same</u> number of molecules as a mole of any other substance (6.02×10^{23}; Avogadro's number).
- Allows one to combine substances in fixed ratios of molecules.

B. Acids, Bases and pH (*p. 47–49*)

1. Dissociation of Water Molecules (*p. 47*)

When water dissociates, what is actually transferred from one water molecule to another?

Occasionally, the hydrogen atom that is shared in a hydrogen bond between two water molecules, shifts from the oxygen atom to which it is covalently bonded to the unshared orbitals of the oxygen atom to which it is hydrogen bonded.
- Only a <u>hydrogen ion</u> (proton with a +1 charge) is actually transferred.
- Transferred proton binds to an unshared orbital of the second water molecule creating a *hydronium ion* (H_3O^+).
- Water molecule that lost a proton has a net negative charge and is called a *hydroxide ion* (OH^-).

$$H_2O + H_2O \rightleftharpoons H_3O^+ + OH^-$$

What is the conventional way to express the dissociation of water?

- By convention, ionization of H_2O is expressed as the *dissociation* into H^+ and OH^-.

$$H_2O \rightleftharpoons H^+ + OH^-$$

- Reaction is reversible.
- At equilibrium, most of the H_2O is <u>not</u> ionized.

2. Acids and Bases (*p. 47–48*)

At equilibrium in pure water at 25°C:
- Number of H^+ ions = number of OH^- ions.
- $[H^+] = [OH^-] = \dfrac{1}{10,000,000} M = 10^{-7} M$
- Note that brackets indicate molar concentration.

<u>Molarity</u>

<u>Hydronium ion</u>
<u>Hydroxide ion</u>

<u>Dissociation</u>

Acid, Base

How do acids and bases directly and indirectly affect the hydrogen ion concentration of a solution?

ACID	BASE
Substance that <u>increases</u> the relative [H$^+$] of a solution.	Substance that <u>reduces</u> the relative [H$^+$] of a solution.
Also removes OH$^-$ because it tends to combine with H$^+$ to form H$_2$O.	May alternately increase [OH$^-$].
For example: (in water) HCl $\longrightarrow$ H$^+$ + Cl$^-$	For example: A base may reduce [H$^+$] directly: NH$_3$ + H$^+$ $\rightleftharpoons$ NH$_4^+$ A base may reduce [H$^+$] indirectly: NaOH $\longrightarrow$ Na$^+$ + OH$^-$ OH$^-$ + H$^+$ $\longrightarrow$ H$_2$O

A solution in which:
- ☞ [H$^+$] = [OH$^-$] is a neutral solution.
- ☞ [H$^+$] > [OH$^-$] is an acidic solution.
- ☞ [H$^+$] < [OH$^-$] is a basic solution.

3. The pH Scale (*p. 48*)

In any aqueous solution:

$$[H^+][OH^-] = 10^{-14} M^2$$

For example:
- ☞ In a neutral solution, [H$^+$] = 10^{-7} M and [OH$^-$] = 10^{-7} M.
- ☞ In an acidic solution where the [H$^+$] = 10^{-5} M, the [OH$^-$] = 10^{-9} M.
- ☞ In a basic solution where the [H$^+$] = 10^{-9} M, the [OH$^-$] = 10^{-5} M.

pH scale

<u>pH scale</u> = Scale used to measure degree of acidity. It ranges from 0 to 14.

What is the basis for the pH scale?

pH

<u>pH</u> = Negative log$_{10}$ of the [H$^+$] expressed in moles per liter.
- ☞ pH of 7 is a neutral solution.
  pH < 7 is an acidic solution.
  pH > 7 is a basic solution.
- ☞ Most biological fluids are within the pH range of 6 to 8. There are some exceptions such as stomach acid with pH = 1.5. (See Campbell, Figure 3.10)
- ☞ Each pH unit represents a <u>tenfold</u> difference (scale is logarithmic), so a slight change in pH represents a large change in actual [H$^+$].

4. Buffers (p. 48–49)

How do buffers work? Can you demonstrate with an example?

By minimizing wide fluctuations in pH, buffers help organisms maintain the pH of body fluids within the narrow range necessary for life (usually pH 6–8).

Buffer = Substance that prevents large sudden changes in pH.
- Are combinations of H⁺-donor and H⁺-acceptor forms of weak acids or bases.
- Work by accepting H⁺ ions from solution when they are in excess, and by donating H⁺ ions to the solution when they have been depleted.

For example: Bicarbonate buffer.

$$H_2CO_3 \underset{\text{response to a drop in pH}}{\overset{\text{response to a rise in pH}}{\rightleftharpoons}} HCO_3^- + H^+$$

H⁺ donor   H⁺ acceptor
weak acid   weak base

$$HCl + NaHCO_3 \rightarrow H_2CO_3 + NaCl$$
strong acid   weak acid

$$NaOH + H_2CO_3 \rightarrow NaHCO_3 + H_2O$$
strong base   weak base

IV. ACID RAIN: UPSETTING THE FITNESS OF THE ENVIRONMENT (p. 49–51)

What causes acid rain? How does it adversely affect the fitness of the environment to support life?

Acid rain = Rain more strongly acidic than pH 5.6.
- Has been recorded as low as pH 1.5 in West Virginia.
- Occurs when sulfur oxides and nitrogen oxides in the atmosphere react with water in the air to form acids.
- Major oxide source is the combustion of fossil fuels by industry and cars.

Acid rain affects the fitness of the environment to support life:
- Lowers soil pH which affects mineral solubility. May leach out necessary mineral nutrients and increase the concentration of minerals that are potentially toxic to vegetation in higher concentration (e.g. aluminum).
- Lowers the pH of lakes and ponds, and runoff carries leached out soil minerals into aquatic ecosystems. This adversely affects aquatic life. For example: In the Western Adirondack Mountains, there are lakes with a pH < 5 that have no fish.

What can be done to reduce the problem?
- Add industrial pollution controls.
- Develop and use antipollution devices.

4 CARBON AND MOLECULAR DIVERSITY

CHAPTER OUTLINE

Aside from water, most biologically important molecules are carbon-based (*organic*).

The structural and functional diversity of organic molecules emerges from the ability of carbon to form large, complex and diverse molecules by bonding to itself and to other elements such as H, O, N, S and P.

I. THE FOUNDATIONS OF ORGANIC CHEMISTRY (p. 53–55)

What were the philosophies of vitalism and mechanism? How did they influence the development of organic chemistry, as well as, mainstream biological thought?

Organic chemistry = The branch of chemistry that specializes in the study of carbon compounds.

Organic molecules = Molecules containing carbon.

Vitalism = Belief in a life force outside the jurisdiction of chemical/physical laws.
- ☞ Organic chemistry in early 19th century was built on a foundation of vitalism because organic chemists could not artificially synthesize organic compounds. It was believed that only living organisms could produce organic compounds.

Mechanism = Belief that all natural phenomena are governed by physical and chemical laws.
- ☞ Pioneers of organic chemistry began to synthesize organic compounds from inorganic molecules. This helped shift mainstream biological thought from vitalism to mechanism.
- ☞ For example, Friedrich Wöhler synthesized urea in 1828; Hermann Kolbe synthesized acetic acid.
- ☞ Stanley Miller (1953) demonstrated the possibility that organic compounds could have been produced under the chemical conditions of primordial Earth.

II. THE VERSATILITY OF CARBON IN MOLECULAR ARCHITECTURE (p. 55–56)

How does carbon's electron configuration determine the kinds and numbers of bonds carbon will form?

The carbon atom:
- ☞ Has an atomic number of 6; therefore, it has 4 valence electrons.
- ☞ Completes its outer energy shell by sharing valence electrons in four covalent bonds. (Not likely to form ionic bonds.)

Emergent properties, such as the kinds and number of bonds carbon will form, are determined by its tetravalent electron configuration.
- ☞ It makes large complex molecules possible. The carbon atom is a central point from which the molecule branches off into four directions.
- ☞ It gives carbon covalent compatibility with many different elements. The four major atomic components of organic molecules are:

| Hydrogen | Oxygen | Nitrogen | Carbon |
| (Valence=1) | (Valence=2) | (Valence=3) | (Valence=4) |

- ☞ It determines an organic molecule's 3-dimensional shape, which may affect molecular function. For example, when carbon forms four single covalent bonds, the four valence orbitals hybridize into teardrop-shaped orbitals that angle from the carbon atom towards the corners of an imaginary tetrahedron.

III. VARIATION IN CARBON SKELETONS (p. 56–58)

How do carbon skeletons vary? How does this variation make diverse and complex organic molecules possible?

Covalent bonds link carbon atoms together in long chains that form the skeletal framework for organic molecules. These carbon skeletons may vary in:
- ☞ Length.
- ☞ Shape (straight chain, branched, ring).
- ☞ Number and location of double bonds.
- ☞ Other elements covalently bonded to available sites.

This variation in carbon skeletons contributes to the complexity and diversity of organic molecules. (See Campbell, Figure 4.6)

Hydrocarbons = Molecules containing only carbon and hydrogen.
- ☞ Are major components of fossil fuels produced from the organic remains of organisms living millions of years ago, though they are not prevalent in living organisms.
- ☞ Have a diversity of carbon skeletons which produce molecules of various lengths and shapes.
- ☞ As in hydrocarbons, a carbon skeleton is the framework for the large diverse organic molecules found in living organisms. Also, some biologically important molecules may have regions consisting of hydrocarbon chains (e.g. fats).

Hydrocarbons

32 Carbon and Molecular Diversity

Isomers

Isomers = Compounds with the same molecular formula but with different structures and hence different properties. Isomers are a source of variation among organic molecules.

What are the differences among structural, geometric and stereoisomers?

Different Types of Isomers: (See Campbell, Figure 4.8)

Structural isomers

Structural isomers = Isomers that differ in the covalent arrangement of their atoms.

☞ Number of possible isomers increases as the carbon skeleton size increases.
☞ May also differ in the location of double bonds.

Geometric isomers

Geometric isomers = Isomers which share the same covalent partnerships, but differ in their spatial arrangements.

☞ Result from the fact that double bonds will not allow the atoms they join to rotate freely about the axis of the bonds.
☞ Subtle differences between isomers affects their biological activity.

Stereoisomers

Stereoisomers = Isomers that are mirror images of each other.

l-isomer   *d*-isomer

☞ Can occur when four different atoms or groups of atoms are bonded to the same carbon (*asymmetric carbon*).
☞ There are two different spatial arrangements of the four groups around the asymmetric carbon. These arrangements are mirror images.
☞ Usually one form is biologically active and its mirror image is not.

Asymmetric carbon

Carbon and Molecular Diversity 33

IV. FUNCTIONAL GROUPS (p. 58–61)

What are functional groups? How do they determine the chemical behavior of organic molecules?

Small characteristic groups of atoms (*functional groups*) are frequently bonded to the carbon skeleton of organic molecules. These functional groups:

- ☞ Have specific chemical and physical properties.
- ☞ Are the regions of organic molecules which are commonly chemically reactive.
- ☞ Behave consistently from one organic molecule to another.
- ☞ Depending upon their number and arrangement, determine unique chemical properties of organic molecules in which they occur.

As with hydrocarbons, diverse organic molecules found in living organisms have carbon skeletons. In fact, these molecules can be viewed as hydrocarbon derivatives with functional groups in place of H, bonded to carbon at various sites along the molecule.

What are six of the major functional groups? (Are you able to recognize or draw them?)

What chemical properties does each functional group give to the organic molecules in which they occur?

A. The Hydroxyl Group (p. 58)

<u>Hydroxyl group</u> = A functional group that consists of a hydrogen atom bonded to an oxygen atom, which in turn is bonded to carbon (−OH).

- ☞ Is a <u>polar</u> group.
- ☞ Makes the molecule to which it is attached <u>water soluble</u>. Polar water molecules are attracted to the polar hydroxyl group which can form hydrogen bonds.
- ☞ Organic compounds with hydroxyl groups are called *alcohols*.

B. The Carbonyl Group (p. 58–59)

<u>Carbonyl group</u> = Functional group that consists of a carbon atom double-bonded to oxygen (−CO).

- ☞ Is a <u>polar</u> group. The oxygen can be involved in hydrogen bonding, and molecules with this functional group are <u>water soluble</u>.
- ☞ Is a functional group found in sugars.
- ☞ If the carbonyl is at the end of the carbon skeleton, the compound is an *aldehyde*.

Glyceraldehyde

- ☞ If the carbonyl is in the internal part of the carbon skeleton, the compound is a *ketone*.

Acetone

Notes column:

Functional groups

Hydroxyl group

Alcohols

Carbonyl group

Aldehyde

Ketone

C. The Carboxyl Group (p. 59)

Carboxyl group = Functional group that consists of a carbon atom which is both double-bonded to an oxygen and single-bonded to the oxygen of a hydroxyl group (−COOH).

☞ Is a polar group and water soluble. The covalent bond between oxygen and hydrogen is so polar, that the hydrogen reversibly dissociates as H^+.

$$\begin{array}{ccc} \text{Acetic} & \text{Acetate} & \text{Hydrogen} \\ \text{acid} & \text{ion} & \text{ion} \end{array}$$

☞ Since it donates protons, this group has acidic properties. Compounds with this functional group are called *carboxylic acids*.

D. The Amino Group (p. 60−61)

Amino group = Functional group that consists of a nitrogen atom bonded to two hydrogens and to the carbon skeleton (−NH$_2$).

☞ Is a polar group and soluble in water.
☞ Acts as a weak base. The unshared pair of electrons on the nitrogen can accept a proton.

$$\text{Amine} \qquad \text{Ammonium ion}$$

☞ Organic compounds with this functional group are called *amines*.

E. The Sulfhydryl Group (p. 61)

Sulfhydryl group = Functional group which consists of an atom of sulfur bonded to an atom of hydrogen (−SH).

☞ Help stabilize the structure of proteins. (Disulfide bridges will be discussed with tertiary structure of proteins in Chapter 5: Structure and Function of Macromolecules.)

F. The Phosphate Group (p. 61)

Phosphate group = Functional group which is the dissociated form of phosphoric acid (H_3PO_4).

☞ Loss of two protons by dissociation leaves the phosphate group with a negative charge.

☞ Has acid properties since it loses protons.
☞ Polar group and soluble in water.
☞ Organic phosphates are important in cellular energy storage and transfer. (ATP is discussed with energy for cellular work in Chapter 6: Introduction to Metabolism.)

5 THE STRUCTURE AND FUNCTION OF MACROMOLECULES

CHAPTER OUTLINE

The topic of macromolecules lends itself well to illustrate three integral themes woven throughout the text and course: (1) there is a natural hierarchy of structural level in biological organization; (2) as one moves up the hierarchy, new properties emerge because of interactions among subunits at the lower levels; and (3) form fits function.

What are the levels of biological hierarchy from subatomic particles to macromolecules?

I. POLYMERS (p. 64–66)

<u>Polymer</u> = (Poly=many; mer=part) Large molecule consisting of many identical or similar subunits connected together.

<u>Monomer</u> = Subunit or building block molecule of a polymer.

<u>Macromolecule</u> = (Macro=large) Large organic polymers.
- ☞ Formation of macromolecules from smaller building block molecules represents another level in the hierarchy of biological organization.
- ☞ There are four classes of macromolecules in living organisms:
    1. Carbohydrates.
    2. Lipids.
    3. Proteins.
    4. Nucleic acids.

    A. Polymers and Molecular Diversity (p. 64–65)

How do organic polymers contribute to biological diversity?

Structural variation of macromolecules is the basis for the enormous diversity of life.
- ☞ There is <u>unity</u> in life as there are only about 40–50 common monomers used to construct macromolecules.
- ☞ There is <u>diversity</u> in life as new properties emerge when these universal monomers are arranged in different ways.

    B. Making and Breaking Polymers (p. 65–66)

How are covalent linkages formed and broken in organic polymers?

Most *polymerization reactions* in living organisms are *condensation reactions*.

Polymer

Monomer

Macromolecule

Polymerization reactions

Condensation synthesis

Hydrolysis

Carbohydrates

Monosaccharides

Polymerization reactions = Chemical reactions that link two or more small molecules to form larger molecules with repeating structural units.

Condensation synthesis = Polymerization reactions during which monomers are covalently linked, producing net removal of a water molecule for each covalent linkage.
- ☞ One monomer loses a hydroxyl (−OH), and the other monomer loses a hydrogen (−H).
- ☞ Removal of water is actually indirect, involving the formation of "activated" monomers (to be discussed in Chapter 6: Introduction to Metabolism).
- ☞ Process requires energy.
- ☞ Process requires biological catalysts or enzymes.

Hydrolysis = (Hydro=water; lysis=loosening) A reaction process that breaks covalent bonds between monomers by the addition of water molecules.
- ☞ A hydrogen from the water bonds to one monomer, and the hydroxyl bonds to the adjacent monomer.
- ☞ For example: digestive enzymes catalyze hydrolytic reactions which break apart large food molecules into monomers that can be absorbed into the bloodstream.

II. **CARBOHYDRATES** (p. 66–71)

What are the distinguishing characteristics of carbohydrates? How are they classified?

Carbohydrates = Organic molecules made of sugars and their polymers.
- ☞ Polymers formed by condensation reactions.
- ☞ Are classified based upon the number of simple sugars.

A. Monosaccharides (p. 66)

Monosaccharides = (Mono=single; racchar=sugar) Simple sugar in which C, H and O occur in the ratio of (CH_2O).
- ☞ Are major nutrients for cells. Glucose is the most common.
- ☞ Can be produced (glucose) by photosynthetic organisms from CO_2, H_2O and sunlight.
- ☞ Store energy in their chemical bonds which is harvested by cellular respiration.
- ☞ Their carbon skeletons are raw material for other organic molecules.
- ☞ Can be incorporated as monomers into disaccharides and polysaccharides.

What are four characteristics of a sugar?

1. An —OH group is attached to each carbon except one, which is double bonded to an oxygen (carbonyl).

   <u>Aldehyde</u>　　　　　<u>Ketone</u>
   Terminal carbon　　　Carbonyl group is
   forms a double bond　within the carbon
   with oxygen.　　　　　skeleton.

   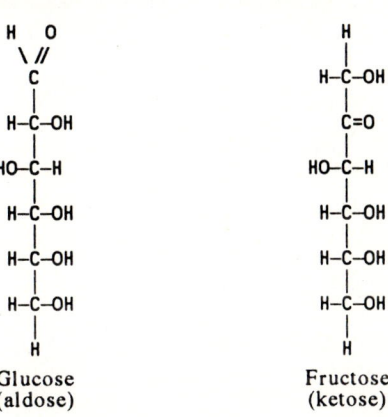

   Glucose　　　　Fructose
   (aldose)　　　　(ketose)

2. Size of the carbon skeleton varies from 3 to 7 carbons. (See Campbell, Figure 5.4) The <u>most common</u> monosaccharides are:

Classification	Number of Carbons	Example
Triose	3	Glyceraldehyde
Pentose	5	Ribose
Hexose	6	Glucose

3. Spatial arrangement around asymmetric carbon may vary. For example, glucose and galactose are stereoisomers.

   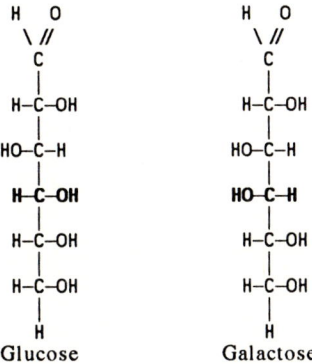

   Glucose　　　Galactose

   The small difference between isomers affects molecular shape which gives these molecules distinctive biochemical properties.

38 The Structure and Function of Macromolecules

4. In aqueous solutions, many monosaccharides form rings. Chemical equilibrium favors the ring structure.

Linear Form of Glucose ⇌ Ring Form of Glucose

B. Disaccharides (p. 66–67)

Disaccharide

Disaccharide = (Di=two; saccharide=sugar) A double sugar that consists of two monosaccharides joined by a *glycosidic linkage*.

How is a glycosidic linkage formed? How would you recognize a glycosidic linkage?

Glycosidic linkage

Glycosidic linkage = Covalent bond formed between two sugar monomers by a condensation reaction (e.g. maltose).

Maltose — 1-4 glycosidic linkage

Examples of disaccharides:

Disaccharide	Monomers	General Comments
Maltose	Glucose + Glucose	Sugar important in brewing beer.
Lactose	Glucose + Galactose	Sugar present in milk.
Sucrose	Glucose + Fructose	Table sugar; most prevalent disaccharide; transport form in plants.

C. **Polysaccharides** (*p. 67–71*)

What are important biological functions of polysaccharides?

<u>Polysaccharides</u> = Macromolecules that are polymers of a few hundred or thousand monosaccharides.
- ☞ Are formed by linking monomers in enzyme-mediated condensation reactions.
- ☞ Have two important biological functions:
    1. Energy storage (starch and glycogen).
    2. Structural support (cellulose and chitin).

1. **Storage Polysaccharides** (*p. 68–69*)

Cells hydrolyze storage polysaccharides into sugars as needed. Two most common storage polysaccharides are starch and glycogen.

<u>Starch</u> = Glucose polymer that is a storage polysaccharide in plants.
- ☞ Helical glucose polymer with α 1-4 linkages. (See Campbell, Figure 5.5)
- ☞ Stored as granules within plant organelles called *plastids*.
- ☞ *Amylose*, the simplest form, is an unbranched polymer.
- ☞ *Amylopectin* is a branched polymer.
- ☞ Most animals have digestive enzymes to hydrolyze starch.
- ☞ Major sources in the human diet are potato tubers and grains (e.g. wheat, corn, rice and fruits of other grasses).

<u>Glycogen</u> = Glucose polymer that is a storage polysaccharide in animals.
- ☞ Large glucose polymer that is more highly branched than amylopectin.
- ☞ Stored in the muscle and liver of humans and other vertebrates.

2. **Structural Polysaccharides** (*p. 69–71*)

Structural polysaccharides include cellulose and chitin.

<u>Cellulose</u> = Linear unbranched polymer of *D*-glucose in β 1-4 linkages.
- ☞ A major structural component of plant cell walls.
- ☞ Differs from starch (also a glucose polymer) in its glycosidic linkages. (See Campbell, Figure 5.9)

Polysaccharides

Starch

Glycogen

Cellulose

40 *The Structure and Function of Macromolecules*

α configuration, β configuration

What is the difference between the glycosidic linkages found in starch and those found in cellulose? Why is this difference biologically important?

STARCH	CELLULOSE
Glucose monomers are in <u>α configuration</u> (−OH group on carbon one is <u>below</u> the ring's plane).	Glucose monomers are in <u>β configuration</u> (−OH group on carbon one is <u>above</u> the ring's plane).
Monomers are connected with α 1-4 linkage.	Monomers are connected with β 1-4 linkage.

☞ Cellulose and starch have different three-dimensional shapes and properties as a result of differences in glycosidic linkages.
☞ Cellulose reinforces plant cell walls. Hydrogen bonds hold together parallel cellulose molecules in bundles of *microfibrils*, which intertwine to form a *fibril*. (See Campbell, Figure 5.9)
☞ Cellulose cannot be digested by most organisms, including humans, because they lack an enzyme that can hydrolyze the β 1-4 linkage. (Exceptions are some symbiotic bacteria, other microorganisms and some fungi.)

Chitin

Chitin = A structural polysaccharide that is a polymer of an amino sugar.
☞ Forms exoskeletons of arthropods.
☞ Found as a building material in the cell walls of some fungi.
☞ Monomer is an *amino sugar*, which is similar to *beta*-glucose with a nitrogen-containing group replacing the hydroxyl on carbon 2.

III. LIPIDS (p. 71–74)

What distinguishes lipids from the other major classes of macromolecules?

Lipids

Lipids = Diverse group of organic compounds that are insoluble in water, but will dissolve in nonpolar solvents (e.g. ether, chloroform, benzene). Important groups are fats, phospholipids and steroids.

A. Fats (p. 71–73)

What are the two building block molecules of a fat?

Fats = Macromolecules constructed from:
1. Glycerol, a three-carbon alcohol.
2. Fatty acid (carboxylic acid).
   - ☞ Composed of a *carboxyl group* at one end and an attached *hydrocarbon chain* ("tail").
   - ☞ Carboxyl functional group has properties of an acid.
   - ☞ Hydrocarbon chain has a long carbon skeleton usually with an even number of carbon atoms (most have 16–18 carbons).
   - ☞ Nonpolar C–H bonds make the chain hydrophobic and not water soluble.

How would you recognize an ester linkage? How is it formed?

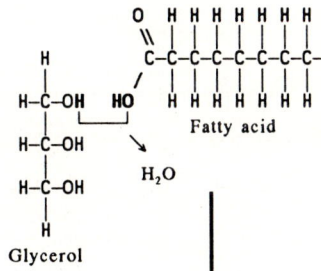

During the formation of a fat, enzyme-catalyzed condensation reactions link glycerol to fatty acids by an *ester linkage*.

Ester linkage = Bond formed between a hydroxyl group and a carboxyl group.

Each of glycerol's three hydroxyl groups can bond to a fatty acid by an ester linkage producing a fat.

Triacylglycerol = A fat composed of three fatty acids bonded to one glycerol by ester linkages (triglyceride).

Some characteristics of fat:
- ☞ Fats are insoluble in water. The long fatty acid chains are hydrophobic because of the many nonpolar C–H bonds.
- ☞ The source of variation among fat molecules is the fatty acid composition.
- ☞ Fatty acids in a fat may all be the same, or some (or all) may differ.
- ☞ Fatty acids may vary in length.

Fats

Ester linkage

Triacylglycerol

42 The Structure and Function of Macromolecules

☞ Fatty acids may vary in the number and location of carbon-to-carbon double bonds:

How do the structures of a saturated and unsaturated fat differ? What unique emergent properties are a consequence of these structural differences?

SATURATED FAT	UNSATURATED FAT
No double bonds between carbons in fatty acid tail.	One or more double bonds between carbons in fatty acid tail.
Carbon skeleton of fatty acid is bonded to maximum number of hydrogens ("saturated" with hydrogens).	Tail kinks at each C=C, so molecules do not pack closely enough to solidify at room temperature.
Usually a solid at room temperature.	Usually a liquid at room temperature.
Most animal fats.	Most plant fats.
e.g. bacon grease, lard and butter.	e.g. corn, peanut and olive oil.

What are some important biological functions of fat?

- ☞ Energy storage. One gram of fat stores twice as much energy as a gram of polysaccharide. (Fat has a higher proportion of energy rich C−H bonds.)
- ☞ More compact fuel reservoir than carbohydrate. Animals store more energy with less weight than plants which use starch, a bulky form of energy storage.
- ☞ Cushions vital organs in mammals (e.g. kidneys).
- ☞ Insulates against heat loss (e.g. marine mammals such as whales and seals).

B. Phospholipids (*p. 73−74*)

What are the building block molecules of a phospholipid? What unique emergent properties does a phospholipid have that are a consequence of combining these monomers into one molecule?

What is the biological importance of phospholipids?

Phospholipids = Compounds with molecular building blocks of glycerol, two fatty acids, a phosphate group and usually an additional small chemical group attached to the phosphate. (See Campbell, Figure 5.14)

- ☞ Differ from fat in that the third carbon of glycerol is joined to a <u>negatively charged</u> phosphate group.
- ☞ Can have small variable molecules (usually charged or polar) attached to phosphate.
- ☞ Show ambivalent behavior towards water. Hydrocarbon tails are hydrophobic, and the polar head (phosphate group with attachments) is hydrophilic.
- ☞ Are diverse depending upon differences in fatty acids and in phosphate attachments.
- ☞ Are major constituents of cell membranes. At the cell surface, phospholipids form a bilayer held together by hydrophobic interactions among the hydrocarbon tails.
- ☞ Phospholipids in water will spontaneously form such a bilayer:

C. **Steroids** (p. 74)

What is the general structure of a steroid? Why is cholesterol important?

<u>Steroids</u> = Lipids which have four fused carbon rings with various functional groups attached.

Cholesterol, an important steroid:
- ☞ Is the precursor of many other steroids including vertebrate sex hormones and bile acids.
- ☞ Is a common component of animal cell membranes.
- ☞ Can contribute to atherosclerosis.

IV. **PROTEINS** (p. 74–83)

What characteristics distinguish proteins from the other major classes of macromolecules? What are some biologically important functions performed by proteins?

Proteins

<u>Proteins</u> = Organic compounds which are complex polymers of *amino acids* linked together by peptide bonds.
- Are abundant, making up 50% or more of cellular dry weight.
- Have important and varied functions in the cell:
  1. Structural support.
  2. Storage (of amino acids).
  3. Transport (e.g. hemoglobin).
  4. Signaling (chemical messengers).
  5. Cellular response to chemical stimuli (receptor proteins).
  6. Movement (contractile proteins).
  7. Defense against foreign substances and disease-causing organisms (antibodies).
  8. Catalysis of biochemical reactions (enzymes).
- Vary extensively in structure; each type has a unique three-dimensional shape.
- Though vary in structure and function, are commonly made of only 20 amino acid monomers.

A. Amino Acids (p. 74–77)

What are the four major components of an amino acid? Which of these components varies from one amino acid to another?

Amino acid

<u>Amino acid</u> = Building block molecule of a protein; most consist of an asymmetric carbon, termed the *alpha carbon*, which is covalently bonded to:
1. Hydrogen atom.
2. Carboxyl group.
3. Amino group.
4. Variable R group (side chain) specific to each amino acid. Physical and chemical properties of the side chain determine the uniqueness of each amino acid.

(At pH's normally found in the cell, both the carboxyl and amino groups are ionized.)

Amino acids contain both carboxyl and amino functional groups. Since one group acts as a weak acid and the other group acts as a weak base, an amino acid can exist in three ionic states. The pH of the solution determines which ionic state predominates.

Cation ⇌ Zwitterion (dipolar ion) ⇌ Anion

How are the physical and chemical properties of side chains responsible for the uniqueness of each amino acid?

The twenty common amino acids can be grouped by properties of side chains: (See Campbell, Figure 5.17)
1. Nonpolar side groups (hydrophobic). Amino acids with nonpolar groups are less soluble in water.
2. Polar side groups (hydrophilic). Amino acids with polar side groups are more soluble in water. Polar amino acids can be grouped further into:
    a. Uncharged polar.
    b. Charged polar.
        ☞ Acidic side groups. Dissociated carboxyl group gives these side groups a net negative charge.
        ☞ Basic side groups. An amino group with an extra proton gives these side groups a net positive charge.

Some proteins have rare amino acids. These are modified from one of the common amino acids <u>after</u> their incorporation into a protein.

B. Polypeptide Chains (p. 77)

What characterizes a peptide bond? How is it formed?

<u>Peptide bond</u> = Covalent bond formed by a condensation reaction that links the carboxyl group of one amino acid to the amino group of the next.
- Has polarity with an amino group on one end (*N-terminus*) and a carboxyl group on the other (*C-terminus*).
- Has a backbone of the repeating sequence, $-N-C-C-N-C-C-$.

Peptide bond

<u>Polypeptide chain</u> = Polymer of many amino acids joined by peptide bonds.
☞ Ranges in length from a few monomers to more than a thousand.
☞ Has a unique linear sequence of amino acids.

Polypeptide chain

C. Protein Conformation (p. 77–78)

What determines protein conformation? Why is protein conformation important?

A protein's function depends upon its unique *conformation*.

<u>Protein conformation</u> = Three-dimensional shape of a protein.

Protein conformation

46 The Structure and Function of Macromolecules

Native conformation

Native conformation = Functional conformation of a protein found under normal biological conditions.
- ☞ Enables a protein to recognize and bind specifically to another molecule (e.g. hormone/receptor, enzyme/substrate and antibody/antigen).
- ☞ Is a consequence of the specific linear sequence of amino acids in the polypeptide.
- ☞ Is produced when a newly formed polypeptide chain coils and folds spontaneously, mostly in response to hydrophobic interactions.
- ☞ Is stabilized by chemical bonds between neighboring regions of the folded protein.

D. Levels of Protein Structure (*p. 78–81*)

The correlation between form and function in proteins is an emergent property resulting from superimposed levels of protein structure:
1. Primary structure.
2. Secondary structure.
3. Tertiary structure.

When a protein has two or more polypeptide chains, it also has:
4. Quaternary structure.

1. Primary Structure (*p. 78–79*)

What is primary structure and what determines it?

Primary structure

Primary structure = Unique sequence of amino acids in a protein.
- ☞ Determined by genes.
- ☞ Slight change can affect a protein's conformation and function (e.g. sickle-cell hemoglobin; See Campbell, Figure 5.22).
- ☞ Can be sequenced in the laboratory. A pioneer in this work was Frederick Sanger who determined the amino acid sequence of insulin (late 1940's and early 1950's). This laborious process involved:
    a. Determination of amino acid composition by complete acid hydrolysis of peptide bonds and separation of resulting amino acids by chromatography. Using these techniques, Sanger identified the amino acids and determined the relative proportion of each.
    b. Determination of amino acid sequence by partial hydrolysis with enzymes and other catalysts to break only specific peptide bonds. Sanger deductively reconstructed the primary structure from fragments with overlapping segments.
- ☞ Most of the sequencing process is now automated.

2. Secondary Structure (p. 79–80)

<u>Secondary structure</u> = Regular, repeated folding of a protein's polypeptide backbone.
- ☞ Contributes to a protein's overall conformation.
- ☞ Stabilized by hydrogen bonds between peptide linkages in the protein's backbone (carbonyl and amino groups).
- ☞ Two types of secondary structure are alpha (α) helix and beta (β) pleated sheet.

What are the two types of secondary structure? What role do hydrogen bonds play in maintaining secondary structure?

a. Alpha (α) Helix

<u>Alpha (α) helix</u> = Secondary structure of a polypeptide that is a helical coil stabilized by hydrogen bonding between every fourth peptide bond (3.6 amino acids per turn).
- ☞ Described by Linus Pauling and Robert Corey in 1951.
- ☞ Found in fibrous proteins (e.g. α-keratin and collagen) for most of their length and in some portions of globular proteins.

b. Beta (β) Pleated Sheet

<u>Beta (β) pleated sheet</u> = Secondary protein structure which is a sheet of antiparallel chains folded into accordion pleats.
- ☞ Held together by interchain hydrogen bonds.
- ☞ Make up the dense core of many globular proteins (e.g. lysozyme) and the major portion of some fibrous proteins (e.g. fibroin, the structural protein of silk).

3. Tertiary Structure (p. 80–81)

<u>Tertiary structure</u> = Irregular contortions of a protein due to bonding between side chains (R groups); third level of protein structure superimposed upon primary and secondary structure.

How do weak interactions and disulfide bridges contribute to tertiary protein structure?

Types of bonds contributing to tertiary structure are weak interactions and covalent linkage (both may occur in the same protein).

a. Weak Interactions

Protein shape is stabilized by the cumulative effect of weak interactions. These weak interactions include:
- ☞ Hydrogen bonding between polar side chains.
- ☞ Ionic bonds between charged side chains.
- ☞ *Hydrophobic interactions* between nonpolar side chains in protein's interior.

<u>Hydrophobic interactions</u> = (Hydro=water; phobos=fear) The clustering of hydrophobic molecules or hydrophobic portions of molecules as a result of their mutual exclusion from water.

48 The Structure and Function of Macromolecules

Disulfide bridges

b. Covalent Linkage

Disulfide bridges form between two cysteine monomers brought together by folding of the protein. This is a strong bond that reinforces conformation.

$$H_3N^+ - \overset{\overset{H}{|}}{C} - \overset{\overset{O}{\parallel}}{C} \diagdown \\ \underset{|}{CH_2} \quad O^- \\ \underset{}{SH}$$

Cysteine

$$-C-CH_2-S-S-CH_2-C-$$

Disulfide Bridge
(S of one cysteine sulfhydryl, bonds to the S of a second cysteine.)

4. Quaternary Structure (p. 81)

Using collagen and hemoglobin as examples, describe quaternary protein structure.

Quaternary structure

Quaternary structure = Structure that results from the interaction among several polypeptides (subunits) in a single protein.
- ☞ For example: collagen, a fibrous protein with three helical polypeptides supercoiled into a triple helix. (See Campbell, Figure 5.26) Found in animal connective tissue, collagen's supercoiled quaternary structure gives it strength.
- ☞ Some globular proteins have subunits that fit tightly together. For example: hemoglobin, a globular protein that has four subunits (two α chains and two β chains).

E. Factors Determining Conformation (p. 81–83)

If a protein's environment is altered, it may lose its native conformation.

Denaturation

Denaturation = A process that alters a protein's native conformation and biological activity.

How can proteins be denatured?

Proteins can be denatured by:
1. Transfer to an organic solvent. Hydrophobic side chains, normally inside the protein's core, move towards the outside. Hydrophilic side chains turn away from the solvent towards the molecule's interior.
2. Chemical agents that disrupt hydrogen bonds, ionic bonds and disulfide bridges.
3. Excessive heat. Increased thermal agitation disrupts weak interactions.

The fact that some denatured proteins return to their native conformation when environmental conditions return to normal is evidence that a protein's amino acid sequence (primary structure) determines conformation. It influences where and which interactions will occur as the molecule arranges into secondary and tertiary structure.

F. The Protein-Folding Problem (p. 83)

It is difficult to predict a protein's conformation from its primary structure:
- ☞ Knowing the final conformation does not reveal the folding intermediates required to get there.
- ☞ A protein's native conformation is dynamic, alternating between several shapes.

Rules of protein folding are important to molecular biologists and the biotechnology industry. This knowledge should allow the design of proteins for specific purposes.

V. NUCLEIC ACIDS (p. 83–87)

What distinguishes nucleic acids from the other major groups of biologically important macromolecules?

A. The Functions of Nucleic Acids: An Overview (p. 83–84)

What are the two types of nucleic acids? What are their functions?

There are two types of nucleic acids:

1. Deoxyribonucleic Acid (DNA)
    - ☞ Can replicate itself.
    - ☞ Is passed from one generation of cells to another.
    - ☞ Is found primarily in the nucleus of eukaryotic cells.
    - ☞ Makes up *genes*, which are units of inheritance that contain instructions for polypeptide synthesis.

2. Ribonucleic Acid (RNA)
    - ☞ Functions in the actual synthesis of proteins coded for by DNA.

Nucleus	mRNA	*Cytoplasm*
Genetic message is transcribed from DNA onto mRNA.	→ moves into cytoplasm	Genetic message translated into a protein.

B. Nucleotides (p. 84–86)

What are the major components of a nucleotide? How are these monomers linked to form a nucleic acid?

Nucleic acid = Polymer of *nucleotides* linked together by condensation reactions.

Nucleotide = Building block molecule of a nucleic acid; made of (1) a five-carbon sugar covalently bonded to (2) a nitrogenous base and (3) a phosphate group.

1. Pentose (5-Carbon Sugar)

There are two pentoses found in nucleic acids: ribose and deoxyribose.

Ribose is the pentose in RNA.

Deoxyribose is the pentose in DNA. (It lacks the −OH group at number two carbon.)

2. Nitrogenous Base

What is the difference between a pyrimidine and a purine?

There are two families of *nitrogenous bases*:

Pyrimidine = Nitrogenous base characterized by a six-membered ring made up of carbon and nitrogen atoms [e.g. cytosine, thymine (DNA only) and uracil (RNA only)].

Purine = Nitrogenous base characterized by a five-membered ring fused to a six-membered ring (e.g. adenine and guanine).

What are the differences among nucleoside, nucleotide and nucleic acid?

Nucleoside = Nitrogenous base + pentose.

Nucleotide = Nitrogenous base + pentose + phosphate.

3. Phosphate

The phosphate group is attached to the number 5 carbon of the sugar.

What are the functions of nucleotides?

☞ Are monomers for nucleic acids.
☞ Transfer chemical energy from one molecule to another (e.g. ATP).
☞ Are electron acceptors in enzyme-controlled redox reactions of the cell (e.g. NAD, NADP).

C. Polynucleotides (*p. 86*)

How are nucleotide monomers linked together to form polynucleotides? Where are the nitrogen bases in this structure? Where are the sugars and phosphates?

<u>Polynucleotide</u> = A polymer of nucleotides joined by *phosphodiester linkages* between the phosphate of one nucleotide and the sugar of the next.
- ☞ Results in a backbone with a repeating pattern of sugar-phosphate-sugar-phosphate.
- ☞ Variable nitrogenous bases are attached to the sugar-phosphate backbone.
- ☞ Each gene contains a unique linear sequence of nitrogenous bases which codes for a unique linear sequence of amino acids in a protein.

D. The Double Helix: An Introduction (*p. 86–87*)

Based upon Watson and Crick's model, how would you briefly describe the three-dimensional structure of DNA?

In 1953, James Watson and Francis Crick proposed the *double helix* as the three-dimensional structure of DNA.
- ☞ Consists of two polynucleotide chains wound in a double helix.
- ☞ Sugar-phosphate backbones are on the outside of the helix.
- ☞ Nitrogenous bases are paired in the interior of the helix and are held together by hydrogen bonds.
- ☞ Base-pairing rules are that adenine (A) always pairs with thymine (T); guanine (G) always pairs with cytosine (C).

VI. DNA AND PROTEINS AS TAPE MEASURES OF EVOLUTION (*p. 87–88*)

Closely related species have more similar sequences of DNA and amino acids, than more distantly related species. Using this type of molecular evidence, biologists can deduce evolutionary relationships among species.

<u>Polynucleotide</u>

<u>Phosphodiester linkages</u>

6

AN INTRODUCTION TO METABOLISM

CHAPTER OUTLINE

I. THE METABOLIC MAP: AN OVERVIEW (p. 91–92)

<u>Metabolism</u> = Totality of an organism's chemical processes.
- ☞ Property emerging from specific molecular interactions within the cell.
- ☞ Concerned with managing cellular resources: material and energy.

How do catabolic and anabolic pathways differ? What is an example of each?

Metabolic reactions are organized into pathways that are orderly series of enzymatically controlled reactions. *Metabolic pathways* are generally of two types:

<u>Catabolic pathways</u> = Metabolic pathways which <u>release energy</u> by breaking down complex molecules to simpler compounds. (e.g. Cellular respiration which degrades glucose to carbon dioxide and water; provides energy for cellular work.)

<u>Anabolic pathways</u> = Metabolic pathways which <u>consume energy</u> to build complicated molecules from simpler ones. (e.g. Photosynthesis which synthesizes glucose from CO_2 and H_2O; any synthesis of a macromolecule from its monomers.)

How are catabolic and anabolic pathways involved in the energy exchanges of cellular metabolism?

Metabolic reactions may be coupled, so that energy released from a catabolic reaction can be used to drive an anabolic one.

II. ENERGY: SOME BASIC PRINCIPLES (p. 92–96)

<u>Energy</u> = Capacity to do work.

  A. Forms of Energy (p. 92)

What is energy? What is the difference between kinetic and potential energy? What are some examples of each?

<u>Kinetic energy</u> = Energy in the process of doing work (energy of motion). For example:
- ☞ Heat (thermal energy) is kinetic energy expressed in random movement of molecules.
- ☞ Light energy from the sun is kinetic energy which powers photosynthesis.

Metabolism

Catabolic pathways

Anabolic pathways

Energy

Kinetic energy

Potential energy = Energy that matter possesses because of its location or arrangement (energy of position). For example:
- In the earth's gravitational field, an object on a hill or water behind a dam have potential energy.
- Chemical energy is potential energy stored in molecules because of the arrangement of nuclei and electrons in its atoms.

B. Energy Transformations (p. 92–93)

Energy can be transformed from one form to another. For example:
- Kinetic energy of sunlight can be transformed into the potential energy of chemical bonds during photosynthesis.
- Potential energy in the chemical bonds of gasoline can be transformed into kinetic mechanical energy which pushes the pistons of an engine.

C. Two Laws of Thermodynamics (p. 93–95)

Explain, in your own words, the First and Second Laws of Thermodynamics.

Thermodynamics = Study of energy transformations.

First Law of Thermodynamics = Energy can be transferred and transformed, but it cannot be created or destroyed (energy of the universe is constant).

Second Law of Thermodynamics = Every energy transfer or transformation makes the universe more disordered (every process increases the *entropy* of the universe).

Entropy = Quantitative measure of disorder that is proportional to randomness (designated by the letter S).

Closed system = Collection of matter under study which is isolated from its surroundings.

Open system = System in which energy can be transferred between the system and its surroundings.

Highly ordered living organisms do not violate the Second Law of Thermodynamics. How is this possible?

Entropy and Living Organisms:

The entropy of a system may decrease, but the entropy of the system _plus its surroundings_ must always increase. Highly ordered living organisms do not violate the second law because they are open systems. For example, animals:
- Maintain highly ordered structure at the expense of increased entropy of their surroundings.
- Take in complex high energy molecules as food and extract chemical energy to create and maintain order.
- Return to the surroundings simpler low energy molecules (CO_2 and water) and heat.

Energy can be transformed, but part of it is dissipated as heat which is largely unavailable to do work. Heat energy can perform work only if there is a heat gradient resulting in heat flow from warmer to cooler.

Combining the first and second laws; the quantity of energy in the universe is constant, but its quality is not.

D. The Free Energy Concept: A Criterion For Spontaneous Change
(p. 95–96)

What portion of a system's energy is available to do work? How is this influenced by a system's entropy, enthalpy and temperature?

Not all of a system's energy is available to do work. The amount of energy that is available to do work is described by the concept of *free energy*. Free energy (G) is related to the system's total energy (H) and its entropy (S) in the following way:

$$G = H - TS$$

where:
- G = Gibbs free energy (energy available to do work)
- H = enthalpy or total energy
- T = temperature in °K
- S = entropy

<u>Free energy (G)</u> = Portion of a system's energy available to do work; is the difference between the total energy (*enthalpy*) and the energy <u>not</u> available for doing work (TS).

Write the Gibbs-Helmholtz equation for free energy change. How do changes in enthalpy, entropy and temperature influence the maximum amount of useable energy that can be harvested from a reaction?

The maximum amount of useable energy that can be harvested from a particular reaction is the system's free energy change from the initial to the final state. This change in free energy (ΔG) is given by the Gibbs-Helmholtz equation at constant temperature and pressure:

$$\Delta G = \Delta H - T\Delta S$$

where:
- ΔG = change in free energy
- ΔH = change in total energy (enthalpy)
- ΔS = change in entropy
- T = absolute temperature (°K)

Why is the concept of free energy useful? What does it indicate about a system under study?

<u>Significance of Free Energy</u>:
1. Indicates the maximum amount of a system's energy which is available to do work.
2. Indicates whether a reaction will occur spontaneously or not.
   - ☞ A *spontaneous reaction* is one that will occur without additional energy.
   - ☞ In a spontaneous process, ΔG or free energy of a system <u>decreases</u> (ΔG < 0).

Free energy

Enthalpy

Spontaneous reaction

What are two major factors that determine whether a process will be spontaneous or have a $-\Delta G$? *How does temperature influence this?*

- A decrease in enthalpy ($-\Delta H$) and an increase in entropy ($+\Delta S$) reduce the free energy of a system and contribute to the spontaneity of a process.
- A higher temperature enhances the effect of an entropy change. Greater kinetic energy of molecules tends to disrupt order as the chances for random collisions increase.
- When enthalpy and entropy changes in a system have an opposite effect on free energy, temperature may determine whether the reaction will be spontaneous or not (e.g. protein denaturation by increased temperature).
- High energy systems, including high energy chemical systems, are unstable and tend to change to a more stable state with a lower free energy.

III. CHEMICAL ENERGY AND LIFE: A CLOSER LOOK (p. 96–97)

How can you classify a reaction based on free energy change? What characterizes exergonic and endergonic reactions, and how do they differ?

Reactions can be classified based upon their free energy changes:

Exergonic reaction = A reaction that proceeds with a net loss of free energy.

Endergonic reaction = An energy-requiring reaction that proceeds with a net gain of free energy.

Exergonic Reaction	Endergonic Reaction
Chemical products have less free energy than the reactant molecules.	Products store more free energy than reactants.
Reaction is energetically downhill.	Reaction is energetically uphill.
Spontaneous reaction.	Non-spontaneous reaction (requires energy input).
ΔG is negative.	ΔG is positive.
$-\Delta G$ is the maximum work the reaction can perform.	$+\Delta G$ is the minimum quantity of work required to drive the reaction.

If a chemical process is exergonic, the reverse process must be endergonic (e.g. cellular respiration, $\Delta G = -686$ kcal/mole; photosynthesis, $\Delta G = +686$ kcal/mole).

Exergonic reaction

Endergonic reaction

What is the relationship between equilibrium and free energy change for a reaction?

There is a relationship between chemical equilibrium and the free energy change (ΔG) of a reaction:
- ☞ As a reaction approaches equilibrium, the free energy of the system decreases (spontaneous and exergonic reaction).
- ☞ When a reaction is pushed away from equilibrium, the free energy of system increases (non-spontaneous and endergonic reaction).
- ☞ When a reaction reaches equilibrium, ΔG = 0 because there is no net change in the system.

In cellular metabolism, endergonic reactions are driven by coupling them to reactions with a greater negative free energy (exergonic). ATP plays a critical role in this energy coupling.

IV. ATP AND CELLULAR WORK (p. 97–100)

What is the function of ATP in the cell?

ATP is the immediate source of energy that drives most cellular work, which includes:
1. *Mechanical work* such as beating of cilia, muscle contraction, cytoplasmic flow, and movement of chromosomes during mitosis and meiosis.
2. *Transport work* such as pumping substances across membranes.
3. *Chemical work* such as the endergonic process of polymerization.

   A. The Structure and Hydrolysis of ATP (p. 98–99)

What are three components of ATP? What is the major class of macromolecules to which it belongs?

ATP (adenosine triphosphate) = Nucleoside triphosphate with unstable phosphate bonds that the cell hydrolyzes for energy to drive endergonic reactions.

- ☞ ATP consists of:
    1. Adenine, a nitrogenous base.
    2. Ribose, a five-carbon sugar.
    3. Chain of three phosphate groups.
- ☞ Unstable bonds between the phosphate groups can be hydrolyzed in an exergonic reaction that releases energy.
- ☞ When the terminal phosphate bond is hydrolyzed, a phosphate group is removed producing ADP (adenosine diphosphate).

    ATP + H$_2$O ⟶ ADP + Ⓟ   ΔG = −7.3 kcal/mol

    This exergonic reaction releases 7.3 kcal/mol under standard conditions (−10 to −12 kcal/mol in the cell).

B. How ATP Performs Work (p. 99)

How does ATP perform cellular work?

Exergonic hydrolysis of ATP is coupled with endergonic processes by transferring a phosphate group to another molecule.
- ☞ Phosphate transfer is enzymatically controlled.
- ☞ The molecule acquiring the phosphate (*phosphorylated* or *activated intermediate*) becomes more reactive.

For example, conversion of glutamic acid to glutamine: (See Campbell, Figure 6.8)

$$\underset{\text{glutamic acid}}{\text{Glu}} + \underset{\text{ammonia}}{NH_3} \longrightarrow \underset{\text{glutamine}}{\text{Gln}} \qquad \Delta G = +3.4 \text{ kcal/mol (endergonic)}$$

Two step process of energy coupling with ATP hydrolysis:

(1) Hydrolysis of ATP and phosphorylation of glutamic acid.

$$\text{Glu} + \text{ATP} \longrightarrow \underset{\substack{\text{unstable}\\\text{phosphorylated}\\\text{intermediate}}}{\text{Glu}-\textcircled{P}} + \text{ADP}$$

(2) Replacement of the phosphate with the reactant ammonia.

$$\text{Glu}-\textcircled{P} + NH_3 \longrightarrow \text{Gln} + \textcircled{P}$$

Overall ΔG:

$$\text{Glu} + NH_3 \longrightarrow \text{Gln} \qquad \Delta G = +3.4 \text{ kcal/mol}$$
$$\text{ATP} \longrightarrow \text{ADP} + \textcircled{P} \qquad \Delta G = \underline{-7.3 \text{ kcal/mol}}$$
$$\text{Net } \Delta G = -3.9 \text{ kcal/mol}$$
(Overall process is exergonic.)

C. The Regeneration of ATP (p. 99)

ATP is continually regenerated by the cell.
- ☞ Process is rapid (10^7 molecules used and regenerated/sec/cell).
- ☞ Reaction is endergonic.

$$\text{ADP} + \textcircled{P} \longrightarrow \text{ATP} \qquad \Delta G = +7.3 \text{ kcal/mol}$$

- ☞ Energy to drive the endergonic regeneration of ATP comes from the exergonic process of cellular respiration.

D. Metabolic Disequilibrium (p. 99–100)

Why is chemical disequilibrium essential for life?

Since many cellular reactions of respiration are reversible, they have the potential to reach equilibrium.
- ☞ At equilibrium, $\Delta G = 0$, so the system can do no work.
- ☞ In the cell, these potentially reversible reactions are pulled forward away from equilibrium, because the products of some reactions become reactants for the next reaction in the metabolic pathway.
- ☞ For example, during cellular respiration a steady supply of high energy reactants such as glucose and removal of low energy products such as CO_2 and H_2O, maintain the disequilibrium necessary for respiration to proceed.

V. ENZYMES (p. 100–106)

Free energy change indicates whether a reaction will occur spontaneously, but does not give information about the speed of reaction.

A chemical reaction will occur spontaneously if it releases free energy ($-\Delta G$), but it may occur too slowly to be effective in living cells. Biochemical reactions require *enzymes* to speed up and control reaction rates.

<u>Catalyst</u> = Chemical agent that accelerates a reaction without being permanently changed in the process, so it can be used over and over.

<u>Enzymes</u> = Biological catalysts, which are usually proteins.

A. Enzymes and the Lowering of Activation Energy (p. 100–102)

Before a reaction can occur, the reactants must absorb energy to break chemical bonds.

<u>Free energy of activation</u> = Amount of energy that reactant molecules must absorb to start a reaction ($\Delta G^{\ddagger}$).

<u>Transition state</u> = Unstable condition of reactant molecules that have absorbed sufficient free energy to react.

Draw and label the energy profile of a chemical reaction including activation energy, free energy change and transition state.

Energy profile of an exergonic reaction:

1. Reactants must absorb enough energy to reach the transition state ($\Delta G^{\ddagger}$) (uphill portion of the curve). Usually the absorption of thermal energy from the surroundings is enough to break chemical bonds.
2. Reaction occurs and energy is released as new bonds form (downhill portion of the curve).
3. ΔG for the overall reaction is the difference in free energy between products and reactants. In an exergonic reaction the free energy of the products is less than reactants.

What is the function of enzymes in biological systems?

Even though a reaction is energetically favorable, there must be an initial investment of activation energy ($\Delta G^{\ddagger}$).

The breakdown of biological macromolecules is exergonic. However, these molecules react <u>very slowly</u> at cellular temperatures because they cannot absorb enough thermal energy to reach transition state.

In order to make these molecules reactive when necessary, cells use biological catalysts called *enzymes*:
- ☞ Are usually proteins.
- ☞ Lower $\Delta G^{\ddagger}$, so the transition state can be reached at cellular temperatures.
- ☞ Do <u>not</u> change the nature of a reaction (ΔG), but only speed up a reaction that would have occurred anyway.
- ☞ Are very selective for which reaction they will catalyze.

B. **The Specificity of Enzymes** (*p. 102*)

Enzymes are specific for a particular *substrate*, and that specificity depends upon the enzyme's three-dimensional shape.

<u>Substrate</u> = The substance an enzyme acts on and makes more reactive.

An enzyme binds to its substrate and catalyzes its conversion to product. The enzyme is released in original form.

Substrate + enzyme ⟶ enzyme-substrate complex ⟶ product + enzyme

The substrate binds to the enzyme's *active site*.

What is the relationship between enzyme structure and enzyme specificity?

<u>Active site</u> = Restricted region of an enzyme molecule which binds to the substrate.
- ☞ Is usually a pocket or groove on the protein's surface.
- ☞ Formed with only a few of the enzyme's amino acids.
- ☞ Determines enzyme specificity which is based upon a compatible fit between the shape of an enzyme's active site and the shape of the substrate.
- ☞ Changes its shape in response to the substrate. This brings its chemical groups into positions that enhance their ability to interact with the substrate and catalyze the reaction.

<u>Induced fit</u> = Change in the shape of an enzyme's active site, which is induced by the substrate.

C. **The Catalytic Cycle of Enzymes** (*p. 102–104*)

What is meant by induced fit? Outline the catalytic cycle of an enzyme.

The entire enzymatic cycle is quite rapid. (See Campbell, Figure 6.15)

<u>Steps in the Catalytic Cycle of Enzymes</u>:
1. Substrate binds to the active site forming an *enzyme-substrate complex*. Substrate is held in the active site by weak interactions (e.g. hydrogen bonds and ionic bonds).
2. *Induced fit* of the active site around the substrate. Side chains of a few amino acids in the active site catalyze the conversion of substrate to product.
3. Product departs active site and the enzyme emerges in its original form.

What are several mechanisms by which enzymes lower activation energy?

Enzymes lower activation energy and speed up reactions by several mechanisms:
- ☞ Active site can hold two or more reactants in the proper position so they may react.
- ☞ Induced fit of the enzyme's active site may distort the substrate's chemical bonds, so less thermal energy (lower $\Delta G^\ddagger$) is needed to break them during the reaction.
- ☞ Active site might provide a micro-environment that is conducive to a particular type of reaction (e.g. localized regions of low pH caused by acidic side chains on amino acids at the active site).
- ☞ Side chains of amino acids in the active site may participate directly in the reaction.

How does substrate concentration affect the rate of an enzyme-controlled reaction?

The initial substrate concentration partly determines the rate of an enzyme controlled reaction.
- ☞ The higher the substrate concentration, the faster the reaction — up to a limit.
- ☞ If substrate concentration is high enough, the enzyme becomes *saturated* with substrate. (The active sites of all enzymes molecules are engaged.)
- ☞ When an enzyme is saturated, the reaction rate depends upon how fast the active sites can convert substrate to product.

D. Factors Affecting Enzyme Activity (*p. 104–106*)

How is enzyme activity controlled or regulated?

1. Environmental Conditions (*p. 104*)

Each enzyme has optimal environmental conditions that favor the most active enzyme conformation.

a. Temperature

Optimal temperature allows the greatest number of molecular collisions without denaturing the enzyme. Optimal temperature range of most human enzymes is 35°–40°C.

b. pH

Optimal pH range for most enzymes is pH 6–8. There are some enzymes that operate best at more extremes of pH (e.g. digestive enzyme, pepsin, found in the acid environment of the stomach has an optimal pH = 2).

c. Ionic Concentration

Inorganic ions can interfere with ionic bonds <u>within</u> the enzyme molecule. With few exceptions, most enzymes cannot tolerate high salt concentrations.

Saturated

2. Cofactors (p. 105)

<u>Cofactors</u> = Small nonprotein molecules that are required for proper enzyme catalysis.
- ☞ May bind tightly to active site.
- ☞ May bind loosely to both active site and substrate.
- ☞ Some are inorganic (e.g. metal atoms of zinc, iron or copper).
- ☞ Some are organic and are called *coenzymes* (e.g. most vitamins).

3. Enzyme Inhibitors (p. 105)

Certain chemicals can selectively inhibit enzyme activity. (See Campbell, Figure 6.17)
- ☞ Inhibition may be <u>irreversible</u> if the inhibitor attaches by covalent bonds.
- ☞ Inhibition may be <u>reversible</u> if the inhibitor attaches by weak bonds.

<u>Competitive inhibitors</u> = Chemicals that resemble an enzyme's normal substrate and compete with it for the active site.
- ☞ Block active site from the substrate.
- ☞ If reversible, the effect of these inhibitors can be overcome by increased substrate concentration.

<u>Noncompetitive inhibitors</u> = Enzyme inhibitors that do not enter the enzyme's active site, but bind to another part of the enzyme molecule.
- ☞ Causes enzyme to change its shape so the active site cannot bind substrate.
- ☞ May act as metabolic poisons (e.g. DDT, many antibiotics).
- ☞ Selective enzyme inhibition is an essential mechanism in the cell for regulating metabolic reactions.

4. Allosteric Regulation (p. 105–106)

<u>Allosteric site</u> = Specific receptor site on some part of the enzyme molecule other than the active site.
- ☞ Most enzymes with allosteric sites have two or more polypeptide chains, each with its own active site. Allosteric sites are located where the subunits join.
- ☞ Allosteric enzymes have two conformations, one catalytically active and the other inactive. (See Campbell, Figure 6.18)
- ☞ Binding of an <u>activator</u> to an allosteric site stabilizes the active conformation.
- ☞ Binding of an <u>inhibitor</u> to an allosteric site stabilizes the inactive conformation.
- ☞ Enzyme activity changes continually in response to changes in the relative proportions of activators and inhibitors (e.g. ATP/ADP).
- ☞ Subunits may interact so that a single activator or inhibitor at one allosteric site will affect the active sites of the other subunits.

Cofactors

Coenzymes

Competitive inhibitors

Noncompetitive inhibitors

Allosteric site

62 An Introduction to Metabolism

Cooperativity

What is the difference between allosteric activation and cooperativity?

Cooperativity = The phenomenon where substrate binding to the active site of one subunit induces a conformational change that enhances substrate binding at the active sites of the other subunits.

VI. THE CONTROL OF METABOLISM (p. 106–107)

How are metabolic pathways regulated?

Metabolic pathways are regulated by controlling enzyme activity.

A. Feedback Inhibition (p. 106–107)

Feedback inhibition

Feedback inhibition = Regulation of a metabolic pathway by its end product, which inhibits an enzyme within the pathway.

threonine $\xrightarrow{\text{Enzyme 1}}$ A $\xrightarrow{\text{Enzyme 2}}$ B $\xrightarrow{\text{Enzyme 3}}$ C $\xrightarrow{\text{Enzyme 4}}$ D $\xrightarrow{\text{Enzyme 5}}$ isoleucine

(initial substrate) — (end product <u>and</u> allosteric inhibitor of enzyme 1)

Feedback Inhibition

Prevents the cell from wasting chemical resources by synthesizing more product than is necessary.

B. Structural Order and Metabolism (p. 107)

Cellular structure orders and compartmentalizes metabolic pathways.

Multienzyme complex

Multienzyme complex = Enzyme assemblage for several steps of a metabolic pathway.
- ☞ Physical arrangement of a multienzyme complex orders the sequence of reactions.
- ☞ Some enzymes have fixed locations in the cell because they are incorporated into a membrane.
- ☞ Dissolved enzymes and their substrates may be localized within organelles such as chloroplasts and mitochondria.

7   A TOUR OF THE CELL

CHAPTER OUTLINE

All organisms are made of cells, the organism's basic unit of structure and function.

The cell as a microcosm can be used to illustrate four themes integral to the text and course:
1. Theme of emergent properties. Life at the cellular level arises from interactions among cellular components.
2. Correlation of structure and function. Ordered cellular processes (e.g. protein synthesis, respiration, photosynthesis, cell-cell recognition, cellular movement, membrane production and secretion) are based upon ordered structures.
3. Interaction of organisms within their environment. Cells are excitable responding to environmental stimuli. In addition, cells are open systems that exchange materials and energy with their environment.
4. Unifying theme of evolution. Evolutionary adaptations are the basis for the correlation between structure and function.

I. HOW CELLS ARE STUDIED (p. 117–121)

What techniques are used to study cell structure and function?

The microscope's invention and improvement in the seventeenth century led to the discovery and study of cells. Modern cell biology integrates the study of cell structure (*cytology*) with biochemistry.
- ☞ *Microscopy* is an indispensable tool for studying cell structure
- ☞ *Cell fractionation* enables researchers to isolate organelles in order to study their function.

A. Microscopy (p. 117–119)

What are the basic operating principles of a light microscope, transmission electron microscope and scanning electron microscope? What are some advantages and limitations of each?

In 1665, Robert Hooke described cells using a *light microscope*. (See Campbell, Figure 7.3) Modern light microscopy is based upon the same principles as microscopy first used by Renaissance scientists.
- ☞ Visible light is focused on a specimen with a *condenser lens*.
- ☞ Light passing through the specimen is refracted with an *objective lens* and an *ocular lens*. The specimen's image is thus magnified and inverted for the observer.

<u>Light microscope</u>

64 A Tour of the Cell

What is the difference between magnification and resolving power? What limits the resolution of a light microscope?

Two important concepts in microscopy are *magnification* and *resolving power*.

<u>Magnification</u> = How much larger an object is made to appear compared to its real size.

<u>Resolving power</u> = Minimum distance between two points that can still be distinguished as two separate points.
- Resolution of a light microscope is limited by the wavelength of visible light. Maximum possible resolution of a light microscope is 0.2μm.
- Highest magnification in a light microscope with maximum resolution is about 1500 times.
- By the early 1900's, optics in light microscopes were good enough to achieve the best resolution, so improvements since then have focused on improving contrast.

In the 1950's, researchers began to use the *electron microscope* which far surpassed the resolving power of the light microscope.
- Resolving power is inversely related to wavelength. Instead of light, electron microscopes use electron beams which have much shorter wavelengths than visible light.
- Modern electron microscopes have a resolving power of about 0.2nm.
- Enhanced resolution and magnification allowed researchers to clearly identify subcellular *organelles* and to study cell *ultrastructure*.
- Two types of electron microscopes are the *transmission electron microscope* (TEM) and the *scanning electron microscope*.

The *transmission electron microscope* (TEM) aims an electron beam at a thin section of specimen which may be stained with metals to absorb electrons and enhance contrast.
- Electrons <u>transmitted</u> through the specimen are focused and the image magnified by using electromagnetic lenses (rather than glass lenses) to bend the trajectories of the charged electrons.
- Image is focused onto a viewing screen or film.
- Used to study internal cellular ultrastructure.

The *scanning electron microscope* (SEM) is useful for studying the surface of a specimen.
- Electron beam <u>scans</u> the surface of the specimen usually coated with a thin film of gold.
- Scanning beam excites secondary electrons on the sample's surface.
- Secondary electrons are collected and focused onto a viewing screen.
- SEM has a great depth of field and produces a three-dimensional image.

<u>Disadvantages of an Electron Microscope</u>:
- Can usually only view dead cells because of the elaborate preparation required.
- May introduce structural artifacts.

Magnification

Resolving power

Transmission electron microscope

Scanning electron microscope

B. Cell Fractionation (p. 120–121)

Why is cell fractionation a useful technique? What are the major steps in the process of fractionation?

Organelles can be separated from cells with *cell fractionation*, a technique gentle enough to preserve organelle function.

<u>Cell fractionation</u> = Technique which involves centrifuging disrupted cells at various speeds and durations to isolate components of different sizes, densities and shapes.

Process of Cell Fractionation:
- Homogenization of tissue and its cells using pistons, blenders or ultrasound devices.
- Centrifugation of the resulting homogenate at a slow speed. Nuclei and other larger particles settle at the bottom of the tube, forming a *pellet*.
- The supernatant is decanted into another tube and centrifuged at a faster speed, separating out smaller organelles.
- The previous step is repeated, increasing the centrifugation speed each time to collect smaller and smaller cellular components from the pellet.
- Once the cellular components are separated and identified, their particular metabolic functions can be determined.

II. THE GEOGRAPHY OF THE CELL: A PANORAMIC VIEW (p. 121–123)

A. Prokaryotic and Eukaryotic Cells (p. 121)

How do prokaryotic and eukaryotic cells differ? In what ways are they similar?

Living organisms are made of either prokaryotic or eukaryotic cells — two major kinds of cells, which can be distinguished by structural organization.

Prokaryotic (pro=before; karyon=kernel)	Eukaryotic (Eu=true; karyon=kernel)
Found only in the Kingdom Monera (bacteria and cyanobacteria).	Found in the Kingdoms Protista, Fungi, Plantae, and Animalia.
No true nucleus; lacks nuclear envelope.	True nucleus; bounded by nuclear envelope.
Genetic material in *nucleoid* region.	Genetic material within nucleus.
No membrane-bound organelles.	Contains cytoplasm with *cytosol* and membrane-bound *organelles*.

<u>Cytoplasm</u> = Entire region between the nucleus and cell membrane.

<u>Cytosol</u> = Semi-fluid medium found in the cytoplasm.

B. Cell Size (p. 121–122)

Size ranges of cells:

Cell Type	Diameter
Mycoplasmas	0.1 – 1.0 μm
Most bacteria	1.0 – 10.0 μm
Most eukaryotic cells	10.0 – 100.0 μm

What determines the upper and lower limits to cell size?

Range of cell size is limited by metabolic requirements. The lower limits are probably determined by the smallest size with:
- ☞ Enough DNA to program metabolism.
- ☞ Enough ribosomes, enzymes and cellular components to sustain life and reproduce.

The upper limits of size are imposed by the surface area to volume ratio. As a cell increases in size, its volume grows proportionately more than its surface area. (See Campbell, Figure 7.7)
- ☞ The surface area of the plasma membrane must be large enough for the cell volume, in order to provide an adequate exchange surface for oxygen, nutrients and wastes.
- ☞ The cytoplasmic volume must be small enough for the nucleus to control.

C. The Importance of Compartmental Organization (p. 122–123)

Why is compartmentalization important in eukaryotic cells?

The average eukaryotic cell has a thousand times the volume of the average prokaryotic cell, but only a hundred times the surface area. Eukaryotic cells compensate for the small surface area to volume ratio by having internal membranes which:
- ☞ Serve as partitions, compartmentalizing the cell.
- ☞ Have unique lipid and protein compositions depending upon their specific functions.
- ☞ May participate in metabolic reactions since many enzymes are incorporated directly into the membrane.
- ☞ Provide localized environmental conditions necessary for specific metabolic processes.
- ☞ Sequester reactions, so they may occur without interference from incompatible metabolic processes elsewhere in the cell.

III. THE NUCLEUS (p. 123–127)

Describe the structure and function of the nucleus including the nuclear envelope, chromosomes and nucleolus.

<u>Nucleus</u> = Generally, a conspicuous membrane-bound cellular organelle in a eukaryote; contains most of the genes that control the entire cell.
- ☞ Averages about 5 μm diameter.
- ☞ Enclosed by a *nuclear envelope*.

Nucleus

A Tour of the Cell 67

<u>Nuclear envelope</u> = A membrane which encloses the nucleus in a eukaryotic cell.
- Is a double membrane separated by a space of about 20–40nm.
- Nuclear side of the inner membrane is associated with a protein layer that helps maintain nuclear shape and helps organize the genetic material (*chromatin*).
- Perforated by *pores* which regulate molecular transport into and out of the nucleus.
- Inner and outer membranes of the nuclear envelope are connected at the margins of the pore. The pores are ordered by a complex of eight protein granules. (See Campbell, Figure 7.11)

The nucleus contains most of the cell's DNA which is organized with proteins into *chromosomes*.

<u>Chromosomes</u> = Long thread-like association of genes, composed of *chromatin* and found in the nucleus of eukaryotic cells.
- Dispersed in nondividing cells.
- Condensed when the cell prepares to divide.
- Each species has a characteristic number of chromosomes (humans have 46).

<u>Chromatin</u> = A complex of DNA and histone proteins, which makes up chromosomes in eukaryotic cells; appears as a mass of stained material in nondividing cells.

The most visible structure within the nondividing nucleus is the *nucleolus*.

<u>Nucleolus</u> = Roughly spherical region in the nucleus of nondividing cells, which consists of *nucleolar organizers* and ribosomes in various stages of production.
- Functions in ribosome synthesis.
- May be two or more per cell.
- Can produce up to 10,000 ribosomes per minute in an actively growing cell.

<u>Nucleolar organizers</u> = Specialized regions of some chromosomes, with multiple copies of genes for ribosome synthesis.

How does the nucleus control protein synthesis in the cytoplasm?

The nucleus controls protein synthesis in the cytoplasm:

Messenger RNA (mRNA) *transcribed* in the nucleus from DNA instructions.
↓
Passes through nuclear pores into cytoplasm.
↓
Attaches to ribosomes where the genetic message is *translated* into primary protein structure.

Nuclear envelope

Chromosomes

Chromatin

Nucleolus

Nucleolar organizers

IV. RIBOSOMES (p. 127–128)

Describe the structure of a eukaryotic ribosome. What is its function? Where is it constructed?

<u>Ribosome</u> = A cytoplasmic organelle which functions in protein synthesis.
- ☞ Constructed in the nucleolus in eukaryotic cells.
- ☞ Cells with high rates of protein synthesis have prominent nucleoli and many ribosomes (e.g. human liver cell has a few million).

Eukaryotic ribosomes are constructed in the nucleolus from two subunits. (See Campbell, Figure 7.12)
- ☞ Subunits are made from: (1) rRNA produced in the nucleolus; (2) RNA produced elsewhere in the nucleus; (3) proteins imported from the cytoplasm.
- ☞ The two subunits are joined in the cytoplasm when they attach to mRNA.
- ☞ Prokaryotic ribosomes are smaller and have slightly different molecular composition. (Some antibiotics such as tetracycline and streptomycin are useful to fight bacterial infections in mammals because they inhibit the prokaryotic ribosome, but do <u>not</u> affect eukaryotic protein synthesis.)

Ribosomes function either free in the cytosol or bound to endomembranes. Bound and free ribosomes are structurally identical and interchangeable.

<u>Free ribosomes</u> = Ribosomes suspended in the cytosol.
- ☞ Most proteins made by free ribosomes will function in the cytosol.

<u>Bound ribosomes</u> = Ribosomes attached to the outside of the endoplasmic reticulum.
- ☞ Generally make proteins that are destined for membrane inclusion or export.

V. THE ENDOMEMBRANE SYSTEM (p. 128–133)

Biologists now consider many membranes of the eukaryotic cell to be part of an *endomembrane system*.
- ☞ Membranes may be interrelated <u>directly</u> through physical contact.
- ☞ Membranes may be related <u>indirectly</u> through *vesicles*.

<u>Vesicles</u> = Membrane-enclosed sacs that are pinched off portions of membranes moving from the site of one membrane to another.

Membranes of the endomembrane system vary in structure and function, and the membranes themselves are dynamic structures changing in composition, thickness and behavior.

What components make up the endomembrane system? What are their individual functions? How do their structures uniquely fit their function?

The endomembrane system includes:
- Nuclear envelope.
- Endoplasmic reticulum.
- Golgi apparatus.
- Lysosomes.
- Vacuoles.
- Plasma membrane (not actually an *endomembrane*, but related to endomembrane system).

A. Endoplasmic Reticulum (*p. 128–129*)

<u>Endoplasmic reticulum (ER)</u> = (Endoplasmic=within the cytoplasm; reticulum= network) Extensive membranous network of tubules and sacs (*cisternae*) which sequesters its internal lumen (*cisternal space*) from the cytosol.
- Most extensive portion of endomembrane system.
- Continuous with the outer membrane of the nuclear envelope; therefore, the space between the membranes of the nuclear envelope is continuous with cisternal space.

There are two distinct regions of ER that differ in structure and function: smooth ER and rough ER.

1. Functions of Smooth ER (*p. 128–129*)

Appears smooth in the electron microscope because its cytoplasmic surface lacks ribosomes. Smooth ER functions in diverse metabolic processes:

a. Synthesizes lipids, phospholipids and steroids.
- For example: mammalian sex hormones and steroids secreted by the adrenal gland.
- Cells that produce and secrete these products are rich in smooth ER (e.g. testes, ovaries, skin oil glands).

b. Participates in carbohydrate metabolism.
- Smooth ER in liver contains an embedded enzyme that catalyzes the final step in the conversion of glycogen to glucose (removes the phosphate from glucose-phosphate).

c. Detoxifies drugs and poisons.
- Smooth ER, especially in the liver, contains enzymes which detoxify drugs and poisons.
- Enzymes catalyze the addition of hydroxyl groups to drugs and poisons. This makes them soluble in the cytosol, so they may be excreted from the body.
- Smooth ER in liver cells proliferates in response to barbiturates, alcohol and other drugs.

Endoplasmic reticulum

Smooth ER

70   A Tour of the Cell

Rough ER

Oligosaccharide

Transport vesicle

    d. **Stores calcium ions necessary for muscle contraction.**
- In a muscle cell, the ER membrane pumps Ca^{++} from the cytosol into the cisternal space.
- In response to a nerve impulse, Ca^{++} leaks from the ER back into the cytosol, which triggers muscle cell contraction.

2. Rough ER and Protein Synthesis (*p. 129*)

Appears rough under an electron microscope because the cytoplasmic side is studded with ribosomes. It is continuous with outer membrane of the nuclear envelope (which may also be studded with ribosomes on the cytoplasmic side). Rough ER manufactures secretory proteins and membrane.

Proteins destined for secretion are synthesized by ribosomes attached to rough ER:

> Ribosomes attached to rough ER synthesize secretory proteins.
> ↓
> Growing polypeptide is threaded through ER membrane into the lumen or *cisternal space*.
> ↓
> Protein folds into its native conformation.
> ↓
> If destined to be a glycoprotein, enzymes localized in the ER membrane catalyze the covalent bonding of an *oligosaccharide* to the secretory protein.
> ↓
> Protein departs in a *transport vesicle* pinched off from *transitional ER* adjacent to the rough ER site of production.

Oligosaccharide = Small polymer of sugar units.

Transport vesicle = Membrane vesicles in transit from one part of the cell to another.

3. Rough ER and Membrane Production (*p. 129*)

Membranes of rough ER grow <u>in place</u> as newly formed proteins and phospholipids are assembled:
- Membrane proteins are produced by the ribosome. As the polypeptide grows it is inserted directly into the rough ER membrane where it is anchored by hydrophobic regions of the proteins.
- Enzymes within the ER membrane synthesize phospholipids from raw materials in the cytosol.
- Newly expanded ER membrane can be transported as a vesicle to other parts of the cell.

B. The Golgi Apparatus (*p. 130–131*)

<u>Golgi apparatus</u> = Organelle made of stacked, flattened membranous sacs (*cisternae*), that modifies, stores and routes products of the endoplasmic reticulum. (See Campbell, Figure 7.15)
- ☞ Membranes of the cisternae sequester cisternal space from the cytosol.
- ☞ Vesicles may transport macromolecules between the Golgi and other cellular structures.
- ☞ Has a distinct polarity. Membranes of cisternae at opposite ends differ in thickness and composition.

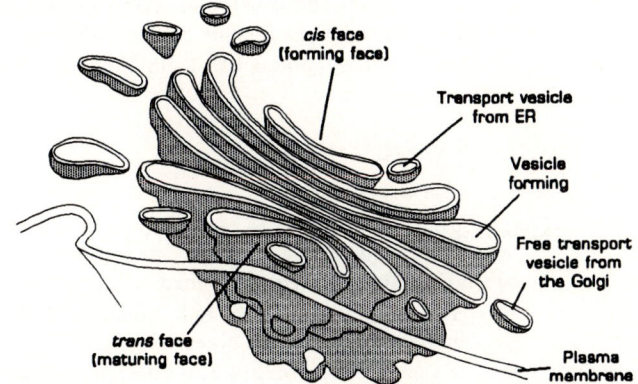

- ☞ Two poles are called the *cis face* (forming face) and the *trans face* (maturing face).
- ☞ *Cis face*, which is closely associated with transi-tional ER, receives products by accepting transport vesicles from the ER. A vesicle fuses its membrane to the cis face of the Golgi and empties its soluble contents into the Golgi's cisternal space.
- ☞ *Trans face* pinches off vesicles from the Golgi and transports molecules to other sites.

Enzymes in the Golgi modify the products of the ER in stages as they move through the Golgi stack from the *cis* to the *trans* face:
- ☞ Alters some membrane phospholipids.
- ☞ Modifies the oligosaccharide portion of glycoproteins.
- ☞ Manufactures certain macromolecules itself (e.g. hyaluronic acid).
- ☞ Targets products for various parts of the cell. Membranous vesicles budded from the Golgi may have external molecules that recognize "docking sites" on the surface of specific organelles.
- ☞ Sorts products for secretion. Phosphate groups or specific oligosaccharides that are added to Golgi products may be molecular identification tags.
- ☞ Products destined for secretion leave the *trans* face in vesicles which eventually fuse with the plasma membrane.

<u>Golgi apparatus</u>

<u>Cis face</u>

<u>Trans face</u>

C. Lysosomes (p. 131–132)

Lysosome = An organelle which is a membrane-enclosed bag of hydrolytic enzymes that digest all major classes of macromolecules.
- Enzymes include lipases, carbohydrases, proteases and nucleases.
- Optimal pH for lysosomal enzymes is about pH 5.
- Lysosomal membrane performs two important functions:
    1. Sequesters potentially destructive hydrolytic enzymes from the cytosol.
    2. Maintains the optimal acidic environment for enzyme activity by pumping H^+s inward from the cytosol to the lumen.
- Hydrolytic enzymes and lysosomal membrane are synthesized in the rough ER and processed further in the Golgi apparatus.
- Lysosomes probably pinch off from the *trans* face of the Golgi apparatus.

Functions of Lysosomes:

1. Intracellular digestion.

    Phagocytosis = (Phago=to eat; cyte=cell) Cellular process of ingestion, where the plasma membrane engulfs particulate substances and pinches off to form a particle-containing *vacuole*.
    - Lysosomes may fuse with food-filled vacuoles, and their hydrolytic enzymes digest the food.
    - Examples are *Amoeba* and other protozoa which eat smaller organisms or food particles.
    - Human cells called *macrophages* phagocytize bacteria and other invaders.

2. Recycle cell's own organic material.
    - Lysosomes may engulf other cellular organelles or part of the cytosol and digest them with hydrolytic enzymes.
    - Resulting monomers are released into the cytosol where they can be recycled into new macromolecules.

3. Programmed cell destruction.

    Destruction of cells by their own lysosomes is important during metamorphosis and development.

Lysosomes and Human Disease:

How are the symptoms of storage diseases caused by impaired lysosomal function?

Symptoms of inherited *storage diseases* result from impaired lysosomal function. Lack of a specific lysosomal enzyme causes substrate accumulation which interferes with lysosomal metabolism and other cellular functions.
- In Pompe's disease, the missing enzyme is a carbohydrase that breaks down glycogen. The resulting glycogen accumulation damages the liver.
- Lysosomal lipase is missing in Tay-Sachs disease, which causes lipid accumulation in the brain.

A Tour of the Cell 73

D. Vacuoles (*p. 132–133*)

Vacuole = Organelle which is a membrane-enclosed sac that is larger than a vesicle (transport vesicle, lysosome or microbody).

What are the types of vacuoles? How do their functions differ?

Vacuole Types and Functions:

Food vacuole = Vacuole formed by phagocytosis which is the site of intracellular digestion in some protozoa and macrophages.

Contractile vacuole = Vacuole, found in some fresh-water protozoa, that pumps excess water from the cell.

Central vacuole = Large vacuole found in most mature plant cells.
- Is enclosed by a membrane called the *tonoplast* which is part of the endomembrane system.
- Develops by the coalescence of smaller vacuoles derived from the ER and Golgi apparatus.
- Is a versatile compartment with many functions:
    1. Stores organic compounds (e.g. protein storage in seeds).
    2. Stores inorganic ions (e.g. K^+ and Cl^-).
    3. Contains hydrolytic enzymes that breakdown macromolecules and recycle organic compounds from worn-out organelles.
    4. Sequesters dangerous metabolic by-products from the cytoplasm.
    5. Contains soluble pigments in some cells (e.g. red and blue pigments in flowers).
    6. May protect the plant from predators by containing poisonous or unpalatable compounds.
    7. Plays a role in plant growth by absorbing water and elongating the cell.
    8. Contributes to the large ratio of membrane surface area to cytoplasmic volume. (There is only a thin layer of cytoplasm between the tonoplast and plasma membrane.)

E. Relationships Among Endomembranes: A Summary (*p. 133*)

What are the relationships among the different components of the endomembrane system?

Components of the endomembrane system are related through direct contact or through vesicles. (See Campbell, Figure 7.19)

Vacuole

Food vacuole

Contractile vacuole
Central vacuole
Tonoplast

74   A Tour of the Cell

Nuclear Envelope —is extension of→ Rough ER ←is confluent with— *Smooth ER*

↓ membrane and secretory proteins produced in ER are transported in

Vesicles

↓ Fuse with the forming face of

Golgi Apparatus

↓ pinches off maturing face

Vesicles

— give rise to → *Lysosomes* and *Tonoplast*

— fuse with and add to plasma membrane and may release cellular products to outside → *Plasma Membrane*

VI. PEROXISOMES (MICROBODIES) (p. 133–134)

Describe the structure of a peroxisome. What is its role in eukaryotic cells?

<u>Peroxisomes</u> = Membrane-bound organelles that contain specialized teams of enzymes for specific metabolic pathways; all contain peroxide-producing oxidases.

- ☞ Bound by a single membrane.
- ☞ Found in nearly all eukaryotic cells.
- ☞ Often have a granular or crystalline core which is a dense collection of enzymes. (See Campbell, Figure 7.20)
- ☞ Contain peroxide-producing oxidases that transfer hydrogen from various substrates to oxygen, producing hydrogen peroxide.

$$RH_2 + O_2 \xrightarrow{oxidase} R + H_2O_2$$

- ☞ Contain catalase, an enzyme that converts toxic hydrogen peroxide to water.

$$2H_2O_2 \xrightarrow{catalase} 2H_2O + O_2$$

- ☞ Peroxisomal reactions have many functions, some of which are:
  1. Breakdown of fats into smaller molecules (acetyl CoA). The products are carried to the mitochondria as fuel for cellular respiration.
  2. Detoxification of alcohol and other harmful compounds. In the liver, peroxisomes enzymatically transfer H from poisons to O_2.

<u>Peroxisomes</u>

☞ Specialized microbodies (*glyoxysomes*) are found in heterotrophic fat-storing tissue of germinating seeds.
   1. Contain enzymes that convert lipid to carbohydrate.
   2. These biochemical pathways make energy stored in seed oils available for the germinating seedling.
☞ Current thought is that peroxisome biogenesis occurs by pinching off from preexisting peroxisomes. Necessary lipids and enzymes are imported from the cytosol.

VII. ENERGY TRANSDUCERS: MITOCHONDRIA AND CHLOROPLASTS
(*p. 134–136*)

Mitochondria and chloroplasts are organelles that transduce energy acquired from the surroundings into forms useable for cellular work.
☞ Enclosed by <u>double</u> membranes.
☞ Membranes are not part of endomembrane system. Rather than being made in the ER, their membrane proteins are synthesized by free ribosomes in the cytosol and by ribosomes located within these organelles.
☞ Contain ribosomes and some DNA that programs a small portion of their own protein synthesis.
☞ Are semiautonomous organelles that grow and reproduce within the cell.

A. Mitochondria (*p. 135*)

<u>Mitochondria</u> = Organelles which are the sites of cellular respiration, a catabolic oxygen-requiring process that uses energy extracted from organic macromolecules to produce ATP.
☞ Found in nearly all eukaryotic cells.
☞ Number of mitochondria per cell varies and directly correlates with the cell's metabolic activity.
☞ Are about 1 μm in diameter and 1–10 μm in length.
☞ Are dynamic structures that move, change their shape and divide.

Describe the structure of a mitochondrion. What is the importance of compartmentalization in mitochondrial function?

<u>Structure of the Mitochondrion:</u>
☞ Enclosed by two membranes that have their own unique combination of proteins embedded in phospholipid bilayers. (See Campbell, Figure 7.23)
☞ Smooth *outer membrane* is highly permeable to small solutes, but it blocks passage of proteins and other macromolecules.
☞ Convoluted *inner membrane* contains embedded enzymes that are involved in cellular respiration. The membrane's many infoldings or *cristae* increase the surface area available for these reactions to occur.

Mitochondria

☞ The inner and outer membranes divide the mitochondrion into two internal compartments:

1. Intermembrane Space
   ☞ Narrow region between the inner and outer mitochondrial membranes.
   ☞ Reflects the solute composition of the cytosol, because the outer membrane is permeable to small solute molecules.

2. Mitochondrial Matrix
   ☞ Compartment enclosed by the inner mitochondrial membrane.
   ☞ Contains enzymes that catalyze many metabolic steps of cellular respiration.
   ☞ Some enzymes of respiration and ATP production are actually embedded in the inner membrane.

B. Chloroplasts (p. 135–136)

What is a plastid? What types of plastids are there and how do they differ?

Plastids = A group of plant and algal membrane-bound organelles that include *amyloplasts*, *chromoplasts* and *chloroplasts*.

Amyloplasts = (Amylo = starch) Colorless plastids that store starch; found in roots and tubers.

Chromoplasts = (Chromo = color) Plastids containing pigments other than chlorophyll; responsible for the color of fruits, flowers and autumn leaves.

Chloroplasts = (Chloro = green) Chlorophyll-containing plastids which are the sites of photosynthesis.
   ☞ Found in eukaryotic algae, leaves and other green plant organs.
   ☞ Are lens-shaped and measure about 2 μm by 5 μm.
   ☞ Are dynamic structures that change shape, move and divide.

What determines the fate of proplastids during plastid development?

Chloroplasts, amyloplasts and chromoplasts can develop from *proplastids*.

Proplastids = Organelles in undifferentiated plant cells, that are destined to become a type of plastid.
- ☞ Their fate depends upon cell location and environment.
- ☞ For example, a proplastid requires light exposure to become a chloroplast. (Chloroplasts may also form from the division of preexisting chloroplasts.)

What are the three functional compartments of a chloroplast? Why is this compartmentalization important in chloroplast function?

Structure of the Chloroplast:

Chloroplasts are lens-shaped organelles measuring about 2–4 μm by 4–7 μm. These organelles are divided into three functional compartments by a system of membranes:

1. **Intermembrane Space.** The chloroplast is bound by a double membrane which partitions its contents from the cytosol. A narrow *intermembrane space* separates the two membranes.

2. **Thylakoid Space.** *Thylakoids* form another membranous system within the chloroplast. The thylakoid membrane segregates the interior of the chloroplast into two compartments: *thylakoid space* and *stroma*.

   Thylakoids = Flattened membranous sacs inside the chloroplast.
   - ☞ Chlorophyll is found in the thylakoid membranes.
   - ☞ Thylakoids function in the steps of photosynthesis that initially convert light energy to chemical energy.

   Thylakoid space = Space inside the thylakoid.

   Grana = (Singular, granum) Stacks of thylakoids in a chloroplast.

3. **Stroma.** Those steps that use chemical energy to convert carbon dioxide to sugar occur in the *stroma*.

   Stroma = Viscous fluid outside the thylakoids.

VIII. THE CYTOSKELETON (*p. 136–143*)

It was originally thought that organelles were suspended in a formless cytosol. Technological advances in both light and electron microscopy (e.g. high voltage E.M.) revealed a three-dimensional view of the cell, which showed a network of fibers throughout the cytoplasm.

Proplastids

Thylakoids

Thylakoid space

Grana

Stroma

A Tour of the Cell

Cytoskeleton

What are probable functions of the cytoskeleton?

Cytoskeleton = A network of fibers throughout the cytoplasm that forms a dynamic framework for support and movement.
- Gives mechanical support to the cell and helps maintain its shape.
- Enables a cell to change shape.
- Associated with motility.
- Constructed from at least three types of fibers: *microtubules* (thickest), *microfilaments* (thinnest) and *intermediate filaments* (intermediate in diameter). (See Campbell, Table 7.2)

What are the monomers, structures and functions of microtubules, microfilaments and intermediate filaments?

A. Microtubules (p. 137–141)

Microtubules

Tubulin

Structure of Microtubules:
- Straight <u>hollow</u> fibers about 25 nm in diameter.
- Constructed from globular proteins called *tubulin* (α-tubulin and β-tubulin).
- Begins as a two-dimensional sheet of tubulin units, which rolls into a tube.
- Tubulin tube can elongate by adding tubulin units to one end. (See Campbell, Table 7.2)

Monomers of Microtubules:
- Building block molecule is a dimer that consists of one α-tubulin molecule and one β-tubulin molecule.
- May be disassembled and the tubulin units recycled to build microtubules elsewhere in the cell.

Function of Microtubules:

Found in cytoplasm of all eukaryotic cells, microtubules may have the following functions:

1. Cellular support.
   - May radiate from the *centrosome*, a *microtubule-organizing center* near the nucleus, and form a framework for cellular support.
   - Microtubular bundles near plasma membrane reinforce cell shape.
2. Tracks for organelle movement. (See Campbell, Figure 7.26)
   - Protein *motor molecules* (e.g. kinesin) interact with microtubules to translocate organelles.

Motor molecules

3. Separation of chromosomes during cell division.
4. Make up *centrioles* in animal cells. (See Campbell, Figure 7.27)

Centriole

Centriole = Pair of cylindrical structures in animal cells, composed of nine sets of triplet microtubules arranged in a ring.
- Are about 150 nm in diameter and are arranged at right angles to each other.
- Pair of centrioles located within the centrosome, replicate during cell division.
- May organize microtubule assembly during cell division, but must not be mandatory for this function since plants lack centrioles.

5. Cell motility.

Cilia and flagella = Locomotor organelles found in eukaryotes, which are formed from a specialized arrangement of microtubules.
- ☞ May propel single-celled organisms (Protista) and motile sperm cells through an aquatic medium.
- ☞ May function to draw fluid across the surface of stationary cells (e.g. ciliated cells lining trachea).

Cilia (singular, cilium)	Flagella (singular, flagellum)
Occur in large numbers on cell surface.	One or a few per cell.
Shorter; 2-20 μm in length.	Longer; 10-200 μm in length.
Work like oars, with a power stroke alternating with a recovery stroke. Moves cell or fluid in a direction perpendicular to the axis of the cilium.	Undulating motion that drives the cell in the same direction as the axis of the flagellum.

How does the ultrastructure of cilia and flagella relate to their function?

Ultrastructure of Cilia and Flagella:
- ☞ Are extensions of plasma membrane with a core of microtubules. (See Campbell, Figure 7.30)
- ☞ Microtubular core is made of nine doublets of microtubules arranged in a ring with two single microtubules in the center (*9 + 2 pattern*).
- ☞ Each doublet is a pair of attached microtubules. One of the pair shares a portion of the other's wall.
- ☞ Each doublet is connected to the center of the ring by *radial spokes* that end near the central microtubules.
- ☞ Each doublet is attached to the neighboring doublet by a pair of *side arms*. Many pairs of sidearms are evenly spaced along the doublet's length.

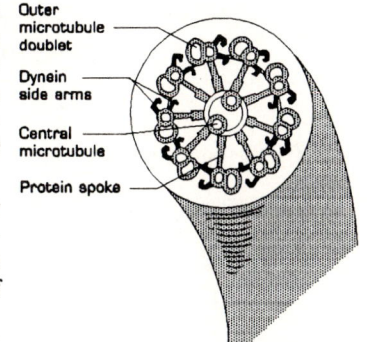

9+2 Pattern in Cross Section

Cilia, Flagella

This unique ultrastructure is necessary for cilia and flagella to function:
- ☞ Sidearms are made of *dynein*, a large protein motor molecule that changes its conformation in the presence of ATP as an energy source.
- ☞ A complex cycle of movements caused by dynein's conformational changes, makes the cilium or flagellum bend: (See Campbell, Figure 7.31)

Sidearms of one doublet attach to the adjacent doublet.
↓
Sidearms swing and the two doublets slide past one another.
↓
Sidearms release.
↓
Sidearms reattach to the adjacent doublet farther along its length.
↓
Cycle is repeated.

- ☞ In cilia and flagella, linear displacement of dynein sidearms is translated into a *bending* by the resistance of the *radial spokes*. Working against this resistance, the "dynein-walking" distorts the microtubules, causing them to bend.

Basal Body = A cellular structure, identical to a centriole, that anchors the microtubular assembly of cilia and flagella.
- ☞ Can convert into a centriole and vice versa.
- ☞ May be a template for ordering tubulin into the microtubules of <u>newly</u> forming cilia or flagella. (As cilia and flagella continue to grow, new tubulin subunits are added to the tips, rather than to the bases.)

B. Microfilaments and Movement (*p. 141–142*)

Structure of Microfilaments:
- ☞ Solid rods about 7 nm in diameter.
- ☞ Built from globular protein monomers, *G-actin*.
- ☞ G-actin monomers are linked into long chains of *F-actin*.
- ☞ Two F-actin chains are wound into a helix.

Function of Microfilaments:

1. Participate in muscle contraction.
   - ☞ Along the length of a muscle cell, parallel actin microfilaments are interdigitated with thicker filaments made of the protein *myosin*, a motor molecule.
   - ☞ With ATP as the energy source, a muscle cell shortens as the thin actin filaments slide across the myosin filaments. Sliding results from the swinging of myosin cross-bridges intermittently attached to actin.

2. Provide support (e.g. bundles of microfilaments in the core of intestinal microvilli).

3. Responsible for localized contraction of cells. Small actin-myosin aggregates exist in some parts of the cell and cause localized contractions. Examples include:
   - ☞ Contracting ring of microfilaments pinches a cell in two during cell division. (See Campbell, Figure 11.10)
   - ☞ Elongation and contraction of *pseudopodia* during *amoeboid movement*.
   - ☞ Involved in *cytoplasmic streaming* (*cyclosis*) found in plant cells.
     - Cytoplasmic streaming (cyclosis) = Flowing of the entire cytoplasm around the space between the vacuole and plasma membrane in a plant cell.

C. Intermediate Filaments (*p. 142–143*)

Structure of Intermediate Filaments:
- ☞ Filaments that are intermediate in diameter (8–12 nm) between microtubules and microfilaments.
- ☞ Diverse class of cytoskeletal elements that differ in diameter and composition depending upon cell type.
- ☞ More permanent than microfilaments and microtubules.

Function of Intermediate Filaments:

1. Specialized for bearing tension; may function as the framework for the cytoskeleton.

2. Reinforce cell shape (e.g. nerve axons).

3. Probably fix organelle position (e.g. nucleus).

IX. THE CELL SURFACE (*p. 143–144, 145*)

Most cells produce coats that are external to the plasma membrane.

Myosin

Cytoplasmic streaming

Intermediate filaments

A Tour of the Cell

Cell walls

A. Cell Walls (p. 143)

Describe the basic design of a cell wall. What is its function?

Plant cells can be distinguished from animal cells by the presence of a *cell wall*:
- Thicker than the plasma membrane.
- Chemical composition varies from cell to cell and species to species.
- Basic design includes strong *cellulose* fibers embedded in a matrix of other polysaccharides and proteins.
- Functions to protect plant cells, maintain their shape, and prevent excess water uptake.
- Has membrane-linked channels, *plasmodesmata*, that connect the cytoplasm of neighboring cells.

How do plant cell walls develop?

Plant cells develop as follows:

Primary cell wall
- Young plant cell secretes a thin, flexible *primary cell wall*. Between primary cell walls of adjacent cells is a *middle lamella* made of *pectins*, a sticky polysaccharide that cements cells together.
- Cell stops growing and strengthens its wall. Some cells:
  1. Secrete hardening substances into primary wall.
  2. Add a *secondary cell wall* between plasma membrane and primary wall.

Secondary cell wall
- *Secondary cell wall* is often deposited in layers with a durable matrix that supports and protects the cell. (See Campbell, Figure 7.33)

B. Glycocalyx of Animal Cells (p. 144)

What is the glycocalyx? What are some functions of the glycocalyx in animal cells?

Glycocalyx
Glycocalyx = Fuzzy coat found outside the plasma membrane of animal cells, made of sticky oligosaccharides covalently bonded to proteins and lipids of the plasma membrane.
- Strengthens cell surface.
- Helps glue cells together.
- May serve as identification tags for specific cell types and aid in cell-cell recognition.

C. Intercellular Junctions (p. 144, 145)

What types of intercellular junctions are found in plant and animal cells? How do their respective structures relate to their function?

Neighboring cells often adhere and interact through special patches of direct physical contact.

Intercellular Junctions in Plants:

    Plasmodesmata (singular, plasmodesma) = Channels that perforate plant cell walls, through which cytoplasmic strands communicate between adjacent cells.
- Lined by plasma membrane. (Plasma membranes of adjacent cells are continuous through a plasmodesma.)
- Allows free passage of water and small solutes. This transport is enhanced by *cytoplasmic streaming*.

Intercellular Junctions in Animals: (See Campbell, Figure 7.34)

    Tight junctions = Intercellular junctions that hold cells together tightly enough to block transport of substances through the intercellular space.
- Specialized membrane proteins in adjacent cells bond directly to each other allowing no space between membranes.
- Usually occur as belts all the way around each cell, that block intercellular transport.
- Frequently found in epithelial layers that separate two kinds of solutions.

    Desmosomes = Intercellular junctions that rivet cells together into strong sheets, but still permit substances to pass freely through intracellular spaces. The desmosome is made of:
1. Intercellular glycoprotein filaments that penetrate and attach the plasma membrane of both cells.
2. A dense disk inside the plasma membrane that is reinforced by *intermediate filaments* made of *keratin* (a strong structural protein).

    Gap junctions = Intercellular junctions specialized for material transport between the cytoplasm of adjacent cells.
- Formed by two connecting protein rings (*connexon*), each embedded in the plasma membrane of adjacent cells. The proteins protrude from the membranes enough to leave an intercellular gap of 2–4 nm.
- Have pores with diameters (1.5 nm) large enough to allow cells to share smaller molecules (e.g. inorganic ions, sugars, amino acids, vitamins), but not macromolecules such as proteins.
- Common in animal embryos and cardiac muscle where chemical communication between cells is essential.

Plasmodesmata

Tight junctions

Desmosomes

Gap junctions

8 MEMBRANE STRUCTURE AND FUNCTION

CHAPTER OUTLINE

What is the function of the plasma membrane?

The *plasma membrane* is the boundary that separates the living cell from its nonliving surroundings. It makes life possible by its ability to discriminate in its chemical exchanges with the environment. This membrane:

- Is about 8 nm thick.
- Surrounds the cell and controls chemical traffic into and out of the cell.
- Is *selectively permeable*; it allows some substances to cross more easily than others.
- Has a unique structure which determines its function and solubility characteristics.

I. MODELS OF MEMBRANE STRUCTURE (p. 151–158)

    A. Two Generations of Membrane Models: A Case Study in the Scientific Process (p. 152–155)

How did scientists use early experimental evidence to make deductions about membrane structure and function?

Membrane function is determined by its structure. Early models of the plasma membrane were deduced from <u>indirect</u> evidence:

1. Evidence: Lipid and lipid soluble materials enter cells more rapidly than substances that are insoluble in lipids (C. Overton, 1895).

   Deduction: Membranes are made of lipids.

   Deduction: Fat-soluble substance move through the membrane by dissolving in it ("like dissolves like").

2. Evidence: *Amphipathic* phospholipids will form an artificial membrane on the surface of water with only the hydrophilic heads immersed in water (Langmuir, 1917).

   <u>Amphipathic</u> = Condition where a molecule has both a hydrophilic region and a hydrophobic region.

   Deduction: Because of their molecular structure, phospholipids can form membranes.

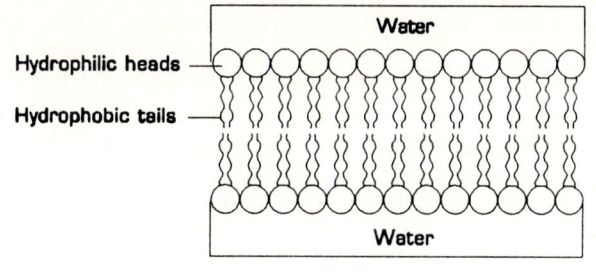

3. Evidence: Phospholipid content of membranes isolated from red blood cells is just enough to cover the cells with two layers (Gorter and Grendel, 1925).

   Deduction: Cell membranes are actually phospholipid bilayers, two molecules thick.

4. Evidence: Membranes isolated from red blood cells contain proteins as well as lipids.

   Deduction: There is protein in biological membranes.

5. Evidence: Wettability of the surface of an actual biological membrane is greater than the surface of an artificial membrane consisting only of a phospholipid bilayer.

   Deduction: Membranes are coated on both sides with proteins, which generally absorb water.

What is the Davson-Danielli model of membrane structure? How did it influence the development of the current model of membrane structure?

Incorporating results from these and other solubility studies, J.F. Danielli and H. Davson (1935) proposed a model of cell membrane structure:

☞ Cell membrane is made of a phospholipid bilayer sandwiched between two layers of globular protein.

☞ The polar (hydrophilic) heads of phospholipids are oriented towards the protein layers forming a hydrophilic zone.

☞ The nonpolar (hydrophobic) tails of phospholipids are oriented in between polar heads forming a hydrophobic zone.

☞ The membrane is approximately 8 nm thick.

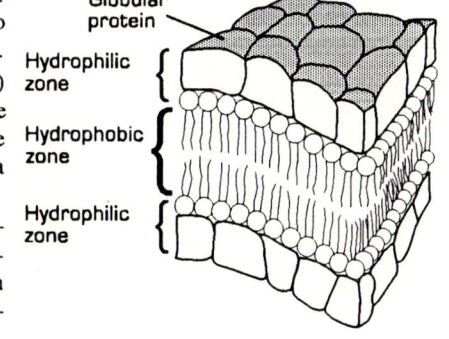

In the 1950's, electron microscopy allowed biologists to visualize the plasma membrane for the first time and provided support for the Davson-Danielli model. Evidence from electron micrographs:

1. Confirmed the plasma membrane was 7 to 8 nm thick (close to the predicted size if the Davson-Danielli model was modified by replacing globular proteins with protein layers in pleated-sheets).

J.F. Danielli, H. Davson

Phospholipid bilayer

2. Showed the plasma membrane was trilaminar, made of two electron-dense bands separated by an unstained layer. It was assumed that the heavy metal atoms of the stain adhered to the hydrophilic proteins and heads of phospholipids and not to the hydrophobic core.

3. Showed internal cellular membranes that looked similar to the plasma membrane. This led biologists (J.D. Robertson) to propose that all cellular membranes were symmetrical and virtually identical.

What were some problems with the Davson-Danielli model?

Though the phospholipid bilayer is probably accurate, there are problems with the Davson-Danielli model:

1. Not all membranes are identical or symmetrical.
   - ☞ Membranes with different functions also differ in chemical composition and structure.
   - ☞ Membranes are bifacial with distinct inside and outside faces.

2. A membrane with an outside layer of proteins would be an unstable structure.
   - ☞ Membrane proteins are not soluble in water, and, like phospholipid, they are *amphipathic*.
   - ☞ Protein layer not likely because its hydrophobic regions would be in an aqueous environment, and it would also separate the hydrophilic phospholipid heads from water.

How did the work of J.D. Robertson, S.J. Singer and G.L. Nicolson clarify current understanding of membrane structure?

In 1972, S.J. Singer and G.L. Nicolson proposed the *fluid mosaic model* which accounted for the amphipathic character of proteins. They proposed:

- ☞ Proteins are individually embedded in the phospholipid bilayer, rather than forming a solid coat spread upon the surface.
- ☞ Hydrophilic portions of both proteins and phospholipids are maximally exposed to water resulting in a stable membrane structure.
- ☞ Hydrophobic portions of proteins and phospholipids are in the nonaqueous environment inside the bilayer.
- ☞ Membrane is a mosaic of proteins bobbing in a fluid bilayer of phospholipids.
- ☞ Evidence from freeze fracture techniques have confirmed that proteins are embedded in the membrane. Using these techniques, biologists can delaminate membranes along the middle of the bilayer. When viewed with an electron microscope, proteins appear to penetrate into the hydrophobic interior of the membrane. (See Campbell, Methods Box)

B. The Fluid Mosaic Model: A Closer Look (*p. 155–158*)

    1. The Fluid Quality of Membranes (*p. 155–156*)

How are membrane molecules free to move in the two-dimensional fluid of the membrane?

    Membranes are held together by hydrophobic interactions, which are weak attractions. (See Campbell, Figure 8.5)
- Most membrane lipids and some proteins can drift laterally within the membrane.
- Molecules rarely flip transversely across the membrane, because hydrophilic parts would have to cross the membrane's hydrophobic core.
- Phospholipids move quickly along the membrane's plane, averaging 2 µm per second.
- Membrane proteins drift more slowly than lipids. The fact that proteins drift laterally was established experimentally by fusing a human and mouse cell (Frye and Edidin, 1970):

    Membrane proteins of a human and mouse cell were labeled with different green and red fluorescent dyes.
    ↓
    Cells were fused to form a hybrid cell with a continuous membrane.
    ↓
    Hybrid cell membrane had initially distinct regions of green and red dye.
    ↓
    In less than an hour, the two colors were intermixed.

- Some membrane proteins are tethered to the cytoskeleton and cannot move far.

How does membrane composition influence a membrane's fluidity?

    Membranes must be fluid to work properly. Solidification may result in permeability changes and enzyme deactivation.
- Unsaturated hydrocarbon tails enhance membrane fluidity, because kinks at the carbon-to-carbon double bonds hinder close packing of phospholipids.
- Membranes solidify if the temperature decreases to a critical point. Critical temperature is lower in membranes with a greater concentration of unsaturated phospholipids.
- Cholesterol, found in plasma membranes of eukaryotes, modulates membrane fluidity by making the membrane:
  1. Less fluid at warmer temperatures (e.g. 37°C body temperature) by restraining phospholipid movement.
  2. More fluid at lower temperatures by preventing close packing of phospholipids.

☞ Cells may alter membrane lipid concentration in response to changes in temperature. Many cold tolerant plants (e.g. winter wheat) increase the unsaturated phospholipid concentration in autumn, which prevents the plasma membranes from solidifying in winter.

2. **Membranes as Mosaics of Structure and Function**
   (p. 156–157)

How are proteins spatially arranged in the cell membrane? How do they contribute to membrane function?

A membrane is a *mosaic* of different proteins embedded and dispersed in the phospholipid bilayer. These proteins vary in both structure and function, and they occur in two spatial arrangements:

a. *Integral proteins*, which are inserted into the membrane so their hydrophobic regions are surrounded by hydrocarbon portions of phospholipids. They may be:
   1. *unilateral*, reaching only partway across the membrane.
   2. *transmembrane*, with hydrophobic midsections between hydrophilic ends exposed on both sides of the membrane.

b. *Peripheral proteins*, which are not embedded but attached to the membrane's surface.
   ☞ May be attached to integral proteins.
   ☞ On cytoplasmic side, may be held by filaments of cytoskeleton.

Membranes are bifacial. The membrane's synthesis and modification by the ER and Golgi determines this asymmetric distribution of lipids, proteins and carbohydrates:
☞ Two lipid layers may differ in lipid composition.
☞ Membrane proteins have distinct directional orientation.
☞ When present, carbohydrates are restricted to the membrane's exterior.
☞ Side of the membrane facing the lumen of the ER, Golgi and vesicles is topologically the same as the plasma membrane's outside face. (See Campbell, Figure 8.9)
☞ Side of the membrane facing the cytoplasm has always faced the cytoplasm, from the time of its formation by the endomembrane system to its addition to the plasma membrane by the fusion of a vesicle.

3. **Membrane Carbohydrates and Cell-Cell Recognition**
   (p. 157–158)

Cell-cell recognition = The ability of a cell to determine if other cells it encounters are alike or different from itself.

Cell-cell recognition is crucial in the functioning of an organism. It is the basis for:
☞ Sorting of an animal embryo's cells into tissues and organs.
☞ Rejection of foreign cells by the immune system.

The way cells recognize other cells is probably by keying on cell markers found on the external surface of the cell membrane. Because of their diversity and location, likely candidates for such cell markers are membrane carbohydrates:
- ☞ Usually branched *oligosaccharides* (<15 monomers).
- ☞ Some covalently bonded to lipids (*glycolipids*).
- ☞ Most covalently bonded to proteins (*glycoproteins*).
- ☞ Vary from species to species, between individuals of the same species and among cells in the same individual.

II. THE TRAFFIC OF SMALL MOLECULES (p. 158–167)

The *selectively permeable* plasma membrane regulates the type and rate of molecular traffic into and out of the cell.

A. Selective Permeability (p. 158–159)

Selective permeability = Property of biological membranes which allows some substances to cross more easily than others.

What factors affect selective permeability of membranes?

The selective permeability of a membrane depends upon (1) membrane solubility characteristics of the phospholipid bilayer and (2) presence of specific integral transport proteins.

1. Permeability of the Lipid Bilayer (p. 158)

The ability of substances to cross the hydrophobic core of the plasma membrane can be measured as the rate of transport through an artificial phospholipid bilayer:

a. Nonpolar (Hydrophobic) Molecules
- ☞ Dissolve in the membrane and cross it with ease (e.g. hydrocarbons and O_2).
- ☞ If two molecules are equally lipid soluble, the smaller of the two will cross the membrane faster.

b. Polar (Hydrophilic) Molecules
- ☞ Small, polar uncharged molecules (e.g. H_2O, CO_2) that are small enough to pass between membrane lipids, will easily pass through synthetic membranes.
- ☞ Larger, polar uncharged molecules (e.g. glucose) will <u>not</u> easily pass through synthetic membranes.
- ☞ All ions, even small ones (e.g. Na^+, H^+) have difficulty penetrating the hydrophobic layer.

Based upon the previous discussion of the fluid mosaic model and selective permeability, explain how hydrophobic interactions influence membrane structure and function.

Selective permeability

2. Transport Proteins (p. 158)

Water, CO_2 and nonpolar molecules rapidly pass through the plasma membrane as they do an artificial membrane.

Unlike artificial membranes, however, biological membranes <u>are</u> permeable to specific ions and certain polar molecules of moderate size (e.g. sugars). These hydrophilic substances avoid the hydrophobic core of the bilayer by passing through *transport proteins*.

<u>Transport proteins</u> = Integral membrane proteins that transport specific molecules or ions across biological membranes. (See Campbell, Figure 8.10)
- May provide a hydrophilic tunnel through the membrane.
- May bind to a substance and physically move it across the membrane.
- Are specific for the substance they translocate.

What are three types of transport proteins? What are their respective patterns of solute transport?

<u>Types of Transport Proteins</u>:

a. *Uniport* — carries a single solute across the membrane.

b. *Symport* — translocates two different solutes simultaneously in the <u>same</u> direction. Both solutes must bind to the protein for transport to occur.

c. *Antiport* — exchanges two solutes by transporting them in opposite directions.

B. Diffusion and Passive Transport (p. 159–160)

<u>Concentration gradient</u> = Regular, graded concentration change over a distance in a particular direction.

<u>Net directional movement</u> = Overall movement away from the center of concentration, which results from random molecular movement in <u>all</u> directions.

What causes diffusion? Why is it a spontaneous process?

<u>Diffusion</u> = The *net* movement of a substance down a *concentration gradient*.
- Results from the intrinsic kinetic energy of molecules (also called thermal motion, or heat).
- Results from random molecular motion, even though the *net* movement may be directional.
- Diffusion continues until a dynamic equilibrium is reached.
- Much of the traffic across cell membranes occurs by diffusion.
- In the absence of other forces, a substance will diffuse from where it is more concentrated to where it is less concentrated. A substance diffuses down its *concentration gradient*.
- Because it decreases free energy, diffusion is a spontaneous process ($-\Delta G$). It increases entropy of a system by producing a more random mixture of molecules.
- A substance diffuses down its <u>own</u> concentration gradient and is not affected by the gradients of other substances.

What regulates the rate of passive transport? Why does a concentration gradient across a membrane represent potential energy?

Passive transport = Diffusion of a substance across a biological membrane.
- Spontaneous process which is a function of a concentration gradient when a substance is more concentrated on one side of the membrane.
- Passive process which does not require the cell to expend energy. It is the potential energy stored in a concentration gradient that drives diffusion.
- Rate of diffusion is regulated by the permeability of the membrane.
- Water diffuses freely across most cell membranes.

C. Osmosis: A Special Case of Passive Transport (p. 160–163)

Hyperosmotic solution = A solution with a greater solute concentration compared to another solution.

Hypoosmotic solution = A solution with a lower solute concentration compared to another solution.

Isosmotic solution = A solution with an equal solute concentration compared to another solution.

What is osmosis? What determines the direction of osmosis? How does bound water affect the osmotic behavior of dilute biological fluids?

Osmosis = Diffusion of water across a selectively permeable membrane.
- Water diffuses down its concentration gradient.
- For example, if two solutions of different concentrations are separated by a selectively permeable membrane that is permeable to water but not the solute, water will diffuse from the hypoosmotic solution to the hyperosmotic solution.
- Some solute molecules can reduce the proportion of water molecules that can freely diffuse. Water molecules form a *hydration shell* around hydrophilic solute molecules, and this bound water cannot freely diffuse across a membrane.
- In dilute solutions including most biological fluids, it is the difference in the proportion of unbound water that causes osmosis, rather than the actual difference in water concentration.
- Direction of osmosis is determined by the difference in <u>total</u> solute concentration, regardless of the type or diversity of solutes in the solutions.

92 Membrane Structure and Function

☞ If two <u>isosmotic</u> solutions are separated by a selectively permeable membrane, water molecules diffuse across the membrane in both directions at an equal rate. There is no <u>net</u> movement of water.

Do water molecules stop moving when two solutions separated by a membrane reach osmotic equilibrium?

Even though there is no <u>net</u> movement of water across the membrane (or osmosis), the water molecules do not stop moving. At equilibrium, the water molecules move in both directions at the same rate.

<u>Osmotic concentration</u> = Total solute concentration of a solution.

<u>Osmotic pressure</u> = Measure of the tendency for a solution to take up water when separated from pure water by a selectively permeable membrane.
  ☞ Osmotic pressure of pure water is zero.
  ☞ Osmotic pressure of a solution is proportional to its osmotic concentration. (The greater the solute concentration, the greater the osmotic pressure.)

How can osmotic pressure be measured?

Osmotic pressure can be measured by an *osmometer*:
  ☞ In one type of osmometer, pure water is separated from a solution by a selectively permeable membrane that is permeable to water but not solute.
  ☞ The tendency for water to move into the solution by osmosis is counteracted by applying enough pressure with a piston so the solution's volume will stay the same.
  ☞ The amount of pressure required to prevent net movement of water into the solution is the *osmotic pressure*.

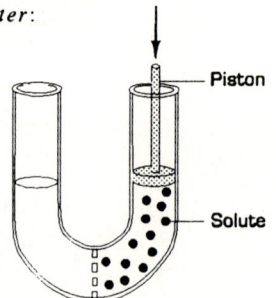

How do living cells regulate water balance? How does the presence or absence of a cell wall affect a cell's response to a change in environmental osmotic concentration?

  1. Water Balance of Cells Without Walls (*p. 162*)

  a. In an isosmotic environment, the volume of an animal cell will remain stable with no net movement of water across the plasma membrane.

  b. In a hyperosmotic environment, an animal cell will lose water by osmosis and *crenate* (shrivel).

  c. In a hypoosmotic environment, an animal cell will gain water by osmosis, swell and perhaps *lyse* (cell destruction).

Organisms without cell walls prevent excessive loss or uptake of water by:
  ☞ Living in an isosmotic environment (e.g. many marine invertebrates are isosmotic with sea water).

☞ Osmoregulating in a hypo- or hyperosmotic environment. Organisms can regulate water balance by removing water in a hypoosmotic environment (e.g. *Paramecium* with contractile vacuoles in fresh water) or conserving water and pumping out salts in a hyperosmotic environment (e.g. bony fish in seawater).

2. **Water Balance of Cells With Walls** (*p. 162*)

Cells of prokaryotes, some protists, fungi and plants have cell walls outside the plasma membrane.

a. In a hyperosmotic environment, walled cells will lose water by osmosis and will *plasmolyze*, which is usually lethal.

Plasmolysis = Phenomenon where a walled cell shrivels and the plasma membrane pulls away from the cell wall as the cell loses water to a hypertonic environment.

b. In a hypoosmotic environment, water moves by osmosis into the plant cell, causing it to swell until internal pressure against the cell wall equals the osmotic pressure of the cytoplasm. A dynamic equilibrium is established (water enters and leaves the cell at the same rate and the cell becomes *turgid*).

Turgid = Firmness or tension such as found in walled cells that are in a hypoosmotic environment where water enters the cell by osmosis.
☞ Ideal state for most plant cells.
☞ Turgid cells provide mechanical support for plants.
☞ Requires cells to be hyperosmotic to their environment.

c. In an isosmotic environment, there is no net movement of water into or out of the cell.
☞ Plant cells become *flaccid* or limp.
☞ Loss of structural support from turgor pressure causes plants to wilt.

D. **Facilitated Diffusion** (*p. 163–164*)

Facilitated diffusion = Diffusion of solutes across a membrane, with the help of transport proteins.
☞ Passive transport because solute is moved down its concentration gradient.
☞ Helps the diffusion of many polar molecules and ions that are impeded by the membrane's phospholipid bilayer.

How are transport proteins similar to enzymes? How do they differ?

Transport proteins share some properties of enzymes:
☞ Transport proteins are specific for the solutes they transport. There is probably a specific binding site analogous to an enzyme's active site.
☞ Transport proteins can be saturated with solute, so the maximum transport rate occurs when all binding sites are occupied with solute.
☞ Transport proteins can be inhibited by molecules that resemble the solute normally carried by the protein (similar to competitive inhibition in enzymes).

Plasmolysis

Turgid

Facilitated diffusion

Transport proteins differ from enzymes in they do not usually catalyze chemical reactions.

Describe one model for facilitated diffusion.

One Model for Facilitated Diffusion:
- ☞ Transport protein most likely remains in place in the membrane and translocates solute by alternating between two conformations.
- ☞ Transport protein might bind to solute in one conformation and deposit it on the other side of the membrane in another conformation.
- ☞ The solute's binding and release may trigger the transport protein's conformational change.
- ☞ In some inherited disorders, transport proteins are missing or are defective (e.g. cystinuria, a kidney disease caused by missing carriers for cystine and other amino acids which are normally reabsorbed from the urine).

E. Active Transport (p. 164–166)

How does active transport differ from diffusion?

Active transport = Energy-requiring process during which a transport protein pumps a molecule across a membrane, against its concentration gradient.
- ☞ Is energetically uphill ($+\Delta G$) and requires the cell to expend energy.
- ☞ Helps cells maintain steep ionic gradients across the cell membrane (e.g. Na^+, K^+, Mg^{++}, Ca^{++} and Cl^-).
- ☞ Transport proteins involved in active transport harness energy from ATP to pump molecules against their concentration gradients.

An example of an active transport system that translocates ions against steep concentration gradients is the *sodium-potassium pump*. Major features of the pump are:

1. The transport protein oscillates between two conformations:
   a. High affinity for Na^+ with binding sites oriented towards the cytoplasm.
   b. High affinity for K^+ with binding sites oriented towards the cell's exterior.
2. ATP phosphorylates the transport protein and powers the conformational change from Na^+ receptive to K^+ receptive.
3. As the transport protein changes conformation, it translocates bound solutes across the membrane.
4. Na^+K^+-pump translocates three Na^+ ions out of the cell for every two K^+ ions pumped into the cell. (Refer to Campbell, Figure 8.19 for the specific sequence of events.)

Membrane Structure and Function

F. The Special Case of Ion Transport (p. 166)

Because anions and cations are unequally distributed across the plasma membrane, all cells have voltages across their plasma membranes.

Membrane potential = Voltage across membranes.
- Ranges from -50 to -200 mv. As indicated by the negative sign, the cell's inside is negatively charged with respect to the outside.
- Affects traffic of charged substances across the membrane.
- Favors diffusion of cations into cell and anions out of the cell (because of electrostatic attractions).

Two forces drive passive transport of ions across membranes:
1. Concentration gradient of the ion.
2. Effect of membrane potential on the ion.

Electrochemical gradient = Diffusion gradient resulting from the combined effects of membrane potential and concentration gradient.
- Ions always diffuse down their electrochemical gradient.
- Ions may not always diffuse down their concentration gradients.
- At equilibrium, the distribution of ions on either side of the membrane may be different from the expected distribution when charge is not a factor.
- Uncharged solutes diffuse down concentration gradients because they are unaffected by membrane potential.

What factors contribute to the creation and maintenance of a membrane potential or electrochemical gradient?

Factors which contribute to a cell's membrane potential (net negative charge on the inside):
1. Negatively charged proteins in the cell's interior.
2. Plasma membrane's selective permeability to various ions. For example, there is a net loss of positive charges as K^+ leaks out of the cell faster than Na^+ diffuses in.
3. The sodium-potassium pump. This *electrogenic pump* translocates 3 Na^+ out for every 2 K^+ in — a net loss of one positive charge per cycle.

Electrogenic pump = A transport protein that generates voltage across a membrane.
- Na^+/K^+ ATPase is the major electrogenic pump in animal cells.
- A *proton pump* is the major electrogenic pump in plants, bacteria and fungi. Also, mitochondria and chloroplasts use a proton pump to drive ATP synthesis.
- Voltages created by electrogenic pumps are sources of potential energy available to do cellular work.

G. Cotransport (p. 166–167)

How can potential energy generated by transmembrane solute gradients be harvested by the cell and used to transport substances across the membrane?

Cotransport = Process where a single ATP-powered pump actively transports one solute and indirectly drives the transport of other solutes against their concentration gradients.

One mechanism of cotransport involves two transport proteins:
1. ATP-powered pump actively transports one solute and creates potential energy in the gradient it creates.
2. Another transport protein couples the solute's downhill diffusion as it leaks back across the membrane with a second solute's uphill transport against its concentration gradient.

For example, plants use a proton pump coupled with sucrose-H⁺ symport.

III. THE TRAFFIC OF LARGE MOLECULES: EXOCYTOSIS AND ENDOCYTOSIS (p. 167–170)

Small molecules cross membranes by:
1. Passing through the phospholipid bilayer.
2. Being translocated by a transport protein.

How are large molecules transported across the cell membrane?

Large molecules (e.g. proteins and polysaccharides) cross membranes by the processes of *exocytosis* and *endocytosis*.

Exocytosis	Endocytosis
Process of exporting macromolecules from a cell by fusion of vesicles with the plasma membrane.	Process of importing macromolecules into a cell by forming vesicles derived from the plasma membrane.
Vesicle usually budded from the ER or Golgi and migrates to plasma membrane.	Vesicle forms from a localized region of plasma membrane that sinks inward; pinches off into the cytoplasm.
Used by secretory cells to export products (e.g. insulin in pancreas, or neurotransmitter from neuron).	Used by cells to incorporate extra-cellular substances.

Even though exocytosis and endocytosis occur, the amount of plasma membrane remains relatively constant in a nongrowing cell.
- ☞ Vesicle fusion with the plasma membrane offsets membrane loss through endocytosis.
- ☞ Vesicles provide a mechanism to rejuvenate or remodel the plasma membrane.

There are three types of endocytosis: (1) *phagocytosis*, (2) *pinocytosis* and (3) *receptor-mediated endocytosis*.

Phagocytosis = (cell eating) Endocytosis of solid particles.
- ☞ Cell engulfs particle with *pseudopodia* and pinches off a food vacuole.
- ☞ Vacuole fuses with a lysosome containing hydrolytic enzymes that will digest the particle.

Pinocytosis = (cell drinking) Endocytosis of fluid droplets.
- ☞ Droplets of extracellular fluid are taken into small vesicles.
- ☞ The process is not discriminating. The cell takes in all solutes dissolved in the droplet.

Receptor-mediated endocytosis = The process of importing specific macromolecules into the cell by the inward budding of vesicles formed from *coated pits*; occurs in response to the binding of specific *ligands* to receptors on the cell's surface.
- ☞ More discriminating process than pinocytosis.
- ☞ Enables cells to acquire bulk quantities of specific substances, even if they are in low concentration in extracellular fluid (e.g. cholesterol).
- ☞ Membrane-embedded proteins with specific receptor sites exposed to the cell's exterior, cluster in regions called *coated pits*.
- ☞ A layer of *clathrin*, a fibrous protein, lines and reinforces the *coated pit* on the cytoplasmic side.
- ☞ A molecule that binds to a specific receptor site of another molecule is called a *ligand*.

Progressive Stages of Receptor-Mediated Endocytosis: (See Campbell, Figure 8.25c)

Extracellular *ligand* binds to *receptors* in a *coated pit*. → Causes inward budding of the coated pit. → Forms a *coated vesicle* inside a *clathrin* cage.
↓
Ingested material is liberated from the vesicle.
↓
Protein receptors are recycled to the plasma membrane.

In a nongrowing cell, the amount of plasma membrane remains relatively constant.
- ☞ Vesicle fusion with the plasma membrane offsets membrane loss through endocytosis.
- ☞ Vesicles provide a mechanism to rejuvenate or remodel the plasma membrane.

Phagocytosis

Pinocytosis

Receptor-mediated endocytosis

CELLULAR RESPIRATION: HARVESTING CHEMICAL ENERGY

CHAPTER OUTLINE

As open systems, cells require outside energy sources to perform cellular work (e.g. chemical, transport and mechanical).

How does energy flow through most ecosystems?

Energy flows into most ecosystems as sunlight.
↓
Photosynthetic organisms trap a portion of the light energy and transform it into chemical bond energy of organic molecules. O_2 is released as a byproduct.
↓
Cells use some of the chemical bond energy in organic molecules to make ATP — the energy source for cellular work.
↓
Energy leaves living organisms as it dissipates as heat.

The products of respiration (CO_2 and H_2O) are the raw materials for photosynthesis. Photosynthesis produces glucose and oxygen, the raw materials for respiration.

Chemical elements essential for life are recycled, but energy is not.

How do cells harvest chemical energy?

Complex organic molecules —Catabolic pathways→ Simpler waste products with <u>less energy</u>
↓ energy
Some energy used to do work & some energy dissipated as heat

Cellular Respiration: Harvesting Chemical Energy 99

How does fermentation differ from cellular respiration?

<u>Fermentation</u> = An ATP-producing catabolic pathway in which both electron donors and acceptors are organic compounds.
- Can be an anaerobic process.
- Results in a partial degradation of sugars.

<u>Cellular respiration</u> = An ATP-producing catabolic process in which the ultimate electron acceptor is an inorganic molecule, such as oxygen.
- Most prevalent and efficient catabolic pathway.
- Is an exergonic process ($\Delta G = -686$ kcal/mol).
- Can be summarized as:

  Organic compounds + Oxygen $\longrightarrow$ Carbon dioxide + Water + Energy
  (food)

What is the overall summary equation for cellular respiration?
- Carbohydrates, proteins and fats can all be metabolized as fuel, but cellular respiration is most often described as the oxidation of glucose:

  $C_6H_{12}O_6 + 6\ O_2 \longrightarrow 6\ CO_2 + 6\ H_2O +$ Energy (ATP + Heat)

I. ATP AND CELLULAR WORK: A REVIEW (p. 174)

What is ATP? How does the cell tap the energy stored in ATP?

The catabolic process of respiration transfers the energy stored in food molecules to *ATP*.

<u>ATP (adenosine triphosphate)</u> = Nucleoside triphosphate with unstable phosphate bonds that the cell hydrolyzes for energy to drive endergonic reactions.
- The cell taps energy stored in ATP by enzymatically transferring terminal phosphate groups from ATP to other compounds. (Recall that direct hydrolysis of ATP would release energy as heat, a form unavailable for cellular work. See Chapter 6.)
- The compound receiving the phosphate group from ATP is said to be *phosphorylated* and becomes more reactive in the process.
- The phosphorylated compound loses its phosphate group as cellular work is performed; inorganic phosphate and ADP are formed in the process. (See Campbell, Figure 9.3)
- Cells must replenish the ATP supply to continue cellular work. Respiration provides the energy to synthesize ATP from ADP and inorganic phosphate.

II. CHEMICAL BACKGROUND: RESPIRATION AS AN OXIDATION-REDUCTION PROCESS (p. 174–178)

A. An Introduction to Redox Reactions (p. 174–175)

<u>Oxidation-reduction reactions</u> = Chemical reactions which involve a partial or complete transfer of electrons from one reactant to another; called *redox reactions* for short.

<u>Oxidation</u> = Partial or complete loss of electrons.

100 Cellular Respiration: Harvesting Chemical Energy

Reduction

Reduction = Partial or complete gain of electrons.

In terms of electron transfer, what happens in an oxidation-reduction reaction?

Generalized Redox Reaction:

Electron transfer requires both a donor and acceptor, so when one reactant is oxidized the other is reduced.

$$Xe^- + Y \longrightarrow X + Ye^-$$

(oxidation on top, reduction on bottom)

Where:
- X = Substance being oxidized; acts as a *reducing agent* because it reduces Y.
- Y = Substance being reduced; acts as an *oxidizing agent* because it oxidizes X.

How are redox reactions involved in energy exchanges?

Not all redox reactions involve a complete transfer of electrons, but, instead, may just change the degree of sharing in covalent bonds. For example:

$$CH_4 + 2\,O_2 \longrightarrow CO_2 + 2\,H_2O + \text{energy}$$

methane   oxygen   carbon dioxide   water

(oxidation on top, reduction on bottom)

- ☞ Covalent electrons of methane are equally shared, because carbon and hydrogen have similar electronegativities.
- ☞ As methane reacts with oxygen to form carbon dioxide, electrons shift away from carbon and hydrogen to the more electronegative oxygen.
- ☞ Since electrons lose potential energy when they shift toward more electronegative atoms, redox reactions that move electrons closer to oxygen release energy.
- ☞ Oxygen is a powerful oxidizing agent because it is so electronegative.

B. Respiration: A Stepwise Redox Reaction (p. 176–178)

What is oxidized in cellular respiration? What is reduced? What happens to the energy released in this exergonic reaction?

Cellular respiration is a redox process that transfers hydrogen from sugar to oxygen.

$$C_6H_{12}O_6 + 6\,O_2 \longrightarrow 6\,CO_2 + 6\,H_2O + \text{energy (used to make ATP)}$$

(oxidation on top, reduction on bottom)

- ☞ Valence electrons of carbon and hydrogen lose potential energy as they shift toward electronegative oxygen.
- ☞ Released energy is used by cells to produce ATP.
- ☞ Carbohydrates and fats are excellent energy stores, because they are rich in C to H bonds.

Without the activation barrier, glucose would combine spontaneously with oxygen.
- ☞ Igniting glucose provides the activation energy for the reaction to proceed; a mole of glucose yields 686 kcal of heat when burned in air.
- ☞ Cellular respiration does not oxidize glucose in one explosive step, as the energy could not be efficiently harnessed in a form available to perform cellular work.
- ☞ Enzymes lower the activation energy in cells, so glucose can be slowly oxidized in a stepwise fashion during glycolysis and Krebs cycle.
- ☞ Hydrogens stripped from glucose are not transferred directly to oxygen, but are first passed to a special electron acceptor — NAD^+.

What is a coenzyme? What coenzymes are involved in the redox reactions of respiration?

Nicotinamide adenine dinucleotide (NAD^+) = A *dinucleotide* that functions as a *coenzyme* in the redox reactions of metabolism. (See Campbell, Figure 9.5)
- ☞ Found in all cells.
- ☞ Assists enzymes in electron transfer during redox reactions of metabolism.

Coenzyme = Small nonprotein organic molecule that is required for certain enzymes to function.

Dinucleotide = A molecule consisting of two nucleotides.

How do coenzymes function in redox reactions?

During the oxidation of glucose, NAD^+ functions as an oxidizing agent by trapping energy-rich electrons from glucose or food. These reactions are catalyzed by enzymes called *dehydrogenases*, which:
- ☞ Remove a pair of hydrogen atoms (2 electrons and 2 protons) from substrate.
- ☞ Deliver the two electrons and <u>one</u> proton to NAD^+.
- ☞ Release the remaining proton into the surrounding solution.

$$\underset{\text{reduction}}{\overset{\text{oxidation}}{\underset{OH}{\overset{H}{R-C-R'}} + NAD^+ \xrightarrow{\text{dehydrogenase}} \underset{O}{\overset{H}{R-\underset{\|}{C}-R'}} + NADH + H^+}}$$

Where:

$R-\underset{OH}{\overset{H}{C}}-R'$ = Substrate that is oxidized by enzymatic transfer of electrons to NAD^+.

NAD^+ = Oxidized coenzyme (net positive charge).

NADH = Reduced coenzyme (electrically neutral).

Cellular Respiration: Harvesting Chemical Energy

In general terms, how is energy from food transferred to ATP? What are the roles of NAD⁺ and the electron transport chain in this process? Why is oxygen necessary for cellular respiration?

These high energy electrons transferred from substrate to NAD⁺ are then passed down the *electron transport chain* to oxygen, powering ATP synthesis (*oxidative phosphorylation*).

Electron transport chains convert some of the chemical energy extracted from food to a form that can be used to make ATP. These transport chains:
- Are composed of electron-carrier molecules built into the inner mitochondrial membrane. Structure of this membrane correlates with its functional role (form fits function).
- Accept energy-rich electrons from reduced coenzymes (NADH and FADH$_2$); and during a series of redox reactions, pass these electrons down the chain to oxygen, the final electron acceptor. The electronegative oxygen accepts these electrons, along with hydrogen nuclei, to form water.
- Release energy from energy-rich electrons in a controlled stepwise fashion; a form that can be harnessed by the cell to power ATP production. If the reaction between hydrogen and oxygen during respiration occurred in a single explosive step, much of the energy released would be lost as heat, a form unavailable to do cellular work.

Electron transfer from NADH to oxygen is exergonic, having a free energy change of −53 kcal/mol.
- Since electrons lose potential energy when they shift toward a more electronegative atom, this series of redox reactions releases energy.
- Each successive carrier in the chain has a higher electronegativity than the carrier before it, so the electrons are pulled downhill towards oxygen, the final electron acceptor and the molecule with the highest electronegativity.

III. AN OVERVIEW OF CELLULAR RESPIRATION (p. 178–179)

What are the three metabolic stages of cellular respiration? Where do they occur? What do they produce?

There are three metabolic stages of cellular respiration:
1. Glycolysis
2. Krebs Cycle
3. Electron transport chain (ETC) and oxidative phosphorylation

Glycolysis is a catabolic pathway that:
- Occurs in the cytosol.
- Partially oxidizes glucose (6C) into two *pyruvate* (3C) molecules.

The *Krebs Cycle* is a catabolic pathway that:
- Occurs in the mitochondrial matrix.
- Completes glucose oxidation by breaking down a *pyruvate* derivative (acetyl CoA) into carbon dioxide.

Glycolysis and the *Krebs Cycle* produce:
- A small amount of ATP by *substrate-level phosphorylation*.
- NADH by transferring electrons from substrate to NAD^+.

The *electron transport chain*:
- Is located at the inner membrane of the mitochondrion.
- Accepts energized electrons from reduced coenzymes (NADH and $FADH_2$) that are harvested during glycolysis and Krebs cycle. Oxygen pulls these electrons down the electron transport chain to a lower energy state.
- Couples this exergonic slide of electrons to ATP synthesis or *oxidative phosphorylation*. This process produces most (90%) of the ATP.

What is the difference between substrate-level phosphorylation and oxidative phosphorylation?

Substrate-level phosphorylation = ATP production by direct enzymatic transfer of phosphate from an intermediate substrate in catabolism to ADP.

Oxidative phosphorylation = ATP production that is coupled to the exergonic transfer of electrons from food to oxygen.

IV. GLYCOLYSIS: A CLOSER LOOK (p. 179–181)

Glycolysis = (Glyco=sweet, sugar; lysis=to split) Catabolic pathway during which six-carbon glucose is split into two three-carbon sugars, which are then oxidized and rearranged by a step-wise process that produces two pyruvate molecules.
- Each reaction is catalyzed by specific enzymes dissolved in the cytosol.
- No CO_2 is released as glucose is oxidized to pyruvate; all carbon in glucose can be accounted for in the two molecules of pyruvate.
- Occurs whether or not oxygen is present.

The reactions of glycolysis occur in two phases:
- *Energy-investment phase* during which the cell uses ATP to phosphorylate the intermediates of glycolysis.
- *Energy-yielding phase* during which two three-carbon intermediates are oxidized. For each glucose molecule entering glycolysis:
  1. A net gain of two ATPs is produced by substrate-level phosphorylation.
  2. Two molecules of NAD^+ are reduced to NADH. Energy conserved in the high-energy electrons of NADH can be used to make ATP by oxidative phosphorylation.

Why is the first phase in glycolysis called the "energy-investment" phase? Why is ATP required for these preparatory steps? In general terms, what is the role of ATP in coupled reactions?

The *energy investment phase* includes five preparatory steps that split glucose in two. This process actually consumes ATP.

Substrate-level phosphorylation

Oxidative phosphorylation

Glycolysis

104 Cellular Respiration: Harvesting Chemical Energy

Step 1: Glucose enters the cell, and carbon six is phosphorylated. This ATP-coupled reaction:
- Is catalyzed by *hexokinase*. (*Kinase* is an enzyme involved in phosphate transfer.)
- Requires an initial <u>investment</u> of ATP.
- Makes glucose more chemically reactive.
- Produces glucose-6-phosphate. Since the plasma membrane is relatively impermeable to ions, the addition of an electrically charged phosphate group traps the sugar in the cell.

How does the carbon skeleton of glucose change as it proceeds through glycolysis?

Step 2: An *isomerase* catalyzes the rearrangement of glucose-6-phosphate to its isomer, fructose-6-phosphate.

Step 3: Carbon one of fructose-6-phosphate is phosphorylated. This reaction:
- Requires an investment of another ATP.
- Is catalyzed by *phospho-fructokinase*, an allosteric enzyme that controls the rate of glycolysis.

Step 4: *Aldolase* cleaves the six-carbon sugar into two isomeric three-carbon sugars.
- This is the reaction for which glycolysis is named.
- For each glucose molecule that begins glycolysis, there are <u>two</u> product molecules for this and each succeeding step.

Step 5: An isomerase catalyzes the reversible conversion between the two three-carbon sugars. This reaction:
- Never reaches equilibrium because only one isomer, *glyceraldehyde phosphate*, is used in the next step of glycolysis.
- Is thus pulled towards the direction of glyceraldehyde phosphate, which is removed as fast as it forms.
- Results in the net effect that, for each glucose molecule, <u>two</u> molecules of glyceraldehyde phosphate progress through glycolysis.

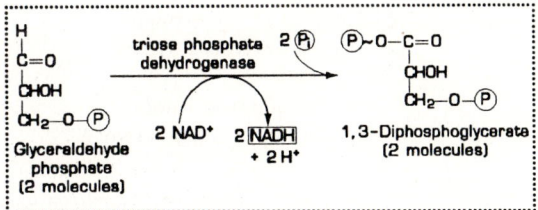

Where in glycolysis do sugar oxidation, substrate-level phosphorylation and reduction of coenzymes occur?

The *energy-yielding phase* occurs after glucose is split into two three-carbon sugars. During these reactions, sugar is oxidized, and ATP and NADH are produced.

Step 6: An enzyme catalyzes two sequential reactions:

1. Glyceraldehyde phosphate is oxidized and NAD^+ is reduced to $NADH + H^+$.
   - This reaction is very exergonic ($\Delta G = -10.3$ kcal/mol) and is coupled to the endergonic phosphorylation phase.
   - For every glucose molecule, 2 NADH are produced.

2. Glyceraldehyde phosphate is phosphorylated on carbon one.
   - The phosphate source is inorganic phosphate, which is always present in the cytosol.
   - The new phosphate bond is a high energy bond with even more potential to transfer a phosphate group than ATP.

Step 7: ATP is produced by *substrate-level phosphorylation*.

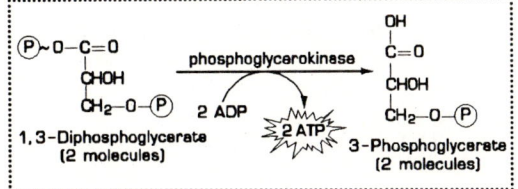

- In a very exergonic reaction, the phosphate group with the high energy bond is transferred from 1,3-diphosphoglycerate to ADP.
- For each glucose molecule, two ATP molecules are produced. The ATP ledger now stands at zero as the initial debt of two ATP from steps one and three is repaid.

106 *Cellular Respiration: Harvesting Chemical Energy*

Step 8: In preparation for the next reaction, a phosphate group on carbon three is enzymatically transferred to carbon two.

Step 9: Enzymatic removal of a water molecule:
- ☞ Creates a double bond between carbons one and two of the substrate.
- ☞ Rearranges the substrate's electrons, which transforms the remaining phosphate bond into an unstable bond.

Step 10: ATP is produced by *substrate-level phosphorylation*.
- ☞ In a highly exergonic reaction, a phosphate group is transferred from PEP to ADP.
- ☞ For each glucose molecule, this step produces two ATP.

What is the summary equation for glycolysis? At the end of glycolysis, where is most of the harnessed energy conserved?

<u>Summary Equation For Glycolysis</u>:

$$C_6H_{12}O_6 + 2\ NAD^+ + 2\ ADP + 2\ \text{P} \longrightarrow 2\ C_3H_4O_3\ (\text{Pyruvate}) + 2\ NADH + 2\ H^+ + 2\ ATP + 2\ H_2O$$

- ☞ Glucose has been oxidized into two pyruvate molecules.
- ☞ The process is exergonic ($\Delta G = -140$ kcal/mol); most of the energy harnessed is conserved in the high-energy electrons of NADH and in the phosphate bonds of ATP.

V. THE KREBS CYCLE: A CLOSER LOOK (*p. 182–184*)

Where is pyruvate oxidized to acetyl CoA? What molecules are produced? How does it link glycolysis to the Krebs cycle?

Most of the chemical energy originally stored in glucose still resides in the two pyruvate molecules produced by glycolysis. The fate of pyruvate depends upon the presence or absence of oxygen. If oxygen is present, pyruvate <u>enters the mitochondrion</u> where it is <u>completely oxidized</u> by a series of enzyme-controlled reactions.

Formation of Acetyl CoA. The junction between glycolysis and the Krebs Cycle is the oxidation of pyruvate to acetyl CoA:
- Pyruvate molecules are translocated from the cytosol into the mitochondrion by a carrier protein in the mitochondrial membrane.
- This step is catalyzed by a *multienzyme complex* which:
    1. Removes CO_2 from the carboxyl group of pyruvate, changing it from a three-carbon to a two-carbon compound. This is the first step where CO_2 is released.
    2. Oxidizes the two-carbon fragment to acetate, while reducing NAD^+ to NADH. Since glycolysis produces two pyruvate molecules per glucose, there are <u>two</u> NADH molecules produced.
    3. Attaches coenzyme A to the acetyl group, forming acetyl CoA. This bond is unstable, making the acetyl group very reactive.

Where does the Krebs cycle occur? Why is it sometimes called the citric acid cycle?

<u>Krebs Cycle</u>: The Krebs Cycle reactions oxidize the remaining acetyl fragments of Acetly-CoA to CO_2. Energy released from this exergonic process is used to reduce coenzyme (NAD^+ and FAD) and to phosphorylate ATP (substrate-level phosphorylation).
- A German-British scientist, Hans Krebs, elucidated this catabolic pathway in the 1930's.
- The Krebs cycle, which is also known as the *citric acid cycle* or *TCA cycle*, has eight enzyme-controlled steps that occur in the *mitochondrial matrix*.

For every turn of the Krebs cycle, what molecules enter and what molecules are produced?

For every turn of Krebs Cycle:
- Two carbons enter in the acetyl fragment of Acetyl CoA.
- Two different carbons are oxidized and leave as CO_2.
- Coenzymes are reduced; three NADH and one $FADH_2$ are produced.
- One ATP molecule is produced by substrate-level phosphorylation.
- Oxaloacetate is regenerated.

<u>FAD</u>

<u>Hans Krebs</u>

108 Cellular Respiration: Harvesting Chemical Energy

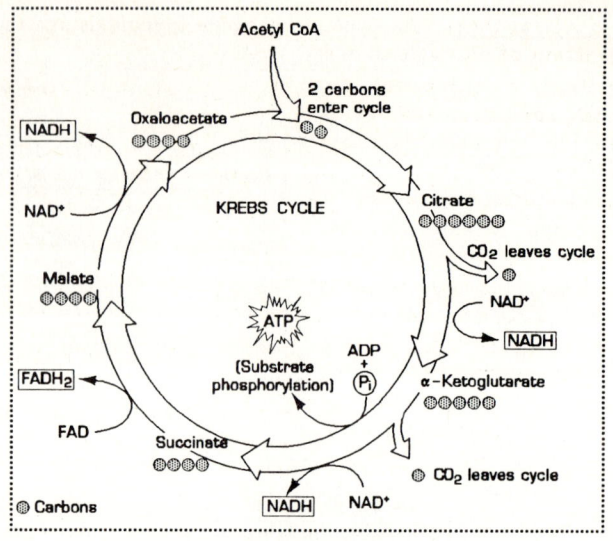

For every glucose molecule split during glycolysis:
- ☞ Two acetyl fragments are produced.
- ☞ It takes two turns of Krebs Cycle to complete the oxidation of glucose.

At what point during cellular respiration is glucose completely oxidized?

The Steps of the Krebs Cycle: (See Campbell, Figure 9.13)

Step 1: The unstable bond of acetyl CoA breaks, and the <u>two-carbon</u> acetyl group bonds to the <u>four-carbon</u> oxaloacetate to form <u>six-carbon</u> citrate.

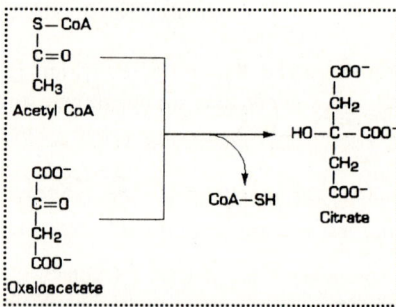

Step 2: Citrate is isomerized to isocitrate.

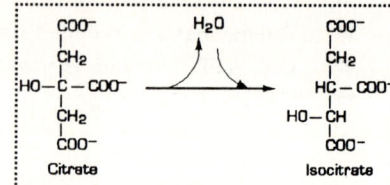

Step 3: Two major events occur during this step:
- ☞ Isocitrate loses CO_2 leaving a <u>five-carbon</u> molecule.
- ☞ The five-carbon compound is oxidized and NAD^+ is reduced.

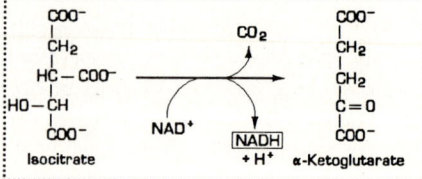

Step 4: A multienzyme complex catalyzes:
- ☞ Removal of CO_2.
- ☞ Oxidation of the remaining <u>four-carbon</u> compound and reduction of NAD^+.
- ☞ Attachment of CoA with a high energy bond to form succinyl CoA.

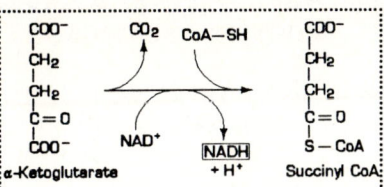

Step 5: Substrate-level phosphorylation occurs in a series of enzyme catalyzed reactions:
- ☞ The high energy bond of succinyl-CoA breaks, and some energy is conserved as CoA is displaced by a phosphate group.
- ☞ The phosphate group is transferred to GDP to form GTP and succinate.
- ☞ GTP donates a phosphate group to ADP to form ATP.

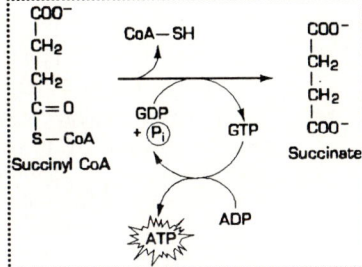

Step 6: Succinate is oxidized to fumarate and FAD is reduced.
- ☞ Two hydrogens are transferred to FAD to form $FADH_2$.
- ☞ The dehydrogenase that catalyzes this reaction is bound to the inner mitochondrial membrane.

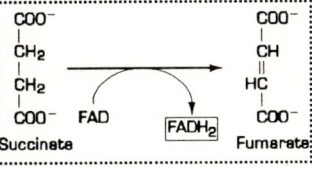

Step 7: Water is added to fumarate which rearranges its chemical bonds to form malate.

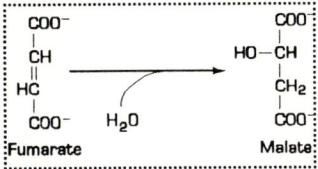

Step 8: Malate is oxidized and NAD^+ is reduced.
- ☞ A molecule of NADH is produced.
- ☞ Oxaloacetate is regenerated to begin the cycle again.

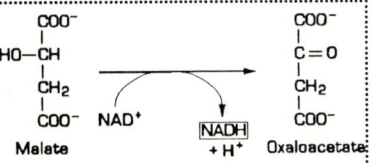

Two turns of the Krebs Cycle produces two ATPs by substrate-level phosphorylation. However, <u>most</u> ATP output of respiration results from *oxidative phosphorylation*.
- ☞ Reduced coenzymes produced by the Krebs Cycle (6 NADH and 2 $FADH_2$ per glucose) carry high energy electrons to the electron transport chain.
- ☞ The ETC couples electron flow down the chain to ATP synthesis.

110 Cellular Respiration: Harvesting Chemical Energy

VI. THE ELECTRON TRANSPORT CHAIN AND OXIDATIVE PHOSPHORYLATION: A CLOSER LOOK (p. 184–189)

Only a few molecules of ATP are produced by substrate-level phosphorylation:
- ☛ 2 net ATPs per glucose from glycolysis.
- ☛ 2 ATPs per glucose from the Krebs Cycle.

How is the exergonic oxidation of glucose coupled to the endergonic synthesis of ATP?

Most molecules of ATP are produced by oxidative phosphorylation.
- ☛ At the end of the Krebs Cycle, most of the energy extracted from glucose is in molecules of NADH and $FADH_2$.
- ☛ These reduced coenzymes link glycolysis and the Krebs Cycle to oxidative phosphorylation by passing their electrons down the electron transport chain to oxygen. (Though the Krebs Cycle occurs only under aerobic conditions, it does not use oxygen directly. The ETC and oxidative phosphorylation require oxygen as the final electron acceptor.)
- ☛ This exergonic transfer of electrons down the ETC to oxygen is coupled to ATP synthesis.

A. Pathway of Electron Transport (p. 184–185)

What is the electron transport chain and where is it located? How do the electron carriers of the chain participate in redox reactions?

The *electron transport chain* is made of electron carrier molecules embedded in the inner mitochondrial membrane.
- ☛ Each successive carrier in the chain has a higher electronegativity than the carrier before it, so the electrons are pulled downhill towards oxygen, the final electron acceptor and the molecule with the highest electronegativity.
- ☛ Except for ubiquinone (Q), most of the carrier molecules are proteins and are tightly bound to *prosthetic groups* (nonprotein cofactors).
- ☛ Prosthetic groups alternate between reduced and oxidized states as they accept and donate electrons.

Protein Electron Carriers	Prosthetic Group
flavoproteins	flavin mononucleotide (FMN)
iron-sulfur proteins	iron and sulfur
cytochromes	heme group

Heme group = Prosthetic group composed of four organic rings surrounding a single iron atom.

Cytochrome = Type of protein molecule that contains a heme prosthetic group and that functions as an electron carrier in the electron transport chains of mitochondria and chloroplasts.
- ☛ There are several cytochromes, each a slightly different protein with a heme group.
- ☛ It is the iron of cytochromes that transfers electrons.

Electron transport chain

Electron carrier

Heme group

Cytochrome

What is the pathway of electrons down the electron transport chain?

Sequence of Electron Transfers Along the Electron Transport Chain:

NADH is oxidized and *flavoprotein* is reduced as high energy electrons from NADH are transferred to FMN.

↓

Flavoprotein is oxidized as it passes electrons to an *iron-sulfur protein*, Fe·S.

↓

Iron-sulfur protein is oxidized as it passes electrons to *ubiquinone (Q)*.

↓

Ubiquinone passes electrons on to a succession of electron carriers, most of which are cytochromes.

↓

Cyt a_3, the last cytochrome passes electrons to *molecular oxygen*, O_2.

↓

As molecular oxygen is reduced it also picks up two protons from the medium to form water. For every two NADHs, one O_2 is reduced to two H_2O molecules.

- ☞ $FADH_2$ also donates electrons to the electron transport chain, but those electrons are added at a lower energy level than NADH.
- ☞ The electron transport chain does not make ATP directly. It generates a proton gradient across the inner mitochondrial membrane, which stores potential energy that can be used to phosphorylate ADP.

    B.   **Chemiosmosis: Coupling Electron Flow to ATP Synthesis**
         (p. 185–189)

How is the exergonic slide of electrons down the electron transport chain coupled to the endergonic production of ATP? What is this coupling process called?

The mechanism for coupling exergonic electron flow from the oxidation of food to the endergonic process of oxidative phosphorylation is *chemiosmosis*.

Cellular Respiration: Harvesting Chemical Energy

Chemiosmosis

Peter Mitchell

Proton gradient

Cristae

Mitochondrial matrix

Intermembrane space

ATP synthases

<u>Chemiosmosis</u> = The coupling of exergonic electron flow down an electron transport chain to endergonic ATP production by the creation of a proton gradient across a membrane. The proton gradient drives ATP synthesis as protons diffuse back across the membrane.
- Proposed by British biochemist, Peter Mitchell (1961).
- The term *chemiosmosis* emphasizes a coupling between (1) chemical reactions (phosphorylation) and (2) transport processes (proton transport).
- Process involved in oxidative phosphorylation and photophosphorylation.

The site of oxidative phosphorylation is the inner mitochondrial membrane, which has many copies of a protein complex, *ATP synthase*. This complex:
- Is an enzyme that makes ATP.
- Uses an existing *proton gradient* across the inner mitochondrial membrane to power ATP synthesis.

Cristae or infoldings of the inner mitochondrial membrane, increase the surface area available for chemiosmosis to occur.

How is membrane structure related to membrane function in chemiosmosis? What enzyme catalyzes oxidative phosphorylation? Where is it located?

Membrane structure correlates with the prominent functional role membranes play in chemiosmosis:
- Using energy from exergonic electron flow, the *electron transport chain* creates the proton gradient by pumping H^+s from the *mitochondrial matrix*, across the *inner membrane* to the *intermembrane space*.
- This proton gradient is maintained, because the membrane's phospholipid bilayer is impermeable to H^+s and prevents them from leaking back across the membrane by diffusion.
- *ATP synthases* use the potential energy stored in a proton gradient to make ATP by allowing H^+ to diffuse down the gradient, back across the membrane. Protons diffuse through the ATP synthase complex, which causes the phosphorylation of ADP.

How does the electron transport chain pump hydrogen ions from the matrix to the intermembrane space? The process is based on spatial organization of the electron transport chain in the membrane. Note that:

- Some electron carriers of the transport chain transport only electrons.
- Some electron carriers accept and release protons along with electrons. These carriers are spatially arranged so that protons are picked up from the matrix and are released into the intermembrane space.

Most of the electron carriers are organized into three complexes: 1) NADH dehydrogenase complex; 2) cytochrome b-c_1 complex; and 3) cytochrome oxidase complex.

- Each complex is an asymmetric particle that has a specific orientation in the membrane.
- As complexes transport electrons, they also harness energy from this exergonic process to pump protons across the inner mitochondrial membrane.

Mobile carriers transfer electrons between complexes. These mobile carriers are:

1. Ubiquinone (Q). Near the matrix, Q accepts electrons from the NADH dehydrogenase complex, diffuses across the lipid bilayer, and passes electrons to the cytochrome b-c_1 complex.

2. Cytochrome c (Cyt c). Cyt c accepts electrons from the cytochrome b-c_1 complex and conveys them to the cytochrome oxidase complex.

When the transport chain is operating:

- The pH in the intermembrane space is one or two pH units lower than in the matrix.
- The pH in the intermembrane space is the same as the pH of the cytosol because the outer mitochondrial membrane is permeable to protons.

The H^+ gradient that results is called a *proton-motive force* to emphasize that the gradient represents potential energy.

Why does the proton gradient across the inner mitochondrial membrane represent potential energy? What is the proton-motive force?

Proton-motive force = Potential energy stored in the proton gradient created across biological membranes that are involved in chemiosmosis.

- This force is an <u>electrochemical gradient</u> with two components:
    1. Concentration gradient of protons (chemical gradient).
    2. Voltage across the membrane because of a higher concentration of positively charged protons on one side (electrical gradient).
- It tends to drive protons across the membrane back into the matrix.

Ubiquinone

Proton-motive force

Electrochemical gradient

Effects of three classes of respiratory poisons provide evidence for chemiosmosis and its dependence upon the structural organization of the mitochondrial membrane.

1. Poisons that block electron flow.
   - ☞ Cyanide blocks electron flow from the Cyt a_3 to oxygen. This stops the electron transport chain so it cannot pump protons. Without the proton gradient, ATP is not produced.
2. Poisons that inhibit ATP synthase.
   - ☞ This class of poisons, which includes the antibiotic oligomycin, directly inhibits ATP synthase.
   - ☞ The proton gradient becomes greater than normal and yet the potential energy of the gradient cannot be tapped to produce ATP.
3. Poisons that make the inner mitochondrial membrane leaky to protons.
   - ☞ These poisons, such as dinitrophenol, are called *uncouplers*. By allowing protons to leak back across the membrane, they uncouple the process of proton pumping with ATP production.
   - ☞ O_2 consumption increases as the ETC continues to work, but no ATP is produced.
   - ☞ This uncoupling mechanism generates metabolic heat as the intensified pace of the ETC continues. Some organisms have evolved a mechanism that provides body heat by naturally uncoupling electron transport from ATP synthesis.
   - ☞ For example, bats use metabolic heat generated this way to break torpor. Some plants use this mechanism to heat flowers, which attracts pollinators. (See Campbell, Figure 9.19)

Chemiosmosis couples exergonic chemical reactions to endergonic H^+ transport, which creates the proton-motive force used to drive cellular work, such as:
- ☞ ATP synthesis in mitochondria (*oxidative phosphorylation*). The energy to create the proton gradient comes from the oxidation of glucose and the ETC.
- ☞ ATP synthesis in chloroplasts (*photophosphorylation*). The energy to create the proton gradient comes from light trapped during the energy-capturing reactions of photosynthesis.
- ☞ ATP synthesis, transport processes and rotation of flagella in bacteria. The proton gradient is created across the plasma membrane. Peter Mitchell first postulated chemiosmosis as an energy-coupling mechanism based on experiments with bacteria.

C. **The Chemiosmotic Model: Incorporating Many Biological Themes** (*p. 189*)

The working model of how mitochondria harvest the energy of food, illustrates many of the text's integrative themes in the study of life:
- ☞ Energy conversion and utilization.
- ☞ Emergent properties. Oxidative phosphorylation is an emergent property of the intact mitochondrion that uses a precise interaction of molecules.
- ☞ Correlation of structure and function. The chemiosmotic model is based upon the spatial arrangement of membrane proteins.

☞ Evolution. In an effort to reconstruct the origin of oxidative phosphorylation and the evolution of cells, biologists compare similarities in the chemiosmotic machinery of mitochondria to that of chloroplasts and bacteria.

VII. SUMMARIZING CELLULAR RESPIRATION (p. 189–190)

Summarize the net ATP yield from the oxidation of a glucose molecule by constructing an ATP ledger which includes coenzyme production during the different stages of glycolysis and cellular respiration.

The net ATP yield from the oxidation of one glucose molecule to six carbon dioxide molecules can be estimated by adding:

1. ATP produced directly by substrate-level phosphorylation during glycoloysis and the Krebs cycle.
   ☞ A net of two ATPs is produced during glycolysis. The debit of two ATPs used during the investment phase is subtracted from the four ATPs produced during the energy-yielding phase.
   ☞ Two ATPs are produced during the Krebs cycle.

2. ATP produced when chemiosmosis couples electron transport to oxidative phosphorylation.
   ☞ The electron transport chain creates enough proton-motive force to produce a maximum of three ATPs for each electron pair that travels from NADH to oxygen. The average yield is actually between two and three ATPs per NADH (2.7).
   ☞ $FADH_2$ is worth a maximum of only two ATPs, since it donates electrons at a lower energy level to the electron transport chain.
   ☞ In most eukaryotic cells, the ATP yield is lower from an NADH produced during glycolysis. The mitochondrial membrane is impermeable to NADH, so its electrons must be shuttled across the membrane. These electrons are received inside the mitochondrion by FAD, a process which downgrades the energy level of those electrons.
   ☞ Maximum ATP yield for each glucose oxidized during cellular respiration:

Process	ATP Produced Directly by Substrate-Level Phosphorylation	Reduced Coenzyme	ATP Produced by Oxidative Phosphorylation	Total
Glycolysis	Net 2 ATP	2 NADH	4 to 6 ATP	6–8
Oxidation of Pyruvate	———	2 NADH	6 ATP	6
Krebs Cycle	2 ATP	6 NADH 2 $FADH_2$	18 ATP 4 ATP	24
			Total	36–38

116   *Cellular Respiration: Harvesting Chemical Energy*

☞ This tally only <u>estimates</u> the ATP yield from respiration. Some variables that affect ATP yield include:
1. Mitochondrial membranes may differ in permeability to protons and resulting proton-motive force.
2. The proton-motive force may be used to drive other kinds of cellular work such as active transport.
3. The total ATP yield is inflated by rounding off the number of ATPs produced per NADH to three.
4. Cellular respiration in prokaryotes produces slightly higher average yields of ATP, since there is no mitochondrial membrane separating glycolysis from electron transport (maximum yield = 38 ATP).

Cellular respiration is remarkably efficient in the transfer of chemical energy from glucose to ATP.
☞ Estimated efficiency in eukaryotic cells is about 40 to 60%.
☞ Energy lost in the process is released as heat.

VIII. FERMENTATION: THE ANAEROBIC ALTERNATIVE (p. 190–192)

<u>Aerobic</u> = (Aer=air; bios=life) Existing in the presence of oxygen.

<u>Anaerobic</u> = (An=without; aer=air) Existing in the absence of free oxygen.

<u>Fermentation</u> = The anaerobic catabolism of organic nutrients.

What is the fate of pyruvate in the absence of oxygen?

Glycolysis oxidizes glucose to two pyruvate molecules. The oxidizing agent for this process is NAD^+, <u>not</u> oxygen.
☞ Some energy released from the exergonic process of glycolysis drives the production of 2 net ATPs by substrate-level phosphorylation.
☞ Glycolysis produces a net of 2 ATPs whether conditions are aerobic or anaerobic.
1. <u>Aerobic conditions</u>: Pyruvate is <u>oxidized</u> further, and more ATP is made as NADH passes electrons removed from glucose to the electron transport chain. NAD^+ is regenerated in the process.
2. <u>Anaerobic conditions</u>: Pyruvate is <u>reduced</u>, and NAD^+ is regenerated. This prevents the cell from depleting the pool of NAD^+, which is the oxidizing agent necessary for glycolysis to continue. No additional ATP is produced.

Why is fermentation necessary?

Fermentation recycles NAD^+ from NADH. This process consists of anaerobic glycolysis plus subsequent reactions that regenerate NAD^+ by reducing pyruvate. Two of the most common types of fermentation are (1) alcohol fermentation and (2) lactic acid fermentation. (See Campbell, Figure 9.21)

<u>Aerobic</u>

<u>Anaerobic</u>

<u>Fermentation</u>

1. Alcohol Fermentation

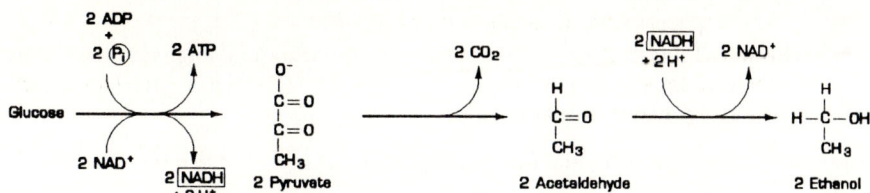

Pyruvate is converted to ethanol in two steps:
   a. Pyruvate loses carbon dioxide and is converted to the two-carbon compound acetaldehyde.
   b. NADH is oxidized to NAD⁺ and acetaldehyde is reduced to ethanol.

Many bacteria and yeast carry out alcohol fermentation under anaerobic conditions.

2. Lactic Acid Fermentation

NADH is oxidized to NAD⁺ and pyruvate is reduced to lactate.
- ☞ Commercially important products of lactic acid fermentation include cheese, yogurt, acetone and methyl alcohol.
- ☞ When oxygen is scarce, human muscle cells switch from aerobic respiration to lactic acid fermentation. Lactate accumulates, but it is gradually carried to the liver where it is converted back to pyruvate when oxygen becomes available.

IX. A COMPARISON OF AEROBIC AND ANAEROBIC CATABOLISM
(p. 192–193)

What are the three major catabolic processes? How do they differ?

There are three major catabolic processes for harvesting food's chemical energy: (1) aerobic respiration, (2) anaerobic respiration, and (3) fermentation. These processes:
- ☞ Are similar in that the high energy electrons from substrate (e.g. glucose) oxidation are transferred to NAD⁺.
- ☞ Differ in the ultimate fate of high energy electrons stored in NADH.

Cellular respiration includes all types of catabolism that use electron transport chains to make ATP, regardless of the substance used as the final electron acceptor. It also includes the Krebs cycle or some modification of the Krebs cycle.

Aerobic respiration occurs only in the presence of oxygen. This process:
- ☞ Uses an electron transport chain to make ATP.
- ☞ Uses oxygen as the final electron acceptor.
- ☞ Produces most ATP by oxidative phosphorylation and some ATP by substrate-level phosphorylation.
- ☞ Is used by plants and animals.

Anaerobic respiration is a catabolic process that does not require the presence of oxygen. This process:
- ☞ Uses an electron transport chain to make ATP.
- ☞ Uses a substance <u>other</u> than free oxygen as a final electron acceptor (e.g. NO_3^- or SO_4^{2-}).
- ☞ Produces most ATP by oxidative phosphorylation and some by substrate-level phosphorylation.
- ☞ Occurs in only a few bacterial groups that exist in anaerobic environments.

Fermentation operates not only without electron transport chains, but without the Krebs cycle as well.
- ☞ No oxygen is required.
- ☞ The final electron acceptor is an organic substrate such as pyruvate or some derivative of pyruvate.
- ☞ ATP is produced by substrate-level phosphorylation only.
- ☞ This process is less efficient than respiration. Respiration yields 18 times more ATP per glucose than fermentation.

Organisms can be classified based upon the effect oxygen has on growth and metabolism.

<u>Strict (obligate) aerobes</u> = Organisms that require oxygen for growth and as the final electron acceptor for aerobic respiration.

<u>Strict (obligate) anaerobes</u> = Microorganisms that only grow in the absence of oxygen and are, in fact, poisoned by it.

<u>Facultative anaerobes</u> = Organisms capable of growth in either aerobic or anaerobic environments.
- ☞ Yeasts, many bacteria and mammalian muscle cells are facultative anaerobes.
- ☞ Can make ATP by fermentation in the absence of oxygen or by respiration in the presence of oxygen.
- ☞ Glycolysis is common to both fermentation and respiration, so pyruvate is a key juncture in catabolism.

```
                        glucose
                           ↓ glycolysis
          no O₂  ——— pyruvate ——— O₂
             ↓                      ↓
      reduced to ethanol       oxidized to
        or lactate &           acetyl CoA &
      NAD⁺ is recycled       oxidation continues
      as NADH is oxidized.     in Krebs cycle.
```

- ☞ Cellular respiration yields 18 times more ATP than does fermentation. Without oxygen, much of the energy in pyruvate is unavailable to the cell.

Strict aerobes

Strict anaerobes

Facultative anaerobes

X. THE EVOLUTIONARY SIGNIFICANCE OF GLYCOLYSIS (p. 193)

What evidence supports the idea that the first prokaryotes produced ATP by glycolysis?

The first prokaryotes probably produced ATP by glycolysis. Evidence includes the following:
- ☞ Glycolysis does not require oxygen, and the oldest known bacterial fossils date back to three-and-a-half billion years ago when oxygen was not present in the atmosphere.
- ☞ Glycolysis is the most widespread metabolic pathway, so it probably evolved early.
- ☞ Glycolysis occurs in the cytoplasm and does not require membrane-bound organelles. Eukaryotic cells with organelles probably evolved about two billion years after prokaryotic cells.

X. THE CATABOLISM OF OTHER MOLECULES (p. 193–194)

How are food molecules other than glucose oxidized to make ATP?

Respiration can oxidize organic molecules other than glucose to make ATP. Organisms obtain most calories from fats, proteins, disaccharides and polysaccharides. These complex molecules must be enzymatically hydrolyzed into simpler molecules or monomers that can enter an intermediate reaction of glycolysis or the Krebs cycle. (See Campbell, Figure 9.24)

Glycolysis can accept a wide range of carbohydrates for catabolism.
- ☞ Starch is hydrolyzed to glucose in the digestive tract of animals.
- ☞ In between meals, the liver hydrolyzes glycogen to glucose.
- ☞ Enzymes in the small intestine break down disaccharides to glucose or other monosaccharides.

Proteins are hydrolyzed to amino acids.
- ☞ Organisms synthesize new proteins from some of these amino acids.
- ☞ Excess amino acids are enzymatically converted to intermediates of glycolysis and the Krebs cycle. Common intermediates are pyruvate, acetyl CoA and α-ketoglutarate.
- ☞ This conversion process deaminates amino acids, and the resulting nitrogenous wastes are excreted.

Fats are excellent fuels because they are rich in hydrogens with high energy electrons. Oxidation of one gram of fat produces twice as much ATP as a gram of carbohydrate.
- ☞ Fat sources may be from the diet or from storage cells in the body.
- ☞ Fats are digested into glycerol and fatty acids.
- ☞ Glycerol can be converted to glyceraldehyde phosphate, an intermediate of glycolysis.
- ☞ Most energy in fats is in fatty acids, which are converted into acetyl CoA by *beta oxidation*. The resulting two-carbon fragments can enter the Krebs cycle.

120 *Cellular Respiration: Harvesting Chemical Energy*

XI. BIOSYNTHESIS (p. 194)

Some organic molecules of food provide the carbon skeletons or raw materials for the synthesis of new macromolecules.

- Some organic monomers from digestion can be used <u>directly</u> in anabolic pathways.
- Some precursors for biosynthesis do not come directly from digested food, but instead come from glycolysis or Krebs cycle intermediates which are diverted into anabolic pathways.
- These anabolic pathways <u>consume</u> ATP produced by catabolic pathways of glycolysis and respiration.
- Glycolysis and the Krebs cycle are metabolic interchanges that can convert one type of macromolecule to another in response to the cell's metabolic demands.

XII. THE CONTROL OF RESPIRATION (p. 194–195)

How is ATP production controlled by the cell? What role does the allosteric enzyme, phosphofructokinase, play in this process?

Cells respond to changing metabolic needs by controlling reaction rates.

Anabolic pathways are switched off when their products are in ample supply. The most common mechanism of control is *feedback inhibition*. (See Campbell, Chapter 6)

Catabolic pathways, such as glycolysis and Krebs cycle, are controlled by regulating enzyme activity at strategic points.

- A key control point is the third step of glycolysis, which is catalyzed by an allosteric enzyme, *phosphofructokinase*.
- The ratio of ATP to ADP and AMP reflects the energy status of the cell, and phosphofructokinase is sensitive to changes in this ratio.
- Citrate (produced in Krebs cycle) and ATP are *allosteric inhibitors* of phosphofructokinase, so when their concentrations rise, the enzyme slows glycolysis. As the rate of glycolysis slows, Krebs cycle also slows since the supply of acetyl CoA is reduced. This synchronizes the rates of glycolysis and Krebs cycle.
- ADP and AMP are *allosteric activators* for phosphofructokinase, so when their concentrations relative to ATP rise, the enzyme speeds up glycolysis which speeds up the Krebs cycle.
- There are other allosteric enzymes that also control the rates of glycolysis and the Krebs cycle.

Phosphofructokinase

10   PHOTOSYNTHESIS

CHAPTER OUTLINE

Photosynthesis transforms solar light energy trapped by chloroplasts into chemical bond energy stored in sugar and other organic molecules. This process:
- Synthesizes energy-rich organic molecules from the energy-poor molecules, CO_2 and H_2O.
- Uses CO_2 as a carbon source and light energy as the energy source.
- Directly or indirectly supplies energy to most living organisms.

How do organisms acquire organic molecules for energy and carbon skeletons? What is the difference between autotrophic and heterotrophic nutrition?

Organisms acquire organic molecules used for energy and carbon skeletons by one of two nutritional modes: 1) *Autotrophic Nutrition* or 2) *Heterotrophic Nutrition*.

1. Autotrophic Nutrition

    <u>Autotrophic nutrition</u> = (Auto=self; trophos=feed) Nutritional mode of synthesizing organic molecules from inorganic raw materials.
    - Examples of autotrophic organisms are plants, which require only CO_2, H_2O and minerals as nutrients.
    - Autotrophic organisms require an energy source to synthesize organic molecules. That energy source may be from light (*photoautotrophic*) or from the oxidation of inorganic substances (*chemoautotrophic*).

How do photoautotrophs and chemoautotrophs acquire the energy needed to synthesize organic molecules? What are some examples of each?

      <u>Photoautotrophs</u> = Autotrophic organisms that use light as an energy source to synthesize organic molecules. Examples are photosynthetic organisms such as plants, algae and some prokaryotes.

      <u>Chemoautotrophs</u> = Autotrophic organisms that use the oxidation of inorganic substances, such as sulfur or ammonia, as an energy source to synthesize organic molecules. Unique to some bacteria, this is a rarer form of autotrophic nutrition.

Autotrophic nutrition

Photoautotrophs

Chemoautotrophs

122 Photosynthesis

2. Heterotrophic Nutrition

   Heterotrophic nutrition = (Heteros=other; trophos=feed) Nutritional mode of acquiring organic molecules from compounds produced by other organisms; heterotrophs are unable to synthesize organic molecules from inorganic raw materials.
   - ☞ Examples are animals that eat plants or other animals.
   - ☞ Examples also include *decomposers*, heterotrophs that decompose and feed on organic litter. Most fungi and many bacteria are decomposers.
   - ☞ Most heterotrophs depend on photoautotrophs for food and oxygen (a by-product of photosynthesis).

I. **CHLOROPLASTS: SITES OF PHOTOSYNTHESIS** (*p. 200–201*)

Where are chloroplasts located?

Although all green plant parts have chloroplasts, leaves are the major organs of photosynthesis in most plants.
- ☞ *Chlorophyll* is the green pigment in chloroplasts that gives a leaf its color and that absorbs the light energy used to drive photosynthesis.

LEAF CROSS SECTION

Stomata *Mesophyll* *Vascular bundle*

- ☞ Chloroplasts are primarily in cells of *mesophyll*, green tissue in the leaf's interior.
- ☞ CO_2 enters and O_2 exits the leaf through microscopic pores called *stomata*.
- ☞ Water absorbed by the roots is transported to leaves through veins or *vascular bundles* which also export sugar from leaves to nonphotosynthetic parts of the plant.

How does chloroplast structure relate to its function?

Chloroplasts are lens-shaped organelles measuring about 2–4 μm by 4–7 μm. These organelles are divided into three functional compartments by a system of membranes:

1. **Intermembrane Space.** The chloroplast is bound by a double membrane which partitions its contents from the cytosol. A narrow *intermembrane space* separates the two membranes.

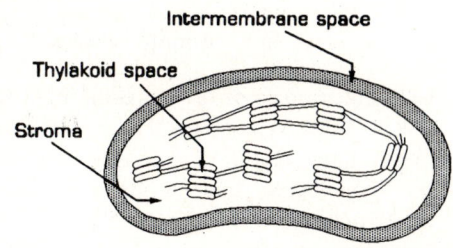

2. **Thylakoid Space.** *Thylakoids* form another membranous system within the chloroplast. The thylakoid membrane segregates the interior of the chloroplast into two compartments: *thylakoid space* and *stroma*.

   Thylakoids = Flattened membranous sacs inside the chloroplast.
   - ☞ Chlorophyll is found in the thylakoid membranes.
   - ☞ Thylakoids function in the steps of photosynthesis that initially convert light energy to chemical energy.

   Thylakoid space = Space inside the thylakoid.

   Grana = (Singular, granum) Stacks of thylakoids in a chloroplast.

3. **Stroma.** Those steps that use chemical energy to convert carbon dioxide to sugar occur in the *stroma*, viscous fluid outside the thylakoids.

Photosynthetic prokaryotes lack chloroplasts, but have chlorophyll built into the plasma membrane or into membranes of numerous vesicles within the cell. (See Campbell, Figure 10.4)
- ☞ These membranes function in a manner similar to the thylakoid membranes of chloroplasts.
- ☞ Photosynthetic membranes of cyanobacteria are usually arranged in parallel stacks of flattened sacs similar to the thylakoids of chloroplasts.

II. HOW PLANTS MAKE FOOD: AN OVERVIEW (p. 201–204)

What is the summary equation for photosynthesis?

Some steps in photosynthesis are not yet understood, but the following summary equation has been known since the early 1800's:

$$6 \, CO_2 + 12 \, H_2O + \text{light energy} \longrightarrow C_6H_{12}O_6 + 6 \, O_2 + 6 \, H_2O$$

- ☞ Glucose ($C_6H_{12}O_6$) is shown in the summary equation, though the main products of photosynthesis are other carbohydrates.
- ☞ Water is on both sides of the equation, because photosynthesis consumes 12 molecules and forms 6.

Indicating the net consumption of water simplifies the equation:

$$6\ CO_2 + 6\ H_2O + \text{light energy} \longrightarrow C_6H_{12}O_6 + 6\ O_2$$

☞ In this form, the summary equation for photosynthesis is the reverse of that for cellular respiration.

☞ Photosynthesis and cellular respiration both occur in plant cells, but plants do not simply reverse the steps of respiration to make food.

The simplest form of the equation is: $CO_2 + H_2O \longrightarrow CH_2O + O_2$

☞ CH_2O symbolizes the general formula for a carbohydrate.

☞ In this form, the summary equation emphasizes the production of a sugar molecule, one carbon at a time. Six repetitions produces a glucose molecule.

A. The Splitting of Water (p. 202–203)

What was van Niel's hypothesis, and how did it contribute to our current understanding of photosynthesis?

The discovery that O_2 released by plants is derived from H_2O and not from CO_2, was one of the earliest clues to the mechanism of photosynthesis.

☞ In the 1930's, C.B. van Niel from Stanford University challenged an early model, which predicted that:

a. O_2 released during photosynthesis comes from CO_2.

$$CO_2 \longrightarrow C + O_2$$

b. CO_2 is split and water is added to the carbon.

$$C + H_2O \longrightarrow CH_2O$$

☞ Van Niel studied bacteria that use hydrogen sulfide (H_2S) rather than H_2O for photosynthesis and that produce yellow sulfur globules as a by-product.

$$CO_2 + 2\ H_2S \longrightarrow CH_2O + H_2O + 2\ S$$

☞ Van Niel deduced that these bacteria split H_2S and use H to make sugar. He generalized that all photosynthetic organisms require hydrogen, but that the source varies:

general: $CO_2 + 2\ H_2X \longrightarrow CH_2O + H_2O + 2\ X$

sulfur bacteria: $CO_2 + 2\ H_2S \longrightarrow CH_2O + H_2O + 2\ S$

plants: $CO_2 + 2\ H_2O \longrightarrow CH_2O + H_2O + O_2$

☞ Van Niel thus hypothesized that plants split water as a source of hydrogen and release oxygen as a by-product.

Scientists later confirmed van Niel's hypothesis by using a heavy isotope of oxygen (^{18}O) as a tracer to follow oxygen's fate during photosynthesis.

☞ If water was labeled with tracer, released oxygen was ^{18}O:

Experiment 1: $CO_2 + H_2O^* \longrightarrow CH_2O + H_2O + O_2^*$

☞ If the ^{18}O was introduced to the plant as CO_2, the tracer did not appear in the released oxygen:

Experiment 2: $CO_2^* + H_2O \longrightarrow CH_2O^* + H_2O^* + O_2$

C.B. van Niel

An important result of photosynthesis is the extraction of hydrogen from water and its incorporation into sugar.
- ☞ Electrons associated with hydrogen have more potential energy in organic molecules than they do in water, where the electrons are closer to electronegative oxygen.
- ☞ Energy is stored in sugar and other food molecules in the form of these high-energy electrons.

B. Photosynthesis as a Redox Process (p. 203)

Respiration is an exergonic redox process; energy is released from the oxidation of sugar.
- ☞ Electrons associated with sugar's hydrogens lose potential energy as carriers transport them to oxygen, forming water.
- ☞ Electronegative oxygen pulls electrons down the electron transport chain, and the potential energy released is used by the mitochondrion to produce ATP.

What is the role of redox reactions in photosynthesis?

Photosynthesis is an endergonic redox process; energy is required to reduce carbon dioxide.
- ☞ Light is the energy source that boosts potential energy of electrons as they are moved from water to sugar.
- ☞ When water is split, electrons are transferred from the water to carbon dioxide, reducing it to sugar.

C. The Two Stages of Photosynthesis (p. 203–204)

Photosynthesis occurs in two stages: the *light reactions* and the *Calvin cycle*.

Light reactions = In photosynthesis, the reactions that convert light energy to chemical bond energy in ATP and NADPH. These reactions:
- ☞ Occur in the thylakoid membranes of chloroplasts.
- ☞ Reduce $NADP^+$ to NADPH. Light absorbed by chlorophyll provides the energy to reduce $NADP^+$ to NADPH, which temporarily stores the energized electrons transferred from water. $NADP^+$ (nicotinamide adenine dinucleotide phosphate), a coenzyme similar to NAD^+ in respiration, is reduced by adding a pair of electrons along with a hydrogen nucleus, or H^+.
- ☞ Give off O_2 as a by-product from the splitting of water.
- ☞ Generate ATP. The light reactions power the addition of a phosphate group to ADP in a process called *photophosphorylation*.

Calvin cycle = In photosynthesis, the carbon-fixation reactions that assimilate atmospheric CO_2 and then reduce it to a carbohydrate; named for Melvin Calvin. These reactions:
- ☞ Occur in the stroma of the chloroplast.
- ☞ First incorporate atmospheric CO_2 into existing organic molecules by a process called *carbon fixation*, and then reduce fixed carbon to carbohydrate.

　　Carbon fixation = The process of incorporating CO_2 into organic molecules.

The Calvin cycle reactions do not require light directly, but reduction of CO_2 to sugar requires the products of the light reactions:
- ☞ NADPH provides the reducing power.
- ☞ ATP provides the chemical energy.

Chloroplasts thus use light energy to make sugar by coordinating the two stages of photosynthesis. (See Campbell, Figure 10.5)
- ☞ Light reactions occur in the thylakoids of chloroplasts.
- ☞ Calvin cycle reactions occur in the stroma.
- ☞ As $NADP^+$ and ADP contact thylakoid membranes, they pick up electrons and phosphate respectively, and then transfer their high-energy cargo to the Calvin cycle.

III. HOW THE LIGHT REACTIONS CAPTURE SOLAR ENERGY (p. 204–212)

To understand how the thylakoids of chloroplasts transform light energy into the chemical energy of ATP and NADPH, it is necessary to know some important properties of light.

A. The Nature of Sunlight (p. 204–205)

The sun emits energy from fusion reactions:
- ☞ Four hydrogen atoms fuse to form one helium atom, but the mass of the newly formed helium is slightly less than the total mass of the hydrogen atoms.
- ☞ The lost mass has been converted to energy (predicted by Albert Einstein in the equation $E = mc^2$).
- ☞ About 120 million tons of solar matter are converted to an enormous amount of energy each minute.

This energy radiates out from the sun into space.
- ☞ A very small fraction of this energy reaches Earth in only a few minutes, as light travels at about 300,000 km/sec.
- ☞ Some of this energy strikes the leaves of plants and is absorbed by the chloroplasts.

In what ways is the behavior of light wavelike? What are the particlelike properties of light?

Sunlight is *electromagnetic energy*. The quantum mechanical model of electromagnetic radiation describes light as having a behavior that is both wavelike and particlelike.

1. Wavelike properties of light.
   - ☞ *Electromagnetic energy* is a form of energy that travels in rhythmic waves which are disturbances of electric and magnetic fields.
   - ☞ A *wavelength* is the distance between the crests of electromagnetic waves.
   - ☞ The electromagnetic spectrum ranges from wavelengths that are less than a nanometer (gamma rays) to those that are more than a kilometer (radio waves). (See Campbell, Figure 10.6)

- *Visible light*, which is detectable by the human eye, is only a small portion of the electromagnetic spectrum and ranges from about 380 to 750 nm. The wavelengths most important for photosynthesis are within this range of visible light.
 2. Particlelike properties of light.
    - Light also behaves as if it consists of discrete particles or quanta called *photons*.
    - Each photon has a fixed quantity of energy which is <u>inversely</u> proportional to the wavelength of light. For example, a photon of violet light has nearly twice as much energy as a photon of red light.

Photons

The sun radiates the full spectrum of electromagnetic energy.
- The atmosphere acts as a selective window that allows visible light to pass through while screening out a substantial fraction of other radiation.
- The visible range of light is the radiation that drives photosynthesis.
- Blue and red, the two wavelengths most effectively absorbed by chlorophyll, are the colors most useful as energy for the light reactions.

B. Photosynthetic Pigments (*p. 205–207*)

Light may be reflected, transmitted or absorbed when it contacts matter. (See Campbell, Figure 10.7)

<u>Pigments</u> = Substances that absorb visible light.

Pigments

- Different pigments absorb different wavelengths of light.
- Wavelengths that are absorbed disappear, so a pigment that absorbs all wavelengths appears black.
- When white light, which contains all the wavelengths of visible light, illuminates a pigment, the color you see is the color most reflected or transmitted by the pigment. For example, a leaf appears green because chlorophyll absorbs red and blue light but transmits and reflects green light.

How is the action spectrum for photosynthesis related to the absorption spectrum for chlorophyll a? Why are they not exactly the same?

Each pigment has a characteristic *absorption spectrum* or pattern of wavelengths that it absorbs. It is expressed as a graph of absorption versus wavelength.

Absorption spectrum

- The absorption spectrum for a pigment in solution can be determined by using a *spectrophotometer*, an instrument used to measure what proportion of a specific wavelength of light is absorbed or transmitted by the pigment. (See Campbell Methods Box, p. 206)

Spectrophotometer

- Since chlorophyll *a* is the light-absorbing pigment that participates directly in the light reactions, the absorption spectrum of chlorophyll *a* provides clues as to which wavelengths of visible light are most effective for photosynthesis.

Chlorophyll a

Photosynthesis

Which wavelengths of light are most effective for photosynthesis?

A graph of wavelength versus rate of photosynthesis is called an *action spectrum* and profiles the relative effectiveness of different wavelengths of visible light for driving photosynthesis.

☞ The action spectrum of photosynthesis can be determined by illuminating chloroplasts with different wavelengths of light and measuring some indicator of photosynthetic rate, such as oxygen release or carbon dioxide consumption.

☞ It is apparent from the action spectrum of photosynthesis that blue and red light are the most effective wavelengths for photosynthesis and green light is the least effective.

The *action spectrum* for photosynthesis does not exactly match the *absorption spectrum* for chlorophyll *a*.

☞ Since chlorophyll *a* is not the only pigment in chloroplasts that absorb light, the absorption spectrum for chlorophyll *a* underestimates the effectiveness of some wavelengths.

☞ Even though only special chlorophyll *a* molecules can participate directly in the light reactions, other pigments, called *accessory pigments*, can absorb light and transfer the energy to chlorophyll *a*.

The *accessory pigments* expand the range of wavelengths available for photosynthesis. These pigments include:

☞ *Chlorophyll b*, a yellow-green pigment with a structure similar to chlorophyll *a*. This minor structural difference gives the pigments slightly different absorption spectra.

☞ *Carotenoids*, yellow and orange hydrocarbons that are built into the thylakoid membrane with the two types of chlorophyll.

C. The Photooxidation of Chlorophyll (p. 207–208)

What happens when chlorophyll or accessory pigments absorb photons?

☞ Colors of absorbed wavelengths disappear from the spectrum of transmitted and reflected light.

☞ The absorbed photon boosts one of the pigment molecule's electrons in its lowest-energy state (*ground state*) to an orbital of higher potential energy (*excited state*).

The only photons absorbed by a molecule are those whose energy state is equal to the difference in energy between the ground state and excited state.

☞ This energy difference varies from one molecule to another. Pigments have unique absorption spectra because pigments only absorb photons corresponding to specific wavelengths.

☞ The photon energy absorbed is converted to potential energy of an electron elevated to the excited state.

The excited state is unstable, so excited electrons quickly fall back to the ground state orbital, releasing excess energy in the process. This released energy may be:
- Dissipated as heat.
- Reradiated as a photon of lower energy and longer wavelength than the original light that excited the pigment. This afterglow is called *fluorescence*.

Fluorescence

Pigment molecules do not fluoresce when in the thylakoid membranes, because nearby *primary electron acceptor* molecules trap excited state electrons that have absorbed photons.

Primary electron acceptor

- In this redox reaction, chlorophyll is photooxidized by the absorption of light energy and the electron acceptor is reduced.
- Because no primary electron acceptor is present, *isolated* chlorophyll fluoresces in the red part of the spectrum and dissipates heat.

The transfer of excited state electrons from chlorophyll to primary electron acceptor molecules is the first step of the light reactions. The energy stored in the trapped electrons powers the synthesis of ATP and NADPH in subsequent steps.

D. The Two Photosystems (p. 208–209)

What are the components of a photosystem? How do they each contribute to the transformation of light energy to chemical energy?

Chlorophyll *a*, chlorophyll *b* and the carotenoids are clustered in assemblies of a few hundred pigment molecules within the thylakoid membrane.
- Only one of the many chlorophyll *a* molecules in each assembly can transfer an excited electron to initiate the light reactions. This specialized chlorophyll *a* molecule is located in the *reaction center*.

Reaction center

- The other chlorophyll *a* molecules, chlorophyll *b* molecules and carotenoids function as light-gathering antennae that absorb photons and pass the energy from molecule to molecule to the reaction center. (See Campbell, Figure 10.11)
- The entire complex of light-gathering molecules and the reaction center is called a *photosystem*.

Photosystem

Two types of photosystems are located in the thylakoid membranes, photosystem I and photosystem II.
- The reaction center of photosystem I has a specialized chlorophyll *a* molecule known as *P700*, which absorbs best at 700 nm (the far red portion of the spectrum).

P700

- The reaction center of photosystem II has a specialized chlorophyll *a* molecule known as *P680*, which absorbs best at a wavelength of 680 nm.

P680

- P700 and P680 are identical chlorophyll *a* molecules, but each is associated with a different protein. This affects their electron distribution and results in slightly different absorption spectra.
- P700 and P680 molecules differ from other chlorophyll *a* molecules in their particular location in the thylakoid membrane, where they are bound to specific proteins and are in closer proximity to their primary electron acceptors.

E. Cyclic Electron Flow (p. 209–210)

What path do electrons follow as they flow through photosystems II and I?

There are two possible routes for electron flow during the light reactions: *cyclic flow* and *noncyclic flow*.

Cyclic electron flow is the simplest pathway, but involves only photosystem I and generates ATP without producing NADPH or evolving oxygen.

☞ It is cyclic because excited electrons that leave from chlorophyll *a* at the reaction center return to the reaction center.

☞ At some points during this pathway, electrons travel in pairs. The cyclic pathway may be visualized as starting with Photosystem I absorbing two photons of light followed by a series of redox reactions which return the electrons along an *electron transport chain* in the thylakoid membrane to chlorophyll.

☞ With each redox reaction along the electron transport chain, electrons lose potential energy until they return to their ground-state orbital in the P700 reaction center.

☞ Absorption of another two photons of light by the pigments send a second pair of electrons through the cyclic pathway.

As excited electrons give up energy along the transport chain to P700, the thylakoid membrane couples the exergonic flow of electrons to the endergonic reactions that phosphorylate ADP to ATP.

☞ This coupling mechanism is *chemiosmosis*.

☞ Some electron carriers can only transport electrons in the company of protons.

☞ The protons are picked up on one side of the thylakoid membrane and deposited on the opposite side as the electrons move to the next member of the transport chain.

☞ The electron flow thus stores energy in the form of a proton gradient across the thylakoid membrane — a *proton-motive force*.

☞ An ATP synthase enzyme in the thylakoid membrane uses the proton-motive force to make ATP. This process is called *photophosphorylation* because the energy required is light.

☞ This form of ATP production is called *cyclic photophosphorylation* as it occurs during the cyclic pathway.

Note that ATP is produced without the production of NADPH or the release of oxygen.

F. **Noncyclic Electron Flow** (p. 210–211)

How does the end result of cyclic electron flow differ from that of noncyclic electron flow? What is the relationship between these two electron pathways in the light reactions?

Both photosystem I and photosystem II function and cooperate in noncyclic electron flow which transforms light energy to chemical energy stored in the bonds of NADPH and ATP. This process:
- ☞ Occurs in the thylakoid membrane.
- ☞ Passes electrons continuously from water to NADP$^+$.
- ☞ Produces ATP by *noncyclic photophosphorylation*.
- ☞ Produces NADPH.
- ☞ Produces O$_2$.

Light excites electrons from P700, the reaction center in photosystem I.
- ☞ These electrons do not return to the reaction center chlorophyll, but are stored as high-energy electrons in NADPH which will later function as the electron donor in the Calvin cycle.
- ☞ The oxidized chlorophyll becomes an oxidizing agent and its electron "holes" must be filled.
- ☞ Photosystem II supplies the electrons to P700 to fill these holes.

When the antenna assembly of photosystem II absorbs light, the energy is transferred to the P680 reaction center.

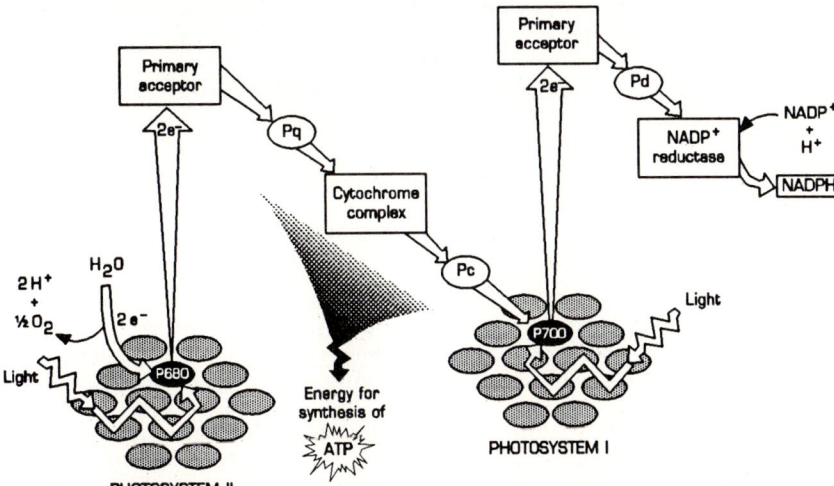

- ☞ Electrons ejected from the P680 are trapped by the photosystem II primary electron acceptor.
- ☞ The electrons are then transferred from this primary electron acceptor to the same transport chain used in cyclic electron flow.
- ☞ As these electrons pass down the electron transport chain, they lose potential energy until they reach the ground state of P700.
- ☞ These electrons then fill the electron vacancies left in photosystem I when NADP$^+$ was reduced.

Noncyclic electron flow

As electrons move from photosystem II to photosystem I, the transport chain pumps H⁺ across the thylakoid membrane.
- ☞ The resulting proton-motive force drives ATP synthesis.
- ☞ When noncyclic electron flow produces ATP, the process is called *noncyclic photophosphorylation*. The mechanism, however, is the same as in cyclic photophosphorylation.

Noncyclic photophosphorylation

Electrons from P680 flow to P700 during noncyclic electron flow, and this process restores the missing electrons in P700. This, however, leaves the P680 reaction center of photosystem II with missing electrons.
- ☞ A water-splitting enzyme extracts electrons from water and passes them to oxidized P680, which has a great affinity for electrons.
- ☞ As water is oxidized, the removal of electrons splits water into two hydrogen ions and an oxygen atom.
- ☞ The oxygen atom immediately combines with a second oxygen atom to form O_2. It is this water-splitting step of photosynthesis that releases O_2.

Summarize the light reactions. Where do they occur? What substances are required? What are the products of these reactions?

Summary of the Light Reactions:

During *noncyclic electron flow*, the photosystems of the thylakoid membrane transform light energy to the chemical energy stored in NADPH and ATP. This process:
- ☞ Pushes low energy-state electrons from water to NADPH, where they are stored at a higher state of potential energy. NADPH, in turn, is the electron donor used to reduce carbon dioxide to sugar (Calvin Cycle).
- ☞ Produces ATP from this light driven electron current.
- ☞ Produces oxygen as a byproduct.

During *cyclic electron flow*, electrons ejected from P700 reach ferredoxin and flow <u>back</u> to P700. This process:
- ☞ Produces ATP.
- ☞ Unlike noncyclic electron flow, does <u>not</u> produce O_2 or NADPH.

G. **Review: A Comparison of Chemiosmosis in Chloroplasts and Mitochondria** (p. 211–212)

Chemiosmosis

Chemiosmosis = The coupling of exergonic electron flow down an electron transport chain to endergonic ATP production by the creation of an electrochemical proton gradient across a membrane. The proton gradient drives ATP synthesis as protons diffuse back across the membrane.

Chemiosmosis in chloroplasts and chemiosmosis in mitochondria are similar in several ways:
- ☞ An electron transport chain assembled in a membrane translocates protons across the membrane as electrons pass through a series of carriers that are progressively more electronegative.
- ☞ An ATP synthase complex built into the same membrane, couples the diffusion of hydrogen ions down their gradient to the phosphorylation of ADP.
- ☞ The ATP synthase complexes and some electron carriers (including quinones and cytochromes) are very similar in both chloroplasts and mitochondria.

What are some important differences between chemiosmosis in oxidative phosphorylation and chemiosmosis in photophosphorylation?

Oxidative phosphorylation in mitochondria and photophosphorylation in chloroplasts differ in the following ways:

1. Electron Transport Chain
    - Mitochondria transfer chemical energy from food molecules to ATP. The high-energy electrons that pass down the transport chain are extracted by the oxidation of food molecules.
    - Chloroplasts transform light energy into chemical energy. Photosystems capture light energy and use it to drive electrons to the top of the transport chain.

2. Spatial Organization
    - The inner mitochondrial membrane pumps protons from the matrix out to the intermembrane space, which is a reservoir of protons that power ATP synthase.
    - The chloroplast's thylakoid membrane pumps protons from the stroma into the thylakoid compartment, which functions as a proton reservoir. ATP is produced as protons diffuse from the thylakoid compartment back to the stroma through ATP synthase complexes that have catalytic heads on the membrane's stroma side. Thus, ATP forms in the stroma where it drives sugar synthesis during the Calvin cycle.

There is a large proton or pH gradient across the thylakoid membrane.
- When chloroplasts are illuminated, there is a thousand-fold difference in H^+ concentration. The pH in the thylakoid compartment is reduced to about 5 while the pH in the stroma increases to about 8.
- When chloroplasts are in the dark, the pH gradient disappears, but can be reestablished if chloroplasts are illuminated.
- Andre Jagendorf (1960's) produced compelling evidence for chemiosmosis when he induced chloroplasts to produce ATP in the dark by using artificial means to create a pH gradient. His experiments demonstrated that during photophosphorylation, the function of the photosystems and the electron transport chain is to create a proton-motive force that drives ATP synthesis.

A tentative model for the organization of the thylakoid membrane includes the following: (See Campbell, Figure 10.16)
- Proton pumping by the thylakoid membrane depends on an asymmetric placement of electron carriers that accept and release protons (H^+).
- There are three steps in the light reactions that contribute to the proton gradient across the thylakoid membrane:
    1. Water is split by Photosystem II on the thylakoid side, releasing protons in the process.
    2. As plastoquinone (Pq), a mobile carrier, transfers electrons to the cytochrome complex, it translocates protons from the stroma to the thylakoid space.
    3. Protons in the stroma are removed from solution as $NADP^+$ is reduced to NADPH.

134 Photosynthesis

☞ NADPH and ATP are produced on the side of the membrane facing the stroma where sugar is synthesized by the Calvin cycle.

IV. HOW THE CALVIN CYCLE MAKES SUGAR: A CLOSER LOOK
(p. 212–214)

What happens during the carbon-fixing reactions of the Calvin-Benson cycle? What substances are required? What substances are produced? What changes occur in the carbon skeleton of the intermediates?

ATP and NADPH produced by the light reactions are used in the Calvin cycle to reduce carbon dioxide to sugar.
  ☞ The Calvin cycle is similar to the Krebs cycle in that the starting material is regenerated by the end of the cycle.
  ☞ Carbon enters the Calvin cycle as CO_2 and leaves as sugar.
  ☞ ATP is the energy source, while NADPH is the reducing agent that adds high-energy electrons to form sugar.
  ☞ The Calvin cycle actually produces a three-carbon sugar *glyceraldehyde 3-phosphate* (glyceraldehyde phosphate).

For the Calvin cycle to synthesize one molecule of sugar, the cycle must make three turns fixing three molecules of CO_2. (See Campbell, Figure 10.17)

Step 1: The Calvin cycle begins when each molecule of CO_2 is attached to a five-carbon sugar, *ribulose biphosphate (RuBP)*.
  ☞ This reaction is catalyzed by the enzyme *RuBP carboxylase* (rubisco).

Glyceraldehyde phosphate

Ribulose biphosphate
Rubisco

- ☞ The product of this reaction is an unstable six-carbon intermediate that immediately splits into two molecules of 3-phosphoglycerate.
- ☞ For every three molecules of carbon dioxide that enter the Calvin cycle via rubisco, three molecules of RuBP are carboxylated and a total of six molecules of 3-phosphoglycerate are formed.

Step 2: Each molecule of 3-phosphoglycerate is phosphorylated by an enzyme that transfers a phosphate group from ATP.
- ☞ This reaction produces 1, 3-diphosphoglycerate.
- ☞ For every 3 molecules of carbon dioxide entering the cycle, six molecules of ATP must be used to produce six molecules of 1, 3-diphosphoglycerate.

Step 3: The unstable phosphate bonds of 1, 3-diphosphoglycerate prime this compound for the addition of high-energy electrons donated from NADPH.
- ☞ This redox reaction reduces 1, 3-diphosphoglycerate to glyceraldehyde phosphate. The reaction is driven by hydrolysis of the phosphate bond transferred to the substrate from ATP in the previous reaction.
- ☞ Electrons from NADPH reduce the carboxyl group of 3-phosphoglycerate to the carbonyl group of glyceraldehyde phosphate, which stores more potential energy.
- ☞ Glyceraldehyde phosphate is a three-carbon sugar, the same sugar produced from the splitting of glucose during glycolysis.

To this point, the Calvin cycle, which started with three molecules of carbon dioxide, has produced six molecules of glyceraldehyde phosphate.
- ☞ Only one molecule of this three-carbon sugar can be counted as a net gain of carbohydrate. The six molecules of glyceraldehyde phosphate contain 18 carbons, a net gain of three carbons which come from CO_2. Fifteen carbons come from the three molecules of five-carbon RuBP at the beginning of the cycle.
- ☞ One molecule of glyceraldehyde phosphate exits the cycle to be used by the plant. The other five molecules are recycled to regenerate three molecules of RuBP.

Step 4: A complex series of reactions rearrange the carbon skeletons of five molecules of glyceraldehyde phosphate into three molecules of RuBP.
- ☞ These reactions require three molecules of ATP.
- ☞ RuBP is thus regenerated and can receive CO_2 to begin the cycle again.

The Calvin cycle uses nine molecules of ATP and six molecules of NADPH to produce a net synthesis of one glyceraldehyde phosphate molecule.
- ☞ The ATP and NADPH are produced by the light reactions.
- ☞ The glyceraldehyde phosphate produced by the Calvin cycle becomes the starting material for a variety of metabolic pathways.
- ☞ Although the three-carbon glyceraldehyde phosphate is the carbohydrate produced by photosynthesis, two molecules of this compound can be rapidly converted to a molecule of glucose, a six-carbon sugar.
- ☞ To produce one molecule of glucose, the Calvin cycle uses 18 ATPs and 12 molecules of NADPH.
- ☞ The ADP and $NADP^+$ resulting from the Calvin cycle are respectively phosphorylated and reduced by the light reactions to regenerate ATP and NADPH.

V. PHOTORESPIRATION: AN EVOLUTIONARY RELIC? (p. 214–215)

A metabolic pathway called *photorespiration* reduces the yield of photosynthesis.

What are the major consequences of photorespiration?

<u>Photorespiration</u> = In plants, a metabolic pathway that consumes oxygen, evolves carbon dioxide, produces no ATP and decreases photosynthetic output.
- ☞ Occurs because the active site of rubisco can accept O_2 as well as CO_2.
- ☞ Produces no ATP molecules.
- ☞ Decreases photosynthetic output by reducing organic molecules used in the Calvin cycle.

What happens to rubisco when the oxygen concentration is much higher than carbon dioxide?

When the O_2 concentration in the leaf's air spaces is higher than CO_2 concentration, rubisco accepts O_2 and transfers it to RuBP. (The "photo" in photorespiration refers to the fact that this pathway usually occurs in light when photosynthesis reduces CO_2 and raises O_2 in the leaf spaces.)

Rubisco transfers O_2 to RuBP
↓
—— Resulting 5-C molecule splits into ——
↓   ↓
Two-C molecule   Three-C molecule
(glycolate)   (3-phosphoglycerate)
↓   ↓
Leaves chloroplast   Stays in the
&   Calvin cycle
goes to peroxisome
↓
A metabolic pathway begins in
the peroxisome and is completed
in the mitochondrion.
↓ Glycolate is broken down into
CO_2

(The "respiration" in photorespiration refers to the fact that this process uses O_2 and releases CO_2.)

Some scientists believe that photorespiration is a metabolic relic from earlier times when the atmosphere contained less oxygen and more carbon dioxide than is present today.
- ☞ Under these conditions, when rubisco evolved, the inability of the enzyme's active site to distinguish carbon dioxide from oxygen would have made little difference.
- ☞ This affinity for oxygen has been retained by rubisco and some photorespiration is bound to occur.

Photorespiration

Whether photorespiration is beneficial to plants is not known.
- ☞ It is known that some crop plants (e.g. soybeans) lose as much as 50% of the carbon fixed by the Calvin cycle to photorespiration.
- ☞ If photorespiration could be reduced in some agricultural plants, crop yields and food supplies would increase.

Photorespiration is fostered by hot, dry, bright days.
- ☞ Under these conditions, plants close their stomata to prevent dehydration by reducing water loss from the leaf.
- ☞ Photosynthesis then depletes available carbon dioxide and increases oxygen within the leaf air spaces. This condition favors photorespiration.

What are two important photosynthetic adaptations that minimize photorespiration? How do they work?

Certain species of plants, which live in hot arid climates, have evolved alternate modes of carbon fixation that minimize photorespiration. C_4 and CAM are the two most important of these photosynthetic adaptations.

VI. C_4 PLANTS (p. 215)

The Calvin cycle occurs in most plants and produces 3-phosphoglycerate, a three-carbon compound, as the first stable intermediate.
- ☞ These plants are called C_3 *plants*, because the first stable intermediate has three carbons.

Many plant species preface the Calvin cycle with reactions that incorporate carbon dioxide into four-carbon compounds.
- ☞ These plants are called C_4 *plants*.
- ☞ The C_4 pathway is used by several thousand species in at least 17 families including corn and sugarcane which are important agricultural plants in the grass family.
- ☞ This pathway is an evolutionary adaptation that enhances a plants's ability to fix CO_2 under those conditions that favor photorespiration.

The leaf anatomy of C_4 plants spatially segregates the Calvin cycle from the initial incorporation of CO_2 into organic compounds. There are two distinct types of photosynthetic cells:

1. Bundle-sheath cells.
    - ☞ Are arranged into tightly packed sheaths around the veins of the leaf.
    - ☞ Thylakoids in the chloroplasts of bundle-sheath cells are not stacked into grana.
    - ☞ The Calvin cycle is confined to the chloroplasts of the bundle sheath.

2. Mesophyll cells.
    - ☞ Mesophyll cells are more loosely arranged in the area between the bundle sheath and the leaf surface.

C_4 pathway

Bundle-sheath cells

Mesophyll cells

138 *Photosynthesis*

The Calvin cycle of C_4 plants is preceded by incorporation of CO_2 into organic compounds in the mesophyll. (See Campbell, Figure 10.18)

Step 1: CO_2 is added to phosphoenolpyruvate (PEP) to form oxaloacetate, a four-carbon product.
- *PEP carboxylase* is the enzyme that adds CO_2 to PEP. Compared to rubisco, it has a much greater affinity for CO_2 and has <u>no</u> affinity for O_2.
- Thus, PEP carboxylase can fix CO_2 efficiently when rubisco cannot — under hot, dry conditions that cause stomata to close, CO_2 concentrations to drop and O_2 concentrations to rise.

Step 2: After CO_2 has been fixed by mesophyll cells, they convert oxaloacetate to another four-carbon compound (usually malate).

Step 3: Mesophyll cells then export malate to bundle-sheath cells through plasmodesmata.
- In the bundle-sheath cells, malate releases CO_2, which is then assimilated into organic matter by rubisco and the Calvin cycle.
- The mesophyll cells thus pump CO_2 into the bundle-sheath cells, preventing photorespiration and enhancing sugar production by maintaining a CO_2 concentration sufficient for rubisco to accept CO_2 rather than oxygen.
- This adaptation is advantageous in hot regions with intense sunlight where C_4 are most likely to be found.

Spatial ordering of C_4 photosynthesis is an adaptation that illustrates the correlation between structure and function.

VII. CAM PLANTS (*p. 215–216*)

A second photosynthetic adaptation exists in succulent plants adapted to very arid conditions. These plants open their stomata primarily at night and close them during the day (opposite of most plants).
- This obviously conserves water during the day, but prevents CO_2 from entering the leaves.
- When the stomata are open at night, CO_2 is taken up and incorporated into a variety of organic acids.
- This mode of carbon fixation is called *crassulacean acid metabolism (CAM)*.
- The organic acids made at night are stored in vacuoles of mesophyll cells until morning, when the stomata close.
- During daytime, the light reactions can supply ATP and NADPH for the Calvin cycle. At this time, CO_2 is released from the organic acids made the previous night and is incorporated into sugar in the chloroplasts.

CAM and C_4 pathways are similar in that CO_2 is first incorporated into organic intermediates before it enters the Calvin cycle.

The pathways differ in that the initial steps of carbon fixation are structurally separate from the Calvin cycle, in C_4 plants, whereas the two steps occur at separate times in CAM plants.

Regardless of whether the plant uses a C_3, C_4 or CAM pathway, all plants use the Calvin cycle to produce sugar from CO_2.

Photosynthesis 139

VIII. SUMMARY: THE FATE OF PHOTOSYNTHETIC PRODUCTS
(p. 216–218)

Light reactions capture light energy and use it to produce ATP and transfer electrons from water to NADP$^+$ to form NADPH.

The Calvin cycle uses the ATP and NADPH to produce sugar from carbon dioxide. (See Campbell, Figure 10.20)

What happens to the products of photosynthesis?

These processes convert light energy that entered the chloroplasts as sunlight to chemical energy in the bonds of sugar molecules.
- ☞ Sugars made in the chloroplasts supply the entire plant with chemical energy and carbon skeletons to synthesize all the major organic molecules produced by the cells.
- ☞ About 50% of the organic material produced by photosynthesis is consumed as fuel for cellular respiration.
- ☞ In some cases, photorespiration causes a loss of photosynthetic products.

Green cells are the only autotrophic parts of a plant with the other parts depending on organic molecules exported from the leaves in vascular bundles.
- ☞ Carbohydrate is transported out of the leaves as the disaccharide sucrose in most plants.
- ☞ The sucrose provides raw material for cellular respiration and many anabolic pathways after arriving at nonphotosynthetic cells.

A large amount of sugar in the form of glucose is used to produce cellulose, the most abundant organic molecule in a plant.

During a 24 hour day, most plants make more organic material than needed for respiratory fuel and precursors for biosynthesis.
- ☞ Extra sugars are synthesized into starch and stored in storage cells of roots, tubers and fruits (some is also stored in chloroplasts).
- ☞ Heterotrophs also consume parts of plants as food materials.

On a global scale, it is estimated that photosynthesis makes about 160 billion metric tons of carbohydrate per year. No other chemical process on Earth is more productive than photosynthesis or as important to the welfare of life.

Sucrose

11 THE REPRODUCTION OF CELLS

CHAPTER OUTLINE

"Where a cell exists, there must have been a preexisting cell, just as the animal arises only from an animal and the plant only from a plant." "*Omnis cellula e cellula*" or "All cells from cells." — Rudolf Virchow (1855).

The ability to reproduce distinguishes living organisms from nonliving objects; this ability has a cellular basis. The perpetuation of life is based on the reproduction of cells or *cell division*.

- In unicellular organisms, the division of one cell to form two reproduces an entire organism (e.g. *Amoeba*).
- In multicellular organisms, cell division allows:
    1. Growth and development from the fertilized egg.
    2. Replacement of damaged or dead cells.

What is the genome? What major events must occur during cell division for the entire genome to be passed on to daughter cells?

Cell division is not just a simple pinching in half, but is a complex process that faithfully passes along the *genome* from one generation of cells to the next.

Genome = All of the genes carried by a set of chromosomes corresponding to the haploid set of a species.

Cell division involves:
- Precise replication of DNA.
- The allocation of this DNA to opposite ends of the cells.
- The separation into two identical daughter cells.

I. BACTERIAL REPRODUCTION (*p. 222–223*)

How does binary fission occur in prokaryotes?

Bacteria and cyanobacteria (prokaryotes) reproduce by the process of *binary fission*:
- A circular DNA molecule with associated proteins forms the single chromosome that carries most bacterial genes.
- This chromosome is about 500 times longer than the cell and is highly folded within.
- The chromosome is copied, and each copy is attached to the plasma membrane.
- Growth of the membrane between the attachment sites separates the two copies of the chromosome.

The Reproduction of Cells 141

- ☞ The bacterium grows to about twice its initial size and the plasma membrane pinches inward.
- ☞ A cell wall forms across the bacterium between the two chromosomes, dividing the original cell into two daughter cells.

II. AN INTRODUCTION TO EUKARYOTIC CHROMOSOMES (p. 223–224)

In a dividing cell:
- ☞ Most cellular components are divided about equally between daughter cells.
- ☞ Genetic material, however, is precisely equally divided between daughter cells, so that each inherits all the genes present in the original cell.

Replicating and distributing extensive sets of genes is simplified by grouping them into multiple *chromosomes*.

<u>Chromosomes</u> = Threadlike structures in eukaryotic nuclei that are composed of DNA and protein.
- ☞ They are duplicated in advance of cell division.
- ☞ Duplicated chromosomes are then distributed by the process of *mitosis*.

<u>Mitosis</u> = (Mitos=thread) Nuclear division during which duplicated chromosomes are evenly distributed into two daughter nuclei; results in two daughter cells that are the genetic equivalent of the parent cell.

All *somatic cells* (all body cells except sperm and ova) in most organisms have the same number of chromosomes in their nuclei.
- ☞ Sperm and ova have exactly one-half the number of chromosomes of somatic cells.
- ☞ The chromosome number in somatic cells is characteristic of a given species (e.g. humans have 46, goldfish have 94).

What is the composition of chromosomes? How does chromosomal structure change in preparation for cell division?

Chromatin, a DNA-protein complex, is organized into a long, thin fiber that is folded and coiled to form the chromosome. Each chromosome thus contains:
1. A long DNA molecule, with thousands of genes.
2. Various proteins that maintain chromosomal structure and help control gene activity.

Before dividing, a cell copies its genome by duplicating every chromosome, each of which forms two identical *sister chromatids*. (See Campbell, Figure 11.4)
- ☞ For much of mitosis, sister chromatids remain joined at a specialized region called the *centromere*.
- ☞ As mitosis progresses, sister chromatids are pulled apart forming two complete chromosome sets, one at each end of the cell.
- ☞ Mitosis may be followed by *cytokinesis*.

<u>Cytokinesis</u> = Cytoplasmic division that forms two separate daughter cells, each containing a single nucleus.

142   *The Reproduction of Cells*

How does chromosomal number change throughout the human life cycle?

HUMAN LIFE CYCLE:

- Individual inherits 46 chromosomes, 23 from each parent.
- *Meiosis* in gonads reduces the chromosome number by one half.
- Sperm cell (23 chromosomes)
- Ovum (23 chromosomes)
- *Fertilization* restores the chromosome number to 46
- Zygote (46 chromosomes)
- *Mitosis* produces genetically identical daughter cells. Process is responsible for growth, development and repair.

III. THE CELL CYCLE: AN OVERVIEW (p. 224)

What are the phases of the cell cycle, and what generally occurs during each phase?

<u>Cell cycle</u> = Well-ordered sequence of events between the time a cell divides to form two daughter cells and the time those daughter cells divide.
- ☞ Includes a doubling of a cell's cytoplasm and reproduction of cellular organelles, precise duplication of DNA, mitosis and cytokinesis.
- ☞ Cell cycle duration varies with the type of cell. Some cells divide each hour, others take more than 24 hours.
- ☞ Nerve and muscle cells never or rarely divide once they are formed.

The *cell cycle* alternates between *M phase*, or dividing phase, and *Interphase*, the nondividing phase:
- ☞ *M phase*, the shortest part of the cell cycle and the phase during which the cell divides, includes:
    1. *Mitosis* — division of the nucleus.
    2. *Cytokinesis* — division of the cytoplasm.

What major events occur during the G_1, S and G_2 periods of interphase?

- ☞ *Interphase*, nondividing phase which is about 90% of the cell cycle and includes most of a cell's growth and metabolic activities. It is a period of intense biochemical activity during which the cell grows and copies its chromosomes in preparation for cell division. Consists of three periods:
    1. G_1 *phase* — first growth phase.
    2. *S phase* — synthesis phase when DNA is synthesized as chromosomes are duplicated.
    3. G_2 *phase* — second growth phase.

IV. THE MECHANICS OF CELL DIVISION: A CLOSER LOOK
(p. 225–230)

Mitosis is unique to eukaryotes and may be an evolutionary adaptation for distributing a large amount of genetic material.
- ☞ Details may vary, but overall process is similar in most eukaryotes.
- ☞ It is a reliable process with experimental evidence showing only one error per 100,000 cell divisions.

A. The Stages of Mitotic Cell Division (p. 226–227)

Mitosis and cytokinesis form a continuum, but for ease of description, mitosis is usually divided into five stages: prophase, prometaphase, metaphase, anaphase and telophase. (See Campbell, Figure 11.6)

What characterizes a G_2 interphase cell?

G_2 of Interphase

In G_2 of interphase:
- ☞ The nucleus is well defined and bounded by the nuclear envelope.
- ☞ One or more nucleoli are present.
- ☞ Two pairs of centrioles are adjacent to nucleus (formed earlier by replication of original pair).
- ☞ Around each pair of centrioles, microtubules form in a radial array, called an *aster*.
- ☞ Chromosomes have been duplicated (occurred in S phase), but cannot be distinguished individually due to loosely packed chromatin fiber arrangement.

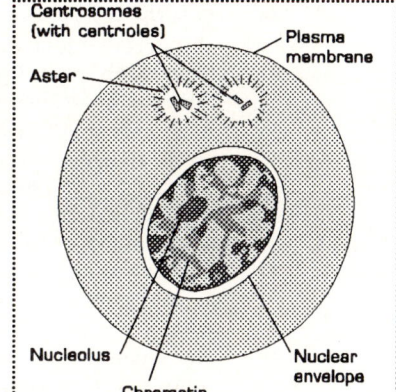

Aster

What are the phases of mitosis proper and what events are characteristic of each phase?

Prophase

During prophase, changes occur in both the nucleus and cytoplasm.

In the nucleus:
- ☞ Nucleoli disappear.
- ☞ Chromatin fibers become tightly coiled and folded into discrete, observable chromosomes.
- ☞ Each duplicated chromosome is composed of two identical sister chromatids joined at the centromere.

In the cytoplasm:
- ☞ Mitotic spindle forms. It is composed of microtubules and associated proteins which are arranged between the two *centrosomes* or microtubule-organizing center.
- ☞ Centrosomes move apart, apparently propelled along the surface of the nucleus by lengthening of the microtubule bundles between them.

Prometaphase

During prometaphase:
- ☞ Nuclear envelope fragments.
- ☞ Absence of nuclear envelope allows microtubules to interact with the more highly condensed chromosomes.
- ☞ *Spindle fibers* (bundles of microtubules) extend from each pole toward the equator of the cell.
- ☞ Each chromatid now has a specialized structure called the *kinetochore* located at the centromere region.
- ☞ Bundles of microtubules (*kinetochore microtubules*) are attached to the kinetochores and put the chromosomes into agitated motion.
- ☞ Many more microtubules (*nonkinetochore microtubules*) radiate from each centrosome toward the metaphase plate without attaching to chromosomes.

Metaphase

During metaphase:
- ☞ Centrosomes are positioned at opposite ends (poles) of the cell.
- ☞ Chromosomes move to the *metaphase plate*, the plane equidistant between the spindle poles.
- ☞ Centromeres of all chromosomes are aligned on the metaphase plate.
- ☞ The long axis of each chromosome is roughly at a right angle to the spindle axis.
- ☞ Kinetochore fibers of sister chromatids face opposite poles, so identical chromatids are attached to kinetochore fibers radiating from opposite ends of the parent cell.
- ☞ Entire structure formed by nonkinetochore microtubules plus kinetochore microtubules is called the *spindle*.

Anaphase

Anaphase begins when paired centromeres of each chromosome move apart.
- ☞ Sister chromatids separate and are considered chromosomes.
- ☞ The spindle apparatus starts moving the separate chromosomes (once joined as sister chromatids) toward opposite poles of the cell. Because kinetochore fibers are attached to the centromeres, the chromosomes move centromeres first in a "V" shape.
- ☞ As chromosomes approach the poles, the kinetochore microtubules shorten at the end attached to the kinetochores.
- ☞ The poles of the cell move farther apart slightly elongating the cell.

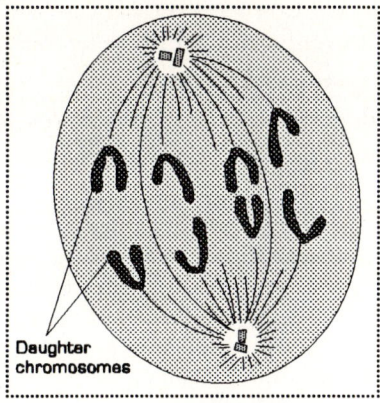

At the end of anaphase, the two poles have complete and equivalent collections of chromosomes.

Telophase

During telophase:
- ☞ Nonkinetochore microtubules further elongate the cell.
- ☞ Daughter nuclei begin to form at the two poles.
- ☞ Nuclear envelopes are formed around the chromosomes from fragments of the parent cell's nuclear envelope and other portions of the endomembrane system.
- ☞ Nucleoli reappear.
- ☞ Chromatin fiber of each chromosome uncoils and the chromosomes become less distinct.

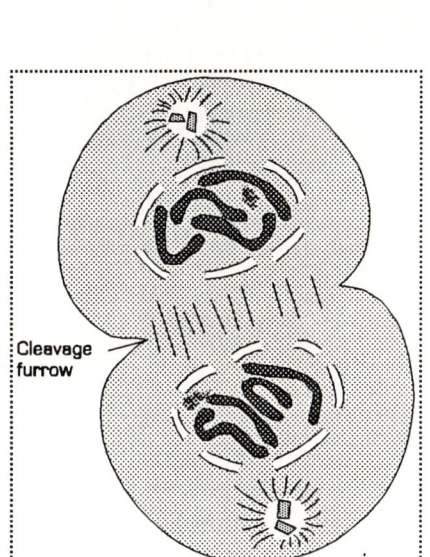

By the end of telophase:
- ☞ Mitosis, the equal division of one nucleus into two genetically identical nuclei, is complete.
- ☞ Cytokinesis has begun and the appearance of two separate daughter cells occurs shortly after mitosis is completed.

The Reproduction of Cells 145

Anaphase

Telophase

146 The Reproduction of Cells

Mitotic spindle

Centrosome

B. **The Structure and Function of the Mitotic Spindle** (*p. 225, 228*)

Describe the structure of the spindle apparatus including centrosomes, nonkinetochore microtubules, kinetochore microtubules, asters and centrioles (in animal cells). How is the spindle apparatus' function dependent upon its structure?

The mitotic spindle is important to the events occurring in mitosis. It forms in the cytoplasm during prophase and is composed of fibers formed from *microtubules* and associated proteins.
- Microtubules of the cytoskeleton are partially disassembled during formation of the mitotic spindle.
- Spindle microtubules are aggregates of two proteins, α- and β-tubulin.
- Elongation of spindle microtubules is accomplished by the addition of more individual protein subunits at one end.
- Parallel microtubules form bundles called *spindle fibers* that are visible under a light microscope.

Spindle microtubules begin assembly in the *centrosome* or microtubule organizing center.
- Microtubules are polar, with distinct + and − ends.
- Microtubule length changes by the addition or removal of tubulin protein units only at the end away from the centrosome, the + end. (The − end is the nongrowing end.)

In animal cells, a pair of centrioles is in the center of the centrosome, but there is evidence that they are not essential for cell division:
- If the centrioles of animal cells are destroyed with a laser microbeam, spindles still form and function during mitosis.
- Plant centrosomes lack centrioles.

What characteristic changes occur in the spindle apparatus during each phase of mitosis? How do current models explain the poleward movement of chromosomes and elongation of the cell's polar axis? What role does dynein play in both events?

Interphase: The centrosome replicates to form two centrosomes located just outside the nucleus.

Prophase and Prometaphase: The two centrosomes move farther apart; spindle microtubules radiate from them, elongating at their + end.

Late Prometaphase: The two centrosomes are at opposite poles of the cell; some of their spindle microtubules attach to chromosomes at the *kinetochores*, specialized regions of the centromeres.
- Each of the two chromatids of a chromosome has its own kinetochore.
- There is a check-and-balance system that equalizes the number of microtubules attached to the two kinetochores of a chromosome and moves the chromosome to the midline of the cell.

☞ Microtubules remain attached to a kinetochore only when there is a force exerted on the chromosome from the cell's opposite end. Kinetochore attachment is thus stabilized only when microtubules from the opposite pole hook to the other kinetochore.

Metaphase: All the duplicated chromosomes are aligned on the cell's midline, or metaphase plate.
☞ About 15 to 35 *kinetochore microtubules*, are attached to each kinetochore.
☞ Many more *nonkinetochore microtubules*, radiate from each centrosome toward the metaphase plate without attaching to chromosomes. Some are too short to reach the metaphase plate, but others extend across the plate and overlap with nonkinetochore microtubules from the opposite pole of the cell.

Anaphase: Begins when the chromosome's centromeres divide and the sister chromatids move as separate chromosomes toward opposite ends of the cell.

How do the kinetochore microtubules function in this poleward movement of chromosomes?
☞ Experimental evidence is that kinetochore microtubules shorten during anaphase by depolymerizing at their kinetochore or + ends. This shortening pulls the chromosomes poleward.
☞ The exact mechanisms of this interaction between kinetochores and microtubules is unknown, but researchers have recently found the motor protein dynein in kinetochores.
☞ Just as dynein drives microtubule movement in flagella, kinetochore dynein may "walk" a chromosome along the shortening microtubules in a similar fashion.

Dynein

What is the function of the nonkinetochore microtubules?
☞ Nonkinetochore tubules elongate the whole cell along the polar axis during anaphase.
☞ These tubules overlap at the middle of the cell and slide past each other away from the cell's equator, reducing the degree of overlap.
☞ It is hypothesized that dynein cross-bridges may form between overlapping tubules to slide them past one another. Alternatively, motor molecules may link the microtubules to other cytoskeletal elements to drive the sliding.
☞ ATP provides the energy for this endergonic process.

Telophase: At the end of anaphase, the duplicate sets of chromosomes are clustered at opposite ends of the elongated parent cell.
☞ Nuclei reform during telophase.
☞ Cytokinesis usually divides the cell's cytoplasm during late telophase.

148 The Reproduction of Cells

C. The Mechanisms of Cytokinesis (p. 228–230)

What is cytokinesis? When does it occur? How does the process differ in animal and plant cells?

Cytokinesis, the process of cytoplasmic division, begins in telophase and occurs differently in animal and plant cells. In animal cells, cytokinesis occurs as *cleavage*:

- First a *cleavage furrow* forms as a shallow groove in the cell surface near the old metaphase plate. (See Campbell, Figure 11.10)
- A *contractile ring* of actin microfilaments forms on the cytoplasmic side of the furrow.
- These microfilaments contract, reducing the diameter of the contractile ring, until the parent cell is pinched in two.
- The remains of the mitotic spindle, which is the last connection between the daughter cells, breaks and the two cells are completely separate.

Cytokinesis in plant cells differs from animal cells, because plants have cell walls:

- No cleavage furrow forms; instead, a *cell plate* forms across the midline of the parent cell (old metaphase plate).
- This cell plate forms from fusing vesicles derived from the Golgi apparatus that are moved along microtubules to the cell's center.
- Vesicle fusion forms two membranes which grow laterally until they unite with the existing plasma membrane. This produces two daughter cells, each with its own plasma membrane.
- A new cell wall forms between the two membranes of the cell plate. (See Campbell, Figure 11.12)

In some exceptional cases, mitosis is not followed by cytokinesis (e.g. certain slime molds form multinucleated masses called *plasmodia*).

V. THE CONTROL OF CELL DIVISION (p. 230–235)

Normal growth, development and maintenance depends upon the proper timing and rate of mitosis in different parts of an organism. Different types of cells exhibit different patterns of cell division. For example:

- Human skin cells divide frequently throughout life.
- Liver cells retain the ability to divide, but do so only in appropriate situations, such as wound repair.
- More specialized cells (e.g. nerve and muscle cells) do not divide in mature humans.

What are several factors identified from cell tissue-culture studies, that stimulate or inhibit cell growth?

Using the technique of tissue culture, researchers have identified many factors that stimulate or inhibit cell division. For example, cells in culture will not divide if:

- Essential nutrients are left out of the culture medium.
- Poisons that inhibit protein synthesis are added to the culture medium.

Margin notes:
- Cleavage
- Cleavage furrow
- Cell plate
- Tissue culture

Even if all conditions are favorable, some mammalian cells will divide in culture only in the presence of specific regulatory substances called *growth factors*. For example:
- Binding of *platelet-derived growth factor* (PDGF) to cell membrane receptors, stimulates cell division in fibroblasts. This regulation occurs not only in cell culture, but in the animal's body as well.
- Each cell type probably responds specifically to a certain growth factor or combination of factors.

Growth factors

Crowding inhibits cell division in a phenomenon called *density-dependent inhibition*.
- Cultured cells normally divide until they form a single layer of cells on the inner surface of the culture container, at which point cell division stops. If some cells are removed, those bordering the open space divide again until the vacancy is filled.
- When cells in culture reach a certain density, the quantities of nutrients and growth regulators become insufficient to support cell division and growth stops.
- This regulatory mechanism probably functions in the body's tissues as well as in cell culture. Cancer cells, however, do not exhibit density-dependent inhibition.
- For normal cell division, cells must also adhere to a substratum: the inside of a culture jar or the extracellular matrix of a tissue; cells will stop dividing if they become detached.

Density-dependent inhibition

At what point in the cell cycle is it determined whether a cell is destined to divide? What is the most important indicator of whether a cell will pass this point?

Whether a cell is destined to divide is determined at the *restriction point* which occurs late in the G_1 phase of the cell cycle.
- If a cell is destined to divide, it progresses beyond the restriction point into the S phase when DNA synthesis begins, and then proceeds through cell division.
- If the cell is not destined to divide, it may exit from the cell cycle at the restriction point and switch to a nondividing state called the G_0 phase.
- Most cells of the body are actually in the G_0 phase. The most specialized cells, such as nerve and muscle, will stay in G_0 and never divide again. Other cells, such as liver cells, can be induced by certain environmental cues to continue through the cell cycle and divide.

Restriction point

For actively dividing cells, the ratio of cytoplasmic volume to genome size is the most important indicator of whether a cell will pass the restriction point.
- Once a cell's volume-to-genome ratio reaches a critical threshold value, it may pass the restriction point and copy its DNA.
- In other words, a cell must add enough cytoplasm to attain a certain size, before DNA synthesis can begin. This prevents daughter cells from becoming progressively smaller with each cell cycle.

Volume-to-genome ratio

Once the cell passes the restriction point, it is destined to divide.
- The onset of the S phase commits the cell to continue through the G_2 and M phases and divide.
- Biologists are only beginning to work out the switches that control the precise sequence of events in cell division. We do know that this sequence probably depends on the completion of each task before the cell can progress to the next stage.

What substance is required for a cell to progress from late G_2 to mitosis?

A high concentration of *MPF* (*maturation promoting factor*) is required for a cell to progress from late interphase (G_2) to mitosis.
- MPF is a complex of proteins that rises and falls in concentration during the cell cycle. It appears in late interphase, reaches its highest concentration during mitosis, and disappears by the end of mitosis.
- During G_2, MPF reaches a threshold concentration, prophase commences, and the cell goes on to divide.
- The high concentration of MPF during mitosis is required for the cell to progress through the mitotic stages, and the decline of MPF signals the end of mitosis and the transition to G_1 of the cell cycle's next interphase.

How does MPF induce the changes that occur in mitosis?

MPF induces mitosis by acting as an enzyme that is probably required for the correct performance of the entire mitotic sequence.
- MPF is a protein *kinase*, an enzyme that activates other proteins by catalyzing the transfer of a phosphate group from ATP to a target protein.
- Some of these targets are themselves protein kinases, which phosphorylate still other proteins in a cascade of activation steps. (See Campbell Figure 11.15) This mechanism enables just one protein, MPF, to elicit many different changes throughout the cell.
- For example, activation of MPF leads to phosphorylation of chromatin proteins, causing chromosomes to condense during prophase. During prometaphase, the nuclear envelope disperses when some of its membrane proteins are phosphorylated.

What causes the cyclical change in MPF concentration?

The rhythmic change in concentration during the cell cycle is a property that emerges from the structure of MPF.

Active MPF consists of two associated proteins: 1) *cdc2* and 2) *cyclin*.
- *cdc2* is the protein kinase responsible for the transition from interphase to M phase. It is inactive unless attached to the second component of MPF, the protein *cyclin*.
- *Cyclin* concentration fluctuates rhythmically throughout the cell cycle, whereas *cdc2* concentration is constant.

Cyclin's rhythmic changes in concentration regulate MPF activity, and thus act as a mitotic clock that regulates the sequential changes in a dividing cell.
- Cyclin is actually produced at a uniform rate throughout the cell cycle, and it accumulates during interphase.
- Cyclin combines with *cdc2* to form active MPF, so as the cyclin concentration in the cell rises and falls, the amount of active MPF changes in a similar way.
- Active MPF acts as a kinase, activating many other proteins that participate in mitosis.
- Near the end of mitosis, cyclin is destroyed by an enzyme that is activated by MPF.
- The destruction of cyclin causes the decline in active MPF at the end of mitosis. Since MPF is activated by cyclin, it is indirectly responsible for its own decline.
- Continuing cyclin synthesis raises the concentration again during interphase. This newly synthesized cyclin binds to *cdc2* to form active MPF, and mitosis begins again.

To summarize the control of the cell cycle:
- The G_1 phase of the cycle is the most variable phase in duration and in the variety of external and internal controls over cell division.
- Nutritional status, growth factors, cell density, and developmental state of the cell affect the length of G_1 and whether the cell will pass the restriction point and divide.
- A cell that does not pass the restriction point will enter the G_0 phase as a nondividing cell.
- If conditions favor cell division, the cell will proceed through the restriction point. When the cell acquires enough cytoplasm to reach the threshold volume-to-genome ratio, the cell will be irreversibly committed to divide.
- The cell duplicates its chromosomes in the S phase and continues to grow in the G_2 phase, the last period of interphase.
- The transition from interphase to mitosis requires a threshold concentration of MPF to activate the proteins required for mitosis.

VI. ABNORMAL CELL DIVISION: CANCER CELLS (p. 235–237)

How does abnormal cell division of cancer cells differ from normal cell division?

Cancer cells do not respond normally to the body's control mechanisms. They divide excessively, invade other tissues and, if unchecked, can kill the whole organism.
- Cancer cells in culture do not respond to the normal signals that stop growth such as density-dependent inhibition.
- They continue to grow until nutrients in the growth medium are exhausted.

Cancer

152 The Reproduction of Cells

Other differences between normal and cancer cells reflect abnormalities in the cell cycle of cancer cells.
- ☞ Cancer cells that stop dividing do so at random points in the cycle instead of at the restriction point in G_1.
- ☞ Cancer cells in culture are immortal in that they continue to divide indefinitely, as long as nutrients are available. Normal mammalian cells in culture divide only about 20 to 50 times before they stop.

The abnormal behavior of cancer cells can be catastrophic when it occurs in the body. Cancer cells have lost the normal restraints on growth and have the ability to invade surrounding tissue.

Cells in tissue culture that have lost the normal controls on growth are said to be *transformed*.

Transformation

Transformation = The conversion of eukaryotic cells in tissue culture to a condition of unregulated growth.

The immune system normally destroys abnormal cells that have converted from normal to cancer cells.
- ☞ If abnormal cells evade destruction, they may proliferate to form a *tumor*, an unregulated growing mass of cells within otherwise normal tissue.
- ☞ If the cells remain at this original site, the mass is *benign* and can be completely removed by surgery.
- ☞ A tumor is *malignant* if the cells have the ability to spread to other parts of the body. Only a malignant tumor is said to be cancer.

Besides having anomalous cell cycles, malignant cells are abnormal in other ways. For example, they may have:
- ☞ Unusual numbers of chromosomes.
- ☞ Aberrant metabolism.
- ☞ Lost attachments to neighboring cells and substratum. This is usually a consequence of abnormal cell surface changes.

Detached cancer cells may spread into other tissues surrounding the original tumor and may even enter the blood and lymph vessels of the circulatory system.
- ☞ These circulating cells can invade other parts of the body and proliferate to form more tumors.
- ☞ This spread of cancer cells beyond their original sites is called *metastasis*.
- ☞ If a tumor metastasizes, it is usually treated with radiation and chemotherapy, which is especially harmful to actively dividing cells.

Researchers are beginning to understand how a normal cell is transformed into a cancer cell.
- ☞ Michael Bishop and Harold Varmus won the Nobel Prize for demonstrating that genetic alterations can cause cancer.
- ☞ Knowledge of how changes in the genome lead to the heritable abnormalities of cancer cells is still rudimentary and depends upon our increased understanding of how normal cells work.

12

MEIOSIS AND SEXUAL LIFE CYCLES

CHAPTER OUTLINE

Why do organisms only reproduce their own kind? Why do offspring more closely resemble their parents than unrelated individuals of the same species?

Reproduction is an emergent property associated with life. The fact that organisms reproduce their own kind is a consequence of *heredity*.

Heredity = Continuity of biological traits from one generation to the next.
- ☞ Results from the transmission of hereditary units, or *genes*, from parents to offspring.
- ☞ Because they share similar genes, offspring more closely resemble their parents or close relatives than unrelated individuals of the same species.

Variation = Inherited differences among individuals of the same species.
- ☞ Though offspring resemble their parents and siblings, they also diverge somewhat as a consequence of inherited differences among them.
- ☞ The development of *genetics* in this century has increased our understanding about the mechanisms of variation and heredity.

Genetics = The scientific study of heredity and variation.

I. GENES, DNA, AND CHROMOSOMES: A BRIEF ORIENTATION (p. 244–245)

DNA = Type of nucleic acid that is a polymer of four different kinds of nucleotides.

Genes = Units of hereditary information that are made of *DNA* and are located on *chromosomes*.
- ☞ Have specific sequences of nucleotides, the monomers of DNA.
- ☞ Most genes program cells to synthesize specific proteins; the action of these proteins produce an organism's inherited traits.

What makes heredity possible?

Inheritance is possible because:
- ☞ DNA is precisely replicated producing copies of genes that can be passed along from parents to offspring.
- ☞ Sperm and ova carrying each parent's genes are combined in the nucleus of the fertilized egg.

Heredity

Variation

Genetics

DNA

Genes

154 *Meiosis and Sexual Life Cycles*

The actual transmission of genes from parents to offspring depends on the behavior of *chromosomes*.

Chromosomes

<u>Chromosomes</u> = Threadlike structures in eukaryotic nuclei that are made of DNA and protein.
- ☞ Consist of a single long DNA molecule that is highly folded and coiled along with proteins.
- ☞ Contain genetic information arranged in a linear sequence.
- ☞ Contain hundreds or thousands of genes, each of which is a specific region of the DNA molecule, or *locus*.

Locus

<u>Locus</u> = Specific location on a chromosome that contains a gene.
- ☞ Each species has a characteristic chromosome number; humans have 46.

II. SEXUAL AND ASEXUAL REPRODUCTION: A COMPARISON (*p. 245*)

How is asexual reproduction different from sexual reproduction?

Asexual reproduction
Sexual reproduction

Asexual Reproduction	Sexual Reproduction
Single individual is the sole parent.	Two parents give rise to offspring.
Single parent passes on *all* its genes to its offspring.	Each parent passes on *half* its genes, to its offspring.
Offspring are genetically identical to the parent.	Offspring have a unique combination of genes inherited from both parents.
Results in a *clone*, or genetically identical individual. Rarely, genetic differences occur as a result of mutation.	Results in greater genetic variation; offspring vary genetically from their siblings and parents.

What generates this genetic variation during sexual reproduction? The answer lies in the process of *meiosis*.

III. AN INTRODUCTION TO SEXUAL LIFE CYCLES: THE HUMAN EXAMPLE (*p. 246–248*)

Diagram the human life cycle. Where in the human body does meiosis occur? What cells are produced by meiosis? What cells are produced by mitosis? Which cells in the human life cycle are haploid? What process in the human life cycle restores the diploid condition?

The human *life cycle* follows the same basic pattern found in all sexually reproducing organisms; *meiosis* and *fertilization* result in alternation between the haploid and diploid condition.

Life cycle

<u>Life cycle</u> = Sequence of stages in an organism's reproductive history, from conception to production of its own offspring.

Somatic cell = Any cell other than a sperm or egg cell.
- Human somatic cells contain 46 chromosomes distinguishable by differences in size, position of the centromere, and staining or banding pattern.
- Using these criteria, chromosomes from a photomicrograph can be matched into *homologous* pairs and arranged in a standard sequence to produce a *karyotype*.

Karyotype = A display or photomicrograph of an individual's somatic-cell metaphase chromosomes that are arranged in a standard sequence. (See Campbell, Methods Box: *Preparation of a Karyotype*)
- Human karyotypes are often made with lymphocytes.
- Can be used to screen for chromosomal abnormalities.

Homologous chromosomes (homologues) = A pair of chromosomes that have the same size, centromere position and staining pattern.
- With one exception, homologues carry the same genetic loci.
- Homologous *autosomes* carry the same genetic loci; however, human *sex chromosomes* carry different loci even though they pair during prophase of Meiosis I. (A discussion of meiosis follows.)

Autosome = A chromosome that is not a sex chromosome.

Sex chromosome = Dissimilar chromosomes that determine an individual's sex.
- Females have a homologous pair of X chromosomes.
- Males have one X and one Y chromosome.
- Thus, humans have 22 pairs of autosomes and 1 pair of sex chromosomes.

Chromosomal pairs in the human karyotype are a result of our sexual origins.
- One homologue is inherited from each parent.
- Thus, the 46 somatic-cell chromosomes are actually two sets of 23 chromosomes; one a maternal set and the other a paternal set.
- Somatic cells in humans and most other animals are *diploid*.

Diploid = Condition in which cells contain two sets of chromosomes; abbreviated as $2n$.

Haploid = Condition in which cells contain one set of chromosomes; it is the chromosome number of *gametes* and is abbreviated as n.

Gamete = A haploid reproductive cell.
- *Sperm cells* and *ova* are gametes, and they differ from somatic cells in their chromosome number. Gametes only have one set of chromosomes.
- Human gametes contain a single set of 22 autosomes and one sex chromosome (either an X or a Y).
- Thus, the haploid number of humans is 23.

The diploid number is restored when two haploid gametes unite in the process of *fertilization*. Sexual intercourse allows a haploid sperm cell from the father to reach and fuse with an ovum from the mother.

Fertilization = The union of two gametes to form a *zygote*.

Zygote

<u>Zygote</u> = A diploid cell that results from the union of two haploid gametes.
- Contains the maternal and parental haploid sets of chromosomes from the gametes and is diploid ($2n$).
- As humans develop from a zygote to a sexually mature adult, the zygote's genetic information is passed with precision to all somatic cells by mitosis.

Gametes are the only cells in the body that are <u>not</u> produced by mitosis.
- Gametes are produced in the ovaries or testes by the process of *meiosis*.
- *Meiosis* is a special type of cell division that produces haploid cells and compensates for the doubling of chromosome number that occurs at fertilization.
- Meiosis in humans produces sperm cells and ova which contain 23 chromosomes.
- When fertilization occurs, the diploid condition ($2n = 46$) is restored in the zygote.

Meiosis

IV. THE VARIETY OF SEXUAL LIFE CYCLES (p. 248–250)

How do sexual life cycles vary among animals, fungi and plants?

Alternation of meiosis and fertilization is common to all sexually reproducing organisms; however, the timing of these two events in the life cycle varies among species. There are three basic patterns of sexual life cycles:

Animal: In animals, including humans, gametes are the only haploid cells.
- Meiosis occurs during gamete production. The resulting gametes undergo no further cell division before fertilization.
- Fertilization produces a diploid zygote that divides by mitosis to produce a diploid multicellular animal.

Fungi and Some Algae: In many fungi and some protists, the only diploid stage is the zygote.
- Meiosis occurs immediately after the zygote forms.
- Resulting haploid cells divide by <u>mitosis</u> to produce a haploid multicellular organism.
- Gametes are produced by <u>mitosis</u> from the already haploid organism.

How are plant life cycles different from those of animals and fungi? What distinguishes a sporophyte from a gametophyte?

Plants and Some Algae: Plants and some species of algae alternate between multicellular haploid and diploid generations.
- ☞ This type of life cycle is called an *alternation of generations*.
- ☞ The multicellular diploid stage is called a *sporophyte*, or spore-producing plant. Meiosis in this stage produces haploid cells called *spores*.
- ☞ Haploid spores divide mitotically to generate a multicellular haploid stage called a *gametophyte*, or gamete-producing plant.
- ☞ Haploid gametophytes produce gametes by mitosis.
- ☞ Fertilization produces a diploid zygote which develops into the next sporophyte generation.

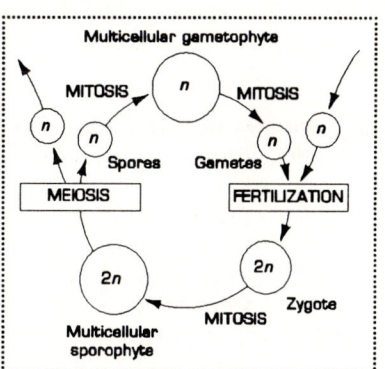

Alternation of generations

Sporophyte

Gametophyte

V. MEIOSIS: A CLOSER LOOK (*p. 250–253*)

Meiosis and sexual reproduction significantly contribute to genetic variation among offspring.

Meiosis includes steps that closely resemble corresponding steps in mitosis.
- ☞ Like mitosis, meiosis is preceded by replication of the chromosomes.
- ☞ Meiosis differs from mitosis in that this single replication is followed by two consecutive cell divisions: *Meiosis I* and *Meiosis II*.
- ☞ These cell divisions produce <u>four</u> daughter cells instead of two as in mitosis.
- ☞ The resulting daughter cells have *half* the number of chromosomes as the original cell; whereas, daughter cells of mitosis have the same number of chromosomes as the parent cell.

What are the phases of meiosis I and meiosis II? What events characterize each phase?

Interphase I: Interphase I precedes meiosis.
- ☞ Chromosomes replicate as in mitosis.
- ☞ Each duplicated chromosome consists of two identical sister chromatids attached at their centromeres.
- ☞ Centriole pairs in animal cells also replicate into two pairs.

Sister chromatids

Meiosis I

Synapsis

Tetrad

Chiasmata

Meiosis I: This cell division reduces the chromosome number by one-half. It includes the following four phases:

What is synapsis? What process causes genetic recombination to occur in prophase I?

Prophase I. This is a longer and more complex process than prophase of mitosis.
- Chromosomes condense and attach at their ends to the nuclear envelope.
- *Synapsis* occurs. During this process, homologous chromosomes come together as pairs.
- Chromosomes condense further until they are distinct structures that can be seen with a microscope. Since each chromosome has two chromatids, each homologous pair in synapsis appears as a complex of four chromatids or a *tetrad*.
- In each tetrad, sister chromatids of the same chromosome are attached at their centromeres. Nonsister chromatids are linked by X-shaped *chiasmata*, sites where homologous strand exchange or *crossing-over* occurs.
- Chromosomes thicken further and detach from the nuclear envelope.

As prophase I continues, the cell prepares for nuclear division.
- Centriole pairs move apart and spindle microtubules form between them.
- Nuclear envelope and nucleoli disperse.
- Chromosomes begin moving to the metaphase plate, midway between the two poles of the spindle apparatus.
- Prophase I typically occupies more than 90% of the time required for meiosis.

Metaphase I. Tetrads are aligned on the metaphase plate.
- Each synaptic pair is aligned so that centromeres of homologues point towards opposite poles.
- Each homologue is thus attached to kinetochore microtubules emerging from the pole it faces, so that the two homologues are destined to separate in anaphase and move towards opposite poles.

Anaphase I. Chromosomes are moved towards the poles by the spindle apparatus.
- ☞ Sister chromatids remain attached at their centromeres and move as a unit towards the same pole, while the homologue moves towards the opposite pole.
- ☞ This differs from mitosis during which chromosomes line up individually on the metaphase plate (rather than in pairs) and sister chromatids are moved apart towards opposite poles of the cell.

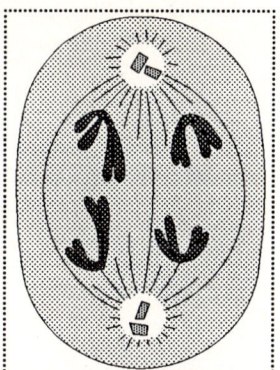

Telophase I and Cytokinesis. The spindle apparatus continues to separate homologous chromosome pairs until the chromosomes reach the poles.
- ☞ Each pole now has a haploid set of chromosomes that are still composed of two sister chromatids attached at the centromere.
- ☞ Cytokinesis occurs simultaneously with Telophase I, forming two <u>haploid</u> daughter cells. *Cleavage furrows* form in animal cells, and cell plates form in plant cells.
- ☞ In some species, nuclear membranes and nucleoli reappear, and the cell enters a period of *interkinesis* before meiosis II. In other species, the daughter cells immediately prepare for meiosis II.
- ☞ Regardless of whether a cell enters interkinesis, <u>no DNA replication occurs before meiosis I</u>.

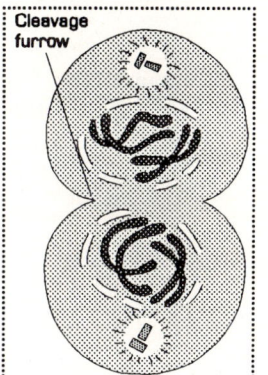

How is mitotic interphase different from meiotic interkinesis?

Meiosis II:

Prophase II.
- ☞ If the cell entered interkinesis, the nuclear envelope and nucleoli disperse.
- ☞ Spindle apparatus forms and chromosomes move towards the metaphase II plate.

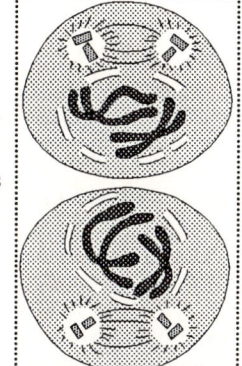

Interkinesis

Meiosis II

Metaphase II. Chromosomes align singly on the metaphase plate.
- Kinetochores of sister chromatids point towards opposite poles.

Anaphase II.
- Centromeres of sister chromatids separate.
- Sister chromatids of each pair (now individual chromosomes) move toward opposite poles of the cell.

Telophase II and Cytokinesis.
- Nuclei form at opposite poles of the cell.
- Cytokinesis occurs producing four haploid daughter cells.

Haploid daughter cells

VI. A COMPARISON OF MITOSIS AND MEIOSIS (p. 253)

What are some key differences between mitosis and meiosis? How does the end result of meiosis differ from that of mitosis?

Though the processes of mitosis and meiosis are similar in some ways, there are some key differences:
- *Meiosis is a reduction division.* Cells produced by mitosis have the same number of chromosomes as the original cell, whereas cells produced by meiosis have half the number of chromosomes as the parent cell.
- *Meiosis creates genetic variation.* Mitosis produces two daughter cells genetically identical to the parent cell and to each other. Meiosis produces four daughter cells genetically different from the parent cell and from each other.

☞ *Meiosis is two successive nuclear divisions*. Mitosis, on the other hand, is characterized by just one nuclear division.

How do prophase, metaphase and anaphase of meiosis I differ from the phases of mitosis? Comparing both processes, how are the chromosomes differently arranged at the metaphase plate?

COMPARISON OF MEIOSIS I AND MITOSIS

	MEIOSIS I	MITOSIS
Prophase	*Synapsis* occurs to form tetrads. *Chiasmata* appear as evidence that crossing over has occurred.	Neither synapsis nor crossing over occurs.
Metaphase	Homologous pairs (tetrads) align on the metaphase plate.	Individual chromosomes align on the metaphase plate.
Anaphase	Meiosis I separates pairs of chromosomes. Centromeres do not divide and sister chromatids stay together. Sister chromatids of each chromosome move to the same pole of the cell; only the homologues separate.	Mitosis separates sister chromatids of individual chromosomes. Centromeres divide and sister chromatids move to opposite poles of the cell.

Meiosis II is virtually identical in mechanism to mitosis, separating sister chromatids.

VII. SEXUAL SOURCES OF GENETIC VARIATION (p. 253–255)

How do independent assortment, crossing over and random fertilization contribute to genetic variation in sexually reproducing organisms?

Meiosis and fertilization are the primary sources of genetic variation in sexually reproducing organisms. Sexual reproduction provides genetic variation by: 1) independent assortment; 2) crossing over during prophase I of meiosis; and 3) random fusion of gametes during fertilization.

A. Independent Assortment of Chromosomes (p. 253–254)

At Metaphase I, each homologous pair of chromosomes aligns on the metaphase plate. Each pair consists of one maternal and one paternal chromosome.

☞ The orientation of the homologous pair to the poles is random, so there is a fifty-fifty chance that a particular daughter cell produced by meiosis I will receive the maternal chromosome of a homologous pair, and a fifty-fifty chance that it will receive the paternal chromosome.

162 Meiosis and Sexual Life Cycles

Independent assortment

Crossing over

- ☞ Each homologous pair of chromosomes orients independently of the other pairs at metaphase I; thus, the first meiotic division results in *independent assortment* of maternal and paternal chromosomes. (See Campbell, Figure 12.9)
- ☞ A gamete produced by meiosis contains just one of all the possible combinations of maternal and paternal chromosomes.

Independent assortment = The random distribution of maternal and paternal homologues to the gametes. (In a more specific sense, assortment refers to the random distribution of genes located on different chromosomes.)
- ☞ Since each homologous pair assorts independently from all the others, the process produces 2^n possible combinations of maternal and paternal chromosomes in gametes, where *n* is the haploid number.
- ☞ In humans, the possible combinations would be 2^{23}, or about eight million.
- ☞ Thus, each human gamete contains one of eight million possible assortments of chromosomes inherited from that person's mother and father.
- ☞ Genetic variation results from this reshuffling of chromosomes, because the maternal and paternal homologues will carry different genetic information at many of their corresponding loci.

B. Crossing Over (p. 245–255)

Another mechanism that increases genetic variation is the process of *crossing over*, during which homologous chromosomes exchange genes.

Crossing over = The exchange of genetic material between homologues; occurs during prophase of meiosis I. This process:
- ☞ Occurs when homologous portions of two nonsister chromatids trade places. During prophase I, X-shaped *chiasmata* become visible at places where this homologous strand exchange occurs.
- ☞ Produces chromosomes that contain genes from <u>both</u> parents.
- ☞ In humans, there is an average of two or three crossovers per chromosome pair.
- ☞ Synapsis during prophase I is precise, so that homologues align gene by gene. The exact mechanism of synapsis is still unknown, but involves the formation of the *synaptonemal complex*, a protein structure that brings the chromosomes into close association.

C. Random Fertilization (p. 255)

Random fertilization is another source of genetic variation in offspring.
- ☞ In humans, an egg cell that is one of eight million different possibilities will be fertilized by a sperm cell that is also one of eight million possibilities. Thus, the resulting zygote can have one of 64 trillion possible diploid combinations.

To summarize, there are three important sources of genetic variability in sexually reproducing organisms:

1. Independent assortment of homologous chromosome pairs during meiosis I.
2. Crossing over between homologous chromosomes during prophase of meiosis I.
3. Random fusion of gametes.

Meiosis and Sexual Life Cycles

VIII. GENETIC VARIATION AND EVOLUTION (p. 255–256)

Why is inheritable variation crucial to Darwin's theory of evolution?

Inheritable variation is the basis for Charles Darwin's theory that natural selection is the mechanism for evolutionary change. Natural selection:

- ☞ Increases the frequency of inheritable variations that favor the reproductive success of some individuals over others.
- ☞ Results in *adaptation*, the accumulation of inheritable variations that are favored by the environment.
- ☞ In the face of environmental change, genetic variation increases the likelihood that some individuals in a population will have inheritable variations that help them cope with the new conditions.

What are the sources of genetic variation?

There are two sources of genetic variation:

1. Sexual reproduction: independent assortment in meiosis I, crossing over in prophase of meiosis I, and random fusion of gametes during fertilization.
2. Mutation, which is rare structural change in a gene.

13 MENDEL AND THE GENE IDEA

CHAPTER OUTLINE

What was the favored model of heredity in the 19th century, prior to Mendel? What was wrong with this model?

Based upon their observations from ornamental plant breeding, biologists in the 19th century realized that both parents contribute to the characteristics of offspring. Before Mendel, the favored explanation of heredity was the *blending theory*.

Blending theory of heredity

Blending theory of heredity = Pre-Mendelian theory of heredity proposing that hereditary material from each parent mixes in the offspring; once blended like two liquids in solution, the hereditary material is inseparable and the offspring's traits are some intermediate between the parental types. According to this theory:
- Individuals of a population should reach a uniform appearance after many generations.
- Once hereditary traits are blended, they can no longer be separated out to appear again in later generations.

This blending theory of heredity was inconsistent with the observations that:
- Individuals in a population do not reach a uniform appearance; inheritable variation among individuals is generally preserved.
- Some inheritable traits skip one generation only to reappear in the next.

How did Mendel's particulate theory of heredity differ from the blending theory of inheritance?

Gregor Mendel

Modern genetics began in the 1860's when Gregor Mendel, an Augustinian monk, discovered the fundamental principles of heredity. Mendel's great contribution to modern genetics was to replace the blending theory of heredity with the *particulate theory of heredity*.

Particulate theory of heredity

Particulate theory of heredity = Gregor Mendel's theory that parents transmit to their offspring discrete inheritable factors (now called genes) that remain as separate factors from one generation to the next.

I. **MENDEL'S MODEL: A CASE STUDY IN THE SCIENTIFIC PROCESS**
   (p. 258–267)

While attending the University of Vienna from 1851–1853, Mendel was influenced by two professors:
- Doppler, a physicist, trained Mendel to apply a *quantitative experimental* approach to the study of natural phenomena.
- Unger, a botanist, interested Mendel in the causes of inheritable variation in plants.

Mendel and the Gene Idea 165

These experiences inspired Mendel to use key elements of the scientific process in the study of heredity. Unlike most nineteenth century biologists, he used a quantitative approach to his experimentation.

What are some features of Mendel's methods that contributed to his success?

 A. **Mendel's Experimental Approach** (*p. 259–260*)

In 1857, Mendel was living in an Augustinian monastery, where he bred garden peas in the abbey garden. He probably chose garden peas as his experimental organisms because:

- They were available in many easily distinguishable varieties.
- Strict control over mating was possible to ensure the parentage of new seeds. Petals of the pea flower enclose the pistil and stamens, which prevents cross-pollination. Immature stamens can be removed to prevent self-pollination. Mendel hybridized pea plants by transferring pollen from one flower to another with an artist's brush.

Character = Detectable inheritable feature of an organism.

Trait = Variant of an inheritable character.

Mendel chose *characters* in pea plants that differed in a relatively clear-cut manner. He chose seven characters, each of which occurred in two alternative forms:

1. Flower color (purple or white)
2. Flower position (axial or terminal)
3. Seed color (yellow or green)
4. Seed shape (round or wrinkled)
5. Pod shape (inflated or constricted)
6. Pod color (green or yellow)
7. Stem length (tall or dwarf)

True breeding = Always producing offspring with the same *traits* as the parents when the parents are self-fertilized.

Mendel started his experiments with *true-breeding* plant varieties, which he hybridized (cross-pollinated) in experimental crosses.

- The true-breeding parental plants of such a *cross* are called the *P generation* (parental).
- The hybrid offspring of the P generation are the *F_1 generation* (first filial).
- Allowing F_1 generation plants to self-pollinate, produces the next generation, the *F_2 generation* (second filial).

Mendel observed the transmission of selected traits for at least three generations and arrived at two principles of heredity that are now known as the *law of segregation* and the *law of independent assortment*.

B. Mendel's Law of Segregation (p. 260-263)

When Mendel crossed true-breeding plants with different forms of a trait, he found that the traits did not blend.

- ☞ Using the scientific process, Mendel designed experiments in which he used large sample sizes and kept accurate quantitative records of the results.
- ☞ For example, a cross between true-breeding varieties, one with purple flowers and one with white flowers, produced F_1 *progeny* (offspring) with only purple flowers.

Hypothesis: Mendel hypothesized that if the inheritable factor for white flowers had been lost, then a cross between F_1 plants should produce only purple-flowered plants.

Experiment: Mendel allowed the F_1 plants to self-pollinate.

Results: There were 705 purple-flowered and 224 white-flowered plants in the F_2 generation — a ratio of 3:1. The inheritable factor for white flowers was not lost, so the hypothesis was rejected.

Conclusions: From these types of experiments and observations, Mendel concluded that since the inheritable factor for white flowers was not lost in the F_1 generation, it must have been masked by the presence of the purple-flower factor. Mendel's factors are now called *genes*; and in Mendel's terms, purple flower is the *dominant trait* and white flower is the *recessive trait*.

What are four components of Mendel's hypothesis that led him to deduce the law of segregation?

Mendel repeated these experiments with the other six characters and found similar 3:1 ratios in the F_2 generations. From these observations he developed a hypothesis that can be divided into four parts:

1. *Alternative forms of genes are responsible for variations in inherited characters.*
    - ☞ For example, the gene for flower color in pea plants exists in two alternative forms; one for purple color and one for white color.
    - ☞ Alternative forms for a gene are now called *alleles*.

2. *For each character, an organism inherits two genes, one from each parent.*
    - ☞ Mendel deduced that each parent contributes one "factor," even though he did not know about chromosomes or meiosis.
    - ☞ We now know that Mendel's factors are genes. Each genetic locus is represented twice in diploid organisms, which have homologous pairs of chromosomes, one set for each parent. Homologous loci may have the same allele as in Mendel's true-breeding organisms or they may differ as in the F_1 hybrids.

3. *If the two alleles differ, one is fully expressed (dominant allele); the other is completely masked (recessive allele).*
    - ☞ Dominant alleles are designated by a capital letter: P = purple flower color.
    - ☞ Recessive alleles are designated by a lowercase letter: p = white flower color.

4. *The two genes for each character segregate during gamete production.*

☞ Without any knowledge of meiosis, Mendel deduced that a sperm cell or ovum carries only one allele for each inherited characteristic, because allele pairs separate (segregate) from each other during gamete production.
☞ Gametes of true-breeding plants will all carry the same allele. If different alleles are present in the parent, there is a 50% chance that a gamete will receive the dominant allele, and a 50% chance that it will receive the recessive allele.
☞ This sorting of alleles into separate gametes is known as Mendel's *law of segregation*.

In your own words, what is Mendel's law of segregation? How many types of gametes can be produced by the F_1 hybrids?

Mendel's law of segregation = Allele pairs segregate during gamete formation (meiosis), and the paired condition is restored by the random fusion of gametes at fertilization.

☞ This law predicts the 3:1 ratio observed in the F_2 generation of a monohybrid cross.
☞ F_1 hybrids (Pp) produce two classes of gametes when allele pairs segregate during gamete formation. Half receive a purple-flower allele (P) and the other half the white-flower allele (p).
☞ During self-pollination, these two classes of gametes unite randomly. Ova containing purple-flower alleles have equal chances of being fertilized by sperm carrying purple-flower alleles or sperm carrying white-flower alleles.
☞ Since the same is true for ova containing white-flower alleles, there are four equally likely combinations of sperm and ova.

How can you use a Punnett square to predict the results of a monohybrid cross? What are the phenotypic and genotypic ratios of the F_2 generation?

The combinations resulting from a genetic cross may be predicted by using a *Punnett Square*.

The F_2 progeny would include:
☞ One-fourth of the plants with two alleles for purple flowers.
☞ One-half of the plants with one allele for purple flowers and one allele for white flowers. Since the purple-flower allele is dominant, these plants have purple flowers.
☞ One-fourth of the plants with two alleles for white flower color which will have white flowers since no dominant allele is present.

The pattern of inheritance for all seven of the characteristics studied by Mendel was the same: one parental trait disappeared in the F_1 generation and reappeared in one-fourth of the F_2 generation.

Mendel's law of segregation

168 *Mendel and the Gene Idea*

Homozygous

Heterozygous

Phenotype

Genotype

Testcross

Some Useful Genetic Vocabulary:

What are the differences between genotype and phenotype; heterozygous and homozygous?

Homozygous = Having two identical alleles for a trait (e.g. PP or pp).
- All gametes carry that allele.
- Homozygotes are *true-breeding*.

Heterozygous = Having two different alleles for a trait (e.g. Pp).
- One-half of the gametes carry one allele (P) with the other half carrying the other allele (p).
- Heterozygotes are not true-breeding.

Phenotype = An organism's expressed traits (e.g. purple or white flowers).
- In Mendel's experiment above, the F_2 generation had a 3:1 phenotypic ratio of plants with purple flowers to plants with white flowers.

Genotype = An organism's genetic makeup (e.g. PP, Pp or pp).
- The F_2 generation had a 1:2:1 genotypic ratio (1 PP:2 Pp:1 pp).

The Testcross

Because some alleles are dominant over others, the genotype of an organism may not be apparent. For example:
- A recessive phenotype (white) is only expressed when the organism is homozygous recessive (pp).
- A pea plant with purple flowers may be either homozygous dominant (PP) or heterozygous (Pp).

How can it be determined experimentally, whether an organism with the dominant phenotype is homozygous dominant or heterozygous?

To determine whether an organism with a dominant phenotype (e.g. purple flower color) is homozygous dominant or heterozygous, you use a *testcross*.

Testcross = The breeding of an organism of unknown genotype with a homozygous recessive.
- For example, if a cross between a purple-flowered plant of unknown genotype (P__) produced only purple-flowered plants, the parent was probably homozygous dominant since a PP × pp cross produces all purple-flowered progeny that are heterozygous (Pp).
- If the progeny of the testcross contains both purple and white phenotypes, then the purple-flowered parent was heterozygous since a Pp × pp cross produces Pp and pp progeny in a 1:1 ratio.

C. Inheritance as a Game of Chance (p. 263–265)

What is a random event? Why is it significant that allele segregation during meiosis and fusion of gametes at fertilization are random events?

Segregation of alleles during gamete formation and fusion of gametes at fertilization are random events. Thus, if we know the genotypes of the parents, we can predict the most likely genotypes of their offspring by using the simple *laws of probability*:

- The probability scale ranges from 0 to 1; an event that is certain to occur has a probability of 1, and an event that is certain <u>not</u> to occur has a probability of 0.
- The probabilities of all possible outcomes for an event must add up to 1.
- For example, when tossing a coin or rolling a six-sided die:

Event	Probability	
Tossing heads with a two-headed coin	1	$1 + 0 = 1$
Tossing tails with a two-headed coin	0	
Tossing heads with a normal coin	$\frac{1}{2}$	$\frac{1}{2} + \frac{1}{2} = 1$
Tossing tails with a normal coin	$\frac{1}{2}$	
Rolling 3 on a six-sided die	$\frac{1}{6}$	$\frac{1}{6} + \frac{5}{6} = 1$
Rolling a number other than 3	$\frac{5}{6}$	

Random events are *independent* of one another.
- The outcome of a random event is unaffected by the outcome of previous such events.
- For example, it is possible that five successive tosses of a normal coin will produce five heads; however, the probability of heads on the sixth toss is still 1/2.

Two basic rules of probability are helpful in solving genetics problems: the *rule of multiplication* and the *rule of addition*.

Mendel and the Gene Idea

How can you determine from a Mendelian cross, the probability that an F₂ individual will be homozygous recessive?

<u>Rule of multiplication</u> = The probability that independent events will occur simultaneously is the product of their individual probabilities. For example:

Question: In a Mendelian cross between pea plants that are heterozygous for flower color (Pp), what is the probability that the offspring will be homozygous recessive?

Answer:

Probability that an egg from the F_1 (Pp) will receive a p allele = 1/2.

Probability that a sperm from the F_1 will receive a p allele = 1/2.

The overall probability that two recessive alleles will unite at fertilization: 1/2 × 1/2 = <u>1/4</u>.

How can you determine from a Mendelian cross, the probability that an F₂ individual will be heterozygous?

<u>Rule of addition</u> = The probability of an event that can occur in two or more independent ways is the sum of the separate probabilities of the different ways. For example:

Question: In a Mendelian cross between pea plants that are heterozygous for flower color (Pp), what is the probability of the offspring being a heterozygote?

Answer: There are two ways in which a heterozygote may be produced: the dominant allele (P) may be in the egg and the recessive allele (p) in the sperm, or the dominant allele may be in the sperm and the recessive in the egg. Consequently, the probability that the offspring will be heterozygous is the sum of the probabilities of those two possible ways:

Probability that the dominant allele will be in the egg with the recessive in the sperm is 1/2 × 1/2 = 1/4.

Probability that the dominant allele will be in the sperm and the recessive in the egg is 1/2 × 1/2 = 1/4.

Therefore, the probability that a heterozygous offspring will be produced is 1/4 + 1/4 = <u>1/2</u>.

<u>The Statistical Nature of Inheritance</u>

If a seed is planted from the F_2 generation of a monohybrid cross, we cannot predict with absolute certainty that the plant will grow to produce white flowers (pp).

We can say that there is a 1/4 chance that the plant will have white flowers.

- Stated in statistical terms: among a large sample of F_2 plants, 25% will have white flowers.
- The larger the sample size, the closer the results will conform to predictions.

D. **Mendel's Law of Independent Assortment** (p. 265–267)

How did Mendel set up his dihybrid crosses?

Mendel deduced the law of segregation from experiments with *monohybrid crosses*, breeding experiments that used parental varieties differing in a single trait. He then performed crosses between parental varieties that differed in two characters or *dihybrid crosses*.

Dihybrid cross = A mating between parents that are heterozygous for two characters (dihybrids).
- For example, Mendel began his experiments by crossing true-breeding parent plants that differed in two characters such as seed color (yellow or green) and seed shape (round or wrinkled). From previous monohybrid crosses, Mendel knew that the allele for yellow seeds (Y) was dominant to the allele for green (y), and that round (R) was dominant to wrinkled (r).
- Plants homozygous for round yellow seeds (RRYY) were crossed with plants homozygous for wrinkled green seeds (rryy).
- The resulting F₁ dihybrid progeny were heterozygous for both traits (RrYy) and had round yellow seeds, the dominant phenotypes.

What two alternate hypothesis did Mendel consider for how two characters might segregate during gamete formation?

- From the F₁ generation, Mendel could not tell if the two characters were inherited independently or not, so he allowed the F₁ progeny to self-pollinate. In the following experiment, Mendel considered two alternate hypotheses:

Hypothesis 1: If the two characters segregate <u>together</u>, the F₁ hybrids can only produce the same two classes of gametes (RY and ry) that they received from the parents, and the F₂ progeny will show a 3:1 phenotypic ratio.

If two characters segregate independently, how many classes of gametes would be produced by the F₁ dihybrids? How would you set up a Punnett square to predict the results of such a dihybrid cross? What would be the phenotypic and genotypic ratios of the F₂ generation?

Monohybrid crosses

Dihybrid cross

172 Mendel and the Gene Idea

Hypothesis 2: If the two characters segregate <u>independently</u>, the F$_1$ hybrids will produce four classes of gametes (RY, Ry, rY, ry), and the F$_2$ progeny will show a 9:3:3:1 ratio.

Experiment: Mendel performed a dihybrid cross by allowing self-pollination of the F$_1$ plants (RrYy × RrYy).

What were the experimental results from Mendel's dihybrid crosses? What conclusions could Mendel draw from these results?

Results: Mendel categorized the F$_2$ progeny and determined a ratio of 315:108:101:32, which approximates 9:3:3:1.

These results were repeatable. Mendel performed similar dihybrid crosses with all seven characters in various combinations and found the same 9:3:3:1 ratio in each case.

He also noted that the ratio for each individual gene pair was 3:1, the same as that for a monohybrid cross.

Conclusions: The experimental results supported the hypothesis that <u>each allele pair segregates independently during gamete formation</u>.

This behavior of genes during gamete formation is referred to as *Mendel's law of independent assortment*.

In your own words, what is Mendel's law of independent assortment?

<u>Mendel's law of independent assortment</u> = Each allele pair segregates independently of other gene pairs during gamete formation.

The rules of probability can be used to solve complex genetics problems. For example, Mendel crossed pea varieties that differed in three characters (*trihybrid crosses*).

Question: What is the probability that a trihybrid cross between two organisms with the genotypes AaBbCc and AaBbCc will produce an offspring with the genotype aabbcc?

Answer: Because segregation of each allele pair is an independent event, we can treat this as three separate monohybrid crosses:

Aa × Aa: probability for aa offspring = 1/4
Bb × Bb: probability for bb offspring = 1/4
Cc × Cc: probability for cc offspring = 1/4

The probability that these independent events will occur simultaneously is the product of their independent probabilities (rule of multiplication). So the probability that the offspring will be aabbcc is:

1/4 aa × 1/4 bb × 1/4 cc = <u>1/64</u>

<u>Mendel's law of independent assortment</u>

II. EXTENDING MENDELIAN GENETICS (p. 267–272)

As mendel described it, characters are determined by one gene with two alleles; one allele completely dominant over the other. There are other patterns of inheritance not described by Mendel, but his laws of segregation and independent assortment can be extended to these more complex cases.

A. Incomplete Dominance (p. 267)

In cases of *incomplete dominance*, one allele is not completely dominant over the other, so the heterozygote has a phenotype that is intermediate between the phenotypes of the two homozygotes.

What is an example of incomplete dominance? Why is it not evidence for the blending theory of inheritance?

Incomplete dominance = Pattern of inheritance in which the dominant phenotype is not fully expressed in the heterozygote, resulting in a phenotype intermediate between the homozygous dominant and homozygous recessive.

☞ For example, when red snapdragons (RR) are crossed with white snapdragons (rr), all F_1 hybrids (Rr) have pink flowers. (The heterozygote produces half as much red pigment as the homozygous red-flower plant.)

☞ Since the heterozygotes can be distinguished from homozygotes by their phenotypes, the phenotypic and genotypic ratios from a monohybrid cross are the same — 1:2:1.

☞ Incomplete dominance is **not** support for the blending theory of inheritance, because alleles maintain their integrity in the heterozygote and segregate during gamete formation. Red and white phenotypes reappear in the F_2 generation.

B. What Is a Dominant Allele? (p. 268–269)

How is the phenotypic expression of the heterozygote affected by complete dominance, incomplete dominance and codominance?

Incomplete dominance

Dominance/recessiveness relationships among alleles vary in a continuum from *complete dominance* on one end of the spectrum to *codominance* on the other, with various degrees of incomplete dominance in between these extremes.

Complete Dominance (A is dominant)	← Incomplete Dominance (A is incompletely dominant) →	Codominance (no dominance)
AA and Aa have the same phenotype	Aa = Intermediate phenotype. Phenotype is intermediate between the two homozygotes (AA & aa).	Aa = Both alleles are equally expressed in phenotype

<u>Complete dominance</u> = Inheritance characterized by an allele that is fully expressed in the phenotype of a heterozygote and that masks the phenotypic expression of the recessive allele; state in which the phenotypes of the heterozygote and dominant homozygote are indistinguishable.

<u>Codominance</u> = Inheritance characterized by full expression of both alleles in the heterozygote.

☞ For example, the MN blood-group locus codes for the production of surface molecules (glycoproteins) on the red blood cell. In this system, there are three blood types: M, N and MN.

Blood Type	Genotype
M	MM
N	NN
MN	MN

☞ The MN blood type is the result of full phenotypic expression of <u>both</u> alleles in the heterozygote; both molecules, M and N, are produced on the red blood cell.

Why is it difficult to distinguish dominance/recessiveness relationships in some situations?

Apparent dominance/recessiveness relationships among alleles reflect the level at which the phenotype is studied. For example:

☞ Tay-Sachs disease is a recessively inherited disease in humans; only children who are homozygous recessive for the Tay-Sachs allele have the disease.

☞ Brain cells of Tay-Sachs babies lack a crucial lipid-metabolizing enzyme. Consequently, lipids accumulate in the brain, causing the disease symptoms and ultimately leading to death.

☞ At the <u>organismal level</u>, since heterozygotes are symptom free, it appears that the normal allele is completely dominant and the Tay-Sachs allele is recessive.

☞ At the <u>biochemical level</u>, inheritance of Tay-Sachs seems to be incomplete dominance of the normal allele, since there appears to be an intermediate phenotype. Heterozygotes have an enzyme activity level that is intermediate between individuals homozygous for the normal allele and individuals with Tay-Sachs Disease.

☞ At the <u>molecular level</u>, the normal allele and the Tay-Sachs allele are actually codominant. Heterozygotes produce equal numbers of normal and dysfunctional enzymes. They lack disease symptoms, because half the normal amount of functional enzyme is sufficient to prevent lipid accumulation in the brain.

Dominance/recessiveness relationships are a consequence of the mechanism that determines phenotypic expression, not the ability of one allele to subdue another at the level of the DNA.

Additionally, dominance/recessiveness relationships among alleles do not determine the relative abundance of alleles in a population.
- ☞ In other words, dominant alleles are not necessarily more common and recessive alleles more rare.
- ☞ For example, the allele for polydactyly is quite rare in the U.S. (1 in 400 births), yet it is caused by a dominant allele. (Polydactyly is the condition of having extra fingers or toes.)

C. **Multiple Alleles** (*p. 269*)

How is it possible that multiple alleles may exist for some characters, when each individual can only carry two alleles? How are multiple alleles inherited in the ABO blood system? Why are the I^A and I^B alleles considered to be codominant?

ABO blood system

Multiple alleles

Some genes may have *multiple alleles*; that is, more than just two alternative forms of a gene. The inheritance of the ABO blood group is an example of a locus with three alleles.

Paired combinations of three alleles produce four possible phenotypes:
- ☞ Blood type A, B, AB, or O.
- ☞ A and B refer to two genetically determined polysaccharides (A and B antigens) which are found on the surface of red blood cells.

There are three alleles for this gene: I^A, I^B, and i.
- ☞ The I^A allele codes for the production of A antigen, the I^B allele codes for the production of B antigen, and the i allele codes for <u>no</u> antigen production on the red blood cell (neither A or B).
- ☞ Alleles I^A and I^B are *codominant* since both are expressed in heterozygotes.
- ☞ Alleles I^A and I^B are dominant to allele i, which is recessive.
- ☞ Even though there are three possible alleles, every person carries only two alleles which specify their ABO blood type; one allele is inherited from each parent.

Since there are three alleles, there are six possible genotypes:

Blood Type	Possible Genotypes	Antigens on the red blood cell	Antibodies in the serum
A	$I^A I^A$ $I^A i$	A	anti-B
B	$I^B I^B$ $I^B i$	B	anti-A
AB	$I^A I^B$	A, B	----
O	ii	----	anti-A anti-B

Foreign antigens usually cause the immune system to respond by producing *antibodies*, globular proteins that bind to the foreign molecules causing a reaction that destroys or inactivates it. In the ABO blood system:
- The antigens are located on the red blood cell and the antibodies are in the serum.
- A person produces antibodies against foreign blood antigens (those not possessed by the individual). These antibodies react with the foreign antigens causing the blood cells to clump or *agglutinate*, which may be lethal.
- For a blood transfusion to be successful, the red blood cell <u>antigens of the donor</u> must be compatible with the <u>antibodies of the recipient</u>.

D. Pleiotropy (p. 269)

<u>Pleiotropy</u> = The ability of a single gene to have multiple phenotypic effects.
- There are many hereditary diseases in which a single defective gene causes complex sets of symptoms (e.g. sickle-cell anemia).
- One gene can also influence a combination of seemingly unrelated characteristics. For example, in tigers and Siamese cats, the gene that controls fur pigmentation also influences the connections between a cat's eyes and the brain. A defective gene causes both abnormal pigmentation and cross-eye condition.

E. Epistasis (p. 269–270)

Different genes can interact to control the phenotypic expression of a single trait. In some cases, a gene at one locus alters the phenotypic expression of a second gene, a condition known as *epistasis*.

In your own words, what is meant by "one gene is epistatic to another"?

<u>Epistasis</u> = (Epi=upon; stasis=standing) Interaction between two nonallelic genes in which one modifies the phenotypic expression of the other.
- If one gene suppress the phenotypic expression of another, the first gene is said to be *epistatic* to the second.
- If epistasis occurs between two nonallelic genes, the phenotypic ratio resulting from a dihybrid cross will deviate from the 9:3:3:1 Mendelian ratio.
- For example, in mice and other rodents, the gene for pigment deposition (C) is epistatic to the gene for pigment (melanin) production. In other words, whether the pigment can be deposited in the fur determines whether the coat color can be expressed. Homozygous recessive for pigment deposition (cc) will result in an albino mouse regardless of the genotype at the black/brown locus (BB, Bb or bb):

How does epistasis affect the phenotypic ratio for a dihybrid cross?

CC, Cc = Melanin deposition
cc = Albino
BB, Bb = Black coat color
bb = Brown coat color

☞ Even though both genes affect the same trait (coat color), they are inherited separately and will assort independently during gamete formation. A cross between black mice that are heterozygous for the two genes results in a 9:3:4 phenotypic ratio:

9 Black (B__C__)
3 Brown (bbC__)
4 Albino (__cc)

BbCc × BbCc

	BC	bC	Bc	bc
BC	BBCC	BbCC	BBCc	BbCc
bC	BbCC	bbCC	BbCc	bbCc
Bc	BBCc	BbCc	BBcc	Bbcc
bc	BbCc	bbCc	Bbcc	bbcc

F. Polygenic Inheritance (p. 270)

Mendel's characters could be classified on an either-or basis, such as purple versus white flower. Many characters, however, are *quantitative characters* that vary in a continuum within a population.

<u>Quantitative characters</u> = Characters that vary by degree in a continuous distribution rather than by discrete (either-or) qualitative differences.
  ☞ Usually, continuous variation is determined not by one, but by many segregating loci or *polygenic inheritance*.

What is polygenic inheritance? Why are most polygenic characters continuous or quantitative?

<u>Polygenic inheritance</u> = Mode of inheritance in which the additive effect of two or more genes determines a single phenotypic character.

For example, skin pigmentation in humans appears to be controlled by at least three separately inherited genes. The following is a simplified model for the polygenic inheritance of skin color:
  ☞ Three genes with the dark-skin allele (A, B, C) contribute one "unit" of darkness to the phenotype. These alleles are incompletely dominant over the other alleles (a, b, c).
  ☞ An AABBCC person would be very dark and an aabbcc person would be very light.
  ☞ An AaBbCc person would have skin of an intermediate shade.
  ☞ Because the alleles have a cumulative effect, genotypes AaBbCc and AABbcc make the same genetic contribution (three "units") to skin darkness. (See Campbell, Figure 13.13)
  ☞ Environmental factors, such as sun exposure, could also affect the phenotype.

G. Nature Versus Nurture: The Environmental Impact on Phenotype (*p. 270–271*)

How can environmental conditions influence the phenotypic expression of a character?

Environmental conditions can influence the phenotypic expression of a gene, so that a single genotype may produce a range of phenotypes. This environmentally-induced phenotypic range is the *norm of reaction* for the genotype.

Norm of reaction = Range of phenotypic variability produced by a single genotype under various environmental conditions. Norms of reaction for a genotype:
- May be quite limited, so that a genotype only produces a specific phenotype, such as the blood group locus that determines **ABO** blood type.
- May also include a wide range of possibilities. For example, an individual's blood cell count varies with environmental factors such as altitude, activity level or infection.
- Are generally broadest for polygenic characters, including behavioral traits.

The expression of most polygenic traits, such as skin color, is *multifactorial*; that is, it depends upon many factors — a variety of possible genotypes, as well as a variety of environmental influences.

H. Integrating a Mendelian View of Heredity and Variation (*p. 272*)

These patterns of inheritance that are departures from Mendel's original description, can be integrated into a comprehensive theory of Mendelian genetics.
- Taking a holistic view, an organism's entire phenotype reflects its overall genotype and unique environmental history.
- Mendelism has broad applications beyond its original scope; extending the principles of segregation and independent assortment helps explain more complex hereditary patterns such as epistasis and quantitative characters.

III. MENDELIAN INHERITANCE IN HUMANS (*p. 272–277*)

Mendelian inheritance in humans is difficult to study because:
- The human generation time is about 20 years.
- Humans produce relatively few offspring compared to most other species.
- Well-planned breeding experiments are impossible.

A. Human Pedigrees (*p. 272–273*)

How are family pedigrees used to deduce inheritance patterns and to predict the occurrence of a trait in future generations?

Our understanding of Mendelian inheritance in humans is based on the analysis of family *pedigrees* or the results of matings that have already occurred.

Mendel and the Gene Idea 179

<u>Pedigree</u> = A family tree that diagrams the relationships among parents and children across generations and that shows the inheritance pattern of a particular phenotypic character.

For example, wooly hair results from a dominant allele W, thus homozygous dominant (WW) and heterozygous (Ww) individuals have this trait. (See Campbell, Figure 13.15)

- ☞ Homozygous recessive (ww) individuals have normal hair.
- ☞ The man at the top of the pedigree has normal hair, so his genotype is ww.
- ☞ His wife had wooly hair, but must be a heterozygote (Ww) since three of their six children had normal hair (if she had been homozygous, all the children would have had wooly hair).

A pedigree not only allows a geneticist to understand the past, but to predict the occurrence of a trait in future generations.

- ☞ If a grandson with wooly hair marries a woman with non-wooly hair and they plan on having three children, the probability that all three children will have wooly hair is 1/8.
- ☞ Since the man is heterozygous (Ww) and the wife is homozygous (ww), each child has a 1/2 probability of inheriting the wooly allele.
- ☞ Using the rule of multiplication, the probability of these independent events occurring together is: 1/2 × 1/2 × 1/2 = 1/8.

This type of analysis is important to geneticists and physicians, especially when the trait being analyzed can lead to a disabling or lethal disorder.

B. Recessively Inherited Disorders (p. 273–274)

A recessive allele that causes a disorder is usually a defective version of the normal allele.

- ☞ Defective alleles that cause disorders code for either a malfunctional protein or no protein at all.
- ☞ Heterozygotes can be phenotypically normal, if one copy of the normal allele is all that is needed to produce sufficient quantities of the specific protein.

How are recessive alleles that cause disorders inherited? What role do carriers play in transmitting recessive disorders to offspring?

Recessively inherited disorders range in severity from nonlethal traits (e.g. albinism) to lethal diseases (e.g. cystic fibrosis). Since these disorders are caused by recessive alleles:

- ☞ The phenotypes are expressed only in homozygotes (aa) who inherit one recessive allele from each parent.
- ☞ Heterozygotes (Aa) can be phenotypically normal and act as *carriers*, possibly transmitting the recessive allele to their offspring.

Pedigree

Carriers

The vast majority of people afflicted with recessive disorders are born to normal parents, both of whom are carriers.

☞ The probability is 1/4 that a mating of two carriers (Aa × Aa) will produce a homozygous recessive zygote.
☞ The probability is 2/3 that a normal child from such a mating will be a heterozygote, or a carrier.

Why are some genetic disorders found more frequently in particular cultural groups? How are the recessively inherited disorders cystic fibrosis, Tay-Sachs disease and sickle-cell anemia, inherited and expressed?

Genetic disorders are not usually distributed evenly among all racial and cultural groups due to the different genetic histories of the world's people. Three examples of such recessively inherited disorders are *cystic fibrosis*, *Tay-Sachs disease* and *sickle-cell anemia*.

Cystic fibrosis is the most common lethal genetic disease in the United States, striking 1 in every 2,500 Caucasians (it is much rarer in other races).

☞ Four percent of the Caucasian population are carriers.
☞ The dominant allele codes for a membrane protein that pumps chloride ions out of cells. These pumps are lacking or are defective in recessive homozygotes, so chloride accumulates abnormally in the cells causing the osmotic uptake of water from the surrounding mucus. Disease symptoms result as the thickened mucus builds up in the pancreas, lungs, digestive tract, and other organs.

Tay-Sachs disease occurs in 1 out of 3,600 births. The incidence is about 100 times higher among Ashkenazic (central European) Jews than among Sephardic (Mediterranean) Jews and non-Jews.

☞ Brain cells of babies with this disease are unable to metabolize gangliosides (a type of lipid), because a crucial enzyme does not function properly.
☞ As lipids accumulate in the brain, the infant begins to suffer seizures, blindness and degeneration of motor and mental performance. The child usually dies after a few years.

Sickle-cell anemia is the most common inherited disease among African-Americans. It affects 1 in 400 African-Americans born in the United States.

☞ The disease is caused by a single amino acid substitution in hemoglobin.
☞ The abnormal hemoglobin molecules tend to link together and crystallize, especially when blood oxygen content is lower than normal. This causes red blood cells to deform from the normal disk-shape to a sickle-shape.
☞ The sickled cells clog tiny blood vessels, causing the pain and fever characteristic of a sickle-cell crisis.

Using sickle-cell anemia as an example, how can a lethal recessive gene be maintained in a population?

About 1 in 10 African-Americans are heterozygous for the sickle-cell allele and are said to have the *sickle-cell trait*.

☞ These carriers are usually healthy, although some suffer symptoms after an extended period of low blood oxygen levels.

- ☞ Carriers can function normally because the two alleles are codominant (heterozygotes produce not only the abnormal hemoglobin but also normal hemoglobin).
- ☞ The high incidence of heterozygotes is related to the fact that in tropical Africa where malaria is endemic, heterozygotes have enhanced resistance to malaria compared to normal homozygotes. Thus, heterozygotes have an advantage over both homozygotes — those who have sickle cell anemia and those who have normal hemoglobin.

Consanguinity

Why does consanguinity increase the probability of homozygosity in offspring? Why is this cause for concern?

The probability of inheriting the same rare harmful allele from both parents, is greater if the parents are closely related.

Consanguinity = A genetic relationship that results from shared ancestry.
- ☞ Consanguineous matings are more likely to produce offspring homozygous for a harmful recessive trait, since individuals with recent common ancestors have a greater possibility of inheriting the same recessive alleles than do unrelated persons.
- ☞ The extent to which human consanguinity increases the incidence of inherited diseases is still debated by geneticists. Embryos homozygous for deleterious mutations are affected so severely that most are spontaneously aborted before birth.
- ☞ Most cultures have laws or taboos which prevent marriage between closely related adults. This may be the result of observations that stillbirths and birth defects are more common when parents are closely related.
- ☞ Consanguinity may not always be harmful. Marriage between close relatives in some human populations, show no observable ill effect (e.g. Tamils of India).

C. **Dominantly Inherited Disorders** (p. 274–275)

Some human disorders are dominantly inherited.
- ☞ For example, *achondroplasia* (a type of dwarfism) affects 1 in 10,000 people who are heterozygous for this gene (homozygous dominant condition results in spontaneous abortion of the fetus).
- ☞ Since 99.99% of the population are homozygous recessive for this gene (normal phenotype), it illustrates well that an allele's dominance or recessiveness does not effect its prevalence in a population.

Why are lethal dominant genes much more rare than lethal recessive genes?

Lethal dominant alleles are much less common than lethal recessive alleles and usually result from new mutations in a gene of the egg or sperm.
- ☞ Such mutations usually kill the developing embryo.
- ☞ Also, the effects of lethal dominant alleles are not masked in the heterozygotes, who usually exhibit the disorder and do not survive to maturity and reproduce.

Consanguinity

Lethal dominant genes

Mendel and the Gene Idea

Late-acting lethal dominants

Huntington's disease

What is an example of a late-acting lethal dominant in humans? How has it escaped elimination?

Late-acting lethal dominants can escape elimination if the disorder does not appear until an advanced age (after afflicted individuals may have transmitted the lethal gene to their children).

- *Huntington's disease*, a degenerative disease of the nervous system, is caused by a late-acting lethal dominant allele. The phenotypic effects do not appear until 35 to 40 years of age. It is irreversible and lethal once the deterioration of the nervous system begins.
- Children of an afflicted parent have a 50% chance of inheriting the lethal dominant allele. New techniques of DNA analysis may detect the gene before the onset of symptoms.

D. Genetic Screening and Counseling (p. 275–277)

How do genetic counselors deduce the risk that prospective parents will have a child with an inherited disease?

Genetic counselors in many hospitals can provide information to prospective parents concerned about a family history for a genetic disorder.

- This preventative approach involves assessing the risk that a particular genetic disorder will occur.
- Risk assessment includes studying the family history for the disease using Mendel's law of segregation to deduce the risk.

For example, a couple is planning to have a child, and both the man and woman had siblings who died from the same recessively inherited disorder. A genetic counselor could deduce the risk of their first child inheriting the disease by using the laws of probability:

Question: What is the probability that the husband and wife are each carriers?

Answer: The genotypic ratio from an Aa × Aa cross is 1 AA:2 Aa:1 aa. Since the parents are normal, they have a 2 out of 3 chance of being carriers.

Question: What is the chance of two carriers having a child with the disease?

Answer: 1/2 (mother's chance of passing on the gene) × 1/2 (father's chance of passing on the gene) = 1/4

Question: What is the probability that their firstborn will have the disorder?

Answer: (Chance that the father is a carrier) × (chance that mother is a carrier) × (chance of two carriers having a child with the disease).

2/3 × 2/3 × 1/4 = 1/9

If the first child is born with the disease, what is the probability that the second child will inherit the disease?

- If the first child is born with the disease, then it is certain that both the man and the woman are carriers. Thus, the probability that other children produced by this couple will have the disease is 1/4.
- The conception of each child is an independent event, because the genotype of one child does not influence the genotype of the other children. So there is a 1/4 chance that any additional child will inherit the disease.

How are carrier recognition, fetal testing and newborn screening used in genetic screening and counseling?

Carrier Recognition

Several tests are available to determine if prospective parents are carriers of genetic disorders.

- ☞ Tests are currently available that can determine heterozygous carriers for the Tay-Sachs allele, cystic fibrosis, and sickle-cell trait.
- ☞ Tests such as these enable people to make informed decisions about having children, but they also could be abused. Ethical dilemmas about how this information could be used points to the immense social implications of such technological advances.

Fetal Testing

A couple that learns they are both carriers for a genetic disease and decide to have a child can determine if the fetus has the disease between the fourteenth and sixteenth weeks of pregnancy.

This test can be done in conjunction with an *amniocentesis*.

Amniocentesis

- ☞ During amniocentesis, a physician inserts a needle into the uterus and extracts about 10 milliliters of amniotic fluid.
- ☞ The presence of certain chemicals in the amniotic fluid are indicative of some genetic disorders.
- ☞ Tests for other disorders (including Tay-Sachs) are performed on cells grown in culture from the fetal cells sloughed off in the amniotic fluid. These cells can also be karyotyped to identify chromosomal defects.

Chorionic villi sampling is a newer technique during which a physician suctions off a small amount of fetal tissue from the chorionic villi of the placenta.

Chorionic villi sampling

- ☞ These cells are embryonic in origin and are proliferating rapidly.
- ☞ Enough cells obtained in this manner are undergoing mitosis. This allows karyotyping to be carried out immediately, usually providing results within 24 hours. The rapidity of this method is a major advantage over amniocentesis; culturing and karyotyping cells obtained by amniocentesis may take several weeks.
- ☞ Another advantage of chorionic villi sampling is that it can be performed at only 8 to 10 weeks of pregnancy.

Other techniques such as *ultrasound* and *fetoscopy* allow physicians to examine a fetus for major abnormalities.

Ultrasound, Fetoscopy

- ☞ Ultrasound is a non-invasive procedure which produces no known risk to fetus or mother.
- ☞ Fetoscopy involves inserting a thin fiber-optic scope into the uterus.
- ☞ In about 1% of the cases, amniocentesis or fetoscopy causes complications such as maternal bleeding or fetal death. Due to this risk, these techniques are usually reserved for cases where the risk of a genetic disorder or other type of birth defect is relatively high.

184 *Mendel and the Gene Idea*

PKU

Multifactorial disorders

<u>Newborn Screening</u>

In most U.S. hospitals, simple tests are routinely performed at birth, which can detect genetic disorders such as *phenylketonuria* (PKU), a recessively inherited disorder:
- ☞ PKU occurs in about 1 in 15,000 births in the United States.
- ☞ Children with this disease cannot properly break down the amino acid phenylalanine.
- ☞ Phenylalanine and its by-product (phenylpyruvic acid) can accumulate in the blood to toxic levels, causing mental retardation.
- ☞ Fetal screening for PKU can detect the deficiency in a newborn and retardation can be prevented with a special diet (low in phenylalanine) that allows normal development.

E. Multifactorial Disorders (*p. 277*)

Not all hereditary diseases are simple Mendelian disorders; that is, diseases caused by the inheritance of certain alleles at a single locus. More commonly, people are afflicted by *multifactorial* disorders, diseases that have both genetic and environmental influences.
- ☞ Examples include heart disease, diabetes, cancer, alcoholism and some forms of mental illness.
- ☞ The hereditary component is often polygenic and poorly understood.
- ☞ The best public-health strategy is to educate people about the role of environmental and behavioral factors that influence the development of these diseases.

14

THE CHROMOSOMAL BASIS OF INHERITANCE

CHAPTER OUTLINE

I. THE CHROMOSOMAL THEORY OF INHERITANCE (p. 280–284)

How were the observations of cytologists and geneticists combined to provide the basis for the chromosome theory of inheritance?

Genetics	Cytology
1860's: Mendel proposed that discrete inherited factors segregate and assort independently during gamete formation	
	1875: Cytologists worked out process of mitosis
	1890: Cytologists worked out process of meiosis
1900: Three botanists (Correns, von Seysenegg and de Vries) independently rediscovered Mendel's principles of segregation and independent assortment	

1902: Cytology and genetics converged as Walter Sutton, Theodor Boveri and others noticed parallels between the behavior of Mendel's factors and the behavior of chromosomes. For example:
 ☞ Chromosomes and genes are both paired in diploid cells.
 ☞ Homologous chromosomes separate and allele pairs segregate during meiosis.
 ☞ Fertilization restores the paired condition for both chromosomes and genes.

Based upon these observations, biologists developed the *chromosome theory of inheritance*. According to this theory:
 ☞ Mendelian factors or genes are located on chromosomes.
 ☞ It is the chromosomes that segregate and independently assort.

Walter Sutton

A. Morgan and the *Drosophila* School (p. 280–283)

What contributions did Thomas Hunt Morgan and those of the Drosophila school make to current understanding of chromosomal inheritance?

Thomas Hunt Morgan from Columbia University performed experiments in the early 1900's which provided convincing evidence that Mendel's inheritable factors are located on chromosomes.

Why is Drosophila melanogaster a good experimental organism?

Morgan selected the fruit fly, *Drosophila melanogaster*, as the experimental organism because these flies:

- ☞ Are easily cultured in the laboratory.
- ☞ Are prolific breeders.
- ☞ Have a short generation time.
- ☞ Have only four pairs of chromosomes which are easily seen with a microscope.

There are three pairs of autosomes (II, III and IV) and one pair of sex chromosomes. Females have two X chromosomes, and males have one X and one Y chromosome.

Tracing a Gene to a Specific Chromosome:

Wild type = Normal or most frequently observed phenotype.

Mutant phenotypes = Phenotypes that are alternatives to the wild type and which are due to mutations in the wild-type gene.

After a year of breeding *Drosophila* to find variant phenotypes, Morgan discovered a single male fly with white eyes instead of the wild-type red. Morgan mated this mutant white-eyed male with a red-eyed female. The cross is outlined below.

w = white-eye allele

w^+ = red-eye or wild-type allele

} Drosophila geneticists symbolize a recessive mutant allele with one or more lowercase letters. The corresponding wild-type allele has a superscript plus sign.

P generation:

w^+w^+ × w
red-eyed ♀ × white-eyed ♂

F₁ generation:

w^+w × w^+
red-eyed ♀ × red-eyed ♂

} The fact that all the F₁ progeny had red eyes, suggested that the wild-type allele was dominant over the mutant allele.

F₂ generation:

w^+w^+ red-eyed ♀ ww^+ red-eyed ♀ ⎫
w^+ red-eyed ♂ w white-eyed ♂ ⎬ White-eyed trait was expressed only in the male, and all the F₂ females had red eyes.

How did Morgan deduce that the Drosophila eye color gene is located on the X chromosome?

Morgan deduced that eye color is linked to sex and that the gene for eye color is located only on the X chromosome. Premises for his conclusions were:
- ☞ If eye color is located only on the X chromosome, then females (XX) carry two copies of the gene, while males (XY) have only one.
- ☞ Since the mutant allele is recessive, a white-eyed female must have that allele on both X chromosomes which was impossible for F₂ females in Morgan's experiment.
- ☞ A white-eyed male has no wild-type allele to mask the recessive mutant allele, so a single copy of the mutant allele confers white eyes.

<u>Sex-linked genes</u> = Genes located on sex chromosomes. The term is commonly applied only to genes on the X chromosome.

B. **Linked Genes** (*p. 283–284*)

What are linked genes? Why does linkage interfere with independent assortment?

Genes located on the same chromosome tend to be *linked* in inheritance and do not assort independently.

<u>Linked genes</u> = Genes that are located on the same chromosome and that tend to be inherited together.
- ☞ Linked genes do not assort independently, because they are on the same chromosome and move together through meiosis and fertilization.
- ☞ Since independent assortment does not occur, a dihybrid cross following two linked genes will not produce an F₂ phenotypic ratio of 9:3:3:1.

T.H. Morgan and his students performed a dihybrid testcross between flies with autosomal recessive mutant alleles for black bodies and vestigial wings and wild-type flies heterozygous for both traits. (A more detailed description follows in a later section.)

b = black body vg = vestigial wings
b^+ = gray body vg^+ = wild-type wings

$b^+b\ vg^+vg$ × $bb\ vg\ vg$
gray, normal wings   black, vestigial wings

- ☞ Resulting phenotypes of the progeny did not occur in the expected 1:1:1:1 ratio for a dihybrid testcross.

<u>Sex-linked genes</u>

<u>Linked genes</u>

- A disproportionately large number of flies had the phenotypes of the parents: gray with normal wings and black with vestigial wings.
- Morgan proposed that these unusual ratios were due to linkage. The genes for body color and wing size are on the same chromosome and are usually inherited together.

II. THE CHROMOSOMAL BASIS OF RECOMBINATION (284–285)

Genetic recombination = The production of offspring with new combinations of traits different from those combinations found in the parents; results from the events of meiosis and random fertilization.

A. The Recombination of Unlinked Genes: Independent Assortment (p. 284–285)

Mendel discovered that some offspring from dihybrid crosses have phenotypes unlike either parent. An example is the following test cross between pea plants:

YY, Yy = yellow seeds   RR, Rr = round seeds
yy   = green seeds   rr   = wrinkled seeds

P generation:

YyRr   ×   yyrr
yellow round   green wrinkled

Testcross progeny:

¼ YyRr   ¼ yyrr   ⎫ Parental types
yellow, round green, wrinkled ⎭ (50%)

¼ yyRr   ¼ Yyrr   ⎫ Recombinant types
green, round yellow, wrinkled ⎭ (50%)

What are parental and recombinant phenotypes? What ratios of parental and recombinant phenotypes result from a dihybrid cross following two characters that are unlinked in inheritance?

Parental types = Progeny that have the same phenotype as one or the other of the parents.

Recombinants = Progeny whose phenotypes differ from either parent.

In this cross, seed shape and seed color are unlinked.
- One-fourth of the progeny have round yellow seeds, and one fourth have wrinkled green seeds. Therefore, one-half of the progeny are *parental types*.
- The remaining half of the progeny are *recombinants*. One fourth are round green and one fourth are wrinkled yellow — phenotypes not found in either parent.
- When half the progeny are recombinants, there is a 50% frequency of recombination.
- A 50% frequency of recombination usually indicates that the two genes are on different chromosomes, because it is the expected result if the two genes assort randomly.

☞ The genes for seed shape and seed color assort independently of one another because they are located on different chromosomes which randomly align during metaphase of meiosis I.

B. The Recombination of Linked Genes: Crossing Over (p. 285)

If genes are totally linked, some possible phenotypic combinations should not appear. Sometimes, however, the unexpected recombinant phenotypes do appear.

As described earlier, T.H. Morgan and his students performed the following dihybrid testcross between flies with autosomal recessive mutant alleles for black bodies and vestigial wings and wild-type flies heterozygous for both traits.

b = black body $\qquad vg$ = vestigial wings
b^+ = gray body $\qquad vg^+$ = wild-type wings

$$b^+b\,vg^+vg \qquad \times \qquad bb\,vg\,vg$$
gray, normal wings $\qquad\qquad$ black, vestigial wings

Phenotypes	Genotypes	Expected Results If Genes Are Unlinked	Expected Results If Genes Are Totally Linked	Actual Results
Black body, normal wings	$\dfrac{b\,vg^+}{b\,vg}$	575	—	206
Gray body, normal wings	$\dfrac{b^+vg^+}{b\,vg}$	575	1150	965
Black body, vestigial wings	$\dfrac{b\,vg}{b\,vg}$	575	1150	944
Gray body, vestigial wings	$\dfrac{b^+vg}{b\,vg}$	575	—	185

Recombination Frequency = $\dfrac{391 \text{ recombinants}}{2300 \text{ total offspring}} \times 100 = 17\%$

How can crossing over unlink genes?

Morgan's results from this diploid testcross showed that the two genes were neither unlinked nor totally linked.

☞ If the genes for wing type and body color were unlinked, they would assort independently, and the progeny would show a 1:1:1:1 ratio of all possible phenotypic combinations.

☞ If the genes were completely linked, the expected results from the testcross would be a 1:1 phenotypic ratio of parental types only.

☞ Morgan's testcross did not produce results consistent with unlinkage or total linkage. The high proportion of parental phenotypes suggested linkage between the two genes.

☞ Since 17% of the progeny were recombinants, the linkage must be incomplete. Morgan proposed that there must be some mechanism that occasionally breaks the linkage between the two genes.

☞ It is now known that the process of *crossing over* during meiosis accounts for the recombination of linked genes. The exchange of parts between homologous chromosomes breaks linkages in parental chromosomes and forms recombinants with new allelic combinations.

III. GENETIC MAPS BASED ON CROSSOVER DATA (p. 285–288)

Scientists used recombination frequencies between genes to *map* the sequence of linked genes on particular chromosomes.

Morgan's *Drosophila* studies showed that some genes are linked more tightly than others.

- For example, the recombination frequency between the *b* and *vg* loci is about 17%.
- The recombination frequency is only 9% between *b* and *cn*, a third locus on the same chromosome. (The cinnabar gene, *cn*, for eye color has a recessive allele causing "cinnabar eyes.")

How can crossover data (recombination frequencies) from experimental crosses be used to map the linear sequence of genes on a chromosome?

A.H. Sturtevant, one of Morgan's students, assumed that if crossing over occurs randomly, the probability of crossing over between two genes is directly proportional to the distance between them.

- Sturtevant used recombination frequencies between genes to assign them a linear position on a chromosome *map*.
- He defined one *map unit* as 1% recombination frequency. (Map units are now called *centimorgans*, in honor of Morgan.)

Using crossover data, a map may be constructed as follows:

1. Establish the relative distance between those genes farthest apart or with the highest recombination frequency.

   b ———————————— vg
   ←———— 17 ————→

Loci	Recombination Frequency	Approximate Map Units
b vg	17.0%	18.5*
cn b	9.0%	9.0
cn vg	9.5%	9.5

2. Determine the recombination frequency between the third gene (cn) and the first (b).

   cn ———— b
   ←— 9 —→

3. Consider the two possible placements of the third gene:

   ←— 9 —→
   cn ——— b ———————— vg
              ←——— 17 ———→

   ←— 9 —→
   b ——— cn ———————— vg
   ←——— 17 ———→

4. Determine the recombination frequency between the third gene (cn) and the second (vg) to eliminate the incorrect sequence.

   ←— 9 —→←— 9.5 —→
   b ——— cn ——— vg
   ←——— 17 ———→

So, the correct sequence is b — cn — vg.

* Note that there are actually 18.5 map units between b and vg. This is higher than that predicted from the recombination frequency of 17.0%. Because b and vg are relatively far apart, double crossovers occur between these loci and cancel each other out, leading us to underestimate the actual map distance.

Locus

Chromosome map

Map unit

Sturtevant and his coworkers extended this method to map other *Drosophila* genes in linear arrays. (See Campbell, Figure 14.9)

The crossover data allowed them to cluster the known mutations into four major linkage groups.
- ☞ Since *Drosophila* has four sets of chromosomes, this clustering of genes into four linkage groups was further evidence that genes are on chromosomes.
- ☞ If linked genes are so far apart on a chromosome that the recombination frequency is 50%, they are indistinguishable from unlinked genes that assort independently.
- ☞ Linked genes that are far apart can be mapped, if additional recombination frequencies can be determined between intermediate genes and each of the distant genes.

What information does cytological mapping provide that cannot be accurately determined by crossover maps?

Maps based on crossover data only give information about the relative position of linked genes on a chromosome. Another technique, *cytological mapping* pinpoints the actual location of genes and the real distance between them.
- ☞ *Cytological mapping* involves screening offspring for mutant phenotypes and associating mutants with chromosomal defects visible by direct microscopic examination.
- ☞ The location of loci derived from maps based on crossover data differs from the spacing derived from cytological mapping, because the frequency of crossing over is not the same for all chromosomal regions.

IV. **SEX CHROMOSOMES AND SEX-LINKED INHERITANCE: A CLOSER LOOK** (*p. 288–290*)

In most species, sex is determined by the presence or absence of special chromosomes. As a result of meiotic segregation, each gamete has one sex chromosome to contribute at fertilization.

Heterogametic sex = The sex that produces two kinds of gametes and determines the sex of the offspring.

Homogametic sex = The sex that produces one kind of gamete.

   A.   **The Chromosomal Basis of Sex in Humans** (*p. 288–289*)

What is the chromosomal basis for sex determination in humans? Which sex is heterogametic in humans?

Mammals, including humans, have an X-Y mechanism that determines sex at fertilization.
- There are two chromosomes, X and Y. Each gamete has one sex chromosome, so when sperm cell and ovum unite at fertilization, the zygote receives one of two possible combinations: XX or XY.
- Males are the heterogametic sex (XY). Half the sperm cells contain an X chromosome, while the other half contain a Y chromosome.
- Females are the homogametic sex (XX); all ova carry an X chromosome.

Whether an embryo develops into a male or female depends upon the presence of a Y chromosome.
- A British research team has identified a single gene, *Sry* (sex-determining region), on the Y chromosome that is responsible for triggering the complex series of events that lead to normal testicular development.
- *Sry* probably codes for a protein that regulates other genes.

B. Sex-Linked Disorders in Humans (p. 289–290)

Some genes on sex chromosomes play a role in sex determination, but these chromosomes also contain genes for other traits.

Sex-linked traits

In humans, the term *sex-linked traits* usually refers to X-linked traits.
- The human X-chromosome is much larger than the Y. Thus, there are more X-linked than Y-linked traits.
- Most X-linked genes have no homologous loci on the Y chromosome.
- Most genes on the Y chromosome not only have no X counterparts, but they encode traits found only in males (e.g. testis-determining factor).
- Examples of sex-linked traits in humans are color blindness, *Duchenne's muscular dystrophy* and hemophilia.

How are sex-linked genes inherited?

Fathers pass X-linked alleles to <u>only</u> and <u>all</u> of their daughters.
- Males receive their X chromosome only from their mothers.
- Fathers cannot, therefore, pass sex-linked traits to their sons.

Mothers can pass sex-linked alleles to both sons and daughters.

- Females receive two X chromosomes, one from each parent.
- Mothers pass on one X chromosome (either the maternal or paternal homologue) to every daughter and son.

```
        XX'    ×    XY
       carrier     normal
         ♀           ♂
              |
    ┌─────┬───┴───┬─────┐
   XX    X'X     XY    X'Y
 normal carrier normal expressed
   ♀      ♀      ♂    trait ♂
```

If a sex-linked trait is due to a recessive allele, a female will express the trait only if she is homozygous.

- Females have two X chromosomes, therefore they can be either homozygous or heterozygous for sex-linked alleles.
- There are fewer females with sex-linked disorders than males, because even if they have one recessive allele, the other dominant allele is the one that is expressed. A female that is heterozygous for the trait can be a carrier, but not show the recessive trait herself.
- A carrier that mates with a normal male will pass the mutation to half her sons and half her daughters.
- If a carrier mates with a male who has the trait, there is a 50% chance that each child born to them will have the trait, regardless of sex.

```
        XX'    ×    X'Y
       carrier    expressed
         ♀        trait ♂
              |
    ┌─────┬───┴───┬─────┐
   XX'   X'X'    XY    X'Y
 carrier expressed normal expressed
   ♀    trait ♀    ♂    trait ♂
```

Why is a recessive sex-linked gene always expressed in human males?

Because males have only one X-linked locus, any male receiving a mutant allele from his mother will express the trait.

- Far more males than females have sex-linked disorders.
- Males are said to be *hemizygous*.

Hemizygous = A condition where only one copy of a gene is present in a diploid organism.

C. **X-Inactivation in Females** (*p. 290*)

How does an organism compensate for the fact that some individuals have a double dosage of sex-linked genes while others have only one?

In female mammals, most diploid cells have only one fully functional X chromosome.

- The explanation for this process is known as the *Lyon hypothesis*, proposed by the British geneticist Mary F. Lyon.
- In females, each of the embryonic cells inactivates one of the two X chromosomes.
- The inactive X chromosome contracts into a dense object called a *Barr body*.

Hemizygous

Lyon hypothesis

Mary F. Lyon

194 The Chromosomal Basis of Inheritance

Barr body

Barr body = A densely staining object inside the nuclear envelope, that is an inactivated X chromosome in female mammalian cells.

Female mammals are a *mosaic* of two types of cells — those with an active maternal X and those with an active paternal X.
- ☞ Which of the two Xs will be inactivated is determined randomly in embryonic cells.
- ☞ After an X is inactivated, all mitotic descendants will have the same inactive X.
- ☞ As a consequence, if a female is heterozygous for a sex-linked trait, about half of her cells will express one allele and the other cells well express the alternate allele.
- ☞ Examples of this type of mosaicism are coloration in calico cats and normal sweat gland development in humans.

V. CHROMOSOMAL ALTERATIONS (p. 290–295)

Meiotic errors and mutagens can cause major chromosomal changes such as altered chromosome numbers or altered chromosomal structure.

A. Alterations of Chromosome Number (p. 291–292)

How do changes in chromosome number occur, and what are the consequences?

Nondisjunction

Nondisjunction = Meiotic or mitotic accident during which certain homologous chromosomes or sister chromatids fail to separate.
- ☞ May occur during Meiosis I so that a homologous pair does not separate. (See Campbell, Figure 14.13a)
- ☞ May occur during Meiosis II when sister chromatids do not separate. (See Campbell, Figure 14.13b)
- ☞ Results in one gamete receiving two of the same type of chromosome and another gamete receiving no copy. The remaining chromosomes may be distributed normally.

Aneuploidy

Aneuploidy = Condition of having an abnormal number of certain chromosomes.
- ☞ Aneuploid offspring may result if a normal gamete unites with an aberrant one produced as a result of *nondisjunction*.
- ☞ An aneuploid cell that has a chromosome in triplicate is said to be *trisomic* for that chromosome.

Trisomic
- ☞ When an aneuploid zygote divides by mitosis, it transmits the chromosomal anomaly to all subsequent embryonic cells.
- ☞ Abnormal gene dosage in aneuploids causes characteristic symptoms in survivors. An example is Down's syndrome which results from trisomy of chromosome 21.

What is the difference between trisomy and triploidy?

Polyploidy = A chromosome number that is more than two complete chromosome sets.
- *Triploidy* is a polyploid chromosome number with three haploid chromosome sets (3N).
- *Tetraploidy* is polyploidy with four haploid chromosome sets (4N).
- Triploids may be produced by fertilization of an abnormal diploid egg produced by nondisjunction of all chromosomes.
- Tetraploidy may result if a diploid zygote undergoes mitosis without cytokinesis. Subsequent normal mitosis would produce a 4N embryo.
- Polyploidy is common in the plant kingdom and is important in plant evolution.
- Polyploids occur rarely among animals, and they are more normal in appearance than aneuploids. Mosaic polyploids, with only patches of polyploid cells, are more common than complete polyploid animals.

B. Alterations of Chromosome Structure (*p. 292–293*)

How is chromosomal structure altered by deletions, duplications, translocations and inversions?

Chromosome breakage can produce various rearrangements affecting the genes of that chromosome.
- Chromosomes which lose a fragment will have a deficiency or *deletion*.
- Fragments without centromeres are usually lost when the cell divides, or they may:
  1. Join to a homologous chromosome producing a *duplication*.
  2. Join to a nonhomologous chromosome (*translocation*).
  3. Reattach to the original chromosome in reverse order (*inversion*).

Crossing-over error is another source of deletions and duplications.
- Crossovers are normally reciprocal, but sometimes one sister chromatid gives up more genes than it receives in an unequal crossover.
- A nonreciprocal crossover results in one chromosome with a deletion and one chromosome with a duplication.

What are the effects of alterations in chromosome structure? What are the consequences of subtle position effects?

Alterations of chromosome structure, can have various effects:
- Homozygous deletions, including a single X in a male, are usually lethal.
- Duplications and translocations tend to have deleterious effects.
- Even if all genes are present in normal dosages, reciprocal translocations and inversions can alter the phenotype because of subtle *position effects*.

Position effect = Influence on a gene's expression because of its location among neighboring genes.

Polyploidy

Triploidy

Tetraploidy

Deletion

Duplication

Translocation

Inversion

Position effect

C. Chromosomal Alterations in Human Disease (p. 293–295)

What type of chromosomal alterations are implicated in the following human disorders: Down syndrome, Klinefelter syndrome, extra Y, metafemale, Turner syndrome, "cri du chat" syndrome and chronic myelogenous leukemia?

Chromosomal alterations are associated with some serious human disorders.

Aneuploidy, resulting from meiotic nondisjunction during gamete formation, usually prevents normal embryonic development and often results in spontaneous abortion.
- ☞ Some types of aneuploidy cause less severe problems, and individuals may be born with a set of characteristic symptoms or *syndrome*.
- ☞ Aneuploid conditions can be diagnosed before birth by *amniocentesis*.

Down syndrome is the most common serious birth defect in the United States.
- ☞ It is usually an extra chromosome 21 (trisomy 21).
- ☞ This syndrome includes characteristic facial features, short stature, heart defects and mental retardation. Individuals with Down syndrome are susceptible to respiratory infections and are prone to developing leukemia and Alzheimer's disease.
- ☞ Though most are sterile, a few women with Down syndrome have had children. There is a 50% chance of transmitting it to each child, since about half of the ova will have an extra chromosome 21.
- ☞ The incidence of Down syndrome offspring correlates with maternal age. This may be related to the long time lag between the first meiotic division during the mother's fetal life and the completion of meiosis at ovulation.

Other rarer diseases caused by autosomal aneuploidy are *Patau syndrome* (trisomy 13) and *Edwards syndrome* (trisomy 18).

Sex chromosome aneuploidies result in less severe conditions than those from autosomal aneuploidies. This may be because:
- ☞ The Y chromosome carries few genes.
- ☞ Copies of the X chromosome become inactivated as Barr bodies.

The basis of sex determination in humans is illustrated by sex chromosome aneuploidies.
- ☞ A single Y chromosome is sufficient to produce maleness.
- ☞ The absence of Y is required for femaleness.

Amniocentesis

Down syndrome

Examples of sex chromosome aneuploidy in males are:

Klinefelter Syndrome
- Genotype: Usually XXY, but may be associated with XXYY, XXXY, XXXXY, XXXXXY.
- Phenotype: Male sex organs with abnormally small testes; sterile; feminine body contours and perhaps breast enlargement; usually of normal intelligence.

Extra Y
- Genotype: XYY
- Phenotype: Normal male; usually taller than average.

Abnormalities of sex chromosome number in females include:

Metafemales
- Genotype: XXX.
- Phenotype: Limited fertility; possible mental retardation.

Turner Syndrome
- Genotype: XO (only known viable human monosomy).
- Phenotype: Short stature; at puberty, secondary sexual characteristics fail to develop; internal sex organs do not mature; sterile.

Specific disorders in humans are also associated with structural alterations of chromosomes.

Deletions in human chromosomes cause severe defects.
- ☞ This applies even in the heterozygous state.
- ☞ *Cri du chat* syndrome is caused by a deletion on chromosome 5. Symptoms are mental retardation, a small head with unusual facial features and a cry that sounds like a mewing cat.

Translocations are also associated with human disorders.
- ☞ Translocations have bene implicated in certain cancers such as *chronic myelogenous leukemia* (CML). A portion of chromosome 22 has switched places with a small fragment from chromosome 9.
- ☞ Some cases of Down syndrome are caused by a translocation of chromosome 21 attached to a chromosome 15. There are two normal chromosomes 21 plus the translocation.

VI. PARENTAL IMPRINTING OF GENES (*p. 295–296*)

The expression of some traits may depend upon which parent contributes the alleles for those traits.
- ☞ For example, two genetic disorders, *Prader-Willi syndrome* and *Angelman syndrome*, are caused by the same deletion on chromosome 15. The symptoms are different depending upon whether the gene was inherited from the mother or from the father.
- ☞ Prader-Willi syndrome is caused by a deletion from the paternal version of chromosome 15. The syndrome is characterized by mental retardation, obesity, short stature, and unusually small hands and feet.

☞ Angelman syndrome is caused by a deletion from the <u>maternal</u> version of chromosome 15. This syndrome is characterized by uncontrollable spontaneous laughter, jerky movements, and other motor and mental symptoms.
☞ The Prader-Willi/Angelman syndromes imply that the deleted genes normally behave differently in offspring, depending on whether they belong to the maternal or the paternal homologue.
☞ In other words, homologous chromosomes inherited from males and females are somehow differently *imprinted*, which causes them to be functionally different in the offspring.

What is genomic imprinting? What evidence supports this model?

<u>Genomic imprinting</u> = Process that induces intrinsic changes in chromosomes inherited from males and females; causes certain genes to be differently expressed in the offspring depending upon whether the alleles were inherited from the ovum or from the sperm cell.
☞ According to this hypothesis, certain genes are imprinted in some way each generation, and the imprint is different depending on whether the genes reside in females or in males.
☞ The same alleles may have different effects on offspring depending on whether they are inherited from the mother or the father.
☞ In the new generation, both maternal and paternal imprints can be reversed in gamete-producing cells, and all the chromosomes are recoded according to the sex of the individual in which they now reside.
☞ *DNA methylation* may be one mechanism for genomic imprinting. (See chapter 18 for futher discussion.)

Fragile-X syndrome is another disorder that may be evidence for genomic imprinting.
☞ Mental retardation is one symptom of this inherited disorder that affects about one in every 1000 males and one in every 2000 females.
☞ Fragile-X syndrome is caused by an abnormal X chromosome, the tip of which hangs on to the rest of the chromosome by a thin thread.
☞ Maternal imprinting of the damaged X chromosome appears to be involved in the expression of the disorder, since it is more likely to appear if the abnormal chromosome is inherited from the mother rather than from the father.
☞ Fragile-X syndrome is more common in males than in females. If a male (XY) inherits a fragile-X chromosome, it has to be from his mother. Whereas, if a female XX inherits the abnormal chromosome, it may be inherited from either her mother or father.

Uniparental disomy, the inheritance of two homologous chromosomes from one parent, may also be additional proof for genomic imprinting.
☞ Normal development evidently requires that both maternal and paternal versions of imprinted chromosomes are inherited.
☞ About 40% of the individuals with Prader-Willi syndrome have two normal copies of chromosome 15, but <u>both</u> have been inherited from the mother.
☞ Such uniparental disomy probably results from meiotic nondisjunction of chromosome 15 in both gametes; an ovum with two copies of chromosome 15 unites with a sperm cell that lacks chromosome 15.

VII. EXTRANUCLEAR INHERITANCE (p. 296–297)

What are some exceptions to the chromosome theory of inheritance? Why are cytoplasmic genes not inherited in a Mendelian fashion?

There are some exceptions to the chromosome theory of inheritance.
- There are extranuclear genes found in cytoplasmic organelles such as plastids and mitochondria.
- These cytoplasmic genes are not inherited in Mendelian fashion because they are not distributed by segregating chromosomes during meiosis.
- In plants, a zygote receives all its cytoplasm from the egg cell, and none from the sperm. Consequently, offspring receive only maternal cytoplasmic genes.
- Cytoplasmic genes in plants were first described by Karl Corens (1909) when he noticed that plant coloration of an ornamental species was determined by the seed bearing plants and not by the pollen producing plants.
- Maternal inheritance of mitochondrial genes also occurs in mammals.

15 THE MOLECULAR BASIS OF INHERITANCE

CHAPTER OUTLINE

Deoxyribonucleic acid or DNA is the genetic material — Mendel's heritable factors and Morgan's genes on chromosomes. Inheritance has its molecular basis in the precise replication and transmission of DNA from parent to offspring.

I. THE SEARCH FOR THE GENETIC MATERIAL (p. 300–303)

Why did researchers originally think that protein is the genetic material?

By the 1940's, scientists knew that chromosomes carry hereditary material and consist of DNA and protein. Most researchers thought protein was the genetic material because:
- ☞ Proteins are macromolecules with great heterogeneity and functional specificity.
- ☞ Little was known about nucleic acids.
- ☞ The physical and chemical properties of DNA seemed too uniform to account for the multitude of inherited traits.

A. Evidence That DNA Can Transform Bacteria (p. 301–302)

How did the results of Griffith's experiments hint that protein is not the genetic material?

Frederick Griffith

In 1928, Frederick Griffith performed experiments which provided evidence that genetic material is a specific molecule.

Griffith was trying to find a vaccine against *Streptococcus pneumoniae*, a bacterium that causes pneumonia in mammals. He knew that:
- ☞ There are two distinguishable strains of the pneumococcus: one produces smooth colonies (S) and the other rough colonies (R).
- ☞ Cells of the smooth strain are encapsulated with a polysaccharide coat and cells of the rough strain are not.
- ☞ These alternative phenotypes (S and R) are inherited.

Griffith performed four sets of experiments: (See Campbell, Figure 15.2)

Experiment: Griffith injected live S strain of *Streptococcus pneumoniae* into mice.
  Results: Mice died of pneumonia.
  Conclusions: Encapsulated strain is pathogenic.

Experiment: Mice were injected with live R strain.
  Results: Mice survived and were healthy.
  Conclusions: The bacterial strain lacking the polysaccharide coat was non-pathogenic.

Experiment: Mice were injected with heat-killed S strain of pneumococcus.

Results: Mice survived and were healthy.

Conclusions: Polysaccharide coat does not cause pneumonia because it is still present in heat-killed bacteria which proved to be non-pathogenic.

Experiment: Heat-killed S cells mixed with live R cells were injected into mice.

Results: Mice developed pneumonia and died. Blood samples from dead mice contained live S cells.

Conclusions: R cells had acquired from the dead S cells the ability to make polysaccharide coats. Griffith cultured S cells from the dead mice. Since the dividing bacteria produced encapsulated daughter cells, he concluded that this newly acquired trait was inheritable. This phenomenon is now called *transformation*.

Transformation = The assimilation of external genetic material by a cell.

Transformation

What was the chemical nature of the transforming agent?
- ☞ Griffith was unable to answer this question, but other scientists continued the search.
- ☞ Griffith's experiments hinted that protein is not the genetic material. Heat denatures protein, yet it did not destroy the transforming ability of the genetic material in the heat-killed S cells.
- ☞ In 1944, after a decade of research, Oswald Avery, Maclyn McCarty and Colin MacLeod discovered that the transforming agent had to be DNA.
- ☞ The discovery by Avery and his coworkers was met with skepticism by other scientists, because they still believed protein was a better candidate for the genetic material.

B. Evidence That Viral DNA Can Program Cells (*p. 302*)

More evidence that DNA is the genetic material came from studies of *bacteriophages*.

Bacteriophage (phage) = Virus that infects bacteria.

Bacteriophage

In 1952, Alfred Hershey and Martha Chase discovered that DNA is the genetic material of a phage known as T2. They knew that T2:
- ☞ Is one of many phages to infect the enteric bacterium *Escherichia coli* (*E. coli*).
- ☞ Like other viruses, is little more than DNA enclosed by a protein coat.
- ☞ Can quickly reprogram an *E. coli* cell to produce T2 phages and release the viruses when the cell lyses.

Alfred Hershey, Martha Chase

How did Hershey and Chase determine which viral (phage) component — DNA or protein — reprograms host bacterial cells? What conclusions could they draw about the nature of hereditary material?

What Hershey and Chase did not know is which viral component — DNA or protein — was responsible for reprogramming the host bacterial cell. They answered this question by performing the following experiment: (See Campbell, Figure 15.3)

Experiment:

Step 1: Viral protein and DNA were tagged with different radioactive isotopes.
- *Protein Tagging*: T2 and *E. coli* were grown in media with radioactive sulfur (^{35}S) which incorporated only into the phage protein.
- *DNA Tagging*: T2 and *E. coli* were grown in media containing radioactive phosphorus (^{32}P) which was incorporated only into the phage DNA.

Step 2: Protein-labeled and DNA-labeled T2 phages were allowed to infect separate samples of nonradioactive *E. coli* cells.

Step 3: Cultures were agitated to shake loose phages that remained outside the bacterial cells.

Step 4: Mixtures were centrifuged forcing the heavier bacterial cells into a pellet on the bottom of the tubes. The lighter viruses remained in the supernatant.

Step 5: Radioactivity in the pellet and supernatant was measured and compared.

Results:

1. In tubes with *E. coli* infected with protein-labeled T2, most of the radioactivity was in the supernatant with viruses.
2. In tubes with *E. coli* infected with DNA-labeled T2, most of the radioactivity was in the pellet with the bacterial cells.
3. When the bacteria containing DNA-labeled phages were returned to culture medium, the bacteria released phage progeny which contained ^{32}P in their DNA.

Conclusions:

1. Viral proteins remain outside the host cell.
2. Viral DNA is injected into the host cell.
3. Injected DNA molecules cause cells to produce additional viruses with more viral DNA and proteins.
4. These data provided evidence that nucleic acids rather than proteins are the hereditary material.

C. **Additional Evidence That DNA Is the Genetic Material of Cells** (p. 302–303)

Hershey and Chase's experiments provided evidence that DNA is the hereditary material in viruses. Additional evidence pointed to DNA as the genetic material in eukaryotes as well.

Some circumstantial evidence was:
- A eukaryotic cell doubles its DNA content prior to mitosis.
- During mitosis, the doubled DNA is equally divided between two daughter cells.
- An organism's diploid cells have twice the DNA as its haploid gametes.

What are Chargaff's rules? What experimental evidence did Chargaff produce that supports DNA as the hereditary material in eukaryotes?

Experimental evidence for DNA as the hereditary material in eukaryotes came from the laboratory of Erwin Chargaff. In 1947, he analyzed the DNA composition of different organisms.

Erwin Chargaff

Using paper chromatography to separate nitrogenous bases, Chargaff reported the following:

1. DNA composition is species-specific.
   - The amount and ratios of nitrogenous bases vary from one species to another.
   - This source of molecular diversity made it more credible that DNA is the genetic material.
2. In every species he studied, there was a regularity in base ratios.
   - The number of adenine (A) residues approximately equaled the number of thymines (T), and the number of guanines (G) equaled the number of cytosines (C).
   - The A=T and G=C equalities became known later as *Chargaff's rules*. The explanation for these rules came with Watson and Crick's structural model for DNA.

Chargaff's rules

II. DISCOVERY OF THE DOUBLE HELIX (p. 303–306)

Reviewing what you learned about DNA in chapter 7, what are the three components of a nucleotide? What are the nitrogenous bases found in DNA? What is the difference between a pyrimidine and a purine? What kind of chemical bond connects the nucleotides of each DNA strand and what type of bond holds the two strands together?

By the 1950's, DNA was accepted as the genetic material, and the covalent arrangement in a nucleic acid polymer was well established. The three dimensional structure of DNA, however, was yet to be discovered. (See Campbell, Figure 15.4)

Among scientists working on the problem were:
  Linus Pauling – California Institute of Technology
  Maurice Wilkins and Rosalind Franklin – King's College in London
  James D. Watson (American) and Francis Crick – Cambridge University

Rosalind Franklin
James Watson, Francis Crick

204 The Molecular Basis of Inheritance

How did Watson and Crick deduce the structure of DNA? What evidence did they use?

James Watson went to Cambridge to work with Francis Crick who was studying protein structure with *X-ray crystallography*.

Watson saw an X-ray photo of DNA produced by Rosalind Franklin at King's College, London. Watson and Crick deduced from Franklin's X-ray data that:

a. DNA is a helix with a uniform width of 2 nm. This width suggested that it had two strands.

b. Purine and pyrimidine bases are stacked .34 nm apart.

c. The helix makes one full turn every 3.4 nm along its length.

d. There are ten layers of nitrogenous base pairs in each turn of the helix.

Watson and Crick built scale models of a double helix that would conform to the X-ray data and the known chemistry of DNA.

- ☞ One of their unsuccessful attempts placed the sugar-phosphate chains inside the molecule.
- ☞ Watson next put the sugar-phosphate chains on the outside which allowed the more hydrophobic nitrogenous bases to swivel to the interior away from the aqueous medium.
- ☞ Their proposed structure is a ladder-like molecule twisted into a spiral, with sugar-phosphate backbones as uprights and pairs of nitrogenous bases as rungs.
- ☞ The two sugar-phosphate backbones of the helix are *antiparallel*; that is, they run in opposite directions.

Antiparallel

What is the base-pairing rule? Why is it significant?

Watson and Crick finally solved the problem of DNA structure by proposing that there is a specific pairing between nitrogenous bases. After considering several arrangements, they concluded:

- ☞ To be consistent with a 2 nm width, a purine on one strand must pair with a pyrimidine on the other.
- ☞ Base structure dictates which pairs of bases can hydrogen bond. The base pairing rule is that adenine can only pair with thymine, and guanine with cytosine.

Purine

Pyrimidine

The Molecular Basis of Inheritance

Purines	Pyrimidines	Possible Base Pairs	Number of Hydrogen Bonds
Adenine (A)	Thymine (T)	A–T	2
Guanine (G)	Cytosine (C)	G–C	3

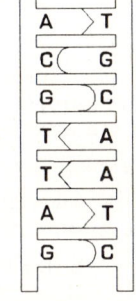

Adenine, Thymine

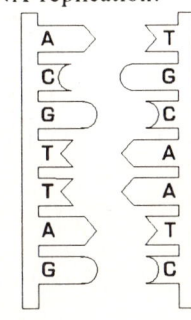

Guanine, Cytosine

The base-pairing rule is significant because:
- It explains Chargaff's rules. Since A must pair with T, their amounts in a given DNA molecule will be about the same. Similarly, the amount of G equals the amount of C.
- It suggests the general mechanisms for DNA replication. If bases form specific pairs, the information on one strand complements that along the other.
- It dictates the combination of complementary base pairs, but places no restriction on the linear sequence of nucleotides along the length of a DNA strand. The sequence of bases can be highly variable which makes it suitable for coding genetic information.
- Though hydrogen bonds between paired bases are weak bonds, collectively they stabilize the DNA molecule. Van der Waals forces between stacked bases also help stabilize DNA.

III. DNA REPLICATION: THE BASIC CONCEPT (p. 307–308)

In April 1953, Watson and Crick's new model for DNA structure, the double helix, was published in the British journal *Nature*. This model of DNA structure suggested a *template* mechanism for DNA replication.
- Watson and Crick proposed that genes on the original DNA strand are copied by a specific pairing of complementary bases, which creates a complementary DNA strand.
- The complementary strand can then function as a template to produce a copy of the original strand.

In a second paper, Watson and Crick proposed that during DNA replication:

1. The two DNA strands separate.

2. Each strand is a template for assembling a complementary strand.

Nucleotides

3. Nucleotides line up singly along the template strand in accordance with the base-pairing rules (A–T and G–C).

4. Enzymes link the nucleotides together at their sugar-phosphate groups.

What is *semiconservative* DNA replication?

Watson and Crick's model is a *semiconservative model* for DNA replication.
- They predicted that when a double helix replicates, each of the two daughter molecules will have one old or *conserved* strand from the parent molecule and one newly created strand.
- In the late 1950's, Matthew Meselson and Franklin Stahl provided the experimental evidence to support the semiconservative model of DNA replication. A brief description of the experimental steps follows.

Matthew Meselson, Franklin Stahl

What three alternate hypotheses did Meselson and Stahl consider to experimentally test the Watson and Crick model for DNA replication? How did their experimental results support the hypothesis that DNA replication is semiconservative?

Hypotheses:

There were three alternate hypotheses for the pattern of DNA replication:

Conservative replication
  a. If DNA replication is *conservative*, then the parental double helix should remain intact and the second DNA molecule should be constructed as entirely new DNA.

Semiconservative replication
  b. If DNA replication is *semiconservative*, then each of the two resulting DNA molecules should be composed of one original or conserved strand (template) and one newly created strand.

Dispersive replication
  c. If DNA replication is *dispersive*, then both strands of the two newly produced DNA molecules should contain a mixture of old and new DNA.

Experiment:

Step 1: Labeling DNA strands with ^{15}N.

  E. coli was grown for several generations in a medium with heavy nitrogen (^{15}N). As *E. coli* cells reproduced, they incorporated the ^{15}N into their nitrogenous bases.

Step 2: Transfer of *E. coli* to a medium with ^{14}N.

  E. coli cells grown in heavy medium were transferred to a medium with light nitrogen, ^{14}N. There were three experimental classes of DNA based upon times after the shift from heavy to light medium:

  a. Parental DNA from *E. coli* cells grown in heavy medium.
  b. First-generation DNA extracted from *E. coli* after one generation of growth in the light medium.
  c. Second-generation DNA extracted from *E. coli* after two replications in the light medium.

Step 3: Separation of DNA classes based upon density differences.

Meselson and Stahl used a new technique to separate DNA based on density differences due to differing proportions of ^{14}N and ^{15}N.

Isolated DNA was mixed with a CsCl solution and placed in an ultracentrifuge.

↓

DNA-CsCl solution was centrifuged at high speed for several days.

↓

Centrifugal force created a CsCl gradient with increased concentration toward the bottom of the centrifuge tube.

↓

DNA molecules moved to a position in the tube where their density equaled that of the CsCl solution. Heavier DNA molecules were closer to the bottom and lighter DNA molecules were closer to the top where the CsCl solution was less dense.

Results:

DNA from *E. coli* grown with ^{15}N was heavier than DNA containing the more common, lighter isotope, ^{14}N.	First-generation DNA after one generation of bacterial growth, was all of intermediate density.	Second generation DNA after two generations of bacterial growth in light medium was of intermediate and light density.

Conclusions:

Their results supported the hypothesis of semiconservative replication for DNA.

☞ The first generation DNA was all hybrid DNA containing one heavy parental strand and one newly synthesized light strand.

☞ These results were predicted by the semiconservative model. If DNA replication were conservative, two classes of DNA would be produced. The heavy parental DNA would be conserved and newly synthesized DNA would be light. There would be no intermediate density hybrid DNA.

☞ The results for first-generation DNA eliminated the possibility of conservative replication, but did not eliminate the possibility of dispersive replication.
☞ Consequently, Meselson and Stahl examined second-generation DNA. The fact that there were two bands, one light and one hybrid band supported the semiconservative model and allowed the investigators to rule out dispersive replication. If DNA replication were dispersive, there would have been only one band intermediate in density between light and hybrid DNA.

IV. A CLOSER LOOK AT DNA REPLICATION (p. 308–313)

The general mechanism of DNA replication is conceptually simple, but the actual process:
☞ Is complex. The helical molecule must untwist while it copies its two antiparallel strands simultaneously. This requires the cooperation of over a dozen enzymes and other proteins.
☞ Is extremely rapid. In prokaryotes, up to 500 nucleotides are added per second. It takes only a few hours to copy the 6 billion bases of a single human cell.
☞ Is accurate. Only about one in a billion nucleotides is incorrectly paired.

What are the major steps in the process of DNA replication? What are the roles of helicase, single-strand binding protein, DNA polymerase, DNA ligase and primase?

A. Getting Started: Origins of Replication (p. 309)

In many organisms DNA replication begins at special sites called *origins of replication* that have a specific sequence of nucleotides.
☞ Specific proteins required to initiate replication bind to each origin.
☞ The DNA double helix opens at the origin and *replication forks* spread in both directions away from the central initiation point creating a *replication bubble*.
☞ Bacterial or viral DNA molecules have only one replication origin.
☞ Eukaryotic chromosomes have hundreds or thousands of replication origins. The many replication bubbles formed by this process, eventually merge forming two continuous DNA molecules.

Replication forks = The Y-shaped regions of replicating DNA molecules where new strands are growing.

B. Elongating a New DNA Strand (p. 309–312)

1. Strand Separation

There are two types of proteins involved in the separation of parental DNA strands:
a. *Helicases* are enzymes which catalyze unwinding of the parental double helix to expose the template.
b. *Single-strand binding proteins* are proteins which keep the separated strands apart and stabilize the unwound DNA until new complementary strands can be synthesized.

2. Synthesis of the New DNA Strands

Enzymes called *DNA polymerases* catalyze synthesis of a new DNA strand.

- ☞ According to base-pairing rules, new nucleotides align themselves along the templates of the old DNA strands.
- ☞ *DNA polymerase* links the nucleotides to the growing strand. These strands grow in the 5' → 3' direction since new nucleotides are added only to the 3' end of the growing strand.

What is the energy source that drives the endergonic synthesis of DNA?

Hydrolysis of *nucleoside triphosphates* provides the energy necessary to synthesize the new DNA strands.

Nucleoside triphosphate = Nucleotides with a triphosphate (three phosphates) covalently linked to the 5' carbon of the pentose.
- ☞ Recall that the pentose in DNA is deoxyribose, and the pentose in RNA is ribose.
- ☞ Nucleoside triphosphates that are the building blocks for DNA, lose two phosphates (pyrophosphate group) when they form covalent linkages to the growing chain.
- ☞ It is the exergonic hydrolysis of this phosphate bond that drives the endergonic synthesis of DNA; it provides the required energy to form the new covalent linkages between nucleotides.

Why is continuous synthesis of both DNA strands not possible? What does antiparallel mean?

Continuous synthesis of both DNA strands at a replication fork is not possible, because:
- ☞ The sugar phosphate backbones of the two complementary DNA strands run in opposite directions; that is, they are *antiparallel*.
- ☞ DNA polymerase can only elongate strands in the 5' to 3' direction.

DNA polymerases

Nucleoside triphosphate
Deoxyribose
Ribose

The problem of antiparallel DNA strands is solved by the continuous synthesis of one strand (*leading strand*) and discontinuous synthesis of the complementary strand (*lagging strand*).

Which DNA strand is the leading strand and which strand is the lagging strand? How is the lagging strand synthesized when DNA polymerase can add nucleotides only to the 3' end?

Leading strand = The DNA strand which is synthesized as a single polymer in the 5' → 3' direction towards the replication fork.

Lagging strand = The DNA strand that is discontinuously synthesized against the overall direction of replication.

- ☞ Lagging strand is produced as a series of short segments called *Okazaki fragments* which are each synthesized in the 5' → 3' direction.
- ☞ Okazaki fragments are 1000–2000 nucleotides long in bacteria and 100 to 200 nucleotides long in eukaryotes.
- ☞ The many fragments are ligated by *DNA ligase*, a linking enzyme that catalyzes the formation of a covalent bond between the 3' end of each new Okazaki fragment to the 5' end of the growing chain.

3. Priming

Before new DNA strands can form, there must be small pre-existing *primers* to start the addition of new nucleotides.

Primer = Short RNA segment that is complementary to a DNA segment and that is necessary to begin DNA replication.
- ☞ Primers are short segments of RNA polymerized by an enzyme called *primase*.
- ☞ A portion of the parental DNA serves as a template for making the primer with a complementary base sequences that is about 10 nucleotides long in eukaryotes.
- ☞ Primer formation must precede DNA replication, because DNA polymerase can only add nucleotides to a polynucleotide that is already correctly base-paired with a complementary strand.

Only one primer is necessary for replication of the leading strand, but many primers are required to replicate the lagging strand.
- ☞ An RNA primer must initiate the synthesis of each Okazaki fragment.
- ☞ The many Okazaki fragments are ligated in two steps to produce a continuous DNA strand:
    1. A DNA polymerase removes the RNA primer and replaces it with DNA.
    2. DNA ligase catalyzes the linkage between the 3' end of each new Okazaki fragment to the 5' end of the growing chain.

C. Proofreading (p. 312)

DNA replication is highly accurate, but this accuracy is not solely the result of base-pairing specificity.
- ☞ Initial pairing errors occur at a frequency of about one in ten thousand, while errors in a complete DNA molecule are only about one in one billion.
- ☞ In bacteria, the initial base-pairing errors are usually corrected by DNA polymerase, which checks each newly added nucleotide against its template. If the polymerase detects an error, it removes the incorrect nucleotide and replaces it.

How do DNA polymerase, ligase and repair enzymes function in DNA proofreading and repair?

V. DNA REPAIR (p. 313)

Mechanisms for repairing DNA help maintain the integrity of the encoded genetic information.
- ☞ Accidental changes in DNA can result from exposure to reactive chemicals, radioactivity, X-rays and ultraviolet light.
- ☞ Cells continuously monitor such accidental changes and repair most.
- ☞ There are more than fifty different types of DNA repair enzymes that repair damage by:
    a. Directly reversing the change.
    b. *Excision repair.* The damaged segment is excised by one *repair enzyme* and the remaining gap is filled in by base-pairing nucleotides with the undamaged strand. *DNA polymerase* and *DNA ligase* are enzymes that catalyze the filling-in process.

Excision repair

16

FROM GENE TO PROTEIN

CHAPTER OUTLINE

Inherited instructions in DNA direct protein synthesis. Thus, proteins are the links between genotype and phenotype, since proteins are directly involved in the expression of specific phenotypic traits.

I. **EVIDENCE THAT GENES SPECIFY PROTEINS** (*p. 316–317*)

What early experimental evidence implicated proteins as the links between genotype and phenotype?

Archibald Garrod was the first to propose the relationship between genes and proteins (1909).
- He suggested that genes dictate phenotypes through enzymes that catalyze reactions.
- As a physician, Garrod was familiar with inherited diseases which he called "inborn errors in metabolism." He hypothesized that such diseases reflected the patient's inability to make particular enzymes.
- One example he studied was alkaptonuria in which the person's urine turns black due to the presence of alkapton which darkens on contact with air.
- Garrod reasoned that normal individuals produced an enzyme which broke down alkapton, while alkaptonuric individuals lacked this enzyme.

A. **How Genes Control Metabolism** (*p. 317*)

Garrod's hypothesis was confirmed several decades later by research which determined that the function of a gene was to direct production of a specific enzyme.
- Biochemists accumulated evidence that cells synthesize and degrade organic compounds via metabolic pathways, with each step in the sequence catalyzed by a specific enzyme.
- Geneticists George Beadle and Boris Ephrussi (1930's) studied eye color in *Drosophila*. They speculated that mutations affecting eye color block pigment synthesis by preventing enzyme production at certain steps in the pigment synthesis pathway.

What types of experiments did Beadle and Tatum perform with Neurospora? What were their experimental controls? How did they demonstrate the relationship between genes and enzymes?

George Beadle and Edward Tatum were later able to demonstrate the relationship between genes and enzymes by studying mutants of a bread mold, *Neurospora crassa*.

Archibald Garrod

George Beadle, Edward Tatum
Neurospora

- ☞ Wild-type *Neurospora* in laboratory colonies can survive on *minimal medium*. All other molecules needed by the mold are produced by its own metabolic pathways from this minimal nutrient source.
- ☞ Beadle and Tatum searched for mutants or *auxotrophs* that could not survive on minimal medium because they lacked the ability to synthesize essential molecules.
- ☞ Mutants were identified by transferring fragments of growing fungi (in complete medium) to vials containing minimal medium. Fragments that didn't grow were identified as auxotrophic mutants.

Auxotroph = (Auxo=to augment; troph=nourishment) Nutritional mutants that can only be grown on *minimal medium* that has been augmented with nutrients not required by the wild type.

Minimal medium = Support medium that is mixed only with molecules that are required for the growth of wild-type organisms.
- ☞ Minimal medium for *Neurospora* contains inorganic salts, sucrose and vitamin, biotin.
- ☞ Nutritional mutants cannot survive only on minimal medium.

Complete growth medium = Minimal medium supplemented with all 20 amino acids and some other nutrients.
- ☞ Nutritional mutants can grow on complete growth medium, since all essential nutrients are provided.

Beadle and Tatum then identified specific metabolic defects (from mutations) by transferring fragments of auxotrophic mutants growing on complete growth medium to vials containing minimal medium each supplemented with only one additional nutrient.
- ☞ Vials where growth occurred indicated the metabolic defect, since the single supplement provided the necessary component.
- ☞ For example, if a mutant grew on minimal medium supplemented with only arginine, it could be concluded that the mutant was defective in the arginine synthesis pathway.

Experiment:

Beadle and Tatum experimented further to more specifically describe the defect in the multistep pathway that synthesizes the amino acid arginine.
- ☞ Arginine synthesis requires three steps each catalyzed by a specific enzyme:

```
              Gene A           Gene B           Gene C
                ↓                ↓                ↓
              Enzyme A         Enzyme B         Enzyme C
Precursor ──────────→ Ornithine ──────→ Citrulline ──────→ Arginine
```

- ☞ They distinguished between three classes of arginine auxotrophs by adding either arginine, citrulline, or ornithine to the medium and seeing if growth occurred.

Results:

Some mutants required arginine, some either arginine or citrulline, and others could grow when any of the three were added.

	Minimal Medium (MM)	MM plus Ornithine	MM plus Citrulline	MM plus Arginine
Wild Type	+	+	+	+
Class I Mutants	−	+	+	+
Class II Mutants	−	−	+	+
Class III Mutants	−	−	−	+

+ = growth, − = no growth

Conclusions:

Beadle and Tatum deduced from their data that the three classes of mutants each lacked a different enzyme and were thus blocked at different steps in the arginine synthesis pathway.

☞ Class I mutants lacked enzyme A; Class II mutants lacked enzyme B; and Class III mutants lacked enzyme C.

☞ Assuming that each mutant was defective in a single gene, they formulated the *one gene–one enzyme* hypothesis, which states that the function of a gene is to dictate the production of a specific enzyme.

In your own words, what was the "one gene–one enzyme" hypothesis? Why was it later changed to "one gene–one polypeptide"?

B. One Gene–One Polypeptide (*p. 317*)

Beadle and Tatum's one gene–one enzyme hypothesis has been slightly modified:

☞ While most enzymes are proteins, many proteins are not enzymes. Proteins that are not enzymes are still, nevertheless, gene products.

☞ Also, many proteins are comprised of two or more polypeptide chains, each chain specified by a different gene.

As a result of this new information, Beadle and Tatum's hypothesis has been restated as *one gene–one polypeptide*.

II. AN OVERVIEW OF PROTEIN SYNTHESIS (*p. 317–319*)

Ribonucleic acid (RNA) links DNA's genetic instructions for making proteins to the process of protein synthesis. It copies or transcribes the message from DNA and then translates that message into a protein.

How does RNA differ from DNA?

☞ RNA, like DNA, is a nucleic acid or polymer of nucleotides.

☞ RNA structure differs from DNA in the following ways:

1. The five-carbon sugar in RNA nucleotides is *ribose* rather than deoxyribose.
2. The nitrogenous base *uracil* is found in place of thymine.

One gene–one enzyme

One gene–one polypeptide

How does information flow from gene to protein? In general terms, what happens during transcription and translation?

The linear sequence of nucleotides in DNA ultimately determines the linear sequence of amino acids in a protein.
- ☞ Nucleic acids are made of four types of nucleotides which differ in their nitrogenous bases. Hundreds or thousands of nucleotides long, each gene has a specific linear sequence of the four possible bases.
- ☞ Proteins are made of twenty types of amino acids linked in a particular linear sequence (the protein's primary structure).
- ☞ Information flows from gene to protein through two major processes, *transcription* and *translation*.

Transcription = The synthesis of RNA using DNA as a template.
- ☞ A gene's unique nucleotide sequence is transcribed from DNA to a complementary nucleotide sequence in messenger RNA (mRNA).
- ☞ The resulting mRNA carries this transcript of protein-building instructions to the cell's protein-synthesizing machinery.

Translation = Synthesis of a polypeptide, which occurs under the direction of messenger RNA (mRNA).
- ☞ During this process, the linear sequence of bases in mRNA is translated into the linear sequence of amino acids in a polypeptide.
- ☞ Translation occurs on *ribosomes*, complex particles composed of ribosomal RNA (rRNA) and protein that facilitate the orderly linking of amino acids into polypeptide chains.

Where do transcription and translation occur in prokaryotes and in eukaryotes? Why is it significant that, in eukaryotes, transcription and translation are separated in space and time?

Prokaryotes and eukaryotes differ in how protein synthesis is organized within their cells.
- ☞ Prokaryotes lack nuclei, so DNA is not segregated from ribosomes or the protein-synthesizing machinery. Thus, transcription and translation occur in rapid succession.
- ☞ Eukaryotes have nuclear envelopes that segregate transcription in the nucleus from translation in the cytoplasm; mRNA, the intermediary, is modified before it moves from the nucleus to the cytoplasm where translation occurs. This *RNA processing* occurs only in eukaryotes.

III. THE GENETIC CODE (p. 319–321)

There is not a one-to-one correspondence between the nitrogenous bases and the amino acids they specify, since there are only 4 nucleotides and 20 amino acids.
- ☞ A two-to-one correspondence of bases to amino acids would only specify 16 (4^2) of the 20 amino acids.
- ☞ A three-to-one correspondence of bases to amino acids would specify 64 (4^3) amino acids.

Transcription
Messenger RNA

Translation

Ribosomes
Ribosomal RNA

RNA processing

From Gene to Protein

What is the basis for information flow from gene to protein? What is a codon?

Researchers have verified that the flow of information from a gene to a protein is based on a triplet code.

- ☞ Triplets of nucleotides are the smallest units of uniform length to allow translation into all 20 amino acids with plenty to spare.
- ☞ These three-nucleotide "words" are called *codons*.

Codon = A three-nucleotide sequence in mRNA that specifies which amino acid will be added to a growing polypeptide or that signals termination; the basic unit of the genetic code.

Genes are not directly translated into amino acids, but are first transcribed as codons into mRNA.

- ☞ For each gene, only one of the two DNA strands (the *coding strand*) is transcribed.
- ☞ The *noncoding strand* is the template for making a new coding strand when DNA replicates.
- ☞ The same DNA strand can be the coding strand for some genes and the noncoding strand for others.

An mRNA is complementary to the DNA template from which it is transcribed.

- ☞ For example, if the triplet nucleotide sequence on the coding DNA strand is CCG; GGC, the codon for glycine, will be the complementary mRNA transcript.
- ☞ Recall that according to the base-pairing rules, uracil (U) in RNA is used in place of thymine (T); uracil thus base pairs with adenine (A).

What is the relationship between the linear sequence of codons on mRNA to the linear sequence of amino acids in a polypeptide?

During translation, the linear sequence of codons along mRNA is translated into the linear sequence of amino acids in a polypeptide.

- ☞ Each mRNA codon specifies which one of 20 amino acids will be incorporated into the corresponding position in a polypeptide.
- ☞ Because codons are base triplets, the number of nucleotides making up a genetic message is three times the number of amino acids making up the polypeptide product.

The first codon was deciphered in 1961 by Marshall Nirenberg of the National Institutes of Health.
- ☞ He synthesized an mRNA by linking only uracil-bearing RNA nucleotides, resulting in UUU codons.
- ☞ Nirenberg added this "poly U" to a test-tube mixture containing the components necessary for protein synthesis. The artificial mRNA (poly U) was translated into a polypeptide containing only one amino acid, phenylalanine.
- ☞ Nirenberg concluded that the mRNA codon UUU specifies the amino acid phenylalanine.
- ☞ These same techniques were used to determine amino acids specified by the codons AAA, GGG, and CCC.

What is the function of the codon, AUG? What are the three stop codons? How many codons are there? Which ones do not code for amino acids?

More elaborate techniques allowed investigators to determine all 64 codons by the mid 1960's. (See Campbell, Figure 16.5)
- ☞ 61 of the 64 triplets code for amino acids.
- ☞ The triplet AUG has a <u>dual</u> function: it provides a signal for the start of translation and codes for methionine.
- ☞ Three codons do not code for amino acids, but signal termination (UAA, UAG and UGA).

In what way is the genetic code redundant? In what way is it unambiguous?

There is redundancy in the genetic code, but no ambiguity.
- ☞ *Redundancy* exists in that two or more codons differing only in their third base, code for the same amino acid (UUU and UUC both code for phenylalanine).
- ☞ *Ambiguity* is absent, since no codon codes for more than one amino acid.

The correct ordering and grouping of nucleotides is important in the molecular language of cells. This ordering is called the *reading frame*.

<u>Reading frame</u> = The correct grouping of adjacent nucleotide triplets into codons that are in the correct sequence on mRNA.
- ☞ For example, the sequence of amino acids −Trp−Phe−Gly−Arg−Phe− can only be assembled in the correct order if the mRNA codons UGGUUUGGCCGUUUU are read in the correct sequence and groups.
- ☞ The cell reads the message in the correct frame as a series of <u>nonoverlapping</u> three-letter words: UGG−UUU−GGC−CGU−UUU.

IV. THE EVOLUTIONARY SIGNIFICANCE OF A COMMON GENETIC LANGUAGE (*p. 321−322*)

What is the evolutionary significance of a nearly universal genetic code?

The genetic code is shared nearly universally among living organisms.
- ☞ For example, the RNA codon CCG is translated into proline in all organisms whose genetic codes have been examined.

Redundancy

Reading frame

218 From Gene to Protein

☞ Experiments show bacteria can translate genetic messages from human cells and human cells can translate bacterial genes.

As with all things, commonality has exceptions.
- ☞ Several ciliates (including *Paramecium* and *Tetrahymena*) have a variation in the standard code. The RNA codons UAA and UAG are not stop signals, but code for the amino acid glutamine.
- ☞ Other exceptions include mitochondria and chloroplasts, which have their own DNA for certain genes and the ability to synthesize proteins.
- ☞ Mitochondrial genetic codes vary among organisms; for example, CUA codes for threonine in yeast mitochondria and leucine in mammalian mitochondria.

The fact that the genetic code is shared universally by all organisms indicates that this code was established very early in life's history.

V. A CLOSER LOOK AT TRANSCRIPTION (p. 322–324)

What is the role of RNA polymerase in transcription?

Transcription of messenger RNA (mRNA) from the coding DNA strand is catalyzed by enzymes called *RNA polymerases*.
- ☞ RNA polymerases separate the two DNA strands and link RNA nucleotides as they base pair along the DNA template.
- ☞ Like DNA polymerase, RNA polymerases can only add nucleotides to the 3' end; thus, mRNA molecules grow in the 5' to 3' direction. (See Campbell, Figure 15.15)
- ☞ Prokaryotes have only one type of RNA polymerase that synthesizes all types of RNA — mRNA, rRNA and tRNA.
- ☞ Eukaryotes have three RNA polymerases that transcribe genes. *RNA polymerase II* is the polymerase that catalyzes mRNA synthesis; it transcribes genes that will be translated into proteins.

Specific DNA nucleotide sequences mark where transcription of a gene begins (*initiation*) and ends (*termination*). Initiation and termination sequences plus the nucleotides in between are called a *transcription unit*.

<u>Transcription unit</u> = Nucleotide sequence on the coding strand of DNA that is transcribed into a single RNA molecule by RNA polymerase; it includes the initiation and termination sequences, as well as the nucleotides in between.
- ☞ In eukaryotes, a transcription unit contains a single gene, so the resulting mRNA codes for synthesis of only one protein.
- ☞ In prokaryotes, a transcription unit can contain several genes, so the resulting mRNA may code for different, but functionally related, proteins.

Transcription occurs in three key steps: a) polymerase binding and initiation; b) elongation; and c) termination.

RNA polymerases

Transcription unit

From Gene to Protein 219

A. RNA Polymerase Binding and Initiation of Transcription
(p. 323–324)

How does RNA polymerase recognize where transcription should begin?

RNA polymerases bind to DNA at regions called *promoters*.

<u>Promoter</u> = Region of DNA that includes the site where RNA polymerase binds and where transcription begins (*initiation site*). In eukaryotes, the promoter is about 100 nucleotides long and consists of:
   a. The initiation site, where transcription begins.
   b. A few nucleotide sequences recognized by specific DNA-binding proteins that help initiate transcription (*transcription factors*).

In eukaryotes, RNA polymerases cannot recognize the promoter without the help of *transcription factors*.

<u>Transcription factors</u> = DNA-binding proteins that bind to specific DNA nucleotide sequences at the promoter and that help RNA polymerase recognize and bind to the promoter region, so that transcription can begin.
   ☞ RNA polymerase II, the enzyme that synthesizes mRNA in eukaryotes, usually cannot recognize a promoter unless a specific transcription factor binds to a region on the promoter called a *TATA box*.

   <u>TATA box</u> = A short nucleotide sequence at the promoter that is rich in thymine (T) and adenine (A) and that is located about 25 nucleotides upstream from the initiation site.
   ☞ RNA polymerase II recognizes the complex between the bound TATA transcription factor and the DNA binding site.
   ☞ Once RNA polymerase recognizes and attaches to the promoter region, it probably associates with other transcription factors before RNA synthesis begins.

In prokaryotes, the RNA polymerase recognizes and binds to DNA by a different mechanism from eukaryotes.
   ☞ Bacterial RNA polymerase has a detachable subunit called sigma (σ) that recognizes the promoter.
   ☞ After RNA polymerase binds to the promoter on DNA and begins transcription, it loses its sigma subunit.

When active RNA polymerase binds to a promoter, the enzyme separates the two DNA strands at the initiation site, and transcription begins.

B. Elongation of the RNA Strand (p. 324)

What are the primary functions of RNA polymerase II?

Once transcription begins, RNA polymerase II moves along DNA and performs two primary functions:
   1. It untwists and opens a short segment of DNA exposing about ten nucleotide bases; one of the exposed DNA strands is the template for base pairing with RNA nucleotides.
   2. It links incoming RNA nucleotides to the 3' end of the elongating strand; thus, RNA grows one nucleotide at a time in the 5' to 3' direction.

Promoter

Transcription factors

TATA box

During transcription, mRNA grows about 30–60 nucleotides per second. As the mRNA strand elongates:
- ☞ It peels away from its DNA template.
- ☞ The noncoding strand of DNA re-forms a DNA–DNA double helix by pairing with the coding strand.

Following in series, several molecules of RNA polymerase II can simultaneously transcribe the same gene.
- ☞ Cells can thus produce particular proteins in large amounts.
- ☞ The growing RNA strands hang free from each polymerase. The length of each strand varies and reflects how far the enzyme has traveled from the initiation site on template DNA.

C. Termination of Transcription (p. 324)

Transcription proceeds until RNA polymerase reaches a *termination site* on the DNA.

<u>Terminator sequence</u> = DNA sequence that signals RNA polymerase to stop transcription and to release the RNA molecule and DNA template.
- ☞ Additional proteins may cooperate with RNA polymerase in termination.

Prokaryotic mRNA is ready for translation as soon as it leaves the DNA template. Eukaryotic mRNA, however, must be processed before it leaves the nucleus and becomes functional.

VI. A CLOSER LOOK AT TRANSLATION (p. 325–331)

Why is tRNA necessary for protein synthesis?

During translation, proteins are synthesized according to a genetic message in the form of sequential codons along mRNA.
- ☞ *Transfer RNA (tRNA)* is the interpreter between the two forms of information — base sequence in mRNA and amino acid sequence in polypeptides.
- ☞ tRNA aligns the appropriate amino acids to form a new polypeptide. To perform this function, tRNA must:
  1. Transfer amino acids from the cytoplasm's amino acid pool to a ribosome.
  2. Recognize the correct codons in mRNA.

Molecules of tRNA are specific for only one particular amino acid. Each type of tRNA associates a distinct mRNA codon with one of the 20 amino acids used to make proteins.
- ☞ One end of a tRNA molecule attaches to a specific amino acid.
- ☞ The other end attaches to an mRNA codon by base pairing with its *anticodon*.

<u>Anticodon</u> = A nucleotide triplet in tRNA that base pairs with a complementary nucleotide triplet (codon) in mRNA.

tRNAs decode the genetic message, codon by codon. For example,
- ☞ The mRNA codon UUU is translated as the amino acid phenylalanine. (See Campbell, Figure 16.5)

- ☞ The tRNA that transfers phenylalanine to the ribosome has an anticodon of AAA.
- ☞ When the codon UUU is presented for translation, phenylalanine will be added to the growing polypeptide.
- ☞ As tRNAs deposit amino acids in the correct order, ribosomal enzymes link them into a chain.

Given a sequence of bases in template DNA (3' ACCAAACCGGCAAAA 5'), what are the corresponding codons transcribed on mRNA, and what are the corresponding anticodons of tRNA? How will this sequence be translated?

A. Transfer RNA (p. 325–326)

All types of RNA in eukaryotes, including tRNA, are transcribed from DNA located within the nucleus.
- ☞ tRNA must travel from the nucleus to the cytoplasm, where translation occurs.
- ☞ Once in the cytoplasm, each tRNA molecule can be used repeatedly.

How does the structure of tRNA fit its function?

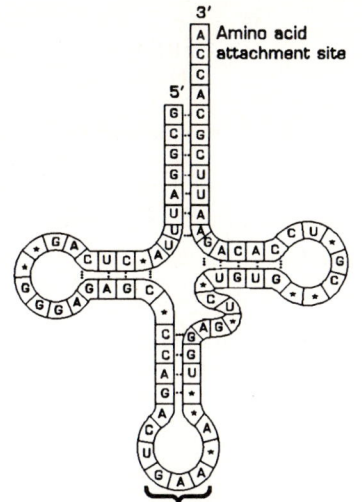

The ability of tRNA to carry specific amino acids and to recognize the correct codons depends upon its structure; its form fits function.
- ☞ tRNA is a single-stranded RNA only about 80 nucleotides long.
- ☞ The strand is folded, forming several double-stranded regions where short base sequences hydrogen bond with other complementary base sequences.
- ☞ A single-plane view reveals a clover leaf shape.

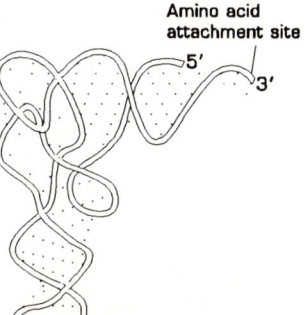

The three-dimensional structure is roughly L-shaped.
- ☞ A loop protrudes at one end of the L and has a specialized sequence of three bases called the *anticodon*.
- ☞ At the other end of the L protrudes the 3' end of the tRNA molecule — the attachment site for an amino acid.

How is it possible that there are only 45 distinct types of tRNA, when there are 64 codons? What is the wobble effect?

There are only about 45 distinct types of tRNA. However, this is enough to translate the 64 codons, since some tRNAs recognize two or three mRNA codons specifying the same amino acid.

☞ This is possible because the base-pairing rules are relaxed between the third base of an mRNA codon and the corresponding base of a tRNA anticodon.

☞ This exception to the base-pairing rules is called *wobble*.

<u>Wobble</u> = The ability of one tRNA to recognize two or three different mRNA codons; occurs when the third base (5' end) of the tRNA anticodon has some play or wobble, so that it can hydrogen bond with more than one kind of base in the third position (3' end) of the codon.

☞ For example, the base U in the wobble position of a tRNA anticodon can pair with either A or G in the third position of an mRNA codon.

☞ Some tRNAs contain a modified base called inosine (I), which is in the anticodon's wobble position and can base pair with U, C or A in the third position of an mRNA codon.

☞ Thus, a single tRNA with the anticodon CCI will recognize three mRNA codons: GGU, GGC or GGA — all of which code for glycine.

B. **Aminoacyl-tRNA Synthetases** (p. 326–327)

How does aminoacyl-tRNA synthetase match a specific amino acid to its appropriate tRNA? Why is the enzyme's three-dimensional conformation important to its function? What is the energy source that drives this endergonic process?

The correct linkage between tRNA and its designated amino acid must occur before the anticodon pairs with its complementary mRNA codon. This process of correctly pairing a tRNA with its appropriate amino acid is catalyzed by an *aminoacyl-tRNA synthetase*.

<u>Aminoacyl-tRNA synthetase</u> = A type of enzyme that catalyzes the attachment of an amino acid to its tRNA.

☞ Each of the 20 amino acids has a specific aminoacyl-tRNA synthetase.

☞ In an endergonic reaction driven by the hydrolysis of ATP, the proper synthetase attaches an amino acid to its tRNA in two steps:

1. *Activation of the amino acid with AMP.* The synthetase's active site binds the amino acid and ATP; the ATP loses two phosphate groups and attaches to the amino acid as AMP (adenosine monophosphate).

2. *Attachment of the amino acid to tRNA.* The appropriate tRNA covalently bonds to the amino acid, displacing AMP from the enzyme's active site.

☞ The aminoacyl-tRNA complex releases from the enzyme and transfers its amino acid to a growing polypeptide on the ribosome.

C. Ribosomes (p. 327)

What is the function of a ribosome? How does a ribosome's structure relate to this function?

Ribosomes coordinate the pairing of tRNA anticodons to mRNA codons.

- ☞ Ribosomes have two subunits (small and large) which are separated when not involved in protein synthesis. (See Campbell, Figure 16.12a)
- ☞ Ribosomes are composed of about 60% *ribosomal RNA* (*rRNA*) and 40% protein.
- ☞ When protein synthesis begins, a small subunit and a large subunit come together and bind the mRNA and tRNAs.

Each ribosome has an mRNA binding site and two (tRNA) binding sites (P and A).

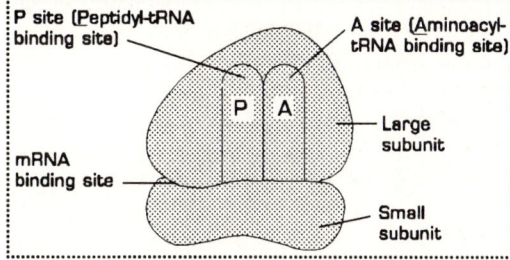

- ☞ The *P site* holds the tRNA carrying the growing polypeptide chain.
- ☞ The *A site* holds the tRNA carrying the next amino acid to be added.

As the ribosome holds the tRNA and mRNA molecules together, enzymes transfer the new amino acid from its tRNA to the carboxyl end of the growing polypeptide.

D. Building a Polypeptide (p. 327–330)

In your own words, what occurs during the process of translation including initiation, elongation and termination? What enzyme, protein factors and energy sources are needed for each stage?

The building of a polypeptide, or translation, occurs in three stages: 1) initiation, 2) elongation, and 3) termination.

All three stages require enzymes and other protein factors; initiation and elongation also require energy provided by GTP (a molecule closely related to ATP).

1. Initiation

Initiation must bring together the mRNA, the first amino acid attached to its tRNA, and the two ribosomal subunits.

a. An initiation complex is assembled as mRNA and a special initiator tRNA bind to a small ribosomal subunit.
   - ☞ First, the small ribosomal subunit binds to a special initiator tRNA carrying methionine, the first amino acid to be added.

P site

A site

☞ Next mRNA aligns on the small ribosomal subunit. mRNA has a specific ribosome-recognition sequence at the 5' upstream end that base pairs with a complementary sequence on rRNA.
☞ Bound initiator tRNA, with the anticodon UAC, finds and base pairs with the initiation codon on mRNA. This initiation codon, AUG, marks the place where translation will begin and is located just downstream from the ribosome-recognition sequence.
☞ Assembly of the initiation complex — small ribosomal subunit, initiator tRNA and mRNA — requires:

1. Protein *initiation factors* that are loosely bound to the small ribosomal subunit.
2. One GTP molecule that probably stabilizes the binding of initiation factors and upon hydrolysis drives the attachment of the large ribosomal subunit.

b. In the second step, a large ribosomal subunit binds to the small one to form a functional ribosome.
☞ The initiator tRNA fits into the P site on the ribosome.
☞ The vacant A site is ready for the next aminoacyl-tRNA.

2. Elongation

Several proteins called *elongation factors* take part in this three-step cycle which adds amino acids one by one to the initial amino acid.

a. *Codon recognition.* The mRNA codon in the A site of the ribosome forms hydrogen bonds with the anticodon of an entering tRNA carrying the next amino acid in the chain.
☞ An elongation factor directs tRNA into the A site.
☞ Hydrolysis of GTP provides energy for this step.

b. *Peptide bond formation.* An enzyme, *peptidyl transferase*, catalyzes the formation of a peptide bond between the polypeptide in the P site and the new amino acid in the A site.
☞ Part of the large ribosomal subunit, peptidyl transferase consists of ribosomal proteins and rRNA.
☞ The polypeptide separates from its tRNA and is transferred to the new amino acid carried by the tRNA in the A site.

c. *Translocation.* The tRNA in the P site releases from the ribosome, and the tRNA in the A site is translocated to the P site.
☞ During this process, the codon and anticodon remain bonded, so the mRNA and the tRNA move as a unit, bringing the next codon to be translated into the A site.
☞ The mRNA is moved through the ribosome only in the 5' to 3' direction.
☞ GTP hydrolysis provides energy for each translocation step.

3. Termination

Each iteration of the elongation cycle takes about 60 milliseconds and is repeated until synthesis is complete and a *termination codon* reaches the ribosome's A site.

Initiation factors

Elongation factors

Peptidyl transferase

Translocation

<u>Termination codon (stop codon)</u> = Base triplet (codon) on mRNA that signals the end of translation.
- Stop codons are UAA, UAG and UGA.
- Stop codons do not code for amino acids.

Termination of translation proceeds as follows:
- When a stop codon reaches the ribosome's A site, a protein *release factor* binds to the codon instead of an aminoacyl-tRNA.
- The release factor causes peptidyl transferase to hydrolyze the bond between the completed polypeptide and the tRNA in the P site. This frees the polypeptide and tRNA, so they can both release from the ribosome.
- The two ribosomal subunits dissociate from the mRNA and separate back into a small and a large subunit.

E. **Polyribosomes** (*p. 331*)

Single ribosomes can make average-sized polypeptides in less than a minute; usually, however, clusters of ribosomes simultaneously translate an mRNA.

<u>Polyribosome</u> = A cluster of ribosomes simultaneously translating an mRNA molecule.
- Once a ribosome passes the initiation codon, a second ribosome can attach to the mRNA.
- Thus, several ribosomes may translate an mRNA at once, making many copies of a polypeptide.

F. **From Polypeptide to Functional Protein** (*p. 331*)

What determines the primary structure of a protein? How must a polypeptide chain be modified, before it can be a fully functional protein?

The biological activity of proteins depends upon a precise folding of the polypeptide chain into a native three-dimensional conformation.
- Genes determine *primary structure*, the linear sequence of amino acids.
- Primary structure determines how a polypeptide chain will spontaneously coil and fold to form a three-dimensional molecule with *secondary* and *tertiary structure*.

Some proteins must be modified further before they become fully functional in the cell.
- Sugars, lipids, phosphate groups, or other additives may be attached to some amino acids.
- One or more amino acids may be enzymatically cleaved from the leading (amino) end of the polypeptide chain.
- Single polypeptide chains may be divided into two or more pieces.
- Two or more polypeptides may join as subunits of a protein that has quaternary structure.

VII. PROTEIN TARGETING (*p. 331–332*)

Eukaryotic ribosomes function either free in the cytosol or bound to endomembranes.
- Bound and free ribosomes are structurally identical and interchangeable.

- Most proteins made by free ribosomes will function in the cytosol.
- Attached to the outside of the endoplasmic reticulum, bound ribosomes generally make proteins that are destined for membrane inclusion or secretory proteins destined for export.

What determines whether a ribosome will be free in the cytosol or attached to the rough ER? How can proteins be targeted for specific sites within the cell?

There is only one type of ribosome, and synthesis of all proteins <u>begins</u> in the cytosol.

- Messenger RNA for some proteins codes for an initial *signal sequence* of 16–20 hydrophobic amino acids at the amino end of the forming polypeptide.
- The signal sequence makes it possible for the ribosome to link to a receptor site on the ER membrane.
- The ribosome continues protein synthesis and the leading end of the new polypeptide (N-terminus) threads into the cisternal space.
- The signal sequence is removed by an enzyme.
- Newly formed polypeptide is released from the ribosome and folds into its native conformation.
- If an mRNA does not code for a signal sequence, the ribosome remains free and synthesizes its protein in the cytosol.

Different signal sequences may dispatch proteins to specific sites other than the ER. For example, newly formed proteins may be targeted for mitochondria or chloroplasts.

VIII. COMPARING PROTEIN SYNTHESIS IN PROKARYOTES AND EUKARYOTES: A REVIEW (p. 332–333)

How is protein synthesis organized in prokaryotes and eukaryotes? What is the significance of a eukaryotic cell's compartmental organization?

Transcription and translation are similar in prokaryotes and eukaryotes, but protein synthesis is organized differently within their cells.

- Prokaryotes lack nuclei, so transcription is not segregated from translation; consequently, translation may begin as soon as the 5' end of mRNA peels away from template DNA, even before transcription is complete.

- The significance of a eukaryotic cell's compartmental organization is that transcription and translation are segregated by the nuclear envelope. This allows mRNA to be modified before it moves from the nucleus to the cytoplasm. Such *RNA processing* occurs only in eukaryotes.

From Gene to Protein 227

IX. RNA PROCESSING IN EUKARYOTES (*p. 333–336*)

How is eukaryotic pre-mRNA processed before it is exported from the nucleus?

Before eukaryotic mRNA is exported from the nucleus, it is processed in two ways: a) both ends are covalently altered and b) intervening sequences are removed and the remainder spliced together.

A. Alteration of the Ends of mRNA (*p. 333*)

During mRNA processing, both the 5′ and 3′ ends are covalently modified.

[Diagram: mRNA structure showing 5′ Cap (G-P-P-P), Leader, Start codon, Coding segment, Stop codon, Trailer, Poly-A tail (AAA··AAA)]

5′ cap = Modified guanine nucleotide (guanosine triphosphate) that is added to the 5′ end of mRNA shortly after transcription begins; has two important functions:
- Protects the growing mRNA from degradation by hydrolytic enzymes.
- Helps small ribosomal subunits recognize the attachment site on mRNA's 5′ end. A *leader* segment of mRNA may also be part of the ribosome recognition signal.
    - Leader sequence = Noncoding (untranslated) sequence of mRNA from the 5′ end to the start codon.

The 3′ end, which is transcribed last, is modified by enzymatic addition of a *poly-A tail*, before the mRNA exits the nucleus.

Poly-A tail = Sequence of 150 to 200 adenine nucleotides added to the 3′ end of mRNA before it exits the nucleus; may function to:
- Help prevent degradation of mRNA in the cytoplasm.
- Regulate protein synthesis by facilitating mRNA's export from the nucleus to the cytoplasm.

The poly-A tail is not attached directly to the stop codon, but to an untranslated *trailer* segment of mRNA.
    Trailer sequence = Noncoding (untranslated) sequence of mRNA from the stop codon to the poly-A tail.

B. RNA Splicing (*p. 333–336*)

The original RNA transcript accurately reflects the complementary base sequence of the gene in template DNA; however, it is much longer than the mRNA that functions in the cytoplasm.
- After RNA processing, only a small portion of the original transcript leaves the nucleus as mRNA.
- The original transcript, or precursor mRNA, is *heterogeneous nuclear RNA*.
    - Heterogeneous nuclear RNA (hnRNA) = Pool of RNA in the nucleus that contains molecules of widely varied sizes; includes primary mRNA transcripts or precursor mRNA (pre-mRNA).

228 From Gene to Protein

Introns

Exons

RNA splicing

Small nuclear ribonucleoproteins

Spliceosome

Genes that code for proteins in eukaryotes are not continuous sequences.

☞ Coding sequences of a gene are interrupted by noncoding segments of DNA called intervening sequences, or *introns*.

Introns = Noncoding sequences in DNA that intervene between coding sequences (exons); are initially transcribed, but not translated, because they are excised from the transcript before mature RNA leaves the nucleus.

☞ Coding sequences of a gene are called *exons*, because they are eventually expressed (translated into protein).

Exons = Coding sequences of a gene that are transcribed and expressed.

Introns and exons are both transcribed to form hnRNA, but the introns are subsequently removed and the remaining exons linked together during the process of *RNA splicing*.

RNA splicing = RNA processing that removes introns and joins exons from eukaryotic pre-mRNA; produces mature mRNA that will move into the cytoplasm from the nucleus.

☞ Enzymes excise introns and splice together exons to form an mRNA with a continuous coding sequence.

☞ RNA splicing also occurs during post-transcriptional processing of tRNA and rRNA.

1. The Mechanisms of RNA Splicing (p. 334–335)

Though there is much left to be discovered, some details of RNA splicing are now known.

☞ Each end of an intron has short boundary sequences that accurately signal the RNA splicing sites.

☞ *Small nuclear ribonucleoproteins (snRNPs)*, play a key role in RNA splicing.

Small nuclear ribonucleoproteins (snRNPs) = Complexes of proteins and small nuclear RNAs that are found only in the nucleus; some participate in RNA splicing. (SnRNPs is pronounced "snurps".)

☞ These small nuclear particles are composed of:

1. *RNA*. Small nuclear RNA (snRNA) is a small RNA molecule with less than 300 nucleotides — much shorter than mRNA.
2. *Protein*. Each snRNP has seven or more proteins.

☞ There are various types of snRNPs with different functions; those involved in RNA splicing are part of a larger, more complex assembly called a *spliceosome*.

Spliceosome = A large molecular complex that catalyzes RNA splicing reactions; composed of pre-mRNA, small nuclear ribonucleoproteins (snRNPs) and other proteins. (See Campbell, Figure 16.21)

☞ As the spliceosome is assembled, one type of snRNP base pairs with a complementary sequence at the 5' end of the intron.

- The spliceosome precisely cuts the RNA transcript at specific sites at either end of the intron, which is excised as a lariat-shaped loop. (See Campbell, Figure 16.21)
- The intron is released and the adjacent exons are immediately spliced together by the spliceosome.

Other kinds of RNA transcripts, such as tRNA and rRNA, are spliced differently; however, as with mRNA splicing, RNA is often involved in catalyzing the reactions.

Are all enzymes proteins?

<u>Ribozymes</u> = RNA molecules that can catalyze reactions by breaking and forming covalent bonds; are called ribozymes to emphasize their enzymelike catalytic activity.
- Ribozymes were first discovered in *Tetrahymena*, a ciliated protozoan that has self-splicing rRNA. That is, intron rRNA catalyzes its own splicing, completely without proteins or extra RNA molecules.
- Since RNA is acting as an enzyme, it can no longer be said that "All enzymes are proteins".
- It has since been discovered that rRNA also functions as an enzyme during translation.

Three-dimensional conformations vary among the types of RNA. These differences in shape give RNA its ability to perform a variety of functions, such as:

1. *Information carrier.* Messenger RNA (mRNA) carries genetic information from DNA to ribosomes; this genetic message specifies a protein's primary structure.
2. *Adaptor molecule.* Transfer RNA (tRNA) acts as an adaptor in protein synthesis by translating information from one form (mRNA nucleotide sequence) into another (protein amino acid sequence).
3. *Catalyst and structural molecule.* During translation, ribosomal RNA (rRNA) plays structural and probably enzymatic roles in ribosomes. Small nuclear RNA (snRNA) in snRNP particles also plays structural and enzymatic roles within spliceosomes that catalyze RNA splicing reactions.
4. *Viral genomes.* Some viruses use RNA as their genetic material.

2. The Functional and Evolutionary Importance of Introns
(p. 335–336)

What are the biological functions of introns and gene splicing?

Introns may play a regulatory role in the cell; for example,
- Intron DNA might include sequences that control gene activity.
- Perhaps the splicing process itself regulates the export of mRNA to the cytoplasm.

<u>Ribozymes</u>

Introns may enable cells in the same organism to make different proteins from the same gene.
- ☞ This can occur if the same RNA transcript is processed differently.
- ☞ For example, all introns may be removed from a particular transcript in one case; but in another, one or more of the introns may be left in place to be translated. Thus, the resulting proteins in each case would be different.

Introns play an important role in the evolution of protein diversity.
- ☞ Exons of a "split gene" may code for different *domains* of a protein that have specific functions, such as, an enzyme's active site or a protein's binding site.
  - Protein domain = Continuous polypeptide sequences that are structural and functional units in proteins with a modular architecture.
- ☞ Genetic recombination can modify protein function by altering just one domain and not the others.
- ☞ Because coding sequences can be separated by considerable distances along the DNA, split genes have higher recombination frequencies than genes with continuous coding regions lacking introns.
- ☞ Thus, introns facilitate recombination of exons between different alleles of a gene, which can produce a novel protein with an altered domain.

The modular arrangement of many genes, with exons separated by introns, allows the shuffling of preexisting DNA segments to produce new proteins. This shuffling may occur by:
- ☞ Homologous strand exchange during crossing over.
- ☞ The action of *transposons*, a type of transposable genetic element to be discussed in Chapter 17.

X. MUTATIONS AND THEIR EFFECTS ON PROTEINS (*p. 336–338*)

Knowing how genes are translated into proteins, scientists can give a molecular description of heritable changes that occur in organisms.

Mutation = A permanent change in DNA that can involve large chromosomal regions or a single nucleotide pair.

Point mutation = A mutation limited to about one or two nucleotides in a single gene.

A. Types of Mutations (*p. 336–338*)

Why do base-pair insertions or deletions usually have a greater effect than base-pair substitutions?

There are two categories of point mutations: 1) base-pair substitutions and 2) base-pair insertions or deletions.

1. **Substitutions** (*p. 336*)

<u>Base-pair substitution</u> = The replacement of one base pair with another; occurs when a nucleotide and its partner from the complementary DNA strand are replaced with another pair of nucleotides according to base-pairing rules.

Depending on how base-pair substitutions are translated, they can result in little or no change in the protein encoded by the mutated gene.
- ☞ Redundancy in the genetic code is why some substitution mutations have no effect. A base pair change may simply transform one codon into another that codes for the same amino acid.
- ☞ Even if the substitution alters an amino acid, the new amino acid may have similar properties to the one it replaces, or it may be in a part of the protein where the exact amino acid sequence is not essential to its activity.

Some base-pair substitutions result in readily detectable changes in proteins.
- ☞ Alteration of a single amino acid in a crucial area of a protein will significantly alter protein activity.
- ☞ On rare occasions, such a mutation will produce a protein that is improved or has capabilities that enhance success of the mutant organism and its descendants.
- ☞ More often, such mutations produce a less active or inactive protein that impairs cell function.

Base-pair substitutions may be *missense mutations* or *nonsense mutations*.

<u>Missense mutation</u> = Base-pair substitution that alters an amino acid codon (sense codon) to a new codon that codes for a different amino acid.
- ☞ Altered codons make sense (are translated), but not necessarily that originally intended.
- ☞ Base-pair substitutions are usually missense mutations.

<u>Nonsense mutation</u> = Base-pair substitution that changes an amino acid codon (sense codon) to a chain termination codon, or *vice versa*.
- ☞ Nonsense mutations can result in premature termination of translation and the production of a shorter than normal polypeptide.
- ☞ Nearly all nonsense mutations lead to nonfunctional proteins.

2. **Insertions or Deletions** (*p. 336–338*)

Base-pair insertions or *deletions* usually have a greater negative effect on proteins than substitutions.

<u>Base-pair insertion</u> = The insertion of one or more nucleotide pairs into a gene.

<u>Base-pair deletion</u> = The deletion of one or more nucleotide pairs from a gene.

Because mRNA is read as a series of triplets during translation, insertion or deletion of nucleotides may alter the reading frame (triplet grouping) of the genetic message. This type of *frameshift mutation* will occur whenever the number of nucleotides inserted or deleted is not 3 or a multiple of 3.

Frameshift mutation = A base-pair insertion or deletion that causes a shift in the reading frame, so that codons beyond the mutation will be the wrong grouping of triplets and will specify the wrong amino acids.
- A frameshift mutation causes the nucleotides following the insertion or deletion to be improperly grouped into codons.
- This results in extensive missense, which will sooner or later end in nonsense (premature termination).
- Frameshift will produce a nonfunctional protein unless the insertion or deletion is very near the end of the gene.

B. **Mutagenesis** (*p. 338*)

How can mutagenesis occur?

Mutagenesis = The creation of mutations.
- Mutations can occur as errors in DNA replication, repair or recombination that result in base-pair substitutions, insertions or deletions.
- Mutagenesis may be a naturally occurring event causing *spontaneous mutations*, or mutations may be caused by exposure to *mutagens*.

Mutagen = Physical or chemical agents that interact with DNA to cause mutations.
- Radiation is the most common physical mutagen in nature and has been used in the laboratory to induce mutations.
- Several categories of chemical mutagens are known including *base analogues*, which are chemicals that mimic normal DNA bases, but base pair incorrectly. (See Campbell, Figure 16.23)
- The Ames test, developed by Bruce Ames, is one of the most widely used tests for measuring the mutagenic strength of various chemicals. Since most mutagens are carcinogenic, this test is also used to screen for chemical carcinogens. (See Campbell, Chapter 16 Methods Box)

17   MICROBIAL MODELS: THE GENETICS OF VIRUSES AND BACTERIA

CHAPTER OUTLINE

Scientists discovered the role of DNA in heredity by studying the simplest of biological systems — viruses and bacteria. Most of the molecular principles discovered through microbe research applies to higher organisms, but viruses and bacteria also have unique genetic features.

- ☞ Knowledge of these unique genetic features has helped scientists understand how viruses and bacteria cause disease.
- ☞ Techniques for gene manipulation emerged from studying genetic peculiarities of microorganisms.

I. THE DISCOVERY OF VIRUSES (p. 344–345)

What contributions did the following scientists make towards the discovery of viruses: A. Mayer, D. Ivanowsky, Martinus Beijerinck and Wendell Stanley?

The discovery of viruses resulted from the search for the infectious agent causing tobacco mosaic disease. This disease stunts the growth of tobacco plants and gives their leaves a mosaic coloration.

1883: A. Mayer, a German scientist demonstrated that the disease was contagious and proposed that the infectious agent was an unusually small bacterium that could not be seen with a microscope.
- ☞ He successfully transmitted the disease by spraying sap from infected plants onto the healthy ones.
- ☞ Using a microscope, he examined the sap and was unable to identify a microbe.

1890's: D. Ivanowsky, a Russian scientist proposed that tobacco mosaic disease was caused by a bacterium that was either too small to be trapped by a filter or that produced a filterable toxin.
- ☞ To remove bacteria, he filtered sap from infected leaves.
- ☞ Filtered sap still transmitted disease to healthy plants.

1897: Martinus Beijerinck, a Dutch microbiologist proposed that the disease was caused by a reproducing particle much smaller and simpler than a bacterium.
- ☞ He ruled out the theory that a filterable toxin caused the disease by demonstrating that the infectious agent in filtered sap could reproduce.

A. Mayer

D. Ivanowsky

Martinus Beijerinck

234 *Microbial Models: The Genetics of Viruses and Bacteria*

```
        Plants were sprayed with filtered sap from
                    diseased plant.
                           ↓
        Sprayed plants developed tobacco mosaic
                       disease.
                           ↓
        Sap from newly infected plants was used to
                    infect others.
```

☞ This experiment was repeated for several generations. He concluded that the pathogen must be reproducing because its ability to infect was undiluted by transfers from plant to plant.
☞ He also noted that unlike bacteria, the pathogen:
    a. Reproduced only within the host it infected.
    b. Could not be cultured on media.
    c. Could not be killed by alcohol.

<u>Wendell M. Stanley</u>

<u>Tobacco mosaic virus</u>

1935: Wendell M. Stanley, an American biologist, crystallized the infectious particle now known as *tobacco mosaic virus* (*TMV*).

II. VIRAL STRUCTURE AND REPLICATION: AN OVERVIEW
(*p. 345–346*)

What are the structural components of viruses? What type of molecules make up each component?

In the 1950's, TMV and other viruses were finally observed with electron microscopes. Viral structure appeared to be unique from the simplest of cells.
☞ The smallest viruses are only 20 nm in diameter.
☞ The virus particle, or *virion*, is just nucleic acid enclosed by a protein coat.

<u>Viruses</u>

<u>Virion</u>

A. Viral Genomes (*p. 345*)

Depending upon the virus, viral genomes:
☞ May be double-stranded DNA, single-stranded DNA, double-stranded RNA or single-stranded RNA.
☞ Are organized as single nucleic acid molecules that are either linear or circular.
☞ May have as few as four genes or as many as several hundred.

B. Capsids and Envelopes (*p. 345–346*)

<u>Capsid</u> = Protein coat that encloses the viral genome.
☞ Its structure may be rod-shaped, polyhedral or complex.
☞ Composed of many protein subunits made from only one or a few types of protein.
☞ Examples are TMV and adenoviruses. (See Campbell, Figure 17.3)

<u>Capsid</u>

Envelope = Membrane that cloaks some viral capsids.
- ☞ Helps viruses infect their host.
- ☞ Is derived from host cell membrane which is usually virus-modified, containing proteins and glycoproteins of viral origin.

The most complex capsids are found among *bacteriophages* or bacterial viruses.
- ☞ Of the first phages studied, seven infected *E. coli*. These were named types 1–7 (T1, T2, T3 ... T7).
- ☞ The T-even phages — T2, T4 and T6 — are structurally very similar.
- ☞ The icosohedral head encloses the genetic material. The protein tailpiece with tail fibers attaches the phage to its bacterial host and injects its DNA into the bacterium.

Structure of the T-Even Phage

III. HOW VIRUSES REPLICATE: VIRAL INFECTION (*p. 346–347*)

Why are viruses considered to be obligate parasites? How do viruses recognize their hosts?

Viral reproduction differs markedly from cellular reproduction, because viruses are *obligate intracellular parasites* which can express their genes and reproduce only within a living cell. Each virus has a specific *host range*.

Host range = Limited number or range of host cells that a parasite can infect.
- ☞ Viruses recognize host cells by a complementary fit between external viral proteins and specific cell surface *receptor sites*.
- ☞ Some viruses have broad host ranges which may include several species (e.g. swine flu and rabies).
- ☞ Some viruses have host ranges so narrow that they can infect only one species (e.g. phages of *E. coli*) or only a single tissue type of one species (e.g. human cold virus which only infects cells of the upper respiratory tract and the AIDS virus which binds only to specific receptors on certain white blood cells).

Depending upon the type of virus, there are several mechanisms used to infect host cells with viral DNA.
- ☞ For example, a T-even phage uses its elaborate tailpiece to inject its DNA into the host cell.
- ☞ Once the viral genome is inside its host cell, it commandeers the host's resources and reprograms the cell to copy the viral genes and manufacture capsid protein.

What are possible patterns of viral genome replication? What are the functions of RNA replicase and reverse transcriptase?

There are three possible patterns of viral genome replication:

1. DNA → DNA. If viral DNA is double-stranded, DNA replication resembles that of cellular DNA, and the virus uses DNA polymerase produced by the host.

2. RNA → RNA. Most RNA viruses contain a gene that codes for *RNA replicase*, an enzyme that uses viral RNA as a template to produce complementary RNA; host cells have no native enzyme to copy RNA.

3. RNA → DNA → RNA. Some RNA viruses encode *reverse transcriptase*, an enzyme that transcribes DNA from an RNA template.

```
                    Viral genomic RNA
                           │
                           │ reverse transcriptase
                           ▼
        transcribes ── Viral DNA ── transcribes
              ▼                          ▼
        messenger RNA               genomic RNA
                                    for new virions
```

Regardless of how viral genomes replicate, all viruses divert host cell resources for viral production.

- Viral genes use the host cell's enzymes, ribosomes, tRNAs, amino acids, ATP and other resources to make copies of the viral genome and produce viral capsid proteins.

- These viral components — nucleic acid and capsids — are assembled into hundreds or thousands of virions, which leave to parasitize new hosts.

Viral nucleic acid and capsid proteins assemble spontaneously into new virus particles, a process called *self-assembly*.

- Since most viral components are held together by weak bonds (e.g. hydrogen bonds and Van der Waals forces), enzymes are not usually necessary for assembly.

- For example, TMV can be disassembled in the laboratory. When mixed together, the RNA and capsids spontaneously reassemble to form complete TMV virions.

IV. **BACTERIAL VIRUSES** (*p. 348 – 350*)

Bacteriophages are the best understood of all viruses, and many of the important discoveries in molecular biology have come from bacteriophage studies.

- In the 1940's, scientists determined how the T phages reproduce within a bacterium. This research helped:

  1. Demonstrate that DNA is the genetic material.
  2. Establish the phage-bacterium system as an important experimental tool.

- Studies on lambda (λ) phage of *E. coli* showed that double-stranded DNA viruses reproduce by two alternative mechanisms: the lytic cycle and the lysogenic cycle.

RNA replicase

Reverse transcriptase

Self-assembly

What characterizes lytic and lysogenic reproductive cycles? How do they differ?

    A.   The Lytic Cycle (*p. 348–349*)

Virulent bacteriophages reproduce by a *lytic* replication cycle.

<u>Virulent phages</u> = Phages that lyse their host cells.

<u>Lytic cycle</u> = A viral replication cycle that results in the death or lysis of the host cell.

The lytic cycle of phage T4 illustrates this type of replication cycle:

    1. Phage attaches to cell surface.
- T4 recognizes a host cell by a complementary fit between proteins on the virion's tail fibers and specific receptor sites on the outer surface of an *E. coli* cell.

    2. Phage contracts sheath and injects DNA.
- ATP stored in the phage tailpiece is the energy source for the phage to: a) pierce the *E. coli* wall and membrane, b) contract its tail sheath, and c) inject its DNA.
- The genome separates from the capsid leaving a capsid "ghost" outside the cell.

    3. Hydrolytic enzymes destroy host cell's DNA.
- The *E. coli* host cell begins to transcribe and translate the viral genome.
- One of the first viral proteins produced is an enzyme that degrades host DNA. The phage's own DNA is protected, because it contains modified cytosine not recognized by the enzyme.

    4. Phage genome directs the host cell to produce phage components: DNA and capsid proteins.
- Using nucleotides from its own degraded DNA, the host cell makes many copies of the phage genome.
- The host cell also produces three sets of capsid proteins and assembles them into phage tails, tail fibers and polyhedral heads.
- Phage components spontaneously assemble into virions.

    5. Cell lyses and releases phage particles.
- Lysozymes specified by the viral genome digest the bacterial cell wall.
- Osmotic swelling lyses the cell which releases hundreds of phages from their host cell.
- Released virions can infect nearby cells.

How can phage concentration in a liquid medium be measured?

Phage reproduction outpaces bacterial growth.
- Lytic cycle takes only 20–30 minutes at 37°C. In that period, a T4 population can increase a hundredfold, whereas a fast-growing *E. coli* population can only double.

Virulent phages

Lytic cycle

238 Microbial Models: The Genetics of Viruses and Bacteria

☞ A measure of phage concentration in a liquid medium is to mix it with a bacterial suspension, spread the mixture onto an agar plate, and after a period of bacterial growth, count the number of *plaques* resulting from cell lysis.

<u>Plaque</u> = A clear spot caused by phage-induced cell lysis in a bacterial culture on solid medium.

What defenses do bacteria have against phage infection?

Bacteria have several defenses against destruction by phage infection.
- ☞ Bacterial mutations can change receptor sites used by phages for recognition, and thus avoid infection.
- ☞ Bacterial *restriction enzymes* recognize and cut up foreign DNA, including certain phage DNA. Bacterial DNA is chemically altered, so it is not destroyed by the cell's own restriction enzymes.

Bacterial hosts and their viral parasites are continually *coevolving*.
- ☞ Most successful bacterial have effective mechanisms for preventing phage entry or reproduction.
- ☞ Most successful phages have evolved ways around bacterial defenses.
- ☞ Many phages check their own destructive tendencies and may coexist with their hosts.

B. The Lysogenic Cycle (*p. 349–350*)

Some viruses can coexist with their hosts by incorporating their genome into the host's genome.

<u>Temperate viruses</u> = Viruses that can reproduce without killing their host cells.
- ☞ They have two possible modes of reproduction, the lytic cycle and the *lysogenic cycle*.
- ☞ An example is phage λ, discovered by E. Lederberg in 1951. (See Campbell, Figure 17.6)

<u>Lysogenic cycle</u> = A viral replication cycle that involves the incorporation of the viral genome into the host cell genome.

Details of the lysogenic cycle were discovered through studies of phage λ life cycle:

1. Phage λ binds to the surface of an *E. coli* cell.
2. Phage λ injects its DNA into the bacterial host cell.
3. λ DNA forms a circle and either begins a lytic cycle or a lysogenic cycle.
4. During a lysogenic cycle, λ DNA inserts by genetic recombination into a specific site on the bacterial chromosome and becomes a *prophage*.

What is a prophage? How can lysogenic conversion affect pathogenicity of host bacterial cells?

> Prophage = A phage genome that is incorporated into a specific site on the bacterial chromosome.
> - Most prophage genes are inactive.
> - One active prophage gene codes for the production of *repressor protein* which switches off most other prophage genes.
> - Prophage genes are copied along with cellular DNA when the host cell reproduces. As the cell divides, both prophage and cellular DNA are passed on to daughter cells.
> - A prophage may be carried in the host cell's chromosomes for many generations.
>
> Occasionally, a prophage may leave the bacterial chromosome.
> - This may be spontaneous or caused by environmental factors (e.g. radiation).
> - The excision process may begin the phage's lytic reproductive cycle.
> - Virions produced during the lytic cycle may begin either a lytic or lysogenic cycle in their new host cells.
>
> Lysogenic cell = Host cell carrying a prophage in its chromosome.
> - It is called lysogenic because it has the potential to lyse.
> - Some prophage genes in a lysogenic cell may be expressed and change the cell's phenotype in a process called *lysogenic conversion*.
> - Lysogenic conversion occurs in bacteria that cause diphtheria, botulism and scarlet fever. Pathogenicity results from toxins coded for by prophage genes.

Prophage

Repressor protein

Lysogenic cell

Lysogenic conversion

V. **ANIMAL VIRUSES** (*p. 350–354*)

　A.　**Diverse Reproductive Cycles of Animal Viruses** (*p. 350–352*)

Replication cycles of animal viruses may show some interesting variations from those of other viruses. Two examples are the replication cycles of 1) viruses with envelopes and 2) viruses with RNA genomes. (See Campbell, Table 17.1 for families of animal viruses grouped by type of nucleic acid.)

　　1.　**Viruses with Envelopes** (*p. 350–351*)

　　Some animal viruses are surrounded by a membranous envelope, which is unique to several groups of animal viruses. This envelope:
> - Is outside the capsid and helps the virus enter host cells.
> - Is a lipid bilayer with glycoprotein spikes protruding from the outer surface.

What characterizes replication cycles of enveloped animal viruses?

Enveloped viruses have replication cycles characterized by: (See Campbell, Figure 17.7)

a. *Attachment.* Glycoprotein spikes protruding from the viral envelope attach to receptor sites on the host's plasma membrane.

b. *Entry.* As the envelope fuses with the plasma membrane, the entire virus (capsid and genome) is transported into the cytoplasm by receptor-mediated endocytosis.

c. *Uncoating.* Cellular enzymes uncoat the genome by removing the protein capsid from viral RNA.

d. *Viral RNA and Protein Synthesis.* Viral enzymes are required to replicate the RNA genome and to transcribe mRNA.
   - Some viral RNA polymerase is packaged in the virion.
   - Viral RNA polymerase (transcriptase) replicates the viral genome and transcribes viral mRNA. Note that the viral genome is a strand complementary to mRNA.

Viral mRNA is translated into viral proteins including:
   - Capsid proteins synthesized in the cytoplasm by free ribosomes.
   - Viral-envelope glycoproteins synthesized by ribosomes bound to rough ER. Glycoproteins produced in the host's ER are sent to the Golgi apparatus for further processing. Golgi vesicles transport the glycoproteins to the plasma membrane, where they cluster at exit sites for the virus.

e. *Assembly and Release.* New capsids surround viral genomes. Once assembled, the virions envelop with host plasma membrane as they bud off from the cell's surface. The viral envelope is derived from:
   - Host cell's plasma membrane lipid.
   - Virus-specific glycoprotein.

Why do herpes infections tend to recur?

__Herpes viruses__

Herpes viruses are double-stranded DNA viruses which:
- Contain envelopes derived from the host cell's nuclear envelope rather than from the plasma membrane.
- Reproduce within the host cell's nucleus.
- Use both viral and cellular enzymes to replicate and transcribe their genomic DNA.
- May integrate their DNA into the cell's genome as a *provirus*. Evidence comes from the nature of herpes infections, which tend to recur. After a period of latency, physical or emotional stress may cause the proviruses to begin a productive cycle again.

Provirus = Viral DNA that inserts into a host cell chromosome.

2. RNA Viruses (p. 351–352)

How are RNA viruses classified? What type of virus is HIV?

All possible viral genomes are represented among animal viruses. Since mRNA is common to all types, DNA and RNA viruses are classified according to the relationship of their mRNA to the genome. (See Campbell, Figure 17.8) In this classification:
- mRNA or the strand that corresponds to mRNA is the *plus (+) strand*; it has the nucleotide sequence that codes for proteins.
- The *minus (−) strand* is a template for synthesis of a plus strand; it is complementary to the sense strand or mRNA.

Animal RNA viruses are classified as following:
- Class III RNA viruses: Double-stranded RNA genome; the minus strand is the template for mRNA. (Reoviruses)
- Class IV RNA viruses: Single plus strand genome; the plus strand can function directly as mRNA, but also is a template for synthesis of minus RNA. (Minus RNA is a template for synthesis of additional plus strands.) Viral enzymes are required for RNA synthesis from RNA templates. (Picornavirus, Togavirus)
- Class V RNA viruses: Single minus strand genome; mRNA is transcribed directly from this genomic RNA. (Rhabdovirus, Paramyxovirus, Orthomyxovirus)
- Class VI RNA viruses: Single plus strand genome; the plus strand is a template for complementary DNA synthesis. *Reverse transcriptase* catalyzes this reverse transcription from RNA to DNA. mRNA is then transcribed from a DNA template. (*Retroviruses*)

Retrovirus = (Retro=backward) RNA virus that uses *reverse transcriptase* to transcribe DNA from the viral RNA genome.
- *Reverse transcriptase* is a type of DNA polymerase that transcribes DNA from an RNA template.
- *HIV (human immunodeficiency virus)*, the virus that causes *AIDS (acquired immunodeficiency syndrome)* is a retrovirus.

How do retroviruses reproduce? What may happen when proviral genes of retroviruses are expressed?

RNA viruses with the most complicated reproductive cycles are the retroviruses, because retroviruses must first carry out *reverse transcription*. (See Campbell, Figure 17.9)

Attachment and entry of the virion.
↓ Enters host cell cytoplasm.

Uncoating of single-stranded RNA genome.
↓ Capsid proteins are enzymatically removed.

Reverse transcription.
↓ Viral RNA is used as a template to produce minus strand DNA.
↓ Reaction is catalyzed by the viral enzyme reverse transcriptase.

Integration.
↓ Newly produced double-stranded viral DNA enters the nucleus.
↓ Viral DNA inserts into chromosomal DNA and becomes a *provirus*.

Viral RNA and protein synthesis.
↓ Proviral DNA is transcribed into mRNA and is translated into proteins. Transcribed RNA may provide genomes for next viral generation.

Expression of proviral genes may:

Produce new virons	Cause expression of *oncogenes*, if present.
Capsid assembly and release of new virions	Transformation of host cell into a cancerous state

B. Viral Diseases in Animals (*p. 352–353*)

How can viruses cause disease symptoms?

It is often unclear how certain viruses cause disease symptoms. Viruses may:
- ☞ Damage or kill cells. In response to a viral infection, lysosomes may release hydrolytic enzymes.
- ☞ Be toxic themselves or cause infected cells to produce toxins.
- ☞ Cause varying degrees of cell damage depending upon regenerative ability of the infected cell. We recover from colds because infected cells of the upper respiratory tract can regenerate by cell division. Poliovirus, however, causes permanent cell damage because the virus attacks nerve cells which cannot divide.
- ☞ Be only indirectly responsible for disease symptoms. Fever, aches and inflammation may result from activities of the immune system.

What are some medical weapons used to fight viral infections?

Medical weapons used to fight viral infections include *vaccines* and *antiviral drugs*.

<u>Vaccines</u> = Harmless variants or derivatives of pathogenic microbes that mobilize a host's immune mechanism against the pathogen.
- ☞ A vaccine has almost completely eradicated smallpox.
- ☞ Effective vaccines exist for polio, rubella, measles, mumps and many other viral diseases.
- ☞ New techniques of modern biotechnology are used to develop antiviral vaccines and to produce natural antiviral agents such as *interferons*. (See Campbell, Chapter 39)

Antiviral drugs have recently been developed.
- ☞ Several are analogs of purine nucleosides that interfere with viral nucleic acid synthesis (e.g. adenine arabinoside and acyclovir).
- ☞ Amantadine and rimantadine are effective in preventing influenza, but the mechanism of action is unknown.

C. **Viruses and Cancer** (*p. 353–354*)

What viruses have been implicated in human cancers?

Some tumor viruses cause cancer in animals.
- ☞ When animal cells grown in tissue culture are infected with tumor viruses, they *transform* to a cancerous state.
- ☞ Examples are members of the retrovirus, papovavirus, adenovirus and herpesvirus groups.
- ☞ Certain viruses are implicated in human cancers:

Viral Group	Examples/Diseases	Cancer Type
Retrovirus	HTLV-1/adult leukemia	Leukemia
Herpesvirus	Epstein-Barr/infectious mononucleosis	Burkitt's lymphoma
Papovavirus	Papilloma/human warts	Cervical cancer
Hepatitis B virus	Chronic hepatitis	Liver cancer

How can tumor viruses transform cells?

Tumor viruses transform cells by inserting viral nucleic aids into host cell DNA.
- ☞ This insertion is permanent as the provirus never excises.
- ☞ Insertion for DNA tumor viruses is straightforward.

Several viral genes have been identified as *oncogenes*.

<u>Oncogenes</u> = Genes found in viruses or as part of the normal eukaryotic genome, that trigger transformation of a cell to a cancerous state.
- ☞ Code for cellular growth factors or for proteins involved in the function of growth factors.
- ☞ Are not unique to tumor viruses, but are found in the normal cells of many species. In fact, some tumor viruses transform cells by activating cellular oncogenes.

244 Microbial Models: The Genetics of Viruses and Bacteria

More than one oncogene must usually be activated to completely transform a cell.
- ☞ Indications are that tumor viruses are effective only in combination with other events such as exposure to carcinogens.
- ☞ Carcinogens probably also act by turning on cellular oncogenes.

VI. PLANT VIRUSES AND VIROIDS (p. 354)

As serious agricultural pests, many of the plant viruses:
- ☞ Stunt plant growth and diminish crop yields.
- ☞ Are RNA viruses.
- ☞ Have rod-shaped capsids with *capsomeres* arranged in a spiral.

Capsomere = Complex capsid subunit consisting of several identical or different protein molecules.

How do plant viruses spread from plant to plant? What is the difference between horizontal and vertical transmission?

Plant viruses spread from plant to plant by two major routes: *horizontal transmission* and *vertical transmission*.

Horizontal transmission = Route of viral transmission in which an organism receives the virus from an external source.
- ☞ Plants are more susceptible to viral infection if their protective epidermal layer is damaged.
- ☞ Insects may be *vectors* that transmit viruses from plant to plant and can inject the virus directly into the cytoplasm.
- ☞ By using contaminated tools, gardeners and farmers may transmit plant viruses.

Vertical transmission = Route of viral transmission in which an organism inherits a viral infection from its parent.
- ☞ Can occur in asexual propagation of infected plants (e.g. by taking cuttings).
- ☞ Can occur in sexual reproduction via infected seeds.

Once a plant is infected, viruses reproduce and spread from cell to cell by passing through plasmodesmata.

Most plant viral diseases have no cure, so current efforts focus on reducing viral propagation and breeding resistant plant varieties.

Another class of plant pathogens called *viroids* are smaller and simpler than viruses.
- ☞ They are small naked RNA molecules with only several hundred nucleotides.
- ☞ It is likely that viroids disrupt normal plant metabolism, development and growth by causing errors in regulatory systems that control gene expression.
- ☞ Viroid diseases affect many commercially important plants such as coconut palms, chrysanthemums, potatoes and tomatoes.

Some scientists believe that viroids originated as escaped introns.
- ☞ Nucleotide sequences of viroid RNA are similar to self-splicing introns found within some normal eukaryotic genes, including rRNA genes.
- ☞ An alternative hypothesis is that viroids and self-splicing introns share a common ancestral molecule.

VII. THE EVOLUTIONARY ORIGIN OF VIRUSES (p. 354–355)

What do viruses and other living organisms have in common? Why do viruses not fit our usual definition of life?

Viruses do not fit our usual definitions of living organisms. They cannot reproduce independently, yet they:
- ☞ Have a genome with the same genetic code as living organisms.
- ☞ Can mutate and evolve.

What evidence suggests that viruses probably evolved from fragments of cellular nucleic acid?

Viruses probably evolved from fragments of cellular nucleic acid. Evidence to support this includes:
- ☞ Genetic material of different viral families is more similar to host genomes than to that of other viral families.
- ☞ Some viral genes are identical to cellular genes (e.g. oncogenes in retroviruses).
- ☞ Viruses of eukaryotes are more similar in genomic structure to their cellular hosts than to bacterial viruses.
- ☞ Viral genomes are similar to certain cellular genetic elements such as plasmids and transposons; they are all *mobile* genetic elements.

VIII. BACTERIAL GENOMES: REPLICATION AND MUTATION (p. 355–356)

What is the structure of a bacterial chromosome? How does it differ from the genomes of viruses and eukaryotes?

The average bacterial genome is larger than a viral genome, but much smaller than that in a typical eukaryotic cell.

Most DNA in a bacterium is found in a single circular *bacterial chromosome* (*genophore*).
- ☞ This chromosome of double-stranded DNA is structurally simpler and has fewer associated proteins than a eukaryotic chromosome.
- ☞ It is found in the *nucleoid* region of the bacterial cell, but is not separate from the rest of the cell. Consequently, transcription and translation can occur simultaneously.
- ☞ Many bacteria also contain extrachromosomal DNA in *plasmids*.

Plasmid = A small double-stranded ring of DNA that carries extrachromosomal genes in some bacteria.

Nucleoid region

Binary fission

How do most bacteria reproduce? Why is replication of the bacterial chromosome considered to be semiconservative?

Most bacteria can rapidly reproduce by *binary fission*. (See Campbell, Chapter 11)
- Semi-conservative replication of the bacterial chromosome begins at a single origin of replication.
- The two replication forks move bidirectionally until they meet and replication is complete. (See Campbell, Figure 17.10)
- Under optimal conditions, some bacteria can divide in twenty minutes. Because of this rapid reproductive rate, bacteria are useful for genetic studies.

Binary fission is asexual reproduction that produces clones — daughter cells that are genetically identical to the parent.
- Though mutations are rare events, they can impact genetic diversity in bacteria because of their rapid reproductive rate.
- Even though mutation can be a major source of genetic variation in bacteria, genetic recombination is responsible for more genetic diversity among populations.

IX. **GENETIC RECOMBINATION AND GENE TRANSFER IN BACTERIA** (p. 356–363)

If binary fission produces clones, how does genetic recombination naturally occur in bacteria?

There are three natural processes of genetic recombination in bacteria: *transformation, transduction* and *conjugation*. These mechanisms of gene transfer occur separately from bacterial reproduction, and are the major source of genetic variation in bacterial populations.

A. Transformation (p. 357)

Transformation

Transformation = Process of gene transfer during which a bacterial cell assimilates genetic material from the surroundings.
- Some bacteria can take up naked DNA from the surroundings. (Refer to Avery's experiments with *Streptococcus pneumoniae* in Chapter 15.)
- Assimilated foreign DNA may be integrated into the bacterial chromosome by recombination (crossing over).
- Progeny of the recipient bacterium will carry a new combination of genes.

How has the biotechnology industry capitalized on the natural mechanism of transformation?

Many bacteria have surface proteins that recognize and import naked DNA from closely related bacterial species.

☞ Though *E. coli* lacks such proteins, it can be artificially induced to take up foreign DNA by incubating it in a culture medium that has a high concentration of calcium ions.
☞ This technique of artificially inducing transformation is used by the biotechnology industry to introduce foreign genes into bacterial genomes, so that bacterial cells can produce proteins characteristic of other species (e.g. insulin and human growth hormone).

B. **Transduction** (*p. 357*)

<u>Transduction</u> = Gene transfer from one bacterium to another by a bacteriophage.

What is the difference between general transduction and specialized transduction?

<u>General transduction</u> = Transduction that occurs when random pieces of host cell DNA are packaged within a phage capsid during the lytic cycle of a phage.
☞ This process can transfer almost any host gene and little or no phage genes.
☞ When the phage particle infects a new host cell, the donor cell DNA can recombine with the recipient cell DNA.

<u>Specialized transduction</u> = Transduction that occurs when a prophage excises from the bacterial chromosome and carries with it some host genes adjacent to the excision site. Also known as *restricted transduction*.
☞ Carried out only by temperate phages.
☞ Differs from general transduction in that:
 1. Specific host genes and most phage genes are packed into the same virion.
 2. Transduced bacterial genes are restricted to specific genes adjacent to the prophage insertion site. In general transduction, host genes are randomly selected and almost any host gene can be transferred.

C. **Conjugation and Plasmids** (*p. 357–361*)

<u>Conjugation</u> = The direct transfer of genes between two cells that are temporarily joined.
☞ Discovered by Joshua Lederberg and Edward Tatum.
☞ Conjugation in *E. coli* is one of the best-studied examples:

A DNA-donating *E. coli* cell extends
external appendages called *sex pili*.
↓
Sex pili attach to a DNA-receiving cell.
↓
A cytoplasmic bridge forms through which
DNA transfer occurs.

The ability to form sex pili and to transfer DNA is conferred by genes in a plasmid called the *F plasmid*.

1. **Plasmids: General Characteristics** (*p. 358–359*)

What are plasmids? How do they replicate?

Plasmid = A small double-stranded ring of DNA that carries extrachromosomal genes in some bacteria.
- ☞ Plasmids have only a few genes, and they are not required for survival and reproduction.
- ☞ Plasmid genes can be beneficial. Examples include the F plasmid, which confers ability to conjugate; and the R plasmid, which confers antibiotic resistance.

These small circular DNA molecules replicate independently:
- ☞ Some plasmids replicate in synchrony with the bacterial chromosome, so only a few are present in the cell.
- ☞ Some plasmids under more relaxed control can replicate on their own schedule, so the number of plasmids in the cell at any one time can vary from only a few to as many as 100.

Some plasmids are *episomes* that can reversibly incorporate in the cell's chromosome.

Episomes = Genetic elements that can replicate either independently as free molecules in the cytoplasm or as integrated parts of the main bacterial chromosome.
- ☞ Examples include some plasmids, as well as, temperate viruses such as lambda phage.
- ☞ Temperate phage genomes replicate separately in the cytoplasm during a lytic cycle, and as an integral part of the host's chromosome during a lysogenic cycle.

How do episomal plasmids and viruses differ?

While plasmids and viruses can both be episomes, they differ in that:
- ☞ Plasmids, unlike viruses, lack an extracellular stage.
- ☞ Plasmids are generally beneficial to the cell, while viruses are parasites that usually harm their hosts.

2. **The F Plasmid and Conjugation** (*p. 359*)

How does the F plasmid control conjugation in bacteria?

The F plasmid (F for fertility) has about 25 genes, most of which are involved in the production of sex pili.
- ☞ Bacterial cells that contain the F factor and can donate DNA ("male") are called *F⁺ cells*.
- ☞ The F factor replicates in synchrony with chromosomal DNA, so the F⁺ factor is inheritable; that is, division of an F⁺ cell results in two F⁺ daughter cells.
- ☞ Cells without the F factor are designated *F⁻* ("female").

For donor and recipient bacterial cells, what are the consequences of conjugation between an F⁺ and F⁻ cell? Hfr and F⁻ cell?

During conjugation between an F⁺ and an F⁻ bacterium:
- The F factor replicates by *rolling circle replication*. The 5' end of the copy peels off the circular plasmid and is transferred in linear form.
- The F⁺ cell transfers a copy of its F factor to the F⁻ partner, and the F⁻ cell <u>becomes</u> F⁺.
- The donor cell remains F⁺, with its original DNA intact.

The F factor occasionally inserts into the bacterial chromosome.
- Integrated F factor genes are still expressed.
- Cells with integrated F factors are called *Hfr cells* (high frequency of recombination).

Conjugation can occur between an Hfr and an F⁻ bacterium.
- As the integrated F factor of the Hfr cell transfers to the F⁻ cell, it pulls the bacterial chromosome behind its leading end.
- The F factor always opens up at the same point for a particular Hfr strain. As rolling circle replication proceeds, the sequence of chromosomal genes behind the leading 5' end is always the same.
- The conjugation bridge usually breaks before the entire chromosome and tail end of the F factor can be transferred. As a result:
    a. Only some bacterial genes are donated.
    b. The recipient F⁻ cell does not become an F⁺ cell, because only part of the F factor is transferred.
    c. The recipient cell becomes a partial diploid.
    d. Recombination occurs between the Hfr chromosomal fragment and the F⁻ cell. Homologous strand exchange results in a *recombinant F⁻ cell*.
    e. Asexual reproduction of the recombinant F⁻ cell produces a bacterial colony that is genetically different from both original parental cells.

How does bacterial conjugation differ from sexual reproduction in eukaryotic organisms?

3. **Interrupting Conjugation to Map Bacterial Chromosomes**
   (p. 359–360)

How can geneticists use several features of conjugation to roughly map a bacterial chromosome?

Scientists can map bacterial chromosomes by utilizing several features of conjugation.
- A specific strain of Hfr bacteria always transfers genes in the same sequence.
- The sequence of gene transfer results from where the F factor is inserted and how it is oriented in the bacterial chromosome. (See Campbell, Figure 17.13c)

Rolling circle replication

Hfr cells

250 Microbial Models: The Genetics of Viruses and Bacteria

☞ The duration of conjugation determines how many chromosomal genes will be transferred.

By artificially interrupting conjugation at different time intervals, geneticists can roughly determine gene location on the bacterial chromosome. The experimental steps are outlined below.

1. Liquid cultures of an Hfr strain and an F⁻ strain with different alleles are mixed together.
2. At successive time intervals, a sample is taken and agitated in a blender to disrupt conjugating pairs.
3. Genetic analysis indicates which genes were transferred during the time period allowed for conjugation.
4. Gene sequence and relative distances between them are deduced from the above information.

4. R Plasmids and Antibiotic Resistance (p. 361)

How has increased antibiotic use contributed to the increase in antibiotic resistant strains of pathogenic bacteria?

One class of nonepisomal plasmids, the *R plasmids* (for resistance), carry genes that confer resistance to certain antibiotics.

☞ Some carry up to 10 genes for resistance to antibiotics.
☞ Some mobilize their own transfer to nonresistant cells.
☞ Increased antibiotic use has selected for antibiotic resistant bacterial strains carrying the R plasmid.
☞ Additionally, R plasmids can transfer resistance genes to bacteria of different species including pathogenic strains. As a consequence, resistant strains of pathogens are becoming more common.

C. Transposons (p. 361–363)

What is a transposon, and who first proposed their existence?

Pieces of DNA called *transposons*, or *transposable genetic elements*, can actually move from one location to another in a cell's genome.

Transposons = DNA sequences that can move from one chromosomal site to another.
☞ Occur as natural agents of genetic change in both prokaryotic and eukaryotic organisms.
☞ Were first proposed in the 1940's by Barbara McClintock, who deduced their existence in maize. Decades later, the importance of her discovery was recognized; in 1983, at the age of 81, she received the Nobel Prize for her work.

There are two patterns of transposition: a) *conservative transposition* and b) *replicative transposition*.

Conservative transposition = Movement of preexisting genes from one genomic location to another; the transposon's genes are not replicated before the move, so the number of gene copies is conserved.

Microbial Models: The Genetics of Viruses and Bacteria

Replicative transposition = Movement of gene copies from their original site of replication to another location in the genome, so the transposon's genes are inserted at some new site without being lost from the original site.

How is transposition different from other mechanisms of genetic recombination?

Transposition is fundamentally different from all other mechanisms of genetic recombination, because transposons may scatter certain genes throughout the genome with no apparent single, specific target.
- ☞ All other mechanisms of genetic recombination depend upon homologous strand exchange: meiotic crossing over in eukaryotes; and transformation, transduction, and conjugation in prokaryotes.
- ☞ Insertion of episomic plasmids into chromosomes is site specific, even though it does not require an extensive stretch of DNA homologous to the plasmid.

1. Insertion Sequences (p. 361–362)

What are insertion sequences? What two types of nucleotide sequences do they include?

The simplest transposons are *insertion sequences*.

Insertion sequences (IS) = The simplest transposons, which only contain genes necessary for the process of transposition. Insertion sequence DNA includes two essential types of nucleotide sequences:
   a. Nucleotide sequence coding for *transposase*.
   b. *Inverted repeats*.

Transposase = Enzyme that catalyzes insertion of transposons into new chromosomal sites.
- ☞ The transposase gene in an insertion sequence is flanked by *inverted repeats*.

Inverted repeats (IR) = Short noncoding nucleotide sequences of DNA that are repeated in reverse order on opposite ends of a transposon. For example:

| DNA strand #1 | ... A T C C G G T ... | ... A C C G G A T ... |
| DNA strand #2 | ... T A G G C C A ... | ... T G G C C T A ... |

Note that each base sequence (IR) is repeated in reverse, on the DNA strand <u>opposite</u> the inverted repeat at the other end. Inverted repeats:
- ☞ Contain only 20 to 40 nucleotide pairs.
- ☞ Are recognition sites for transposase.

Replicative transposition

Insertion sequences

Transposase

Inverted repeats

252 Microbial Models: The Genetics of Viruses and Bacteria

How does transposase catalyze transposition? What other enzyme is required?

Transposase catalyzes the recombination by:
- ☞ Binding to the inverted repeats and holding them close together.
- ☞ Cutting and resealing DNA necessary to insert the transposon at a new site.

Insertion of transposons also requires other enzymes, such as DNA polymerase. For example,
- ☞ At the target site, transposase makes staggered cuts in the two DNA strands, leaving short segments of unpaired DNA at each end of the cut.
- ☞ Transposase inserts the transposon into the open target site.
- ☞ DNA polymerase helps form *direct repeats*, which flank transposons in their target site. Gaps in the two DNA strands fill in when nucleotides base pair with the exposed single-stranded regions.

<u>Direct repeats</u> = Two or more identical DNA sequences in the same molecule.
- ☞ The transposition process creates direct repeats that flank transposons in their target site.

Transposed insertion sequences are likely to somehow alter the cell's phenotype; they may:
- ☞ Cause mutations by interrupting coding sequences for proteins.
- ☞ Increase or decrease a protein's production by inserting within regulatory regions that control transcription rates.

Transposition of insertion sequences probably plays a significant role in bacterial evolution as a source of genetic variation.
- ☞ Though insertion sequences only rarely cause mutations (about one in every 10^6 generations), the mutation rate from transpositions is about the same as the mutation rate from extrinsic causes, such as radiation and chemical mutagens.

Microbial Models: The Genetics of Viruses and Bacteria 253

Why do some categorize insertion sequences as "selfish" DNA?

Because the role of insertion sequences is not known, some categorize them as parasitic or "selfish" DNA that perpetuates without providing any known benefit to the cell. They may be related to certain viruses, such as the phage Mu.
- Phage Mu ("mutator") can cause ten thousand times the normal mutation rate in its host cells, by transposing copies of Mu DNA to several sites.
- As a temperate phage, Mu uses the lysogenic cycle to reproduce. Unlike most temperate phages, it inserts at random in the bacterial genome, rather than at a specific target site.
- Insertion sequences transpose less often than Mu and lack genes for capsid proteins that make an extracellular stage possible.

2. **Complex Transposons and R Plasmids** (*p. 362–363*)

How does a complex transposon differ from an insertion sequence? What are some examples of genetic elements that contain complex transposons? How can transposons generate genetic diversity?

Complex transposons = Transposons which include additional genetic material besides that required for transposition; consist of one or more genes flanked by insertion sequences.
- The additional DNA may have any nucleotide sequence.
- Can insert into almost any stretch of DNA since their insertion is not dependent upon DNA sequence homology.
- Generate genetic diversity in bacteria by moving genes from one chromosome, or even one species, to another. This diversity may help bacteria adapt to new environmental conditions.
- An example is a transposon that carries a bacterial gene for antibiotic resistance.

Examples of genetic elements that contain one or more complex transposons include:
- F factor.
- R plasmids. (See Campbell, Figure 17.18c)
- DNA version of the retrovirus genome.

Complex transposons

X. THE CONTROL OF GENE EXPRESSION IN PROKARYOTES
(p. 363–368)

What are two main strategies cells use to control metabolism?

Genes switch on and off as conditions in the intracellular environment change. Cells have two main ways of controlling metabolism:

1. Regulation of enzyme activity.
   - ☞ The end product of an anabolic pathway may turn off its own production by inhibiting activity of an enzyme at the beginning of the pathway (*feedback inhibition*).
   - ☞ Useful for immediate short-term response.
2. Regulation of gene expression.
   - ☞ Accumulation of product triggers a mechanism that inhibits mRNA production by genes that code for an enzyme at the beginning of the pathway (*gene repression*).
   - ☞ Slower to take effect than feedback inhibition, but is more economical for the cell. It prevents unneeded protein synthesis for enzymes, as well as, unneeded pathway product.

An example illustrating regulation of a metabolic pathway is the tryptophan pathway in *E. coli*. (See Campbell, Figure 17.18) Mechanisms for gene regulation were first discovered for *E. coli*. Current understanding of such regulatory mechanisms at the molecular level is still limited to bacterial systems.

A. Operons: The Basic Concept (p. 364–366)

Regulated genes can be switched on or off depending on the cell's metabolic needs. From their research on the control of lactose metabolism in *E. coli*, Francois Jacob and Jacques Monod proposed a mechanism for the control of gene expression, the *operon* concept.

Structural gene = Gene that codes for a polypeptide.

Operon = A regulated cluster of adjacent *structural genes* with related functions.
   - ☞ Common in bacteria and phages.
   - ☞ Has a single promoter region, so an RNA polymerase will transcribe all structural genes on an all-or-none basis.
   - ☞ Transcription produces a single *polycistronic* mRNA with coding sequences for all enzymes in a metabolic pathway (e.g. tryptophan pathway in *E. coli*).

Polycistronic mRNA = A large mRNA molecule that is a transcript of several genes.
   - ☞ Is translated into separate polypeptides.
   - ☞ Contains codons signaling the termination and initiation of translation for each polypeptide.

Why can grouping genes into an operon be advantageous?

Grouping genes into operons is advantageous because:
   - ☞ Expression of these genes can be coordinated. When a cell needs the product of a metabolic pathway, all the necessary enzymes are synthesized at one time.
   - ☞ The entire operon can be controlled by a single *operator*.

Operator = A DNA segment between an operon's promoter and structural genes, which controls access of RNA polymerase to structural genes.
- ☞ Sometimes overlaps the transcription starting point for the operon's first structural gene.
- ☞ Acts as an on/off switch for movement of RNA polymerase and transcription of the operon's structural genes.

What determines whether an operator is in the "on" or "off" mode?

By itself, the operator is on; it is switched off by a protein *repressor*.

How are repressors similar to enzymes?

Repressor = Specific protein that binds to an operator and blocks transcription of the operon.
- ☞ Blocks attachment of RNA polymerase to the promoter.
- ☞ Is similar to an enzyme, in that it:
    1. Has an active site with a specific conformation, which discriminates among operators. Repressor proteins are specific only for operators of certain operons.
    2. Binds <u>reversibly</u> to DNA.
    3. May have an allosteric site in addition to its DNA-binding site.
- ☞ Repressors are encoded by *regulatory genes*.

What is the difference between a structural gene and a regulatory gene?

Regulatory genes = Genes that code for repressor or regulators of other genes.
- ☞ Are often located some distance away from the operons they control.
- ☞ Are involved in switching on or off the transcription of structural genes by the following process:

Transcription of the regulatory gene
↓ produces
mRNA
↓ translated into
Regulatory protein
↓ binds to
Operator
↓ represses or activates
Transcription of operon's structural genes

Regulatory genes are continually transcribed, so their activity depends upon how efficient their promoters are in binding RNA polymerase.
- ☞ They produce repressor molecules continuously, but slowly.

256 Microbial Models: The Genetics of Viruses and Bacteria

☞ Operons are still expressed even though repressor molecules are always present, because repressors are not always capable of blocking transcription; they alternate between inactive and active conformations.

Using the trp operon as an example, how can a metabolite cue a repressor? What are the functions of the operator, repressor and corepressor?

A repressor's activity depends upon the presence of key metabolites in the cell.

☞ Regulation of the *trp* operon in *E. coli* is an example of how a metabolite cues a repressor: (See Campbell, Figure 17.19)
☞ Repressible enzymes catalyze the anabolic pathway that produces tryptophan, an amino acid.
☞ Tryptophan accumulation represses synthesis of the enzymes that catalyze its production.

```
          absent                    present
         ──────── Tryptophan ────────
            ↓                          ↓
  Repressor protein is in     Repressor protein is in
  inactive conformation.      active conformation.
            ↓                          ↓
  trp operon is turned on.    Binds to operator.
                                       ↓
                              trp operon is switched off.
```

How does tryptophan activate the repressor protein?

☞ The repressor protein, which normally has a low affinity for the operator, has a DNA binding site plus an allosteric site specific for tryptophan.
☞ When tryptophan binds to the repressor's allosteric site, it activates the repressor causing it to change its conformation.
☞ The activated repressor binds to the operator, which switches the *trp* operon off.
☞ Tryptophan functions in this regulatory system as a *corepressor*.

Corepressor = A molecule, usually a metabolite, that binds to a repressor protein, causing the repressor to change into its active conformation.

☞ Only the *repressor-corepressor complex* can attach to the operator and turn off the operon.
☞ When tryptophan concentrations drop, it is less likely to be bound to repressor protein. The *trp* operon, once free from repression, begins transcription.
☞ As concentrations of tryptophan rise, it turns off its own production by activating the repressor.
☞ Enzymes of the tryptophan pathway are said to be *repressible*.

trp operon

Corepressor

B. Repressible Versus Inducible Enzymes: Two Types of Negative Gene Regulation (p. 366–367)

What are the differences between repressible and inducible enzymes?

  Repressible enzymes = Enzymes which have their synthesis inhibited by a metabolite (e.g. tryptophan).

  Inducible enzymes = Enzymes which have their synthesis stimulated or induced by specific metabolites.

How does the lac operon function? How can it operate without a corepressor?

  Some operons can be switched on or *induced* by specific metabolites (e.g. *lac* operon in *E. coli*).

  ☞ *E. coli* can metabolize the disaccharide lactose. Once lactose is transported into the cell, β-galactosidase cleaves lactose into glucose and galactose:

$$\text{lactose} \xrightarrow{\beta\text{-galactosidase}} \text{glucose} + \text{galactose}$$
$$\text{(disaccharide)} \qquad\qquad \text{(monosaccharides)}$$

  ☞ When *E. coli* is in a lactose-free medium, it only contains a few β-galactosidase molecules.

  ☞ When lactose is added to the medium, *E. coli* increases the number of mRNA molecules coding for β-galactosidase. These mRNA molecules are quickly translated into thousands of β-galactosidase molecules.

  ☞ Lactose metabolism in *E. coli* is programmed by the *lac* operon which has three structural genes:

   1. *lac Z* — Codes for β-galactosidase which hydrolyzes lactose.
   2. *lac Y* — Codes for a permease, a membrane protein that transports lactose into the cell.
   3. *lac A* — Codes for transacetylase, an enzyme that has no known role in lactose metabolism.

  ☞ The *lac* operon has a single promoter and operator. The *lac* repressor is innately active, so it attaches to the operon without a corepressor.

  ☞ Allolactose, an isomer of lactose, acts as an *inducer* to turn on the *lac* operon: (See Campbell, Figure 17.20)

Allolactose
↓ binds to repressor
Inactivated repressor loses affinity for *lac* operon.
↓
Operon is transcribed.
↓
Enzymes for lactose metabolism are produced.

Repressible enzymes

Inducible enzymes

lac operon

How do the differences between repressible and inducible enzymes reflect differences in the pathways they control?

Differences between repressible and inducible operons reflect differences in the pathways they control.

Repressible Enzymes	Inducible Enzymes
Their genes are switched on until a specific metabolite activates the repressor.	Their genes are switched off until a specific metabolite inactivates the repressor.
Generally function in anabolic pathways.	Function in catabolic pathways.
Pathway end product switches off its own production by repressing enzyme synthesis.	Enzyme synthesis is switched on by the nutrient the pathway uses.

Repressible and inducible operons share similar features of gene regulation. In both cases:
- ☞ Specific repressor proteins control gene expression.
- ☞ Repressors can assume an active conformation that blocks transcription and an inactive conformation that allows transcription.
- ☞ Which form the repressor assumes depends upon cues from a metabolite.

Both systems are thus examples of *negative control*.
- ☞ Binding of active repressor to an operator always turns off structural gene expression.
- ☞ The *lac* operon is a system with negative control, because allolactose does not interact directly with the genome. The derepression allolactose causes is indirect, by freeing the *lac* operon from the repressor's negative effect.

Positive control of a regulatory system occurs only if an activator molecule interacts directly with the genome to turn on transcription.

C. CAP: An Example of Positive Gene Regulation (*p. 367–368*)

What is the difference between positive and negative control? What is an example of each in the lac operon?

The *lac* operon is under dual regulation which includes negative control by repressor protein and positive control by *catabolite activator protein*.

CAP (catabolite activator protein) = A protein that binds within an operon's promoter region and enhances the promoter's affinity for RNA polymerase.
- ☞ It is necessary for the normal expression of the *lac* operon. (Even if allolactose is present to inactivate the repressor, transcription proceeds slowly because the promoter has such a low affinity for RNA polymerase.)
- ☞ It is a *positive regulator* because it directly interacts with the genome to stimulate gene expression.
- ☞ It can bind to the promoter only if glucose is absent from the cell.

Negative control

Positive control

CAP

E. coli preferentially uses glucose over lactose as a substrate for glycolysis. So, normal expression of the lac operon requires:
- ☞ *Presence* of lactose.
- ☞ *Absence* of glucose.

Most genes required for glucose catabolism are *constitutive genes*.

<u>Constitutive genes</u> = Unregulated genes that are continually transcribed.
- ☞ Examples are genes that produce proteins always needed by the cell (e.g. enzymes of glycolysis).
- ☞ Are not all transcribed at the same rate. Differences in relative transcriptional rates are evolutionary adaptations built into the genome, so an individual *E. coli* cell cannot alter these rates.
- ☞ Differ from inducible genes, which are only active in the presence of specific metabolites (e.g. lac operon requires allolactose).

How is CAP affected by the absence or presence of glucose?

- ☞ When glucose is missing, the cell accumulates *cyclic AMP (cAMP)*, a nucleotide derived from ATP.
- ☞ The cAMP activates CAP so that it can bind to the *lac* promoter.
- ☞ When glucose concentration rises, glucose catabolism decreases the intracellular concentration of cAMP. Thus, cAMP releases CAP.

```
        low                              high
         ──── Glucose concentration ────
          ↓                                ↓
  cAMP concentration rises.         cAMP becomes scarce.
          ↓                                ↓
  cAMP binds to CAP.                CAP loses its cAMP.
          ↓                                ↓
  cAMP-CAP complex binds            CAP disengages from the
  to lac promoter.                  lac promoter.
          ↓                                ↓
  Efficient transcription           Slowed transcription of
  of lac operon.                    lac operon.
```

In this dual regulation of the *lac* operon:
- ☞ Negative control by the repressor determines whether or not the operon will transcribe the structural genes.
- ☞ Positive control by CAP determines the rate of transcription.

How does E. coli use the negative and positive controls of the lac operon to economize on RNA and protein synthesis?

E. coli economizes on RNA and protein synthesis with the help of these negative and positive controls.
- ☞ CAP is an activator of several different operons that program catabolic pathways.

Constitutive genes

Cyclic AMP

- ☞ Glucose's presence deactivates CAP. This, in turn, slows synthesis of those enzymes a cell needs to use catabolites other than glucose.
- ☞ *E. coli* preferentially uses glucose as its primary carbon and energy source, and the enzymes for glucose catabolism are *constitutive*.
- ☞ Consequently, when glucose is present, CAP does not work and the cell's systems for using secondary energy sources are inactive.

When glucose is absent, the cell metabolizes alternate energy sources.
- ☞ The cAMP level rises, CAP is activated and transcription begins of operons that program the use of alternate energy sources.
- ☞ Which operon is actually transcribed depends upon which nutrients are available to the cell. For example, if lactose is present, the *lac* operon will be switched on as allolactose inactivates the repressor.

18   GENOME ORGANIZATION AND EXPRESSION IN EUKARYOTES

CHAPTER OUTLINE

Why is gene regulation more complex in eukaryotes than in prokaryotes?

Eukaryotic gene regulation is more complex than in prokaryotes, because eukaryotes:
- ☞ Have larger, more complex genomes. This requires that eukaryotic DNA be more complexly organized than prokaryotic DNA.
- ☞ Require cell specialization or *differentiation*.

<u>Cellular differentiation</u> = Divergence in structure and function of different cell types, as they become specialized during an organism's development.
- ☞ Cell differentiation requires that gene expression must be regulated on a long-term basis.
- ☞ Highly specialized cells, such as muscle or nerve, express only a small percentage of their genes, so transcription enzymes must locate the right genes at the right time.
- ☞ Uncontrolled or incorrect gene action can cause serious imbalances and disease, including cancer. Thus, eukaryotic gene regulation is of interest in medical as well as basic research.

DNA-binding proteins regulate gene activity in all organisms — prokaryotes as well as eukaryotes.
- ☞ Usually, it is DNA transcription that is controlled.
- ☞ Eukaryotes have more complex chromosomal structure, gene organization and cell structure than prokaryotes, which offer added opportunities for controlling gene expression.

I.   GENOME ORGANIZATION AT THE MICROSCOPIC LEVEL
(p. 373–375)

How does genome organization differ between prokaryotic and eukaryotic cells?

Prokaryotic and eukaryotic cells both contain double-stranded DNA, but their genomes are organized differently.

Prokaryotic DNA is:
- ☞ Usually circular.
- ☞ Much smaller than eukaryotic DNA; it makes up a small nucleoid region only visible with an electron microscope.
- ☞ Associated with only a few protein molecules.
- ☞ Less elaborately structured and folded than eukaryotic DNA; bacterial chromosomes have some additional structure as the DNA-protein fiber forms loops that are anchored to the plasma membrane.

Cellular differentiation

262 Genome Organization and Expression in Eukaryotes

Eukaryotic DNA is:
- ☞ Complexed with a large amount of protein to form *chromatin*.
- ☞ Highly extended and tangled during interphase.
- ☞ Condensed into short, thick, discrete *chromosomes* during mitosis; when stained, chromosomes are clearly visible with a light microscope.

Eukaryotic chromosomes contain an enormous amount of DNA, which requires an elaborate system of DNA packing to fit all of the cell's DNA into the nucleus.

What is the current model for progressive levels of DNA packing?

    A.   Nucleosomes, or "Beads on a String" (*p. 373*)

Histone proteins associated with DNA are responsible for the first level of DNA packing in eukaryotes.

<u>Histones</u> = Small proteins that are rich in basic amino acids and that bind to DNA, forming chromatin.
- ☞ Contain a high proportion of positively charged amino acids (arginine and lysine), which bind tightly to the negatively charged DNA.
- ☞ Are present in approximately equal amounts to DNA in eukaryotic cells.
- ☞ Are similar from one eukaryote to another, suggesting that histone genes have been highly conserved during evolution. There are five types of histones in eukaryotes.

How do histones influence folding in eukaryotic DNA? What is the basic unit of DNA packing?

Partially unfolded *chromatin* (DNA and its associated proteins) resembles beads spaced along the DNA string. Each beadlike structure is a *nucleosome*. (See Campbell, Figure 18.2a).

<u>Nucleosome</u> = The basic unit of DNA packing; it is formed from DNA wound around a protein core that consists of two copies each of four types of histone. A fifth histone (H$_1$) may be present on DNA next to the nucleosome.
- ☞ Nucleosomes may control gene expression by limiting access of transcription proteins to DNA.
- ☞ Nucleosome heterogeneity may also help control gene expression; nucleosomes may differ in the extent of amino acid modification and in the type of nonhistone proteins present.

    B.   Higher Levels of DNA Packing (*p. 373–375*)

The *30-nm chromatin fiber* is the next level of DNA packing. (See Campbell, Figure 18.2b)
- ☞ This structure consists of a tightly wound coil with six nucleosomes per turn.
- ☞ Molecules of histone H$_1$ pull the nucleosomes into a cylinder 30nm in diameter.

In the next level of higher-order packing, the 30-nm chromatin fiber forms *looped domains*.

Genome Organization and Expression in Eukaryotes 263

- Each loop in the 30-nm chromatin fiber contains 20,000 to 100,000 base pairs.
- Looped domains coil and fold, further compacting the chromatin in a mitotic chromosome characteristic of metaphase.

Interphase chromatin is much less condensed than mitotic chromatin, but it still exhibits higher-order packing.
- Its nucleosome string is usually coiled into a 30-nm fiber, which is folded into looped domains.
- Evidence is that interphase looped domains attach to a scaffolding inside the nuclear envelope and that chromatin fibers of different chromosomes do not become entangled.

What is the difference between heterochromatin and euchromatin?

Portions of some chromosomes remain highly condensed throughout the cell cycle, even during interphase. Such *heterochromatin* is not transcribed.

Heterochromatin = Chromatin that remains highly condensed during interphase and that is not actively transcribed.

Euchromatin = Chromatin that is less condensed during interphase and is actively transcribed; euchromatin becomes highly condensed during mitosis.

What is the function of heterochromatin in interphase cells?

- Since most heterochromatin is not transcribed, it may be a coarse control of gene expression.
- For example, Barr bodies in mammalian cells are X chromosomes that are mostly condensed into heterochromatin. In female somatic cells, one X chromosome is a Barr body, so the other X chromosome is the only one transcribed.

II. **GENOME ORGANIZATION AT THE MOLECULAR LEVEL**
    (*p. 375–378*)

DNA in eukaryotic genomes is organized differently from that in prokaryotes.
- In prokaryotes, most DNA codes for protein (mRNA), tRNA or rRNA, and coding sequences are uninterrupted. Small amounts of noncoding DNA consist mainly of control sequences, such as promoters.
- In eukaryotes, most DNA does *not* encode protein or RNA, and coding sequences may be interrupted by long stretches of noncoding DNA (introns). Certain DNA sequences may be present in multiple copies.

   A. **Repetitive Sequences** (*p. 375*)

Where is satellite DNA found? What role might it play in the cell?

About 10–25% of total DNA in higher eukaryotes consists of short (5 to 10 nucleotides) sequences that are tandemly repeated thousands of times. These DNA segments have a different natural density due to base composition differences.
- The density difference allows these segments to be isolated by ultracentrifugation in a cesium chloride density gradient.

264 Genome Organization and Expression in Eukaryotes

☞ DNA separated in this way appears as a separate band from the other DNA in the centrifuge tube, so it is called *satellite DNA*.

☞ Most satellite DNA in chromosomes is located at the tips and the centromere; it probably plays a structural role during chromosome replication and chromatid separation in mitosis and meiosis.

B. Multigene Families (p. 375–376)

Most eukaryotic genes are *unique sequences* present as single copies in the genome, but some genes are part of a *multigene family*.

Multigene family = A collection of genes that are similar or identical in sequence and presumably of common ancestral origin; such genes may be clustered or dispersed in the genome.

How can multigene families of identical genes be advantageous for a cell?

Families of identical genes are almost exclusively genes with a final product of RNA, not protein.

☞ An exception is the multigene family of identical genes that code for histone proteins.

☞ Identical genes are usually clustered.

☞ Genes for the major rRNA molecules are a multigene family of identical genes. These genes are repeated in series several hundred or thousand times, creating huge tandem arrays of genes, which enable cells to make millions of ribosomes when needed during active protein synthesis. (See Campbell, Figure 18.8)

Using α-globin and β-globin genes as examples, how do multigene families of nonidentical genes probably evolve? What is the role of transposition?

Examples of multigene families of nonidentical genes are two related families of genes that code for globin α and β polypeptide subunits of hemoglobin. (See Campbell, Figure 18.4)

☞ α-globins are encoded for by one gene family on chromosome 16 in humans, and β-globins are encoded for by the other family on chromosome 11.

☞ Different versions of each globin subunit are expressed at different times of development, which allows hemoglobin to change with changes in the organism's environment as it progresses from an embryo to a fetus, and then to an adult.

☞ Sequence similarities of the globin genes indicate that the α-like globins and β-like globins evolved from a common ancestral globin.

Families of identical genes probably arise from a single ancestral gene by repeated gene duplication.

☞ *Tandem gene duplication* appears to be frequent and results from mistakes made in DNA replication and recombination.

Families of nonidentical genes probably arise from mutations that accumulate in duplicated genes over time. *Pseudogenes* are evidence for such gene duplication and mutation.

Pseudogene = Nonfunctional gene that has a DNA sequence similar to a particular functional gene, but that lacks sites necessary for expression.

Satellite DNA

Multigene family

Pseudogene

- Several pseudogenes exist in the noncoding DNA between functional genes of the globin gene families.
- Globin pseudogenes are evidence that duplicated genes may move by transposition involving reverse transcription. Two characteristics of pseudogenes implicate an mRNA-intermediate; that is, pseudogenes:
  1. Lack introns.
  2. Have poly-A tails (like mRNA).

In addition to pseudogenes and repetitive DNA, significant amounts of noncoding DNA are found <u>within</u> genes as introns.

C. Organization of a Typical Eukaryotic Gene: A Review (p. 376)

The following is a brief review of the molecular anatomy of a eukaryotic gene and its transcript.

Eukaryotic genes:
- Contain introns, noncoding sequences that intervene within the coding sequence.
- Contain a promoter sequence at the 5' upstream end; RNA polymerase attaches to a promoter sequence and transcribes introns along with the coding sequences, or exons.

The primary RNA transcript (hnRNA) is processed into mature mRNA by:
- Removal of introns.
- Addition of a modified guanosine triphosphate cap at the 5' end.
- Addition of a poly-A tail at the 3' end.

Eukaryotic genes may be regulated by other noncoding control sequences located thousands of bases away from the promoter. These *enhancers* can powerfully influence transcription of the associated gene.

<u>Enhancer</u> = Noncoding DNA control sequence that enhances a gene's transcription and that is located thousands of bases away from the gene's promoter.

D. The Arrangement of Coordinately Controlled Genes (p. 376–378)

How are coordinately controlled genes arranged in prokaryotes and eukaryotes?

Coordinately controlled genes are arranged differently in a eukaryotic chromosome that in prokaryotic genomes.
- Prokaryotic genes that are turned on and off together are often clustered into operons; these adjacent genes share regulatory sites located at one end of the cluster. All genes of the operon are transcribed into one mRNA molecule and are translated together.
- Eukaryotic genes coding for enzymes of a metabolic pathway are often scattered over different chromosomes. Even functionally related genes on the same chromosome have their own promoters and are individually transcribed.

Eukaryotic genes, however, are often coordinately expressed, and a plausible mechanism involves the presence of a specific nucleotide sequence found in the regulatory elements of every gene in the coordinated group.

Enhancer

Genome Organization and Expression in Eukaryotes

☞ This sequence would be recognized by a single regulatory protein that activates or represses all genes in synchrony.

☞ This type of mechanism is known in bacteria; there is evidence for such shared sequences in eukaryotic cells.

III. GENOME PLASTICITY (p. 378–379)

In general terms, how can genome plasticity influence gene expression?

An organism's genome is plastic, or changeable, in ways that affect the availability of specific genes for expression.

☞ Genes may be available for expression in some cells and not others, or at some time in the organism's development and not others.

☞ Genes may, under some conditions, be amplified or made more available than usual.

☞ Changes in the physical arrangement of DNA, such as levels of DNA packing, affect gene expression. For example, genes in heterochromatin and mitotic chromosomes are not expressed.

The structural organization of an organism's genome is also somewhat plastic; movement of DNA within the genome and chemical modification of DNA influence gene expression.

A. Gene Amplification and Selective Gene Loss (p. 378)

What are the effects of gene amplification and selective gene loss?

Gene amplification may temporarily increase the number of gene copies at certain times in development.

Gene amplification = Selective synthesis of DNA, which results in multiple copies of a single gene.

☞ For example, rRNA genes in amphibians are selectively replicated in the oocyte, synthesizing a million or more additional copies of the rRNA genes that exist as extrachromosomal circles of DNA.

☞ Gene amplification permits the oocyte to make huge numbers of ribosomes that will produce the vast amounts of proteins needed when the egg is fertilized.

Genes may also be selectively lost in certain tissues.

Chromosome diminution = Elimination of whole chromosomes or parts of chromosomes from certain cells early in embryonic development.

☞ For example, chromosome diminution occurs in gall midges during early development; all but two cells lose 32 of their 40 chromosomes during the first mitotic division after the 16-cell stage.

☞ The two cells that retain the complete genome are germ cells that will produce gametes in the adult. The other 14 cells become somatic cells with only 8 chromosomes.

B. DNA Methylation (p. 378)

What is DNA methylation? How might it affect gene expression?

DNA methylation is the addition of methyl groups ($-CH_3$) to bases of DNA after DNA synthesis.

- ☞ Most plant and animal DNA contains methylated bases (usually cytosine).
- ☞ About 5% of the cytosine residues are methylated.
- ☞ This may be a cellular mechanism for long-term control of gene expression. When researchers examine different types of cells from different tissues, they find that:
    1. Genes that are not expressed are more heavily methylated than those that are expressed.
    2. Drugs that inhibit methylation can induce gene reactivation, even in Barr bodies.

DNA methylation may affect gene expression by causing structural changes in the DNA double helix.

C. **Rearrangements in the Genome** (p. 378–379)

How can rearrangements in the genome activate or inactivate genes?

Rearrangements of DNA in somatic cells can activate or inactivate specific genes.

1. Transposons (p. 379)

All organisms probably have transposons that move DNA from one location to another within the genome. (See Campbell, Chapter 17) Transposons can rearrange the genome by:
- ☞ Inserting into the middle of a coding sequence of another gene; it can prevent the interrupted gene from functioning normally.
- ☞ Inserting within a sequence that regulates transcription; the transposition may increase or decrease a protein's production.
- ☞ Inserting its own gene just downstream from an active promoter that activates its transcription.

2. Yeast Mating-Type Genes (p. 379)

There are two mating types in yeast, α and a; rearrangements in the genome can change one mating type to another.
- ☞ These unicellular fungi can mate only if they are opposite mating types.
- ☞ Mating types are distinguished by which of two alleles (α or a) is present at the *mating-type locus* (*MAT*).

Mating-type locus

Yeasts can change mating types by switching alleles at the MAT locus; a *cassette mechanism* is involved in this switch.

Cassette mechanism

- ☞ There are unexpressed (silent) copies of both the α and a genes at a site distant from the MAT locus.
- ☞ When a cell changes mating types, the gene in the MAT locus is excised and a copy of the other mating-type gene is inserted.
- ☞ Silent copies of both mating-type alleles are retained in the genome so descendants of a yeast cell can change mating-type repeatedly.
- ☞ This is called the *cassette mechanism* since it involves removal of one gene from the expressed ("play") slot and its replacement with another gene from the genome's silent collection.

268 Genome Organization and Expression in Eukaryotes

Immunoglobulins

3. Immunoglobulin Genes (p. 379)

During cellular differentiation in mammals, permanent rearrangements of DNA segments occur in those genes that encode antibodies, or *immunoglobulins*.

Immunoglobulins = A class of proteins (antibodies) produced by B lymphocytes that specifically recognize and help combat viruses, bacteria, and other invaders of the body. Immunoglobulin molecules consist of: (See Campbell, Figure 18.7)
- ☞ Four polypeptide chains held together by disulfide bridges.
- ☞ Each chain has two major parts:
    a. A *constant region*, which is the same for all antibodies of a particular class.
    b. A *variable region*, which gives an antibody the ability to recognize and bind to a specific foreign molecule.

B lymphocytes, which produce immunoglobulins, are a type of white blood cell found in the mammalian immune system.
- ☞ The human immune system contains millions of subpopulations of B lymphocytes that produce different antibodies.
- ☞ B lymphocytes are very specialized; each differentiated cell and its descendants produce only one specific antibody.
- ☞ Cell specialization of B lymphocytes is a consequence of the antibody gene's unique organization.

What is the genetic basis for antibody diversity?

As an unspecialized embryonic cell differentiates into a B lymphocyte, its antibody gene is pieced together randomly from several DNA segments that are physically separated in the genome.
- ☞ In the genome of an embryonic cell, a long stretch of DNA separates the sequence coding for the constant region, from a site containing hundreds of coding sequences for the variable regions.
- ☞ As a B lymphocyte differentiates, the interrupting stretch of DNA is deleted, connecting a specific DNA sequence coding for a variable region to a DNA sequence coding for a constant region.
- ☞ The joined segments form the continuous nucleotide sequence that functions as the gene for one of the polypeptides.
- ☞ The primary RNA transcript is processed in the usual way to form mRNA that is translated into one of the polypeptides of an antibody molecule.

Antibody variation results from different combinations of variable and constant regions in the polypeptides, as well as, from different combinations of polypeptides that form complete antibody molecules.

IV. THE CONTROL OF GENE EXPRESSION (p. 379–383)

Where in the steps of gene expression can gene regulation occur?

Complexities in chromosome structure, gene organization and cell structure provide opportunities for the control of gene expression in eukaryotic cells. The steps of gene expression where gene regulation can occur are outlined below.

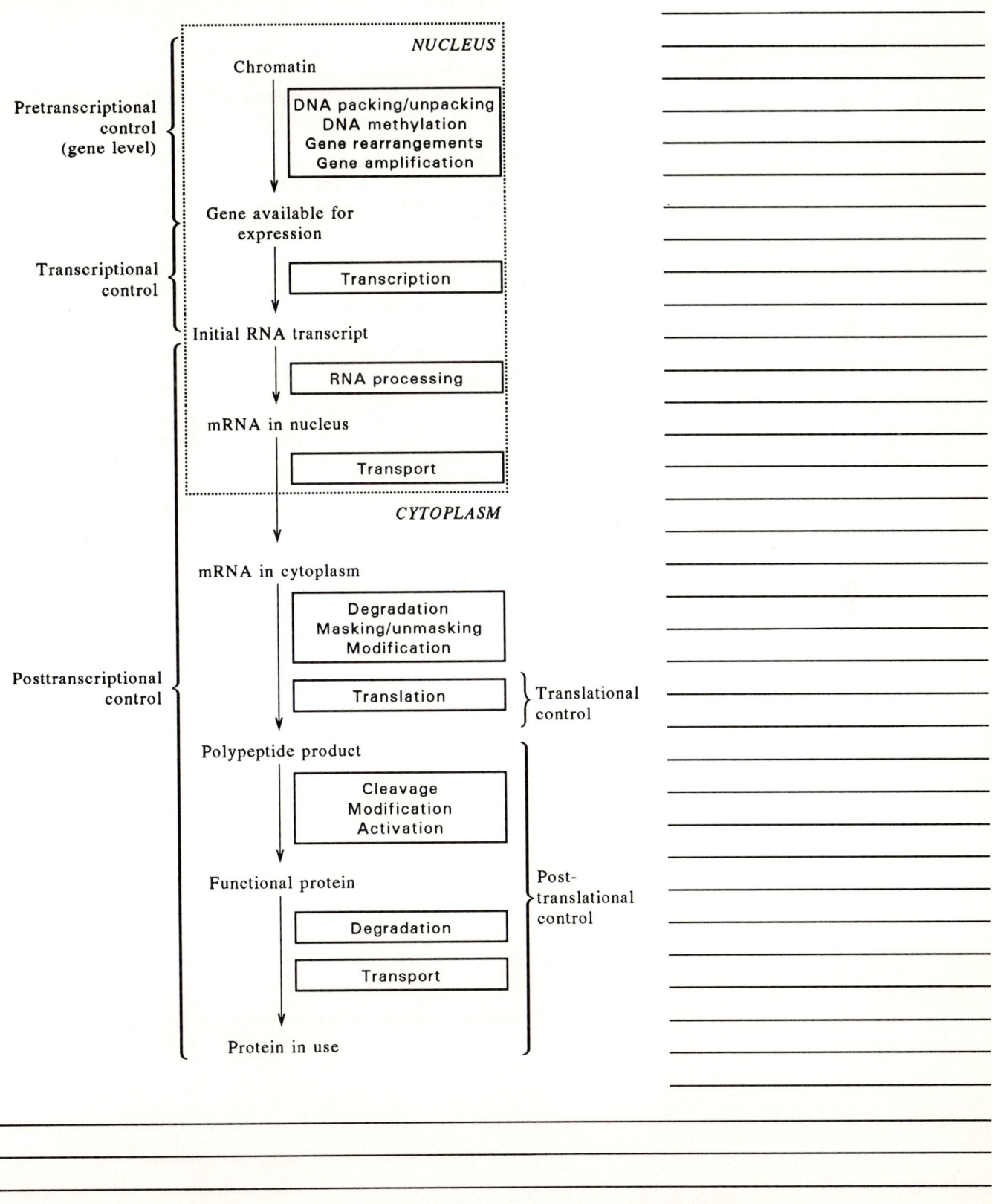

270 Genome Organization and Expression in Eukaryotes

A. Transcriptional Control of Gene Expression (p. 380–381)

What potential role do promoters and enhancers play in transcriptional control?

Transcription in both prokaryotes and eukaryotes requires that RNA polymerase recognize and bind to DNA at the promoter. However, in eukaryotes:

- ☞ RNA polymerase cannot recognize the promoter without the help of a protein *transcription factor* that binds to a specific upstream, noncoding sequence on the promotor.
- ☞ Some transcription factors bind to *enhancer* regions, noncoding control sequences of DNA that may be thousands of bases away from the promoter and coding sequence. (See Campbell, Figure 18.5)

Associations between transcription factors and enhancer sites strongly influence gene expression in eukaryotes. *How do enhancers stimulate transcription of specific genes?*

- ☞ One hypothesis is that a hairpin loop forms in DNA, bringing the transcription factor bound to enhancer into contact with transcription factors and polymerase at the promoter. (See Campbell, Figure 18.9)
- ☞ Diverse enhancer sequences and their transcription factors may selectively activate gene expression at appropriate stages in cell development.

What are three proposed structural motifs for transcription factors that are grouped by DNA-binding domain?

There are three main structural motifs of transcription factors that are grouped by DNA-binding *domain*, the protein's functional region that binds to DNA. (See Campbell, Figure 18.10)

1. *Helix-turn-helix domain.* This domain is two regions of α helix connected by a short turn; the two α helices bind to specific DNA sequences.
2. *Zinc-finger domain.* "Zinc fingers" are folded polypeptide chains that are stabilized by zinc atoms.
3. *Leucine zipper domain.* This domain consists of two polypeptide chains, each with a leucine rich region that zips together to form active DNA-binding protein.

B. Posttranscriptional Control of Gene Expression (p. 381–383)

Gene expression is measured by the types and amounts of proteins, tRNA, or rRNA that a cell makes. Transcription produces a primary transcript, but gene expression may be interrupted or stimulated at any posttranscriptional step.

How does the nuclear envelope in eukaryotes offer a level of post-transcriptional control beyond that found in prokaryotes?

Because eukaryotic cells have a nuclear envelope, translation is segregated from transcription. This offers additional opportunities for controlling gene expression.

1. **RNA Processing and Export** (*p. 381–382*)

   A primary RNA transcript must be processed before it functions as mRNA, tRNA, or rRNA.
   - ☞ A 5' cap and poly-A tail are added to mRNA.
   - ☞ Introns are removed and exons are spliced together.

   The processed transcript is exported to the cytoplasm.
   - ☞ It probably passes through the nuclear envelope by way of a nuclear pore.
   - ☞ In the cytoplasm, mRNA interacts with specific proteins and may associate with ribosomes to be translated.

   Each step of RNA processing and export is an opportunity for controlling gene expression; however, little is known about how these posttranscriptional steps are regulated.

2. **Regulation of mRNA Degradation** (*p. 382*)

Why can the ability to rapidly degrade mRNA be an adaptive advantage for prokaryotes? What is the importance of mRNA degradation in eukaryotes?

   Protein synthesis is also controlled by mRNA's lifespan in the cytoplasm.
   - ☞ Prokaryotic mRNA molecules are degraded by enzymes after only a few minutes. Thus, bacteria can quickly alter patterns of protein synthesis in response to environmental change.
   - ☞ Eukaryotic mRNA molecules can exist for several hours or even weeks.
   - ☞ The longevity of a mRNA affects how much protein synthesis it directs. Those that are viable longer can produce more of their protein.
   - ☞ For example, long-lived mRNAs for hemoglobin are repeatedly translated in developing vertebrate red blood cells.

   In some cases, mRNA molecules accumulate and a mass translation occurs after some control signal is received.
   - ☞ Prior to fertilization, the ovum produces and stores the mRNA that will be used for protein synthesis during the first embryonic cleavage.
   - ☞ The inactive mRNA is stored in the ovum until fertilization, when a control signal triggers translation.
   - ☞ Delayed translation of stored mRNA allows cells to respond quickly with an explosive burst of protein synthesis when it is needed later in development.

3. **Translational and Posttranslational Control** (*p. 382–383*)

How may gene expression be controlled at the translational and posttranslational level?

Compared to prokaryotes, eukaryotes require many more protein factors, especially initiation factors, for translation.
- For example, the sudden appearance of an initiation factor triggers translation of inactive mRNA in fertilized eggs. This protein factor must be present for ribosomes and mRNA to interact.

Posttranslational control is the last level of control for regulating gene expression.
- Many eukaryotic polypeptides must be modified or transported before becoming biologically active. Such modifications include:
  a. Cleavage of the polypeptide into smaller molecules (e.g. insulin).
  b. Adding chemical groups, such as sugars.
  c. Dispatching proteins targeted for specific sites designated by signal sequences.
- Selective degradation of particular proteins and regulation of enzyme activity are also control mechanisms of gene expression.

V. **THE ROLE OF SMALL MOLECULES IN REGULATING EUKARYOTIC GENE EXPRESSION** (*p. 383–384*)

In eukaryotes, certain small organic molecules combine with regulatory proteins to influence transcription.
- Examples include animal steroid hormones, chemical messengers produced in one part of the body that are carried in the bloodstream to affect target organs elsewhere in the body.
- Effect of steroids on gene transcription in target cells was first clearly recognized in insects.

A. **Chromosome Puffs: Evidence for the Regulatory Role of Steroid Hormones in Insects** (*p. 383*)

What experimental evidence demonstrates transcriptional control of eukaryotic gene expression? How do chemical signals such as steroid hormones influence transcriptional control?

Evidence for transcriptional control and the influence of steroids on gene expression has come from studies of giant polytene chromosomes found in the salivary gland and other tissues of insects.
- Chromosome puffs appear at specific sites along polytene chromosomes at characteristic stages in larval development.
- Puff formation occurs when DNA loops extend out from the chromosome axis.
- Autoradiography has shown that puffs correspond to regions of intense RNA synthesis, which indicates that the DNA extending out is more accessible to RNA polymerase.

As a larva develops, puff location along the chromosome changes.
- At molting, some puffs disappear and new puffs form at other sites.
- Shifting puffs indicate that selective switching on and off of specific genes occurs during development.
- *Ecdysone* (molting hormone) can induce changes in puff patterns, which demonstrates that gene regulation responds to specific chemical signals.

B. The Action of Steroid Hormones in Vertebrates (p. 383–384)

What are the key steps of steroid hormone action on gene expression in vertebrates?

Steroid hormones are chemical signals that can activate gene expression in target cells of vertebrates. The key steps of steroid hormone action are outlined below.
- Since steroids are lipid soluble, they readily diffuse across the plasma membrane into the cytoplasm.
- From the cytoplasm, the steroid enters the nucleus, where it binds to a soluble, *steroid-receptor* protein — a DNA-binding protein that can activate transcription of a particular gene.
- In absence of steroid, an *inhibitory protein* binds to the steroid receptor and blocks its DNA-binding domain, preventing the receptor from binding to DNA.
- When steroid is present, its binding to the receptor causes release of inhibitory protein and activates the steroid receptor, so it can attach to DNA at *enhancer* sequences that control steroid-responsive genes.

Steroid-receptor protein

Inhibitory protein

Summary: The Control of Gene Expression in Eukaryotes
- Various cell types of multicellular organisms express different genes.
- Rearrangements in the genome make certain genes available for expression and other genes unavailable.
- Opportunities for the control of gene expression exist at each step in the pathway from gene to functional protein.
- Transcriptional control is important in determining which genes are expressed; selective binding of regulatory proteins to enhancer sequences in DNA, stimulates transcription of specific genes.
- Regulatory activity of some DNA-binding proteins is affected by certain hormones and other chemical cues.

Genome Organization and Expression in Eukaryotes

VI. GENE EXPRESSION AND CANCER (p. 384–386)

What normal control mechanisms limit cell growth and division?

When a normal cell transforms into a cancerous cell, density-dependent inhibition and other controls limiting growth and division are absent. (See Campbell, Chapter 11) Cancer can be induced by *carcinogens* that cause mutations altering gene expression.

<u>Carcinogens</u> = Physical agents such as X-rays and chemical agents that cause cancer by mutating DNA.

Certain viruses can also cause cancer. Whether cancer is caused by physical agents, chemical agents, or viruses, the mechanism involves activation of *oncogenes* that are either native to the cell or introduced in viral genomes.

<u>Oncogene</u> = Gene responsible for a cell becoming cancerous.

Proto-oncogenes (normal cellular genes corresponding to oncogenes) are believed to be key genes in the control of cell growth and differentiation.
- ☞ These key genes are being identified by comparing DNA sequences of known oncogenes with known genes for cell growth factors.
- ☞ Studies of cancer and its causes are providing information into the control of normal development, and vice versa.
- ☞ One oncogene has been found to correspond to a gene that functions as a growth factor in normal wound healing.

How can proto-oncogenes be converted to oncogenes?

Four mechanisms can convert proto-oncogenes to oncogenes.
1. *Gene amplification.* Sometimes more copies of oncogenes are present in a cell than is normal.
2. *Chromosome translocation.* Malignant cells frequently contain chromosomes that have broken and rejoined, placing pieces of different chromosomes side-by-side and possibly separating the oncogene from its normal control regions.
3. *Gene transposition.* Abnormal expression of an oncogene may occur if the oncogene or its regulatory gene are transposed to an unusual locus.
4. *Point mutation.* A slight change in the nucleotide sequence might produce a growth-stimulating protein that is more active or more resistant to degradation than the normal protein.

How can changes in tumor-suppressor genes influence the transforming of normal cells into cancerous cells?

In addition to mutations affecting growth-stimulating proteins, changes in genes whose products <u>inhibit</u> cell division can also be involved in cancer.

Tumor-suppressor genes code for proteins that normally inhibit uncontrolled cell division.
- ☞ Mutations that affect these genes and prevent the production of normal tumor-suppressor protein may contribute to the onset of cancer (e.g. retinoblastoma allele).

- ☞ Children who are homozygous for the retinoblastoma allele develop malignant eye tumors. The mutated genes do not produce normal tumor-suppressor protein.
- ☞ When a normal allele coding for tumor suppressor is inserted into cells with mutated genes, abnormal growth of cancerous cells in culture is stopped.

More than one somatic mutation is probably needed to transform normal cells into cancerous cells. One of the best understood examples is colorectal cancer.
- ☞ Development of metastasizing colorectal cancer is gradual, and the first sign is unusually rapid cell division of apparently normal cells in the colon lining; a benign tumor (polyp) appears, and eventually a malignant tumor may develop.
- ☞ During this process, mutations in oncogenes and tumor-suppressor genes gradually accumulate. After a number of genes have changed, the tumor becomes malignant.
- ☞ In most cases studied, four changes occur at the DNA level: the activation of a cellular oncogene (*ras*) and the inactivation of three tumor-suppressor genes on chromosomes 5, 18, and 17. (See Campbell, Figure 18.13)

How are oncogenes involved in virus-induced cancers?

Virus-induced cancers probably account for 15% of human cancer worldwide.
- ☞ Viruses might add oncogenes to cells or disrupt DNA, affecting proto-oncogenes or a tumor-suppressor gene.
- ☞ However, viral-induced DNA changes are probably not all that is required to cause cancer; additional DNA changes needed must be triggered by environmental agents.

19 DNA TECHNOLOGY

CHAPTER OUTLINE

Recombinant DNA technology

Recombinant DNA technology refers to the set of techniques for recombining genes from different sources *in vitro* and transferring this recombinant DNA into a cell where it may be expressed.
- These techniques were first developed around 1975 for basic research in bacterial molecular biology, but this technology has also led to many important discoveries in basic eukaryotic molecular biology.

Biotechnology

Such discoveries resulted in the appearance of the *biotechnology* industry. Biotechnology refers to the use of living organisms or their components to do practical tasks such as:
- The use of microorganisms to make wine and cheese.
- Selective breeding of livestock and crops.
- Production of antibiotics from microorganisms.
- Production of monoclonal antibodies.

Application of recombinant DNA techniques makes modern biotechnology a more precise and systematic process than earlier research methods. It is also a more powerful tool since it allows genes to be moved across the species barrier.

Human genome project
- The *human genome project* is an important application of this technology. This project's goal is to transcribe and translate the entire human genome in order to better understand the human organism.
- A variety of other applications are possible for this technology, but the practical goal of all applications is the improvement of human health and food production.

I. BASIC STRATEGIES OF GENE MANIPULATION AND ANALYSIS
(*p. 391–404*)

Prior to the discovery of recombinant DNA techniques, procedures for altering the genes of organisms were constrained by the need to find and propagate desirable mutants.
- Geneticists relied on either natural processes, mutagenic radiation, or chemicals to induce mutations.
- In a laborious process, each organism's phenotype was checked to determine the presence of the desired mutation.
- Microbial geneticists developed techniques for screening mutants. For example, bacteria was cultured on media containing an antibiotic to isolate mutants which were antibiotic resistant.

Selection of mutants is not gene manipulation, since it does not involve the transfer of a desired gene from one organism to another. Before 1975, transferring genes between organisms was accomplished by cumbersome and nonspecific breeding procedures. The only exception to this was the use of bacteria and their phages.

Phages

- Genes can be transferred from one bacterial strain to another by the natural processes of transformation, conjugation or transduction.

Transformation

- Geneticists used these processes to carry out detailed molecular studies on the structure and functioning of prokaryotic and phage genes.
- Ideal for laboratory experiments, bacteria and phages are relatively small, have simple genomes, and are easily propagated.
- Although the technique was available to grow plant and animal cells in culture, the workings of their genomes could not be examined using existing methods.

A. **Recombinant DNA and Gene Cloning** (*p. 391–399*)

How has recombinant DNA technology helped scientists study the eukaryotic genome?

Recombinant DNA technology now makes it possible for scientists to examine the structure and function of the eukaryotic genome, because it contains several key components:
- Biochemical tools that allow construction of recombinant DNA.
- Methods for purifying DNA molecules and proteins of interest.
- Vectors for carrying recombinant DNA into cells and replicating it.
- Techniques for determining nucleotide sequences of DNA molecules.

1. **Restriction Enzymes** (*p. 391–393*)

What is the natural function of restriction enzymes? How do bacterial cells protect their own DNA from restriction?

Restriction enzymes are major tools in recombinant DNA technology.

Restriction enzymes

- First discovered in the late 1960's, these enzymes occur naturally in bacteria where they protect the bacterium against intruding DNA from other organisms.
- This protection involves *restriction*, a process in which the foreign DNA is cut into small segments.

Restriction

- Most restriction enzymes only recognize short, specific nucleotide sequences called *recognition sequences*. They only cut at specific points within those sequences.

Bacterial cells protect their own DNA from restriction through *modification* or methylation of DNA.

Modification

- Methyl groups are added to nucleotides within the recognition sequences.
- Modification is catalyzed by separate enzymes that recognize these same DNA sequences.

There are several hundred restriction enzymes and about a hundred different specific recognition sequences.

- ☞ Recognition sequences are symmetric in that the same sequence of four to eight nucleotides is found on both strands, but run in opposite directions.
- ☞ Restriction enzymes usually cut phosphodiester bonds of both strands in a staggered manner, so that the resulting double-stranded DNA fragments have single-stranded ends, called *sticky ends*.
- ☞ The single-stranded short extensions form hydrogen-bonded base pairs with complementary single-stranded stretches on other DNA molecules.

What are sticky ends? How is their creation by restriction enzymes useful in producing a recombinant DNA molecule?

Sticky ends of *restriction fragments* are used in the laboratory to join DNA pieces from different sources (cells or even different organisms).
- ☞ These unions are temporary since they are only held by a few hydrogen bonds.
- ☞ These unions can be made permanent by adding the enzyme *DNA ligase*, which catalyzes formation of covalent phosphodiester bonds.

The outcome of this process is the same as natural genetic recombination, the production of recombinant DNA – a DNA molecule carrying a new combination of genes. (See Campbell, Figure 19.4)

How are restriction enzymes and gel electrophoresis used to isolate DNA fragments?

Gel electrophoresis is used to separate the restriction fragments after large molecules of DNA are treated with a restriction enzyme.
- This technique separates macromolecules on the basis of size and electric charge by measuring rate of movement through an electric field.
- Particular DNA molecules can be identified by the band patterns they produce in gel electrophoresis after digestion with restriction enzymes. Segments of chromosomal DNA, viral DNA, and plasmid DNA can be identified in this manner.
- Gel electrophoresis can also be used to isolate and purify individual fragments with interesting genes. These individual fragments can be recovered from the gel and still retain full biological activity.

2. Gene-Cloning in a Plasmid: A Closer Look (*p. 394*)

Recombinant DNA molecules are only useful if they can be made to replicate and produce a large number of copies.

What are the major steps in a typical gene-cloning procedure?

A typical gene-cloning procedure includes the following steps: (See Campbell, Figure 19.4)

Step 1: *Isolation of two kinds of DNA.*
- Bacterial plasmids and foreign DNA containing the gene of interest are isolated.
- In this example, the plasmid is from *E. coli* and has two genes, *amp*^{R} and *tet*^{R}, that confer antibiotic resistance to ampicillin and tetracycline.
- Note that the recognition sequence for the restriction enzyme used in this example is within the *tet*^{R} gene.

Step 2: *Treatment of plasmid and foreign DNA with the same restriction enzyme.*
- The restriction enzyme cuts plasmid DNA at the *restriction site*, the recognition sequence within the *tet*^{R} gene.
- The foreign DNA is cut into thousands of fragments by the same restriction enzyme; one of the fragments contains the gene of interest.
- When the restriction enzyme cuts, it produces *sticky ends* on both the foreign DNA fragments and the plasmid.

Step 3: *Mixture of foreign DNA with clipped plasmids.*
- Sticky ends of the plasmid base pair with complementary sticky ends of foreign DNA fragments.

Step 4: *Addition of DNA ligase.*
- DNA ligase catalyzes the formation of covalent bonds, joining the two DNA molecules and forming a new plasmid with recombinant DNA.

Step 5: *Introduction of recombinant plasmid into bacterial cells.*
- The naked DNA is added to a bacterial culture.

Gel electrophoresis

Plasmid DNA

Recombinant plasmid

Gene cloning

☞ Some bacteria will take up the plasmid DNA by transformation.

Step 6: *Production of multiple gene copies by gene cloning.*
☞ Bacteria with the recombinant plasmid are allowed to reproduce, cloning the inserted gene in the process.
☞ Recombinant plasmids can be identified by the fact that they are ampicillin resistant but tetracycline sensitive; the foreign gene inserts within the tet^R gene and inactivates it.

3. **Inserting DNA into Cells** (*p. 394*)

How are plasmid and phage vectors produced? How are vectors used in recombinant DNA technology?

Small DNA fragments can be inserted into a bacterial plasmid without interfering with its ability to replicate within the bacterial cell.
☞ Isolated plasmids can then be introduced into cells by transformation.
☞ The plasmids are thus *vectors*, a means to move recombinant DNA from test tubes into cells.

Vectors

Bacteriophage λ

Bacteriophage λ can also be used as a vector.
☞ The middle of the linear genome, which contains nonessential genes, is deleted by using restriction enzymes.
☞ Restriction fragments of foreign DNA are then inserted to replace the deleted area.
☞ The recombinant phage DNA is introduced into an *E. coli* cell.
☞ The phage replicates itself inside the bacterial cell.
☞ Each new phage particle carries the foreign DNA "passenger."

How can foreign DNA be inserted into eukaryotic cells?

Sometimes it is necessary to clone DNA in eukaryotic cells rather than in bacteria. Under the right conditions, yeast and animal cells growing in culture can also take up foreign DNA from the medium.
☞ If the new DNA becomes incorporated into chromosomal DNA or can replicate itself, it can be cloned with the cell.
☞ Since yeast cells have plasmids, scientists can construct recombinant plasmids that combine yeast and bacterial DNA and that can replicate in either cell type.
☞ Viruses can also be used as vectors with eukaryotic cells.

There are more aggressive techniques for inserting foreign DNA into eukaryotic cells:
☞ In *electroporation*, a brief electric pulse applied to a cell solution causes temporary holes in the plasma membrane, through which DNA can enter.
☞ With thin needles, DNA can be injected directly into a eukaryotic cell.
☞ DNA attached to microscopic metal particles can be fired into plant cells with a gun.

Electroporation

4. **Sources of Genes for Cloning** (*p. 394–398*)

What are two major sources of genes for cloning?

There are two major sources of DNA which can be inserted into vectors and cloned:
1. DNA isolated directly from an organism.
2. Complementary DNA made in the laboratory from mRNA templates.

DNA isolated from a particular organism is an obvious source of genes.
- Scientists isolate the organism's DNA which contains the gene of interest.
- Restriction enzymes are used to cut this DNA into thousands of pieces which are slightly larger than a gene.
- All of these pieces are then inserted into plasmids or viral DNA.
- These vectors containing the foreign DNA are introduced into bacteria.
- This produces the *genomic library*: thousands of DNA segments from a genome, each of which is carried by a plasmid or phage.

Genomic library

What is the advantage of using cDNA as a source of genes for cloning? How can reverse transcriptase be used to produce cDNA? What is the natural function of reverse transcriptase in retroviruses?

One problem with direct isolation of DNA from an organism is that eukaryotic genes of interest may be too large to clone easily because they contain introns (noncoding regions). A solution is to make *complementary DNA (cDNA)* in the laboratory, by using mRNA molecules as templates. (See Campbell, Figure 19.6)
- The enzyme *reverse transcriptase* (from retroviruses) is used to catalyze DNA synthesis from an RNA template.
- Even if the gene of interest contains introns, the cDNA does not, because they have been removed from the template during RNA processing.
- Without introns, the cDNA gene is more manageable in size than the original gene.
- The cDNA also has the potential of being translated into protein by bacterial cells which lack RNA-processing capabilities.
- To be transcribed in bacteria, the cDNA is attached to other DNA containing a promoter and other essential transcription signals.

Complementary DNA

Reverse transcriptase, Retroviruses

The cDNA method produces partial genomic libraries because mRNA molecules for a particular gene usually cannot be distinguished from other RNA in a cell.
- By using cells from specialized tissues or a cell culture used exclusively for making one gene product, the majority of mRNA produced is for the gene of interest.

☞ For example, most of the mRNA in precursors of mammalian erythrocytes is for the protein hemoglobin.

5. Using Probes to Find a Gene of Interest (p. 398)

How can scientists identify the bacterial clone that contains the gene of interest? What is a probe and how is it used to locate a gene?

Identifying the bacterial clone containing the gene of interest is difficult regardless of which approach (genomic library or cDNA) is used.

Screening for the protein can be used if the clones translate the gene into a product.
☞ Detection may be based on activity (for enzymes) or structure (using antibodies that combine with the protein).

Screening for the gene itself is most often used.
☞ These methods use base pairing between the gene and a complementary sequence on another nucleic acid (RNA or DNA).
☞ Nucleic acid segments can be synthesized and used to identify a gene if a part of the gene's nucleotide sequence is known or inferred by its protein product.
☞ The synthesized nucleic acid used to find the gene of interest is called a *probe*.
☞ The location of a probe is traced by labeling it with a radioactive isotope. (See Campbell, Figure 19.7)

When a bacterial clone carrying the gene of interest is identified, the gene can be easily isolated in large amounts and used for further studies.

Cloned genes can also be used as probes to identify similar or identical genes.

6. Making Gene Products Using Genetic Engineering (p. 398–399)

Yeast genes, as well as those of more complex eukaryotes, can be expressed in bacteria. This possibility surprised scientists because:
☞ Prokaryotes have different signals that control gene transcription and translation.
☞ The enzymes that recognize these signals are also different.

Bacteria are the organisms of choice for large scale, commercial production of gene products, because:
☞ They can be grown rapidly and cheaply in large fermenters.
☞ Their genomes are relatively simple and easy to manipulate.
☞ Bacteria can be manipulated to secrete a protein as it is made, simplifying purification of the protein.

Probe

How can the expression of eukaryotic genes in bacterial cells be maximized?

Expression of eukaryotic genes in bacterial cells can be maximized by several genetic and biochemical methods. They include:
- Using plasmid vectors that produce many copies per cell.
- Changing the promoter controlling the gene's expression to a highly active one.
- Attaching a eukaryotic gene to the initial portion of a bacterial gene for a protein that is usually produced in large quantities.

Why might it be important to insert foreign eukaryotic genes into eukaryotic cells rather than into bacterial cells? What is the advantage of using yeast cells for this purpose?

Eukaryotic cells must be used to study eukaryotic mechanisms of gene function and regulation.
- While eukaryotic genes are initially cloned in bacteria, they are later returned to eukaryotic cells such as yeast for expression and study.
- Yeast have advantages for use in the production of gene products (commercially or for research).
    1. They are easy to grow.
    2. They are very simple eukaryotes.
    3. It is simple to reinsert genes into yeast since they can be made to take up DNA by transformation and have plasmids for the process.

Bacteria and yeast are not suitable for every purpose. For certain applications, plant or animal cell cultures must be used.
- Cells of more complex eukaryotes carry out certain biochemical processes not found in yeasts (e.g. only animal cells produce antibodies).

B. **DNA Synthesis and Sequencing** (*p. 399*)

How are DNA synthesis and sequencing used in modern studies of eukaryotic genomes?

It is possible to synthesize artificial genes in the laboratory (e.g. genes for insulin).
- This once tedious task can now be done with machines that can rapidly produce genes several hundred nucleotides long.
- This approach can only be used for short genes with known nucleotide sequences.

284 DNA Technology

Sanger Method

Dideoxyribonucleotide

What is the basis of the Sanger method for sequencing DNA?

The most common method for determining the sequence of a DNA molecule is the Sanger Method. (See Campbell, Methods Box, page 400 for a description of DNA sequencing by the Sanger Method.) The method is based on:

- Use of restriction enzymes to cut DNA into discrete, reproducible fragments with unique sequences.
- Synthesizing *in vitro* DNA strands complementary to a strand from the restriction fragment being sequenced.
- Incorporation of a modified nucleotide that blocks further DNA synthesis (*dideo*xyribonucleotide, which lacks 2 OH groups).

Restriction enzymes are used to cut the DNA into restriction fragments. Each fragment is processed in the following manner:

Step 1: A preparation containing one of the strands of the DNA fragment is divided into four portions.
- Each portion is incubated with a radioactively labeled primer, DNA polymerase, and the four deoxyribonucleotide triphosphates (ingredients necessary for complementary strand synthesis).
- Each mixture also contains <u>one</u> of the four nucleotides in the modified dideoxy (dd) form.

Step 2: New strand synthesis begins with the radioactive primer and continues until a dideoxyribonucleotide is incorporated, where synthesis stops.
- Since both deoxy and dideoxy forms of one nucleotide are present, they compete for positions.
- A set of radioactive strands of varying lengths are eventually produced.

Step 3: The new DNA strands are separated by polyacrylamide gel electrophoresis, which can separate strands differing by one nucleotide in length.
- The sequence in the new strands is then read from the radioactive bands, and the sequence of the original template is deduced.

DNA sequencing techniques have enabled scientists to collect thousands of DNA sequences in computer data banks.
- This knowledge greatly improves the understanding of genes and genetic control elements.
- Using a computer, scientists can scan long sequences for shorter sequences known to be protein recognition sites and control sequences. They can also scan for similarities to known sequences of other genes or other organisms.
- Nucleotide sequences can also be automatically translated into amino acid sequences.

DNA Technology   285

C. Amplifying DNA by the Polymerase Chain Reaction (PCR)
(p. 399–402)

What is the polymerase chain reaction (PCR)? What are some advantages to this technique? For what purposes is it currently being used?

<u>Polymerase chain reaction</u>

PCR is another promising new technique which allows any piece of DNA to be quickly amplified (copied many times) in vitro. (See Campbell, Methods Box, page 401)
- DNA is incubated under appropriate conditions with special primers and DNA polymerase molecules.
- Billions of copies of the DNA are produced in just a few hours.
- PCR is highly specific; primers determine the sequence to be amplified.
- Only minute amounts of DNA are needed.

PCR is presently being applied in many ways for analysis of DNA from a wide variety of sources:
- Ancient DNA fragments from a woolly mammoth; DNA is a stable molecule and can be amplified by PCR from sources thousands, even millions, of years old.
- DNA from tiny amounts of tissue or semen found at crime scenes.
- DNA from single embryonic cells for prenatal diagnosis.
- DNA of viral genes from cells infected with difficult to detect viruses such as HIV.

Amplification of DNA by PCR is being used in the human genome project to produce linkage maps without the need for large family pedigree analysis.
- DNA from sperm of a single donor can be amplified to analyze the immediate products of meiotic recombination.
- This process eliminates the need to rely on chance that offspring will be produced with a particular type of recombinant chromosome.
- It makes it possible to study genetic markers that are extremely close together.

D. RFLP Analysis (p. 402–404)

What causes restriction fragment length polymorphisms among homologous DNA segments carrying different alleles of a gene?

Gel electrophoresis of restriction fragments results in a characteristic banding pattern.
- Each band corresponds to a DNA restriction fragment of a certain length.
- DNA segments carrying different alleles of a gene can result in dissimilar banding patterns, since numbers and locations of restriction sites may not be the same in the different nucleotide sequences. (See Campbell, Figure 19.8)
- Similar differences in banding patterns result when *noncoding* segments of DNA are used as starting material.

Differences in restriction fragment length that reflect variations in homologous DNA sequences are called *restriction fragment length polymorphisms* (RFLPs).

<u>Restriction fragment length polymorphisms</u>

☞ DNA sequence differences on homologous chromosomes that result in RFLPs are scattered abundantly throughout genomes, including the human genome.
☞ RFLPs are not only abundant, but can easily be detected whether or not they affect the organism's phenotype; they can be located in an exon, intron, or any noncoding part of the genome.

What techniques are used in RFLP analysis? Why are RFLPs useful to scientists?

☞ RFLP analysis involves five techniques often used in recombinant DNA technology: (See Campbell, Methods Box, pages 402-403)
1. Restriction enzyme digestion of DNA.
2. Gel electrophoresis.
3. Southern blotting.
4. Use of radioactive DNA probes.
5. Autoradiography.

☞ Because RFLPs can be readily detected with recombinant DNA methods, they are extremely useful as *genetic markers* for making linkage maps.
☞ A RFLP marker is often found in numerous variants in a population. RFLPs have provided many markers for mapping the human genome since geneticists are no longer limited to genetic variations that lead to phenotypic differences or protein products.

RFLPs are proving useful in several areas.
☞ Disease genes are being located by examining known RFLPs for linkage to them. RFLP markers inherited at a high frequency with a disease are probably located close to the defective gene on the chromosome.
☞ An individual's RFLP markers provide a "genetic fingerprint" which can be used in forensics, since there is a very low probability that two people would have the same set of RFLP markers.

II. **APPLICATIONS OF RECOMBINANT DNA TECHNOLOGY** (p. 404-412)

A. Uses of DNA Technology in Basic Research (p. 404)

What are some practical applications of recombinant DNA technology in biological research?

Recombinant DNA technology is a refined research tool which allows scientists to study more specific questions in almost all fields of biology.

Study of the molecular details of eukaryotic gene structure and function is possible with the use of identified clones as probes.
☞ A radioactively labeled probe can be used to detect similar DNA segments in the same or different genomes by hydrogen bonding to complementary nucleotide sequences.
☞ These methods do not depend on gene expression, so geneticists do not have to infer genotype from phenotype.

Finding DNA segments with a probe is helpful, because it can:
- ☞ Provide information about evolutionary relationships between the gene of interest within or among species.
- ☞ Be used to examine the natural form of a gene, allowing the study of regulatory and other noncoding sequences adjacent to the gene.
- ☞ Increase the understanding of gene organization and control.

DNA probes can be used to map genes on eukaryotic chromosomes.   *In situ hybridization*
- ☞ This technique (*in situ hybridization*) uses a radioactive DNA probe that base pairs with complementary sequences on intact chromosomes.
- ☞ Autoradiography and chromosome staining reveal to which band of which chromosome the probe has attached.

Recombinant DNA technology allows for the production of large quantities of proteins which are usually present in small amounts.
- ☞ Traditional techniques cannot purify and characterize many scarce molecules that control cell metabolism and development.
- ☞ The production of large amounts of proteins will allow scientists to catalog all proteins produced by particular organs such as the brain.

B. The Human Genome Project (*p. 404–405*)

What approaches are being used to map the human genome? What are some potential benefits of this project?

Four complementary approaches are being used in an attempt to map the entire human genome.

1. *Genetic mapping (linkage mapping) of the human genome.*   *Genetic mapping*
    - ☞ At least 3000 genes or other identifiable loci on the DNA (genetic markers) are to be identified initially. Many of the markers will be RFLPs.
    - ☞ This initial map will allow researchers to more easily map other genes by testing for genetic linkage to known markers.

2. *Physical mapping of the human genome.*   *Physical mapping*
    - ☞ Each chromosome will be broken into a number of identifiable fragments.
    - ☞ The actual order of fragments in the chromosome will then be determined.

3. *Sequencing the human genome.*
    - ☞ The exact order of nucleotide pairs of each chromosome will be determined.
    - ☞ This will be the most time consuming part of the project as each haploid set of human chromosomes contains about 3 billion nucleotide pairs.

4. *Analyzing the genomes of other species.*
    - ☞ Scientists will also map the genomes of species that are particularly useful for genetic research.
    - ☞ Such species will include *E. coli*, yeast, the plant *Arabadopsis* and the mouse.

These four approaches will be used to completely map the human genome and provide an understanding of how the human genome compares to those of other organisms. Potential benefits include:
- ☞ Identification and mapping of genes responsible for genetic diseases will aid diagnosis, treatment and prevention.
- ☞ Detailed knowledge of the genomes of humans and other species will give insight into genome organization, control of gene expression, cellular growth and differentiation, and evolutionary biology.

Methods used to construct the human genetic map combine:
- ☞ Classical pedigree analysis of large families.
- ☞ Many molecular and recombinant DNA techniques.

How are chromosome walking and PCR used in the Human Genome Project? What is the basis for chromosome walking and what information does it provide?

Two valuable techniques used in the human genome project include: 1) *chromosome walking* (overlap hybridization) and 2) *PCR amplification*.

1. *Chromosome walking* (*overlap hybridization*). This process combines several techniques to order numerous restriction fragments that result from treating long eukaryotic DNA molecules with restriction enzymes. (See Campbell, Methods Box, page 406)
    - ☞ Overlapping DNA segments are created by cutting the original DNA with two different restriction enzymes; two libraries are produced with the resulting fragments.
    - ☞ Beginning with a probe for an already known DNA sequence, the researcher "walks" along the chromosome from this point. From overlaps between DNA segments, one can deduce the order in which the segments occur in the original DNA.

2. *PCR amplification*.
    - ☞ PCR can amplify specific portions of DNA from individual sperm cells — the immediate products of meiotic recombination.
    - ☞ Researchers can amplify DNA in amounts sufficient for study, and analyze samples as large as thousands of sperm.
    - ☞ Based on the crossover frequencies between genes, researchers can deduce human linkage maps without having to find large families for pedigree analysis.

Achieving the goals of the Human Genome Project in a timely way will come from advances in automation and utilization of the latest electronic technology.

C. **Medical Uses of DNA Technology** (*p. 405–409*)

What are some medical applications of recombinant DNA technology?

Modern biotechnology has resulted in significant advances in many areas of medicine.

1. **Diagnosis of Diseases** (*p. 405–407*)

Medical scientists currently use DNA technology, in particular PCR and labeled DNA probes, to detect genetic diseases and identify heterozygote carriers.
- This allows early disease detection and identification of carriers for potentially harmful recessive mutations — even before the onset of symptoms.
- Genes have been cloned for many genetic disorders including hemophilia, phenylketonuria, cystic fibrosis, and Duchenne's muscular dystrophy.
- Gene cloning permits direct detection of gene mutations. A cloned normal gene can be used as a probe to find the corresponding gene in cells being tested; the alleles are compared with normal and mutant standards usually by RFLP analysis.

When the normal gene has not been cloned, a closely linked RFLP marker may indicate the presence of an abnormal allele if the RFLP marker is frequently co-inherited with the disease.
- Blood samples from relatives can be used to determine which RFLP marker is linked to the abnormal allele and which is linked to the normal allele.
- The RFLP markers must be different for the normal and abnormal alleles.
- Under these conditions, the RFLP marker variant found in the person being tested can reveal whether the normal or abnormal allele is likely to be present.
- Alleles for cystic fibrosis and Huntington's disease can be detected in this manner.

2. **Human Gene Therapy** (*p. 407–408*)

What is the basic principle behind human gene therapy?

Traceable genetic disorders in individuals may eventually be correctable.
- Theoretically, it should be possible to replace or supplement defective genes with functional normal genes using recombinant DNA techniques.
- Correcting somatic cells of individuals with well-defined, life-threatening genetic defects will be the starting place.

The principle behind human gene therapy is that normal genes are introduced into a patient's own somatic cells. For this therapy to be effective, there are several requirements:
- The few cells receiving the normal gene must actively reproduce, so the normal gene will be replicated.
- The cells' normal protein products can correct the disorder. Gene therapy is especially suited for single-enzyme deficiency diseases.

For example, several children with an immunodeficiency disease are currently being treated successfully by gene therapy; their disorder results from a lack of the enzyme adenosine deaminase (ADA).
- Some of the patient's T lymphocytes (a type of white blood cell) are removed and cultured in vitro with harmless retroviral vectors carrying the normal ADA allele.
- The recombinant retroviruses infect the cells, inserting the normal ADA allele (along with a DNA version of the viral genome).
- The patient's own engineered cells carrying a normal ADA allele are then returned to the patient's body by intravenous injection.

A newer approach using undifferentiated bone marrow cells could offer a broader, more long-lasting treatment. (See Campbell, Figure 19.10)
- The normal ADA allele is inserted into a patient's undifferentiated bone marrow cells rather than into lymphocytes.
- The bone marrow cells multiply and give rise to all the cells of the immune system. Also, the bone marrow cells continue to reproduce throughout life.

What types of technical questions have been raised by the early research in human gene therapy?

Many technical questions have been raised by this possibility:
- Can the proper genetic control mechanisms be made to operate on the transferred gene?
- At what stage of biological development is intervention most effective?
- When is intervention too late?
- Would undifferentiated or differentiated cells be best to use?
- How can the correct gene or gene product be placed in the tissues where it is required?
- How can developmental diseases be treated when a gene's product is only needed during a brief period in development?

What are some difficult social and ethical issues surrounding gene therapy?

Gene therapy raises difficult social and ethical questions:
- Who will have access to gene therapy? The procedures are very expensive and only available in major medical centers.
- Should we try to treat germ cells with the hope of correcting a defect in future generations?

Treating germ cells is possible and has been used in mice for some time.
- Mice defective in a gene for a crucial blood polypeptide have been treated with a corresponding human gene.
- Recipient mice and their descendants contain the active human gene not only in the proper location (erythrocytes) but also during the proper stage of development.
- Expression of foreign genes in other animal experiments has not been so successful.

Some critics feel that tampering with human genes is wrong regardless of the situation, because it will lead to eugenics. Others say that genetic engineering of somatic cells is no different than other conventional medical interventions used to save lives.

3. Vaccines (*p. 408*)

Developing vaccines for prevention of infectious diseases is also possible using recombinant DNA techniques. Prevention by vaccine is the only way to fight diseases for which no treatment exists.

Traditional vaccines for viral diseases are of two types:
- Particles of virulent virus that have been inactivated by chemical or physical means.
- Active virus particles of an attenuated viral strain.

Since the particles in both cases are similar to active virus, both types of vaccine will trigger an animal's immune system to produce antibodies, which react very specifically against invading pathogens.

Biotechnology is being used in several ways to modify current vaccines and to produce new ones. Recombinant DNA techniques can be used to:
- Produce large amounts of specific protein molecules (*subunits*) from the protein coat of disease causing viruses, bacteria, or other microbes. If these protein subunits cause immune responses to the pathogens, they can be used as vaccines.
- Modify genomes of pathogens to directly attenuate them. Vaccination with live, attenuated organisms is more effective than a subunit vaccine. Small amounts of material trigger greater immune response, and pathogens attenuated by gene-splicing may also be safer than using natural mutants.
- Alter a virus to produce a vaccine against several diseases. For example, vaccinia, the virus used to produce smallpox vaccine, may have its genes that induce immunity to smallpox replaced with genes that induce immunity to other diseases. The vaccinia virus can even carry genes required to vaccinate for several diseases at once.

4. Other Pharmaceutical Products (*p. 408–409*)

Recombinant DNA techniques use bacteria to produce hormones and other mammalian proteins for use in treating human patients. Insulin and growth hormone have already been approved.

Treatment of diabetics in the United States has been improved by genetically engineering bacteria to produce insulin.
- Insulin produced this way is chemically identical to that made by the human pancreas, so it causes fewer adverse reactions than previous insulin extracted from pig and cattle pancreas.

The growth hormone molecule is much larger than insulin (almost 200 amino acids long) and more species specific.
- Thus, growth hormone from other animals is not an effective growth stimulator in humans.

Subunits

gene-splicing

Vaccinia

Erythropoietin

☞ Hormone produced by genetically engineered bacteria was approved in 1985 for the clinical treatment of children with hypopituitarism (pituitary dwarfism).
☞ Previously, these individuals were treated with growth hormone obtained from human cadavers.
☞ Genetically engineered growth hormone will eventually be used in other treatments (to promote healing in burn victims and individuals with bone fractures).

Erythropoietin (EPO) was produced in 1989 using recombinant DNA methods.
☞ This protein is normally produced by the kidney and stimulates erythrocyte production in the bone marrow.
☞ EPO produced by biotechnology will provide a source of the protein to treat anemia from various causes including kidney damage.

How is biotechnology used in new approaches to fight viral infection?

Disease-causing viruses are less susceptible to drug treatment than bacterial or fungal pathogens, so recent pharmaceutical research has focused on novel ways to fight viral infection. For example, two new approaches using biotechnology include the production of: a) *antisense nucleic acid* and b) *surface-receptor drugs*.

  a. *Antisense nucleic acid.* Drugs may soon be produced that prevent viral mRNAs from being translated in an infected cell.
  ☞ For example, single-stranded antisense nucleic acid (DNA or RNA) could be used to base pair with viral mRNA and block its translation.

  b. *Surface-receptor drugs.* Recombinant DNA techniques can be used to make drugs that either block or mimic surface receptors on cell membranes.
  ☞ For example, a new experimental drug substitutes for a receptor protein that HIV binds to when it attacks white blood cells. The premise is that the drug will decoy the AIDS virus and prevent it from infecting white blood cells.

D. **Forensic Uses of DNA Technology** (p. 409–410)

What types of forensic evidence can be produced by DNA technology?

Forensic labs can determine blood or tissue type from blood, small fragments of other tissue or semen left at the scene of violent crime. These tests, however, have limitations:
☞ They require fresh tissue in sufficient amounts for testing.
☞ This approach can exclude a suspect but is not evidence of guilt, because many people have the same blood type or tissue type.

Theoretically, DNA testing can identify an individual with certainty, since everyone's DNA base sequence is unique (except for identical twins). The most commonly used DNA technology for this forensic application is the RFLP autoradiograph. (See Campbell, Figure 19.11)

☞ This method is used to compare DNA samples from the suspect, the victim, and a small amount of semen, blood or other tissue found at the scene of the crime.
☞ Restriction fragments from the DNA samples are separated by electrophoresis; radioactive probes mark the bands containing RFLP markers.
☞ Even a small set of RFLP markers from an individual can provide a *DNA fingerprint* that is of forensic use; the probability that two individuals would have the same RFLP markers is quite low.

How reliable is DNA fingerprinting?

☞ Though each individual's DNA fingerprint is unique, most forensic tests do not analyze the entire genome but focus on tiny regions known to be highly variable from one person to another.
☞ Even though the probability is minute (one in a million) that two people will have matching DNA fingerprints for the few regions tested, some courts have thrown out DNA evidence because of uncertainty about its probability of being correct.

As with most new technology, forensic applications of DNA fingerprinting raises important ethical questions such as:
☞ Once collected, what happens to the DNA data?
☞ Should DNA fingerprints be filed or destroyed? Some states now save DNA data from convicted criminals.

E. Agricultural Uses of DNA Technology (p. 410-412)

What are some agricultural uses of DNA technology?

Recombinant DNA techniques are being used to study plants and animals of agricultural importance to improve their productivity.

1. Animal Husbandry (p. 410)

Products produced by recombinant DNA methods, such as vaccines, antibodies and growth hormone, are already used in animal husbandry. For example,
☞ *Bovine growth hormone* (*BGH*). Made by *E. coli*, BGH is injected into milk cows to enhance milk production and into beef cattle to increase weight gain.
☞ *Cellulase*. Also produced by *E. coli*, this enzyme hydrolyzes cellulose making all plant parts useable for animal feed.

Transgenic animals, those that contain DNA from other species, have been commercially produced by injecting foreign DNA into egg nuclei or early embryos.
☞ Transgenic beef and dairy cattle, hogs, sheep and several species of commercially raised fish have been produced for potential agricultural use.

Transgenic animals

2. Manipulating Plant Genes (p. 410–411)

Why are plant cells easier to engineer than animal cells?

Plant cells are easier to engineer than animal cells, because an adult plant can be produced from a single cell growing in tissue culture.

- ☞ This is important since many types of genetic manipulation are easier to perform and assess on single cells than on whole organisms.
- ☞ Asparagus, cabbage, citrus fruits, sunflowers, carrots, alfalfa, millet, tomatoes, potatoes, and tobacco are all commercial plants that can be grown from single somatic cells.

How can plant genes be manipulated using the Ti plasmid as a vector? What is the role of Agrobacterium tumefaciens in this process?

The best developed DNA vector for plant cells is the *Ti plasmid* (tumor inducing), carried by the normally pathogenic bacterium *Agrobacterium tumefaciens*.

- ☞ Ti plasmid usually induces tumor formation in infected plants by integrating a segment of its DNA (called T DNA) into the chromosomes of the host plant cell.
- ☞ Researchers have turned this plasmid into a useful vector by eliminating its disease-causing ability without interfering with its potential to move genetic material into infected plants.

With recombinant DNA methods, foreign genes can be inserted into Ti plasmid.

- ☞ The recombinant plasmid can either be put back into *Agrobacterium tumefaciens*, which is used to infect plant cells in culture, or it can be introduced directly into plant cells.
- ☞ Individual modified plant cells are then used to regenerate whole plants that contain, express and pass on to their progeny the foreign gene. (See Campbell, Figure 19.12)

Since monocots are not infected by Agrobacterium, how is foreign DNA transferred into monocotyledonous plants?

Using Ti plasmid as a vector has one major drawback, only dicotyledons are susceptible to infection by *Agrobacterium*; important commercial plants such as corn and wheat are monocotyledons and cannot be infected.

- ☞ Newer methods that allow researchers to overcome this limitation are *electroporation* and use of the *DNA particle gun*.
- ☞ Electroporation uses high-voltage jolts of electricity to open temporary pores in the cell membrane; foreign DNA can enter the cells through these pores.
- ☞ The DNA gun shoots tiny DNA-coated metal pellets through the cell walls into the cytoplasm, where the foreign DNA becomes integrated into the host cell DNA.

Ti plasmid

Agrobacterium tumefaciens

T DNA

Though cloning plant DNA is straightforward, plant molecular geneticists still face several technical problems:
☞ Identifying genes of interest may be difficult.
☞ Many important plant traits (such as crop yield) are polygenic.

3. **Some Early Successes of Genetic Engineering in Plants** *(p. 411)*

How has DNA technology been used to: a) engineer herbicide resistant crops; b) suppress ripening and retard spoilage in tomatoes; and c) produce crop plants resistant to pathogens and insects?

Genetic engineering of plants has yielded positive results in cases where useful traits are determined by single genes. For example,
☞ A bacterial gene that makes plants resistant to glyphosate (a powerful herbicide) has been successfully introduced into several crop plants; glyphosate-resistant plants makes it easier to grow crops and destroy weeds simultaneously.
☞ The first gene-spliced fruits approved by the FDA for human consumption were tomatoes engineered with antisense RNA to suppress ripening and retard spoilage. The general process is outlined below:

    a. Researchers clone the tomato gene coding for enzymes responsible for ripening.

    b. The complementary (antisense) gene is cloned.

    c. The antisense gene is spliced into the tomato plant's DNA, where it transcribes mRNA complementary to the ripening genes' mRNA.

    d. When the ripening gene produces a normal mRNA transcript, the antisense mRNA binds to it, blocking synthesis of the ripening enzyme.

Crop plants are being engineered to resist pathogens and pest insects. For example,
☞ Tomato and tobacco plants can be engineered to carry certain genes of viruses that normally infect and damage plants; expression of these versions of viral genes confer resistance to viral attack.
☞ Some plants have been engineered to resist insect attack, reducing the need to apply chemical insecticides to crops. Genes for an insecticidal protein have been successfully transferred from the bacterium, *Bacillus thuringiensis*, to corn, cotton and potatoes; field tests show that these engineered plants are insect resistant.

In the near future, crop plants developed with recombinant DNA techniques are likely to:
☞ Be made more productive by enlarging agriculturally valuable parts — whether they be roots, leaves, flowers, or stems.
☞ Have an enhanced food value. For example, corn and wheat might be engineered to produce mixes of amino acids optimal for the human diet.

4. **The Nitrogen Fixation Challenge** (p. 411–412)

What is nitrogen fixation? Why is this process so important to plant growth, including crop yield? How might biotechnology influence this process?

Nitrogen fixation is the conversion of atmospheric, gaseous nitrogen (N_2) into nitrogen-containing compounds.
- Gaseous nitrogen is useless to plants.
- Nitrogen-fixing bacteria live in the soil or are symbiotic within plant roots.
- The nitrogen-containing compounds produced by nitrogen fixation are taken up from the soil by plants and used to make organic molecules such as amino acids and nucleotides.
- A major limiting factor in plant growth and crop yield is availability of useable nitrogen compounds. This is why nitrogenous fertilizers are used in agriculture.

Recombinant DNA technology can possibly be used to increase biological nitrogen fixation of bacterial species living in the soil or in association with plants.

III. SAFETY AND ETHICAL ISSUES (p. 412–413)

How are recombinant DNA studies and the biotechnology industry regulated with regards to safety and policy matters?

When scientists realized the potential power of DNA technology, they also became concerned that recombinant microorganisms could create hazardous new pathogens, which might escape from the laboratory.
- In response to these concerns, scientists developed and agreed upon a self-monitoring approach, which was soon formalized into federal regulatory programs.
- Today, governments and regulatory agencies worldwide are monitoring the biotechnology industry — promoting potential industrial, medical, and agricultural benefits, while ensuring that new products are safe.
- In the U.S., the FDA, National Institutes of Health (NIH) Recombinant DNA Advisory Committee, Department of Agriculture (USDA), and Environmental Protection Agency (EPA) set policies and regulate new developments in genetic engineering.

What are some ethical concerns about new developments in DNA technology?

While genetic engineering holds enormous potential for improving human health and increasing agricultural productivity, new developments in DNA technology raise ethical concerns. For example, mapping the human genome will contribute to significant advances in gene therapy, but:
- Who should have the right to examine someone else's genes?
- Should a person's genome be a factor in their suitability for a job?
- Should insurance companies have the right to examine an applicant's genes?
- How do we weigh the benefits of gene therapy against assurances that the gene vectors are safe?

What are some environmental and health concerns about the potential dangers of genetically engineered organisms and their products?

For environmental problems such as oil spills, genetically engineered organisms may be part of the solution, but what is their potential impact on native species?

For new medical products, what is the potential for harmful side effects, both short-term and long-term?

- New medical products must pass exhaustive tests before the FDA approves it for general marketing.
- Currently awaiting federal approval are hundreds of new genetically engineered diagnostic products, vaccines, and drugs — including some to treat AIDS and cancer.

For agricultural products, what are the potential dangers of introducing new genetically engineered organisms into the environment?

- Some argue that producing transgenic organisms is only an extension of traditional hybridization and should not be treated differently from the production of other hybrid crops or animals. The FDA holds that if products of genetic engineering are not significantly different from products already on the market, testing is not required.
- Others argue that creating transgenic organisms by splicing genes from one species to another is radically different from hybridizing closely related species of plants or animals.
- Some concerns are that genetically altered food products may contain new proteins that are toxic or cause severe allergies; genetically engineered crop plants could become superweeds resistant to herbicides, disease and insect pests; and engineered crop plants may hybridize with native plants and pass their new genes to closely related plants in the wild.

20

DESCENT WITH MODIFICATION: A DARWINIAN VIEW OF LIFE

Evolution

Charles Darwin, On the Origin of Species by Means of Natural Selection

CHAPTER OUTLINE

Evolution includes the processes that have transformed life on Earth from its earliest beginnings to the seemingly infinite diversity that characterizes it today.

Charles Darwin published *On the Origin of Species by Means of Natural Selection* on November 24, 1859.

☞ This book not only presented the first convincing case for evolution but also united the field of biology into a cohesive science.
☞ Darwin's book covered many aspects of living organisms including their diversity, origins and relationships, similarities and differences, geographical distribution, and adaptations to their environment.
☞ None of these aspects of life makes sense except in the context of evolution.

What major points did Darwin make in his book?

Darwin made two major points in his book:

1. Species evolved from ancestral species and were not specially created.
2. *Natural selection* is a mechanism that could result in this evolutionary change.

I. **PRE-DARWINIAN VIEWS** (*p. 420–424*)

The impact of Darwin's ideas was a product of timing as much as logic.

☞ Darwinism, as an intellectual revolution, partially depended upon the historical and social context. (See Campbell, Figure 20.3)
☞ Darwin's view of life contrasted sharply with the accepted viewpoint: the Earth was only a few thousand years old and was populated by unchanging life forms made by the Creator during a single week.
☞ Thus, *On the Origin of Species by Means of Natural Selection* not only challenged prevailing scientific views, but also challenged the roots of Western culture.

A. **The Scale of Life and Natural Theology** (*p. 422*)

Many Greek philosophers believed in the gradual evolution of life. However, the two that influenced Western culture most, Plato and his student Aristotle, held opinions which were inconsistent with any concept of evolution.

What were the premises of Plato's philosophy of idealism or essentialism? Why did this philosophy rule out evolution?

Plato's philosophy of *idealism* (*essentialism*) held that there are two coexisting worlds: an ideal eternal real world and an illusionary imperfect world that humans perceive with their senses.
- ☞ To Plato, variations in plant and animal populations are merely imperfect representatives of ideal forms; only these perfect ideal forms are real.
- ☞ Idealism ruled out evolution which would be counterproductive in a world where ideal organisms are already perfectly adapted to their environments.

How did the views of Aristotle, Plato's student, differ from Platonic philosophy? For Aristotle, what was the basis of the scala naturae?

Aristotle questioned the Platonic philosophy of dual worlds, but his beliefs also excluded evolution.
- ☞ He recognized, being a student of nature, that organisms range from relatively simple to very complex.
- ☞ Aristotle believed that all living organisms could be arranged on a scale of increasing complexity (*scala naturae*); on this ladder of life, each form had its allotted rung and each rung was occupied.
- ☞ In this view of life, species are fixed and do not evolve.
- ☞ The *scala naturae* view of life prevailed for over 2000 years.

How did Judeo-Christian culture influence Western thought about evolution?

The Old Testament account of creation from the Judeo-Christian culture fortified prejudice against evolution, and *creationist-essentialist* dogma that species were individually designed and permanent became embedded in Western thought.
- ☞ Evolutionists before Darwin were unable to overcome the doctrine of fixed species.
- ☞ *Natural Theology*, a philosophy concerned with discovering the Creator's plan by studying his works, dominated European and American biology even as Darwinism emerged.
- ☞ For natural theologians, adaptations of organisms were evidence that the Creator had designed every species for a particular purpose.
- ☞ Natural theology's major objective was to classify species revealing God's created steps on the ladder of life.

What important contribution did Carolus Linnaeus make to biological science? Why is it ironic that Linnaeus' taxonomic system was used as a focal point for Darwin's arguments for evolution?

Carolus Linnaeus (1707–1778), a Swedish physician and botanist, sought order in the diversity of life *ad majorem Dei gloriam* (for the greater glory of God). He was the father of *taxonomy*, the area of biology concerned with naming and classifying organisms.
- ☞ He developed the system of *binomial nomenclature* still used today, which names organisms according to genus and species.

300 *Descent With Modification: A Darwinian View of Life*

> ☞ He adopted a system for grouping species into categories and ranking the categories into a hierarchy. For example, similar species are grouped into a genus; similar genera are grouped into the same order; etc.

Linnaeus found order in the diversity of life with his hierarchy of taxonomic categories.
> ☞ The clustering of certain species in taxonomic groups did not imply evolutionary relationships to Linnaeus, since he believed that species were permanent creations.
> ☞ Linnaeus, a natural theologian, developed his classification scheme only to reveal God's plan and even stated *Deus creavit, Linnaeus disposuit* ("God creates, Linnaeus arranges").
> ☞ It is ironic that in the nineteenth century, the taxonomic system of Linnaeus became a focal point in Darwin's arguments for evolution.

B. Cuvier, Fossils, and Catastrophism (p. 422–423)

What are fossils, and how are they formed? Using evidence from the fossil record, what can scientists deduce about life's history?

Fossils = Relics or impressions of organisms from the past, preserved in rock.

Most fossils are found in *sedimentary rocks*, which form when new layers of sand and mud settle to the bottom of seas, lakes, and marshes, covering and compressing older layers into rock (e.g. sandstone and shale).
> ☞ Sedimentary rock may be deposited in many layers (*strata*) in places where shorelines repeatedly advance and retreat.
> ☞ Later erosion can wear away the upper (younger) strata, revealing older strata which had been buried.
> ☞ The fossil record thus holds and displays graphic and indisputable evidence that Earth has had a succession of flora and fauna.

Paleontology = The study of fossils.
> ☞ This science was largely founded by the French anatomist Georges Cuvier (1769–1832).

What contribution did George Cuvier make to paleontology? How did Cuvier and his followers use the concept of catastrophism to oppose evolution?

Cuvier realized that life's history was recorded in fossil-containing strata and documented the succession of fossil species in the Paris Basin.
> ☞ He noted that each stratum was characterized by a unique set of fossil species and that the older (deeper) the stratum, the more dissimilar the flora and fauna from modern life forms.
> ☞ Cuvier understood that extinction had been a common occurrence in the history of life since, from stratum to stratum, new species appeared and others disappeared.

Although Cuvier had this evidence in hand, he was a serious and effective opponent to the evolutionists of his day.

☞ He reconciled the fossil evidence with his belief that species were immutable, by speculating that boundaries between the fossil strata corresponded in time to catastrophic events such as floods or droughts that destroyed many species then living at that location. Multiple strata thus indicated many catastrophes.
☞ This view of Earth history is known as *catastrophism*.

Catastrophism

Cuvier explained the appearance of new species in younger rock that were absent from older rock by proposing that:
☞ Periodic catastrophes which resulted in mass extinctions were usually confined to local geographical regions.
☞ After the native flora and fauna had become extinct, the region would be repopulated by foreign species immigrating from other areas.

Though Cuvier was winning debates against evolutionists, a new theory of Earth history was gaining popularity among geologists and it would greatly influence Darwin.

C. Gradualism in Geology (*p. 423*)

How did James Hutton's principle of gradualism differ from the concept of catastrophism? What was Charles Lyell's idea of uniformitarianism?

The principle of *gradualism* holds that profound change is the cumulative product of slow but continuous processes.

Gradualism

☞ This very different idea of how geological processes had shaped the Earth's crust competed with Cuvier's theory of catastrophism.
☞ James Hutton, a Scottish geologist, proposed in 1795 that it was possible to explain the various land forms by looking at mechanisms currently operating in the world.

James Hutton

☞ For example, canyons form by erosion from rivers passing through them, and fossil-bearing sedimentary rocks are built of particles that had been eroded from the land and carried by rivers to the sea.

Charles Lyell, a leading geologist of Darwin's time expanded Hutton's gradualism into the theory known as *uniformitarianism*.

Charles Lyell
Uniformitarianism

☞ Uniformitarianism was Lyell's extreme idea that geological processes are so uniform that their rates and effects must balance out through time (e.g. processes that build mountains are eventually balanced by the erosion of mountains).

How was Darwin influenced by Hutton's and Lyell's observations?

Darwin rejected uniformitarianism, but was greatly influenced by conclusions that followed directly from the observations of Hutton and Lyell:
☞ If geological change results from slow, continuous actions rather than sudden events, then the Earth must be much older than the 6000 years indicated by many theologians on the basis of biblical inference.
☞ Very slow and subtle processes persisting over a great length of time can cause substantial change.

Jean Baptiste Lamarck

D. Lamarck's Theory of Evolution (p. 424)

While several eighteenth century naturalists suggested that life had evolved with the Earth's formation, only Jean Baptiste Lamarck (1744–1829) developed a comprehensive model which attempted to explain how life evolved.

Lamarck was in charge of the invertebrate collection at the Natural History Museum in Paris.
- By comparing extant species to fossil forms, Lamarck saw what appeared to be several lines of descent.
- Each line of descent being a chronological series of older to younger fossils leading to a modern species.

Lamarck imagined many ladders of life which organisms could climb (as opposed to Aristotle's single ladder without movement).
- The bottom rungs were occupied by microscopic organisms which were continually generated spontaneously from inanimate material.
- At the top of the ladders were the most complex plants and animals.

What were Jean Baptiste Lamarck's views about evolution? How did this differ from Aristotle's scala naturae?

Lamarck believed that evolution was driven by an innate tendency toward increasing complexity, which he equated with perfection.
- As organisms attained perfection, they became better and better adapted to their environments.
- Thus, Lamarck believed that evolution responded to organisms' *sentiments interieurs* ("felt needs").

What mechanism did Lamarck propose for the way specific adaptations evolve?

Lamarck proposed a mechanism by which specific adaptations evolve, which included two related principles:

1. Use and disuse. Those body organs used extensively to cope with the environment become larger and stronger while those not used deteriorate.

2. Inheritance of acquired characteristics. The modifications an organism acquired during its lifetime could be passed along to its offspring.

The inheritance of acquired characteristics was generally accepted during Lamarck's era, although it is ridiculed by some today.
- To most of Lamarck's contemporaries, the mechanism of evolution was irrelevant since the creationist-essentialist view prevailed and no theory of evolution could be taken seriously.

Even though the inheritance of acquired characteristics was incorrect, what contributions did Lamarck make towards the development of evolutionary theory?

Although his mechanism of evolution was in error, Lamarck deserves credit for his unorthodox theory which:
- ☞ Proposed that evolution is the best explanation for both the fossil record and the extant diversity of life.
- ☞ Emphasized the great age of the Earth.
- ☞ Stressed that adaptation to the environment is a primary product of evolution.

II. ON THE ORIGIN OF DARWINISM (p. 424–426)

Charles Darwin was born in Shrewsbury, England, in 1809 (the same year Lamarck published his theory of evolution).

Charles showed an interest in nature, but was sent by his father (a physician) to the University of Edinburgh to study medicine at the age of 16.
- ☞ Charles found medicine boring and distasteful and left Edinburgh without a degree. He then enrolled at Christ College at Cambridge University to study for the clergy.
- ☞ Nearly all naturalists and other scientists were clergymen, and a majority held to the philosophy of natural theology.
- ☞ Charles studied under the Reverend John Henslow, a professor of botany at Cambridge, and received his B.A. degree in 1831.
- ☞ Professor Henslow recommended him to Captain Robert FitzRoy who was preparing the survey ship HMS *Beagle* for an around the world voyage.

Reverend John Henslow
Captain Robert FitzRoy
HMS Beagle

How did Charles Darwin use his observations from the voyage of the HMS Beagle to formulate and support his theory of evolution?

A. The Voyage of the Beagle (p. 424–425)

The HMS *Beagle*, with Darwin aboard, sailed from England in December 1831.
- ☞ The mission of the voyage was to chart the poorly known South American coastline.
- ☞ Darwin spent most of his time ashore while the ship's crew surveyed the coast.
- ☞ He collected thousands of specimens of the exotic and diverse flora and fauna.

While the ship worked its way around the continent, Darwin observed the various adaptations of plants and animals that inhabited the diverse environments of South America: Brazilian jungles, grasslands of the Argentine pampas, desolate islands of Tierra del Fuego, and the Andes Mountains.
- ☞ While unique adaptations were obvious, Darwin also noted that the South American flora and fauna from different regions were distinct from the flora and fauna of Europe.
- ☞ He also noted temperate species were taxonomically closer to species living in tropical regions of South America than to temperate species of Europe.
- ☞ Additionally, the South American fossils he found (while differing from modern species) were distinctly South American in their resemblance to the living plants and animals of that continent.

304 Descent With Modification: A Darwinian View of Life

Galapagos Islands

☞ This led Darwin to consider the peculiarities of the geographical distribution of species.

Geographical distribution was particularly confusing in the case of the fauna of the Galapagos Islands, which lie on the equator about 900 km west of South America.

☞ Most animal species on the Galapagos are unique to those islands, but resemble species living on the South American mainland.

☞ Darwin collected 13 types of finches from the Galapagos, and although they were similar, they seemed to represent different species.

☞ Some were unique to individual islands, while others were found on two or more islands that were close together.

What was the importance of Charles Lyell's work to Darwin's developing view of life?

Principles of Geology

Darwin had read Lyell's *Principles of Geology* by the time the *Beagle* left the Galapagos.

☞ Lyell's ideas, coupled with his own experiences on the Galapagos, caused Darwin to doubt the church's position that the Earth was static and had been created only a few thousand years before.

☞ When Darwin acknowledged that the Earth was very old and constantly changing, he had taken an important step toward recognizing that life on Earth had also evolved.

B. Darwin Frames His View of Life (*p. 426*)

Darwin was not sure whether the 13 types of finches he collected on the Galapagos were different species or varieties of the same species.

☞ After he returned to England in 1836, an ornithologist indicated that they were actually different species.

☞ Knowing this, he reassessed observations made during the voyage and began the first of several notebooks on the origin of species in 1837.

Darwin perceived the origin of new species and adaptation as closely related processes.

☞ A new species could arise from an ancestral form if, due to fragmentation of the original population, different populations were isolated in different environments by geographical barriers.

☞ The populations would diverge more and more in appearance as each adapted to local conditions.

☞ Gradually, after many generations, the populations would become so dissimilar as to be separate species.

☞ This caused the Galapagos finches to develop different beak forms, which are adapted to specific foods available on their home islands. (See Campbell, Figure 20.7)

☞ Understanding evolution required an explanation for how such adaptations could arise.

By the early 1840's, Darwin had formulated the major features of his theory of natural selection as the mechanism for evolution.

☞ It was not published at this time due to his poor health and the fact that he rarely left home.
☞ Although reclusive, Darwin was well known as a naturalist from the specimens and letters he had sent to England from the voyage on the *Beagle*.
☞ His notoriety led to frequent correspondence and visits with Lyell, Henslow, and other scientists.

Darwin wrote a long essay on the origin of species and natural selection in 1844.

☞ He realized the importance and subversive nature of his work, but did not publish the information because he wished to gather more evidence in support of his theory.
☞ Evolutionary thinking was emerging at this time and Lyell admonished Darwin to publish on the subject before someone else published it first.

How did Alfred Russel Wallace influence Charles Darwin?

In June 1858, Darwin received a letter from Alfred Wallace who was working as a specimen collector in the East Indies.
☞ Accompanying the letter was a manuscript detailing Wallace's own theory of natural selection which was almost identical to Darwin's.
☞ The letter asked Darwin to evaluate the theory and forward the manuscript to Lyell if it was thought worthy of publication.
☞ Darwin did so, although he felt that his own originality would be "smashed."
☞ Lyell and a colleague presented Wallace's paper along with excerpts from Darwin's unpublished 1844 essay to the Linnaean Society of London on July 1, 1858.

Darwin finished *The Origin of Species* and published it the next year.
☞ Darwin is considered the main author of the idea since he developed and supported natural selection much more extensively than Wallace.
☞ Darwin's book and its proponents convinced the majority of biologists that evolution was fact within a decade.
☞ Darwin succeeded where previous evolutionists had failed not only because science was beginning to move away from natural theology, but mainly because he convinced his readers with unmatched logic and an avalanche of evidence.

III. **THE CONCEPTS OF DARWINISM** (*p. 426–431*)

A. **Common Descent** (*p. 427*)

What did Darwin mean by the principle of common descent and "descent with modification"? What evidence convinced Darwin that species change over time?

Darwin used the phrase "descent with modification," not evolution, in the first edition of *The Origin of Species*.

Natural selection

Alfred Wallace

Principle of common descent

Descent with modification

- He perceived a unity in life with all organisms related through descent from some unknown prototype that lived in the remote past.
- Diverse modifications (adaptations) accumulated as descendants from this common ancestor moved into various habitats over millions of years.

Darwin viewed the history of life as a tree with many branches and rebranching from a common trunk to the tips of the twigs, symbolic of the current diversity of organisms.
- A common ancestor at each fork was ancestral to all lines of evolution branching from that fork.
- Closely related species share many characteristics because their lineage of common descent extends to the smallest branches of the evolutionary tree.
- Most branches of evolution are dead ends since about 99% of all species that ever lived are extinct.

B. Natural Selection and Adaptation (p. 427–431)

Most of Darwin's book focused on how populations of individual species become better adapted to their local environments through natural selection.

In your own words, what were three inferences Darwin made from his observations, which led him to propose natural selection as mechanism for evolutionary change?

Ernst Mayr of Harvard University dissected the logic of Darwin's theory of natural selection into three inferences based on five facts:

Fact 1: All species have such great potential fertility that their population size would increase exponentially if all individuals that are born would reproduce successfully.

Fact 2: Most populations are normally stable in size except for seasonal fluctuations.

Fact 3: Natural resources are limited.

  Inference 1: Production of more individuals than the environment can support leads to a struggle for existence among individuals of a population, with only a fraction of offspring surviving each generation.

Fact 4: Individuals of a population vary extensively in their characteristics; no two individuals are exactly alike.

Fact 5: Much of this variation is heritable.

  Inference 2: Survival in the struggle for existence is not random, but depends in part on the hereditary constitution of the surviving individuals. Those individuals whose inherited characteristics fit them best to their environment are likely to leave more offspring than less fit individuals.

  Inference 3: This unequal ability of individuals to survive and reproduce will lead to a gradual change in a population, with favorable characteristics accumulating over the generations.

Ernst Mayr

Why was variation so important to Darwin's theory?

Natural selection is this differential success in reproduction, and its product is adaptation of organisms to their environment.

- Even slight advantages of some variations will accumulate in a population over many generations of being perpetuated by natural selection.
- Natural selection thus occurs from the interaction between the environment and the inherent variability in a population.
- Variations in a population arise by chance, but natural selection is not a chance phenomenon since environmental factors set definite criteria for reproductive success.
- Excessive production of new individuals insures a struggle for life.

How did the Reverend Thomas Malthus' essay influence Charles Darwin?

Darwin was already aware of the struggle for existence when he read an essay on human population written by the Reverend Thomas Malthus in 1798.

- In this essay, Malthus held that much of human suffering was a consequence of the potential for human populations to grow at a faster rate than increased supplies of food and other resources could keep pace.
- This capacity for overproduction is common to all species and only a fraction of new individuals complete development and leave offspring of their own; the rest die or are unable to reproduce in some manner.

Variation and overproduction are the two characteristics of populations that make natural selection possible.

- On the average, the most fit individuals pass their genes on to more offspring than less fit individuals.
- This results from an environmental screening of the variations which favors some over others.

What is the difference between artificial selection and natural selection? Why was Darwin's knowledge of domestic plant and animal breeding crucial to his argument that natural selection is the mechanism of evolutionary change?

Darwin found evidence that this favoritism could cause substantial changes in a population in *artificial selection* (the breeding of domesticated plants and animals).

- Useful species have been modified over many generations by humans who selected individuals with desired traits as breeding stocks.
- The plants and animals we grow for food show little resemblance to their wild ancestors.
- Darwin reasoned that if such change could be achieved by artificial selection in a relatively short period of time, then natural selection should be capable of considerable modifications of species over hundreds of thousands of generations.

Darwin postulated that natural selection operating in differing contexts over vast spans of time could account for the diversity of life.
- ☞ In Darwin's view, life did not evolve suddenly by quantum leaps, but instead by a gradual accumulation of small changes.
- ☞ Gradualism is fundamental to the Darwinian view of evolution.

Darwin's view of life can thus be summarized: the diverse forms of life have arisen by descent with modification from ancestral species, and the mechanism of modification has been natural selection working continuously over enormous tracts of time.

1. Some Subtleties of Natural Selection (p. 428–429)

Populations are important in evolutionary theory, since a population is the smallest unit that can evolve.

<u>Population</u> = A group of interbreeding individuals belonging to a particular species and sharing a common geographic area.

Why is the population the smallest unit that can evolve?

Natural selection centers on interactions between individual organisms and their environment, but individuals do not evolve.
- ☞ Evolution can only be measured as change in relative proportions of variations in a population over several generations.
- ☞ Natural selection can only amplify or diminish heritable variations.
- ☞ Organisms can become modified through experiences in such a way that they become better adapted to their environments, but these acquired characteristics cannot be inherited.
- ☞ Consequently, evolutionists must distinguish between adaptations an organism acquires by its own actions and those that are innate adaptations evolving in a population over many generations as a result of natural selection.

Specifics of natural selection are regional and timely since environmental factors vary from area to area and from time to time.
- ☞ An adaptation under one set of conditions may be useless or detrimental in different circumstances.

2. Natural Selection in Action: Three Examples (p. 429–431)

Using real life examples, how does natural selection result in evolutionary change?

A case history which provides an excellent example of situational natural selection is that of the English peppered moth, *Biston betularia*.
- ☞ This species occurs in two varieties throughout the English midlands. One form is light with dark splotches; the second uniformly dark.
- ☞ These moths feed at night and rest during the day.
- ☞ Resting spots are sometimes trees and rocks encrusted with light-colored lichens.

- Light individuals are camouflaged against this background, while dark moths are obvious and easily seen by predators such as birds. (See Campbell, Figure 20.10)

Before the Industrial Revolution, dark peppered moths were very rare (probably due to predation by birds).
- Industrial pollution killed most of the lichens over much of the countryside resulting in a darkened landscape by the late 1800's.
- Under these "new" conditions, the dark moths were less conspicuous and light colored moths more obvious.
- The frequency of dark moths in populations increased, and by the early 1900's, the population of *Biston* near Manchester consisted entirely of dark moths.
- This phenomenon is known as *industrial melanism* and occurred in hundreds of other species of moths in polluted areas.

Industrial melanism

This is not a case of inheritance of acquired characteristics since the environment did not create favorable characteristics.
- Natural selection acted upon the existing variability in the populations, varying the survival and reproduction of some individuals (dark moths) over others (light moths).
- The dark moths were favored because the light moths were more obvious to predators on the newly darkened background.
- Dark moths thus left more offspring since light moths were eaten before they could reproduce.
- Other factors may have been involved, but predation by birds was a major component.

This example shows that natural selection works in the present and tends to adapt organisms to their local environment by favoring traits that work best for the current situation.
- Recently the case of the peppered moth has taken a new turn.
- In areas where industrial pollution has been curbed, lichens are again becoming established and the light form of *Biston* is becoming more prevalent.

Another example of natural selection in action is protective coloration in a population of house mice (*Mus musculus*).
- In 1962, Larry Brown tested the hypothesis that coat color in house mice (yellow or brown) affects their survival in an environment (dark wood granary) with predators (domestic cats).
- Both brown and yellow mice were sealed in a dark wood granary to protect them from cats.
- Eight months later, Brown censused the mouse population and counted 46% yellow mice.
- A hole was then cut in the granary to allow the cats in to prey on the mice.
- Four months later, the mouse population had declined by 62% and no yellow mice were trapped. Brown concluded that the cats selectively preyed upon the light mice, because they were more visible against the dark wood.

310 *Descent With Modification: A Darwinian View of Life*

There are hundreds of examples of natural selection in laboratory populations of *Drosophila*. A few other examples of natural selection in action include:
- ☞ Changes in beak depth of ground finches (*Geospiza fortis*) on the Galapagos island of Daphne Major. Peter and R. Grand discovered that average beak depth (an inherited trait) oscillated with rainfall. They concluded that in wet years, birds preferentially fed on small seeds. Survival in dry years depended on birds being able to crack the less preferred larger seeds, since smaller seeds were less plentiful than in wet periods. Average beak size increased during dry years. (See Campbell, Figure 20.12)
- ☞ Antibiotic resistance in bacteria.
- ☞ Body size of guppies exposed to different predators. (See Campbell, Chapter 1)

IV. THE SIGNS OF EVOLUTION (*p. 431–435*)

What lines of evidence did Charles Darwin use to support the principle of common descent?

Darwin used several lines of evidence to support his principle of common descent.
- ☞ More recent discoveries, including those from molecular biology, lend even more support to this principle.

A. Biogeography (*p. 431–432*)

It was *biogeographical* evidence that first suggested common descent to Darwin.

Biogeography = The geographical distribution of species.

Islands have many endemic species which are closely related to species on the nearest mainland or neighboring island. Some logical questions follow:
- ☞ Why are two islands with similar environments in different parts of the world not populated by closely related species, but rather by species more closely related to those from the nearest mainland even when that environment is quite different?
- ☞ Why are South American tropical animals more closely related to South American desert animals than to African tropical animals?
- ☞ Why does Australia have such a diversity of marsupial animals and very few placental animals even though the environment can easily support placentals?

B. The Fossil Record (*p. 432*)

The fossil record was quite fragmentary in Darwin's time, and he was troubled by the absence of transitional fossils linking modern life to ancestral forms.
- ☞ Even though the fossil record is still incomplete, paleontologists continue to find important new fossils, and many key links are no longer missing.
- ☞ For example, *Archeopteryx* is a link between birds and their reptilian ancestors.

Although still incomplete, the fossil record provides information that is compatible with other types of evidence about the major branches of descent in the tree of life. For example:
- ☞ Prokaryotes are placed as the ancestors of all life by evidence from cell biology, biochemistry, and molecular biology.
- ☞ Fossil evidence shows the chronological appearance of the vertebrates as being sequential with fishes first, followed by amphibians, reptiles and then birds and mammals. This sequence is also supported by many other types of evidence.

C. Taxonomy (*p. 432–433*)

To Darwin, Linnaeus' taxonomic scheme reflected the branching genealogy of the tree of life.
- ☞ It recognized that the diversity of organisms could be ordered into "groups subordinate to groups" with organisms at the different taxonomic levels related through descent from common ancestors.
- ☞ Classification alone does not confirm the principle of common descent, but when combined with other lines of evidence, the relationships are clear.
- ☞ For example, genetic analysis of species that are thought to be closely related on the basis of anatomical features and other criteria reveals a common hereditary background.

D. Comparative Anatomy (*p. 433–434*)

Anatomical similarities between species grouped in the same taxonomic category show evidence of their common descent.

The skeletal components of the forelimbs of mammals is a good example. (See Campbell, Figure 20.11)
- ☞ Although the limbs are used for different functions, it is obvious that the same skeletal elements are present.
- ☞ It is logical that whether the forelimb is a foreleg, wing, flipper, or arm, the basic similarity is the consequence of descent from a common ancestor and that the limbs have been modified for different functions.

<u>Homologous structures</u> = Structures that are similar because of common ancestry.
- ☞ Other evidence from comparative anatomy supports that evolution is a remodeling process in which ancestral structures that functioned in one capacity have become modified as they take on new functions.

<u>Vestigial organs</u> = Rudimentary structures of marginal or no use to an organism.
- ☞ Vestigial organs are remnants of structures that had important functions in ancestral forms but are no longer essential.
- ☞ An example of vestigial organs are the remnants of pelvic and leg bones in snakes. They show descent from a walking ancestor, but have no function in the snake.
- ☞ Vestigial organs seem to support the Lamarckian concept of use and disuse, but they can be explained by natural selection.

Homologous structures

Vestigial organs

E. Comparative Embryology (p. 434)

Closely related organisms go through similar stages in their embryonic development.

All vertebrate embryos (fishes, amphibians, reptiles, birds, mammals) go through an embryonic stage in which they possess gill slits on the sides of their throats.
- ☞ As development progresses, the gill slits develop into divergent structures characteristic of each vertebrate class.
- ☞ In fish, the gill slits form gills; in humans, they form the eustachian tubes that connect the middle ear with the throat.
- ☞ Development of diverse organs with different functions from the gill slits common to all vertebrate embryos supports the conclusion that all vertebrates descended from aquatic ancestors with gills.

In the late nineteenth century, embryologists developed the view that "ontogeny recapitulates phylogeny."
- ☞ This view held that the embryonic development of an individual organism (*ontogeny*) is a replay of the evolutionary history of the species (*phylogeny*).
- ☞ This is an extreme view; what does occur is a series of similar *embryonic* stages that exhibit the same characteristics, not a sequence of adult-like stages.
- ☞ These characteristics then become modified, possibly by natural selection, since even embryonic processes can ultimately affect the fitness of the adult organism.

Comparative embryology can often establish homology among structures that are so altered in later development that their common origin can not be determined by comparing their fully developed forms.

F. Molecular Biology (p. 434–435)

How can molecular biology be used to study evolutionary relationships among organisms?

An organism's hereditary background is reflected in its genes and their protein products.
- ☞ Siblings have greater similarity in their DNA and proteins than do two unrelated organisms of the same species.
- ☞ Likewise, two species considered to be closely related by other criteria should have a greater proportion of their DNA and proteins in common than more distantly related species.

Molecular taxonomists use a variety of modern techniques to measure the degree of similarity among DNA nucleotide sequences of different species.
- ☞ The closer two species are taxonomically, the higher the percentage of common DNA.
- ☞ This type of evidence supports common descent.
- ☞ Common descent is also supported by the fact that closely related species also have proteins of similar amino acid sequence (resulting from inherited genes).
- ☞ If two species have many genes and proteins with sequences of monomers that match closely, the sequences must have been copied from a common ancestor. (See Campbell, Table 20.1)

Ontogeny, Phylogeny

Molecular biology has also substantiated Darwin's idea that all forms of life are related to some extent through branching descent from the earliest organisms.
- Even taxonomically distant organisms (bacteria and mammals) have some proteins in common.
- For example, cytochrome *c* (the respiratory protein) is found in all aerobic species. Cytochrome *c* molecules of all species are very similar in structure and function even though mutations have substituted amino acids in some areas of the protein during the course of evolution.
- Additional evidence for the unity of life is the common genetic code. This mechanism has been passed through all branches of evolution since its beginning in an early form of life.

Cytochrome c

Mutations

V. JUST A THEORY? (p. 435–436)

What is the problem with the statement that Darwinism is "just a theory"? What is the difference between the colloquial and scientific use of the word "theory"?

Dismissing Darwinism as "just a theory" is flawed because:
- Darwin made *two* claims:
    1. Modern species evolved from ancestral forms.
    2. The mechanism for evolution is natural selection.
- The conclusion that species change or evolve is based on historical fact.

What then is theoretical about evolution?
- Theories are conceptual frameworks with great explanatory power used to interpret facts.
- That species can evolve is fact, but the *mechanism* Darwin proposed for that change — natural selection — is a theory. Darwin used this theory of natural selection to explain facts of evolution documented by fossils, biogeography, and other historical evidence.

In science, "theory" is very different from the colloquial use of the word, which comes closer to what scientists mean by a hypothesis, or educated guess.
- Unifying concepts do not become scientific theories, unless their predictions stand up to thorough and continuous testing by experiment and observation.
- Good scientists, however, do not allow theories to become dogma; many evolutionary biologists now question whether natural selection alone can account for evolutionary history observed in the fossil record.

21
HOW POPULATIONS EVOLVE

CHAPTER OUTLINE

Evolution in a population can be observed as a change in the prevalence of inherited characteristics over generations.
- ☞ The peppered moths (See Campbell, Chapter 20) are excellent examples that show the "sorting" which can occur under changing environmental conditions.
- ☞ It is also important to note that while natural selection works on individuals, it is the population that evolves. Darwin understood this, but was unable to determine its genetic basis.

I. THE MODERN EVOLUTIONARY SYNTHESIS (p. 438–439)

Why did Darwin have difficulty convincing biologists that natural selection is a mechanism of evolutionary change?

Within a few years after the publication of *The Origin of Species*, most biologists were convinced that evolution was a fact. Darwin was less successful in convincing them that natural selection was the mechanism for evolution.
- ☞ The major obstacle to this acceptance was that little was known about how traits were inherited since nothing resembling a theory of genetics had yet been formulated.
- ☞ Darwin could observe that variation occurred, but could not explain how.
- ☞ Since heredity is fundamental to natural selection, Darwin felt compelled to address the problem.
- ☞ To do this, Darwin used a variation of Lamarck's concept of inheritance of acquired characteristics, although this contradicted his own model of evolution.

Gregor Mendel

Gregor Mendel was a contemporary of Darwin's, but his work went unnoticed or unappreciated.
- ☞ It is ironic that Mendel's principles of inheritance give credibility to natural selection, and no one noticed.
- ☞ It is also ironic that when Mendel's work was finally noticed in the early 1900's, many geneticists believed the laws of inheritance were at odds with Darwin's theory of natural selection.

Why was the genetic basis for natural selection not immediately apparent to geneticists who rediscovered Mendel's work in the early 1900s?

We now know the population traits that Darwin emphasized vary in a continuum and are influenced by multiple genetic loci.
- ☞ Early geneticists did not recognize continuous variation and believed in the "either-or" inheritance of discrete traits.

☞ Recognizing only discrete qualitative traits, they found no genetic basis for natural selection to work. The central idea of Darwin's theory was that natural selection acts on subtle variations within a population.

Genetic research focused on mutations during the 1920's and an alternative to Darwin's theory of natural selection was widely accepted.
☞ This alternative was that evolution occurred in rapid leaps as a result of radical phenotypic changes caused by mutations.
☞ This idea contrasted with Darwin's view of gradual evolution due to environmental selection acting on continuous variations among individuals of a population.
☞ Considerable sentiment was also afforded *orthogenesis* (the idea that evolution has been a predictable progression to more and more elite forms of life).

Orthogenesis

Why was the emergence of population genetics an important turning point for evolutionary theory? What is meant by the "modern synthesis"?

The emergence of *population genetics* was an important turning point for evolutionary theory.
☞ Population genetics emphasizes the extensive genetic variation within populations and recognizes the importance of quantitative inheritance.
☞ Progress in this area reconciled Mendelism and Darwinism in the 1930's.

Population genetics

The *modern synthesis*, a comprehensive theory of evolution, was formulated in the 1940's as the genetic basis of variation and natural selection was worked out.
☞ Many scientists (paleontologists, taxonomists, biogeographers) contributed to the modern synthesis, including Theodosius Dobzhansky, Ernst Mayr, G.G. Simpson and G. Ledyard Stebbins.
☞ This theory emphasizes the importance of populations as the units of evolution, the essential role of natural selection, and gradualism.

Modern synthesis

Most of Darwin's ideas persist in the modern synthesis although many evolutionists are challenging some generalizations of the modern synthesis.
☞ This debate focuses on the rate of evolution and on the relative importance of evolutionary mechanisms other than natural selection.
☞ These debates do not question the fact of evolution, only what mechanisms are most important in the process.
☞ Such disagreements indicate that the study of evolution is very lively and that it continues to develop as a science.

II. THE GENETICS OF POPULATIONS (p. 439–442)

Population = Localized group of organisms which belong to the same *species*.

Species = Groups of actually or potentially interbreeding natural populations, which are reproductively isolated from other such groups.

Population

Species

How are most species geographically distributed? Why are individuals likely to be more closely related to others from their own population center, than to those from a different population center?

Most species are not evenly distributed over a geographical range, but are concentrated in several localized population centers.

316 How Populations Evolve

- Each population center is isolated to some extent from other population centers with only occasional gene flow among these groups.
- Obvious examples are isolated populations found on widely separated islands or in unconnected lakes.
- Some populations are not separated by such sharp boundaries. For example, a species with two population centers may be connected by an intermediate sparsely populated range.
- Even though these two populations are not absolutely isolated, individuals are more likely to interbreed with others from their population center. Gene flow between the two population centers is thus reduced by the intermediate range.

A. The Gene Pool and Microevolution (p. 439)

Gene pool = The total aggregate of genes in a population at any one time.

The gene pool consists of all the alleles at all gene loci in all individuals of a population. Alleles from this pool will be combined to produce the next generation.
- In a diploid species, an individual may be homozygous or heterozygous for a locus since each locus is represented twice.
- An allele is said to be *fixed* in the gene pool if all members of the population are homozygous for that allele.
- Normally there will be two or more alleles for a gene, each having a relative frequency in the gene pool.

How can microevolutionary change affect allele frequencies in a gene pool? (Allele frequency is the proportion or percentage of an allele in the population's gene pool.)

For example, in peppered moth populations living in unpolluted areas, the allele for light color has a much higher frequency than the allele for dark color.
- In areas of industrialization, the allele for dark color increased in frequency because of increased predation on light colored moths.
- This shift in the population's relative allelic frequencies is small scale evolutionary change called *microevolution*.

B. The Hardy-Weinberg Theorem (p. 440-442)

In your own words, what is the Hardy-Weinberg theorem?

In the absence of other factors, the segregation and recombination of alleles during meiosis and fertilization will not alter the overall genetic makeup of a population.
- The frequencies of alleles in the gene pool will remain constant unless acted upon by other agents; this is known as the *Hardy-Weinberg Theorem*.
- The Hardy-Weinberg model describes the genetics of nonevolving populations. This theorem can be tested with theoretical population models.

To test the Hardy-Weinberg theorem, imagine an isolated population of wildflowers with the following characteristics: (See Campbell, Figure 21.2)
- It is a diploid species with both red and white flowers.

Gene pool

Microevolution

Hardy-Weinberg Theorem

☞ The population size is 500 plants: 480 plants have red flowers, 20 plants have white flowers.
☞ Red flower color is coded for by the dominant allele "A," white flower color is coded for by the recessive allele "a."
☞ Of the 480 red flowered plants, 320 are homozygous (AA) and 160 are heterozygous (Aa). Since white color is recessive, all white flowered plants are homozygous aa.

Given the above information, how would you calculate the frequency of the dominant allele (A)?

☞ There are 1000 genes for flower color in this population, since each of the 500 individuals has two genes (this is a diploid species).
☞ A total of 320 genes are present in the 160 heterozygotes (Aa): half are dominant (160 A) and half are recessive (160 a).
☞ 800 of the 1000 total genes are dominant.
☞ The frequency of the A allele is 80% or 0.8 (800/1000).

Genotypes	# of plants		# of A alleles per individual		Total # A alleles
AA plants	320	×	2	=	640
Aa plants	160	×	1	=	160
					800

How would you calculate the frequency of the recessive allele (a)?

☞ 200 of the 1000 total genes are recessive.
☞ The frequency of the a allele is 20% or 0.2 (200/1000).

Genotypes	# of plants		# of A alleles per individual		Total # A alleles
aa plants	20	×	2	=	40
Aa plants	160	×	1	=	160
					200

If mating is random in this population, what will happen to the frequencies of "A" or "a" in the next generation?

Assuming that mating in the population is completely random (all male-female mating combinations have equal chances), the frequencies of A and a will remain the same in the next generation.
☞ Each gamete will carry one gene for flower color, either A or a.
☞ Since mating is random, there is an 80% chance that any particular gamete will carry the A allele and a 20% chance that any particular gamete will carry the a allele.

Given the allele frequencies of A (.8) and a (.2), what will be the frequencies of the three possible genotypes (AA, Aa, aa) in the next generation?

The frequencies of the three possible genotypes of the next generation can be calculated using the rule of multiplication: (See Campbell, Chapter 13)
☞ The probability of two A alleles joining is 0.8 × 0.8 = 0.64; thus, 64% of the next generation will be AA.

- ☞ The probability of two a alleles joining is 0.2 × 0.2 = 0.04; thus, 4% of the next generation will be aa.
- ☞ Heterozygotes can be produced in two ways, depending upon whether the sperm or ovum contains the dominant allele (Aa or aA). The probability of a heterozygote being produced is thus (0.8 × 0.2) + (0.2 × 0.8) = 0.16 + 0.16 = 0.32.

In the Hardy-Weinberg equilibrium population, does sexual reproduction (segregation and crossing-over in meiosis and random mating) cause rare genes to become more rare or common genes more common? If the hypothetical wildflower population were not in equilibrium, how many generations of random mating would it take to achieve Hardy-Weinberg equilibrium?

The frequencies of possible genotypes in the next generation are 64% AA, 32% Aa and 4% aa.
- ☞ The frequency of the A allele in the new generation is 0.64 + (0.32/2) = 0.8, and the frequency of the a allele is 0.04 + (0.32/2) = 0.2. Note that the alleles are present in the gene pool of the new population at the <u>same</u> frequencies they were in the original gene pool.
- ☞ Continued sexual reproduction with segregation, recombination and random mating would <u>not alter</u> the frequencies of these two alleles: the gene pool of this population would be in a state of equilibrium referred to as *Hardy-Weinberg equilibrium*.
- ☞ If our original population had not been in equilibrium, only one generation would have been necessary for equilibrium to become established.

What is the general Hardy-Weinberg equation? How can it be used to calculate allele and genotype frequencies?

From this theoretical wildflower population, a general formula, called the *Hardy-Weinberg equation*, can be derived to calculate allele and genotype frequencies.
- ☞ The Hardy-Weinberg equation can be used to consider loci with three or more alleles.
- ☞ By way of example, consider the simplest case with only two alleles with one dominant to the other.
- ☞ In our wildflower population, let p represent allele A and q represent allele a, thus p = 0.8 and q = 0.2.
- ☞ The sum of frequencies from all alleles must equal 100% of the genes for that locus in the population: $p + q = 1$.
- ☞ Where only two alleles exist, only the frequency of one must be known since the other can be derived:

$$1 - p = q \quad \text{or} \quad 1 - q = p$$

When gametes fuse to form a zygote, the probability of producing the AA genotype is p^2; the probability of producing aa is q^2; and the probability of producing an Aa heterozygote is $2pq$ (remember heterozygotes may be formed in two ways).

Hardy-Weinberg equilibrium

☞ The sum of these frequencies must equal 100%, thus:

$$p^2 + 2pq + q^2 = 1$$
$$\text{Frequency of AA} \quad \text{Frequency of Aa} \quad \text{Frequency of aa}$$

What are some practical uses of the Hardy-Weinberg equation? Why is it important to the study of evolution?

The Hardy-Weinberg equation permits the calculation of allelic frequencies in a gene pool, if the genotype frequencies are known. Conversely, the genotype can be calculated from known allelic frequencies.

For example, the Hardy-Weinberg equation can be used to calculate the frequency of inherited diseases in humans (e.g. phenylketonuria):

☞ 1 of every 10,000 babies in the United States is born with phenylketonuria (PKU) which can result in mental retardation if untreated.

☞ The allele for PKU is recessive, so babies with this disorder are homozygous recessive = q^2.

☞ Thus $q^2 = 0.0001$, with $q = 0.01$ (the square root of 0.0001).

☞ The frequency of p can be determined since $p = 1 - q$:

$$p = 1 - 0.01 = 0.99$$

☞ The frequency of carriers (heterozygotes) in the population is 2pq.

$$2pq = 2(0.99)(0.01) = 0.0198$$

☞ Thus, about 2% of the U.S. population are carriers for PKU.

The Hardy-Weinberg equilibrium is important to the study of evolution since it tells us what will happen in a <u>nonevolving</u> population.

☞ This equilibrium model provides a base line from which evolutionary departures take place.

☞ It provides a reference point with which to compare the frequencies of alleles and genotypes of natural populations whose gene pools may be changing.

What are the conditions a population must meet in order to maintain Hardy-Weinberg equilibrium?

For Hardy-Weinberg equilibrium to be maintained, five conditions <u>must</u> be met:

1. *Very large population size.*
2. *Isolation from other populations.* There is no migration of individuals into or out of the population.
3. *No mutations.*
4. *Random mating.*
5. *No natural selection.* All genotypes are equal in reproductive success. Differential reproductive success can alter gene frequencies.

Hardy-Weinberg equilibrium describes the genetics of an ideal population and it is important to note that this situation never exists in nature.

III. CAUSES OF MICROEVOLUTION (p. 442–445)

In real populations, several factors can upset Hardy-Weinberg equilibrium and cause microevolutionary change.

What five factors can cause microevolutionary change in a population? Which of these can lead to adaptive change? Which are nonadaptive, non-Darwinian changes?

Microevolution can be caused by *genetic drift, gene flow, mutation, nonrandom mating,* and *natural selection*; each of these conditions is a deviation from the criteria for Hardy-Weinberg equilibrium.
- ☞ Of these five possible agents for microevolution, only natural selection generally leads to an accumulation of favorable adaptations in a population.
- ☞ The other four are nonadaptive and are usually called non-Darwinian changes.

A. Genetic Drift (p. 442–444)

What is genetic drift? Why is the effect of genetic drift so great in a small population?

Genetic drift = Changes in the gene pool of a small population due to chance.
- ☞ If a population is small, stochastic events have a greater impact on gene frequencies.
- ☞ Chance events may cause the frequencies of alleles to drift randomly from generation to generation, since the existing gene pool may not be accurately represented in the next generation.

For example, assume our theoretical wildflower population contains only 25 plants, and the genotypes for flower color occur in the following numbers: 16 AA, 8 Aa and 1 aa. In this case, a chance event could easily change the frequencies of the two alleles for flower color.
- ☞ A rock slide or passing herbivore which destroys three AA plants would immediately change the frequencies of the alleles from A = 0.8 and a = 0.2, to A = 0.77 and a = 0.23.
- ☞ Although this change does not seem very drastic, the frequencies of the two alleles were changed by a chance event.

The larger the population, the less important is the effect of genetic drift.
- ☞ Even though natural populations are not infinitely large (in which case genetic drift could be completely eliminated as a cause of microevolution), most are so large that the effect of genetic drift is negligible.
- ☞ However, some populations are small enough that genetic drift can play a major role in microevolution, especially when the population has less than 100 individuals.

Two situations which result in populations small enough for genetic drift to be important are the *bottleneck effect* and the *founder effect*.

Nonadaptive

Genetic drift

1. **Bottleneck Effect** (*p. 443*)

What is the bottleneck effect?

The size of a population may be reduced drastically by such natural disasters as volcanic eruptions, earthquakes, fires, floods, etc. which kill organisms nonselectively.

- The small surviving population is unlikely to represent the genetic makeup of the original population.
- Genetic drift which results from drastic reduction in population size is referred to as the *bottleneck effect*.
- By chance some individuals survive. In the small remaining population, some alleles may be overrepresented, some underrepresented, and some alleles may be totally absent.
- Genetic drift which has occurred may continue to affect the population for many generations, until it is large enough for random drift to be insignificant.

How does the bottleneck effect influence genetic variability in a population?

The bottleneck effect reduces overall genetic variability in a population since some alleles may be entirely absent. For example, a population of northern elephant seals was reduced to just 20 individuals by hunters in the 1890's.

- Since this time, these animals have been protected and the population has increased to about 30,000 animals.
- Researchers have found that <u>no</u> variation exists in the 24 loci examined from the present population. A single allele has been fixed at each of the 24 loci due to genetic drift by the bottleneck effect.
- This contrasts sharply with the large amount of genetic variation found in southern elephant seal populations which did not undergo the bottleneck effect.

A lack of genetic variation in South African cheetahs may also have resulted from genetic drift, since the large population was severely reduced during the last ice age and again by hunting to near extinction.

2. **Founder Effect** (*p. 443–444*)

What is the founder effect? Are the gene frequencies in the small founding group likely to be the same as the gene frequencies in the original population? Why do founding populations tend to be influenced by genetic drift?

When a few individuals colonize a new habitat, genetic drift is also likely to occur. Genetic drift in a new colony is called the *founder effect*.

- The smaller the founding population, the less likely its gene pool will be representative of the original population's genetic makeup.
- The most extreme example would be when a single seed or pregnant female moves into a new habitat.

Bottleneck effect

Founder effect

- If the new colony survives, random drift will continue to affect allele frequencies until the population reaches a large enough size for its influence to be negligible.
- No doubt, the founder effect was instrumental in the evolutionary divergence of the Galapagos finches.

The founder effect probably resulted in the high frequency of *retinitis pigmentosa* (a progressive form of blindness that affects humans homozygous for this recessive allele) in the human population of Tristan da Cunha, a group of small Atlantic islands.
- This area was colonized by 15 people in 1814, and one must have been a carrier.
- The frequency of this allele is much higher on this island than in the populations from which the colonists came.

Although inherited diseases are obvious examples of the founder effect, this form of genetic drift can alter the frequencies of any alleles in the gene pool.

B. Gene Flow (*p. 444*)

What is gene flow? What are the consequences of gene flow on between-population differences?

Gene flow = The migration of fertile individuals, or the transfer of gametes, between populations.
- Natural populations may gain or lose alleles by gene flow, since they do not have gene pools which are closed systems required for Hardy-Weinberg equilibrium.
- Gene flow tends to reduce between-population differences which have accumulated by natural selection or genetic drift.
- An example of gene flow would be if our theoretical wildflower population was to begin receiving wind blown pollen from an all white-flower population in a neighboring field. This new pollen could greatly increase the frequency of the white flower allele, thus also altering the frequency of the red flower allele.
- Extensive gene flow can eventually group neighboring populations into a single population.

C. Mutation (*p. 444*)

A new mutation which is transmitted in gametes immediately changes the gene pool of a population by substituting one allele for another.

In our theoretical wildflower population, if a mutation in a white flowered plant caused that plant to begin producing gametes which carried a red flower allele, the frequency of the white flower allele is reduced and the frequency of the red flower allele is increased.

Why does mutation have little quantitative effect on a large population?

Mutation itself has little quantitative effect on large populations in a single generation, since mutation at any given locus is very rare.
- Mutation rates of one mutation per 10^5 to 10^6 gametes are typical, but vary depending on the species and locus.

Gene flow

- An allele with a 0.50 frequency in the gene pool that mutates to another allele at a rate of 10^{-5} mutations per generation would take 2000 generations to reduce the frequency of the original allele from 0.50 to 0.49.
- The gene pool is effected even less, since most mutations are reversible.
- If a new mutation increases in frequency, it is because individuals carrying this allele are producing a larger percentage of offspring in the population due to genetic drift or natural selection, not because mutation is producing the allele in abundance.

If mutation has such a negligible effect on the gene pool, why is mutation an important factor in evolutionary change?

Mutation is important to evolution since it is the original source of genetic variation, which is the raw material for natural selection.

D. Nonrandom Mating (p. 444–445)

How does nonrandom mating affect a population's genotype frequencies? Allele frequencies? What are two kinds of nonrandom mating?

Nonrandom mating increases the number of homozygous loci in a population, but does not in itself alter frequencies of alleles in a population's gene pool. There are two kinds of nonrandom mating: *inbreeding* and *assortative mating*.

1. Inbreeding

Individuals of a population usually mate with close neighbors rather than with more distant members of a population, especially if the members of the population do not disperse widely.
- This violates the Hardy-Weinberg criteria that an individual must choose its mate at random from the population.
- Since neighboring individuals of a large population tend to be closely related, inbreeding is promoted.
- Self-fertilization, which is common in plants, is the most extreme example of inbreeding.

Inbreeding results in relative genotypic frequencies (p^2, $2pq$, q^2) that deviate from the frequencies predicted for Hardy-Weinberg equilibrium, but does not in itself alter frequencies of alleles (p and q) in the gene pool.

Self-fertilization in our theoretical wildflower population would increase the frequencies of homozygous individuals and reduce the frequency of heterozygotes.
- Selfing of AA and aa individuals would produce homozygous plants.
- Selfing of Aa plants would produce half homozygotes and half heterozygotes.
- Each new generation would see the proportion of heterozygotes decrease, while the proportions of homozygous dominant and homozygous recessive plants would increase.
- Inbreeding without selfing would also result in a reduction of heterozygotes, although it would take much longer.

Inbreeding

One effect of inbreeding is that the frequency of homozygous recessive phenotypes increases.

An interesting thing to note is that even if the phenotypic and genotypic ratios change, the values of p and q do not change in these situations, only the way they are combined. A smaller proportion of recessive alleles are masked by the heterozygous state.

2. **Assortative Mating**

Assortative mating is another type of nonrandom mating which results when individuals mate with partners that are like themselves in certain phenotypic characters. For example:
- ☞ Snow geese occur in a blue variety and a white variety, with the blue color allele being dominant. Birds prefer to mate with those of their own color, blue with blue and white with white; this results in a lower frequency of heterozygotes than predicted by Hardy-Weinberg.
- ☞ Blister beetles (*Lytla magister*) in the Sonoran Desert usually mate with a same-size individual.

E. **Natural Selection** (*p. 445*)

The Hardy-Weinberg equilibrium condition that all individuals in a population have equal ability to produce viable, fertile offspring is probably never met.

In your own words, what is natural selection? What are the consequences of natural selection?

In any sexually reproducing population, variation among individuals exists and some variants leave more offspring than others. *Natural selection* is this differential success in reproduction.

Due to selection, alleles are passed on to the next generation in disproportionate numbers relative to their frequencies in the present generation.
- ☞ If in our theoretical wildflower population, white flowers are more visible to herbivores than red flowers, plants with red flowers (both AA and Aa) would leave more offspring on the average.
- ☞ Genetic equilibrium would be disturbed and the frequency of allele A would increase and the frequency of the a allele would decrease.

What is meant by the statement that natural selection is the only agent of microevolution which is adaptive?

Natural selection is the only agent of microevolution which is adaptive, since it accumulates and maintains favorable genotypes.
- ☞ Environmental change would result in selection favoring genotypes present in the population which can survive the new conditions.
- ☞ Variability in the population makes it possible for natural selection to occur.

IV. THE GENETIC BASIS OF VARIATION (p. 445–450)

Members of a population may vary in subtle or obvious ways. It is the genetic basis of this variation that makes natural selection possible.

A. The Nature and Extent of Genetic Variation Within and Between Populations (p. 445–447)

Darwin considered the slight differences between individuals of a population as raw material for natural selection.

While we are more conscious of the variation among humans, an equal if not greater amount of variation exists among the many plant and animals species.
- ☞ Phenotypic variation is a product of inherited genotype and numerous environmental influences.
- ☞ Only the genetic or inheritable component of variation can have adaptive impact as a result of natural selection.

Polygenic characters which vary quantitatively within a population are responsible for much of the inheritable variation.
- ☞ For example, the height of the individuals in our theoretical wildflower population may vary from very short to very tall with all sorts of intermediate heights.

Discrete characters which are determined by only one locus vary categorically, such as flower color in our wildflowers, without intermediates.
- ☞ In our wildflower population, the red and white flowers would be referred to as different *morphs* (contrasting forms of a Mendelian character).

What is a polymorphic character?

- ☞ A population is referred to as *polymorphic* for a character if two or more morphs are present in noticeable frequencies. (See Campbell, Figure 21.6)
- ☞ Polymorphism is found in human populations not only in physical characters (e.g. presence or absence of freckles) but also in biochemical characters (e.g. ABO blood group).

What biochemical technique allowed researchers to detect the great amount of genetic variation in natural populations? Why does this technique underestimate genetic variation?

Darwin did not realize the extent of genetic variation in populations, since much of the genetic variation can only be determined with biochemical methods.
- ☞ Electrophoresis has been used to determine genetic variation among individuals of a population. This technique allows researchers to identify variations in protein products of specific gene loci. (See Campbell, Methods Box, page 396)
- ☞ Electrophoretic studies show that in *Drosophila* populations the gene pool has two or more alleles for about 30% of the loci examined, and each fly is heterozygous at about 12% of its loci.

Polygenic characters

Morphs

Polymorphic

326 *How Populations Evolve*

Electrophoresis

- Thus, there are about 700–1200 heterozygous loci in each fly. Any two flies in a *Drosophila* population will differ in genotype at about 25% of their loci.
- Electrophoretic studies also show comparable variation in the human population.

Note that electrophoresis underestimates genetic variation:
- Proteins produced by different alleles may vary in amino acid composition and still have the same overall charge, which makes them indistinguishable by electrophoresis.
- Also, DNA variation not expressed as protein is not detected by electrophoresis.

What are some factors that can produce geographic variation among closely related populations?

Geographical variation

Geographical variation in allele frequencies exists among populations of most species.
- Natural selection can contribute to geographical variation, since at least some environmental factors are different between two locales. For example, one population of our wildflowers may have a higher frequency of white flowers because of the prevalence in that area of pollinators that prefer white flowers.
- Genetic drift may cause chance variations among different populations.
- Also, subpopulations may appear within a population due to localized inbreeding resulting from a "patchy" environment.

Cline

Cline = One type of geographical variation that is a graded change in some trait along a geographic transect.
- Clines may result from a gradation in some environmental variable.
- It may be a graded region of overlap where individuals of neighboring populations interbreed.
- For example, the average body size of many North American mammal species gradually increases with increasing latitude. It is presumed that the reduced surface area to volume ratio associated with larger size helps animals in cold environments conserve body heat.
- Studies of geographical variation confirm that genetic variation affects spatial differences of phenotypes in some clines. For example, yarrow plants are shorter at higher elevations, and some of this phenotypic variation has a genetic basis. (See Campbell, Figure 21.7)

B. Sources of Genetic Variation (*p. 447–448*)

Genetic variation results from mutation and sexual recombination.

1. Mutation (*p. 447–448*)

Mutations produce new alleles. They are rare and random events which usually occur in somatic cells and are thus not inheritable.
- Only mutations that occur in cell lines which will produce gametes can be passed to the next generation.
- Geneticists estimate that only an average of one or two mutations occur in each human gamete-producing cell line.

Point mutation = Mutation affecting a single base in DNA.
- ☞ Much of the DNA in eukaryotes does not code for proteins, and it is uncertain how a point mutation in these regions affect an organism.
- ☞ Point mutations in structural genes may cause little effect, partly due to the redundancy of the genetic code.

Mutations that alter a protein enough to affect the function are more often harmful than beneficial, since organisms are evolved products shaped by selection and a chance change is unlikely to improve the genome.
- ☞ Occasionally, a mutant allele is beneficial, which is more probable when environmental conditions are changing.
- ☞ The mutation which allowed house flies to be resistant to DDT was present in the population and under normal conditions resulted in reduced growth rate. It became beneficial to the house fly population only after a new environmental factor (DDT) was introduced and tipped the balance in favor of the mutant alleles.

Chromosomal mutations usually affect many gene loci and tend to disrupt an organism's development.
- ☞ On rare occasions, chromosomal rearrangement may be beneficial. These instances (usually by translocation) may produce a cluster of genes with cooperative functions when inherited together.

Duplication of chromosome segments is nearly always deleterious.
- ☞ If the repeated segment does not severely disrupt genetic balance, it may persist for several generations and provide an expanded genome with extra loci.
- ☞ These extra loci may take on new functions by mutation while the original genes continue to function.
- ☞ Shuffling of exons within the genome (single locus or between loci) may also produce new genes.

Mutation can produce adequate genetic variation in bacteria and other microorganisms which have short generation times.
- ☞ Some bacteria reproduce asexually by dividing every 20 minutes, and a single cell can produce a billion descendants in only 10 hours.
- ☞ With this type of reproduction, a beneficial mutation can increase in frequency in a bacterial population very rapidly.
- ☞ A bacterial cell with a mutant allele which makes it antibiotic resistant could produce an extremely large population of clones in a short period, while other cells without that allele are eliminated.
- ☞ Although bacteria reproduce primarily by asexual means, most increase genetic variation by occasionally exchanging and recombining genes through processes such as conjugation, transduction and transformation.

Point mutation

328 How Populations Evolve

2. Recombination (p. 448)

Even though mutation can be a source of genetic variability, why does it only contribute a negligible amount to genetic variation in a population? What process is responsible for nearly all the genetic variation in a population?

The contribution of mutations to genetic variation is negligible.

- Mutations are so infrequent at a single locus that they have little effect on genetic variation in a large gene pool.
- Although mutations produce new alleles, nearly all genetic variation in a population results from new combinations of alleles produced by sexual recombination.

Sexual recombination

Gametes from each individual vary extensively due to crossing over and random segregation during meiosis.

- Thus, each zygote produced by a mating pair possesses a unique genetic makeup.
- Sexual reproduction produces new combinations of old alleles each generation.

Plants and animals depend almost entirely on sexual recombination for genetic variation which makes adaptation possible.

C. How Genetic Variation is Preserved (p. 448–449)

Natural selection tends to produce genetic uniformity in a population by eliminating unfavorable genotypes. This tendency is opposed by several mechanisms that preserve or restore variation.

How is genetic variation preserved in a natural population?

1. Diploidy (p. 448)

Diploidy hides much genetic variation from selection by the presence of recessive alleles in heterozygotes.

- Since recessive alleles are not expressed in heterozygotes, less favorable or harmful alleles may persist in a population.
- This variation is only exposed to selection when two heterozygotes mate and produce offspring homozygous for the recessive allele.
- If a recessive allele has a frequency of 0.01 and its dominant counterpart 0.99, then 99% of the recessive allele copies will be protected in heterozygotes. Only 1% of the recessive alleles will be present in homozygotes and exposed to selection. (See Campbell, Figure 21.8)
- The more rare the recessive allele, the greater its protection by heterozygosity. That is, a greater proportion are hidden in heterozygotes by the dominant allele.
- This type of protection maintains a large pool of alleles which may be beneficial if conditions change.

2. Balanced Polymorphism (p. 448–449)

Selection may also preserve variation at some gene loci.

<u>Balanced polymorphism</u> = The ability of natural selection to maintain diversity in a population.

One mechanism by which selection preserves variation is *heterozygote advantage*.
- Natural selection will maintain two or more alleles at a locus if heterozygous individuals have a greater reproductive success than any type of homozygote.
- An example is the recessive allele that causes sickle-cell anemia in homozygotes. The locus involved codes for one chain of hemoglobin.
- Homozygotes for this recessive allele develop sickle-cell anemia which is often fatal.
- Heterozygotes are resistant to malaria. Heterozygotes thus have an advantage in tropical areas where malaria is prevalent, since homozygotes for the dominant allele are susceptible to malaria and homozygous recessive individuals are incapacitated by the sickle-cell condition.
- In some African tribes from areas where malaria is common, 20% of the hemoglobin loci in the gene pool is occupied by the recessive allele.

Other examples of heterozygote advantage are found in crop plants (e.g. corn) where inbred lines become homozygous at more loci and show stunted growth and sensitivity to diseases.
- Crossbreeding different inbred varieties often produces hybrids which are more vigorous than the parent stocks.
- This *hybrid vigor* is probably due to:
    1. Segregation of harmful recessives that were homozygous in the inbred varieties.
    2. Heterozygote advantage at many loci in the hybrids.

Balanced polymorphism can also result from patchy environments where different phenotypes are favored in different subregions of a populations geographic range. (See Campbell, Figure 21.9)

Frequency-dependent selection also causes balanced polymorphism.
- In this situation the reproductive success of any one morph declines if that phenotype becomes too common in the population.
- For example, in *Papilio dardanus*, an African swallowtail butterfly, males have similar coloration but females occur in several morphs.
- The female morphs resemble other butterfly species which are noxious to bird predators. *Papilio* females are not noxious, but birds avoid them because they look like distasteful species.

Balanced polymorphism

Heterozygote advantage

Sickle-cell anemia

Frequency-dependent selection

Batesian mimicry

☞ This type of protective coloration (*Batesian mimicry*) would be less effective if all the females looked like the same noxious species, because birds would encounter good-tasting mimics as often as noxious butterflies and would not associate a particular color pattern with bad taste.

D. Is All Genetic Variation Adaptive? (*p. 450*)

In your own words, what is the neutral theory of molecular evolution? How can changes in gene frequency be nonadaptive?

Some genetic variations found in populations confer no selective advantage or disadvantage. They have little or no impact on reproductive success. This type of variation is called *neutral variation*.

Neutral variation

☞ Much of the protein variation found by electrophoresis is adaptively neutral.
☞ For example, 99 known mutations affect 71 of 146 amino acids in the beta hemoglobin chain in humans. Some, like the sickle-cell anemia allele, affect the reproductive potential of an individual, while others have no obvious effect.
☞ The *neutral theory* of molecular evolution states than many variant alleles at a locus may confer no selective advantage or disadvantage.
☞ Natural selection would not affect the relative frequencies of neutral variations. Frequency of some neutral alleles will increase in the gene pool and others will decrease <u>due the chance effects of genetic drift</u>.

What is meant by "selfish DNA"?

Variation in DNA which does not code for proteins may also be nonadaptive.

☞ Most eukaryotes contain large amounts of DNA in their genomes which have no known function. Such noncoding DNA can be found in varying amounts in closely related species.
☞ Some scientists speculate that noncoding DNA has resulted from the inherent capacity of DNA to replicate itself and has expanded to the tolerance limits of the each species. The entire genome could exist as a consequence of being self-replicating rather than by providing an adaptive advantage to the organism.

Selfish DNA

☞ Transposons might fit this definition of "selfish DNA," although the degree of influence these sequences have on the evolution of genomes is not known.

Evolutionary biologists continue to debate how much variation, or even whether variation, is neutral.

☞ It is easy to show that an allele is deleterious to an organism.
☞ It is not easily shown that an allele provides no benefits, since such benefits may occur in immeasurable ways.
☞ Also, a variation may appear to be neutral under one set of environmental conditions and not neutral under other conditions.

We cannot know how much genetic variation is neutral, but if even a small portion of a population's genetic variation significantly affects the organisms, there is still a tremendous amount of raw material for natural selection and adaptive evolution.

V. ADAPTIVE EVOLUTION (p. 450–453)

Adaptive evolution results from a combination of both:
- ☞ Chance events that produce new genetic variation (e.g. mutation and sexual recombination).
- ☞ Natural selection that favors propagation of some variations over others.

A. Fitness (p. 450–451)

Darwinian fitness is measured by the relative contribution an individual makes to the gene pool of the next generation.
- ☞ It is not a measure of physical and direct confrontation, but of the success of an organism in producing progeny.
- ☞ Organisms may produce more progeny because they are more efficient feeders, attract more pollinators (as in our wildflowers), avoid predators, etc.

Survival does not guarantee reproductive success, since a sterile organism may outlive fertile members of the population.
- ☞ A long life span may increase fitness if the organism reproduces over a longer period of time (thus leaving more offspring) than other members of the population.
- ☞ Even if all members of a population have the same life span, those that mature early and thus have a longer reproductive time span, have increased their fitness.
- ☞ Every aspect of survival and fecundity are components of fitness.

What is relative fitness? What role does it play in adaptive evolution?

Relative fitness = The contribution of a genotype to the next generation compared to the contributions of alternative genotypes for the same locus.
- ☞ For example, if red flower plants (AA and Aa) in our wildflower population produce more offspring than white flower plants (aa), then AA and Aa genotypes have a higher relative fitness.

Statistical estimates of fitness can be produced by the relative measure of selection <u>against</u> an inferior genotype. This measure is called the *selection coefficient*.
- ☞ For comparison, relative fitness of the most fecund variant (AA or Aa in our wildflower population) is set at 1.0.
- ☞ If white flower plants produce 80% as many progeny on average, then the white variant relative fitness is 0.8.
- ☞ The selection coefficient is the difference between these two values (1.0 − 0.8 = 0.2).
- ☞ The more disadvantageous the allele, the greater the selection coefficient.
- ☞ Selection coefficients can range up to 1.0 for a lethal allele.

Why does the rate of decline for a deleterious allele depend upon whether the allele is dominant or recessive to the more successful allele?

The rate of decline in relative frequencies of deleterious alleles in a population depends on the magnitude of the selection coefficient working against it and whether the allele is dominant or recessive to the more successful allele.

☞ Deleterious recessives are normally protected from elimination by heterozygote protection.
☞ Selection against harmful dominant alleles is faster since they are expressed in heterozygotes.

The rate of increase in relative frequencies of beneficial alleles is also affected by whether it is a dominant or recessive.
☞ New recessive mutations spread slowly in a population (even if beneficial) because selection can not act in its favor until the mutation is common enough for homozygotes to be produced.
☞ New dominant mutations that are beneficial increase in frequency faster since even heterozygotes benefit from the allele's presence (for example, the mutant dark color producing allele in peppered moths).

Most new mutations, whether harmful or beneficial, probably disappear from the gene pool early due to genetic drift.

B. What Does Selection Act On? (p. 451)

What does selection act on? What factors contribute to the overall fitness of a genotype at any one locus?

Selection acts on phenotypes, indirectly adapting a population to its environment by increasing or maintaining favorable genotypes in the gene pool.
☞ Since it is the phenotype (physical traits, metabolism, physiology, and behavior) which is exposed to the environment, selection can only act indirectly on genotypes.

The connection between genotype and phenotype may not be as simple and definite as with our wildflower population where red was dominant to white.
☞ *Pleiotropy* (the ability of a gene to have multiple effects) often clouds this connection. The overall fitness of a genotype depends on whether its beneficial effects exceed any harmful effects on the organism's reproductive success.
☞ Polygenic traits also make it difficult to distinguish the phenotype-genotype connection. Whenever several loci influence the same characteristic, the members of the population will not fit into definite categories, but represent a continuum along a range.

An organism is an integrated composite of many phenotypic features, and the fitness of a genotype at any one locus depends upon the entire genetic context. A number of genes may work cooperatively to produce related phenotypic traits.

C. Modes of Natural Selection (p. 451–452)

What are three modes of natural selection? How are they distinguished from each other?

The frequency of a heritable characteristic in a population may be affected in one of three different ways by natural selection, depending on which phenotypes are favored. (See Campbell, Figure 21.11)

Margin note: Pleiotropy

1. **Stabilizing Selection**

Stabilizing selection favors intermediate variants by selecting against extreme phenotypes.
- ☞ The trend is toward reduced phenotypic variation and greater prevalence of phenotypes best suited to relatively stable environments.
- ☞ For example, human birth weights are in the 3–4 kg range. Much smaller and much higher birth weight babies have a greater infant mortality.

2. **Directional Selection**

Directional selection favors variants of one extreme. It shifts the frequency curve for phenotypic variations in one direction toward rare variants which deviate from the average of that trait.
- ☞ This is most common when members of a species migrate to a new habitat with different environmental conditions or during periods of environmental change.
- ☞ For example, fossils show the average size of European black bears increased after periods of glaciation, only to decrease during warmer interglacial periods.

3. **Diversifying Selection**

In diversifying selection, opposite phenotypic extremes are favored over intermediate phenotypes.
- ☞ This occurs when environmental conditions are variable in such a way that extreme phenotypes are favored.
- ☞ For example, balanced polymorphism of *Papilio* where butterflies with characteristics between two noxious model species (thus not favoring either) gain no advantage from their mimicry.

D. **Sexual Selection** (p. 452–453)

What is sexual dimorphism? How can it influence evolutionary change?

Sexual dimorphism = Distinction between the secondary sexual characteristics of males and females.
- ☞ Often seen as differences in size, plumage, lion's manes, deer antlers, or other adornments in males.
- ☞ In vertebrates it is usually the male that is the "showier" sex.
- ☞ In some species, males use their secondary sexual characteristics in direct competition with other males to obtain female mates (especially where harem building is common). These males may defeat other males in actual combat, but more often they use ritualized displays to discourage male competitors.

Darwin viewed sexual selection as a separate selection process leading to sexual dimorphism.
- ☞ These enhanced secondary sexual characteristics usually have no adaptive advantage other than attracting mates.
- ☞ However, if these adornments increase a males ability to attract more mates, his reproductive success is increased and he contributes more to the gene pool of the next generation.

The evolutionary outcome is usually a compromise between natural selection and sexual selection.
- In some cases the line between these two types of selection is not distinct, as in male deer.
- A stag may use his antlers to defend himself from predators and also to attract females.

E. **Does Evolution Fashion Perfect Organisms?** (*p. 453*)

Why doesn't natural selection breed perfect organisms?

Natural selection <u>cannot</u> breed perfect organisms because:

1. *Organisms are locked into historical constraints.* Each species has a history of descent with modification from ancestral forms.
   - Natural selection modifies existing structures and adapts them to new situations, it does not rebuild organisms.
   - For example, back problems suffered by some humans are in part due to the modification of a skeleton and musculature from the anatomy of four-legged ancestors which are not fully compatible to upright posture.

2. *Adaptations are often compromises.*
   - Each organism must be versatile enough to do many different things.
   - For example, seals spend time in the water and on rocks; they could walk better with legs, but swim much better with flippers.
   - Prehensile hands and flexible limbs allow humans to be very versatile and athletic, but they also make us prone to sprains, torn ligaments, and dislocations. Structural reinforcement would prevent many of these disabling occurrences but would limit agility.

3. *Not all evolution is adaptive.*
   - Genetic drift probably affects the gene pool of populations to a large extent.
   - Alleles which become fixed in small populations formed by the founder effect may not be better suited for the environment than alleles that are eliminated.
   - Similarly, small surviving populations produced by bottleneck effect may be no better adapted to the environment or even less well adapted than the original population.

4. *Selection can only edit variations that exist.*
   - These variations may not represent ideal characteristics.
   - New genes are not formed by mutation on demand.

These limitations thus allow natural selection to operate on a "better than" basis and subtle imperfections are the best evidence for evolution.

22 THE ORIGIN OF SPECIES

CHAPTER OUTLINE

The development of new species is the focal point of evolutionary theory and the mechanism by which biological diversity arises.

What are the two patterns of evolutionary change? How do they differ?

There are two patterns of evolutionary change: *anagenesis* and *cladogenesis*.

Anagenesis (phyletic evolution) = The transformation of an unbranched lineage of organisms, sometimes to a state different enough from the ancestral population to justify renaming it as a new species.

Cladogenesis (branching evolution) = The budding of one or more new species from a parent species that continues to exist.
- ☞ Cladogenesis is more important than anagenesis in the history of life, because it is more common and can promote biological diversity by increasing the number of species.
- ☞ Some evolutionary lineages that appear to be phyletic based on available fossils, may actually be cladogenetic.

I. THE SPECIES PROBLEM (p. 456–459)

Species exist in nature as discrete units, usually distinguishable from other species.
- ☞ Many times trained taxonomists give specific scientific names to flora and fauna which are already recognized by native populations as being different.
- ☞ Although species are discrete in nature, it is difficult to devise a formal definition of a species.

A. Two Concepts of Species (p. 457–458)

What is a morphospecies? How is the morphospecies concept useful to biologists?

Morphospecies = Species defined by their anatomical features.

Linnaeus (founder of modern taxonomy) described species in terms of their physical form (morphology) and this is still the most common method used for describing species.

There may be some difficulties in applying the morphospecies concept to some situations.
- ☞ Sometimes, it is difficult to determine if a set of organisms represents multiple species or a single species with extensive phenotypic variation.

Anagenesis

Cladogenesis

Morphospecies

336 The Origin of Species

Ernst Mayr

Biological species

Gene flow

☞ At other times, two populations morphologically almost indistinguishable may be different species based on other criteria.
☞ Even though there are difficulties with the morphospecies concept, it can be useful in the field, since its basis is observable and measurable anatomy.
☞ Most species have been designated as separate entities based on morphological criteria.

The morphospecies concept does not address the discontinuity that exists between species from the standpoint of evolutionary theory.

According to Ernst Mayr, what is a biological species?

Biological species = A population or group of populations whose members have the potential to interbreed with one another in nature to produce fertile offspring, but cannot successfully interbreed with members of other populations.
☞ This alternative to the morphospecies concept was formulated by Mayr in 1942.
☞ The biological species concept holds that the biological species is the largest unit of population in which gene flow is possible and that each such species is circumscribed by reproductive barriers that preserve its integrity by blocking (genetically isolating) genetic mixing with other species.
☞ Members of a species (called *conspecifics*) are united, at least potentially, by being reproductively compatible.
☞ Biological species are defined by their reproductive isolation from other species in natural environments (hybrids may be possible between two species in the laboratory).

B. Limitations of the Biological Species Concept (*p. 458–459*)

What are some limitations of the biological species concept?

The biological species concept cannot be applied to organisms that are completely asexual in their reproduction. Some protists and fungi, some commercial plants (bananas), and many bacteria are exclusively asexual.
☞ Asexual reproduction effectively produces a series of clones which, genetically speaking, represent a single organism.
☞ Asexual organisms can be assigned to species only by grouping clones with the same morphology and biochemical characteristics.

The biological species concept cannot be applied to extinct organisms represented only by fossils. These must be classified by the morphospecies concept.

It is also very difficult to apply the biological species concept to two populations which are geographically segregated. Can they potentially interbreed in nature even if they are similar enough to be placed into the same species based on morphological characteristics?

In some cases, unambiguous determination of species is not possible, even though the populations are sexual, contemporaneous and contiguous.

☞ Four phenotypically distinct populations of the deer mouse (*Peromyscus maniculatus*) found in the Rocky Mountains are geographically isolated and referred to as *subspecies*. (See Campbell, Figure 22.5)

☞ These populations overlap at certain locations and some interbreeding occurs in these areas of cohabitation, which indicates they are the same species by the biological species criteria.

☞ Two subspecies (*P. m. artemisiae* and *P. m. nebrascensis*) are an exception, since they do not interbreed in the area of cohabitation. However, their gene pools are not completely isolated since they freely interbreed with other neighboring populations.

☞ This could eventually produce gene flow between the populations of *P. m. artemisiae* and *P. m. nebrascensis* through the other populations.

☞ This circuitous route could only produce a very limited gene flow, but the route is open and possible.

☞ If this route was closed by extinction or geographic isolation of the neighboring populations, then *P. m. artemisiae* and *P. m. nebrascensis* could be named separate species without reservation.

More examples are being discovered where there is a blurry distinction between populations with limited gene flow and full biological species with segregated gene pools.

☞ If two populations cannot interbreed when in contact, they are clearly distinct species.

☞ When there is gene flow (even very limited) between two populations that are in contact, it is difficult to apply the biological species concept.

☞ This is equivalent to finding two populations at different stages in their evolutionary descent from a common ancestor, which is to be expected if new species arise by gradual divergence of populations.

Other species concepts have been developed in an effort to accommodate the dynamic, quantitative aspects of speciation; however, the species problem may never be completely resolved as it is unlikely that a single definition of species will apply to all cases.

II. REPRODUCTIVE BARRIERS (p. 459–462)

How can gene flow between closely related species be prevented?

Reproductive barrier = Any factor that impedes two species from producing fertile hybrids, thus contributing to reproductive isolation.

☞ Most species are genetically sequestered from other species by more than one type of reproductive barrier.

☞ Only intrinsic biological barriers to reproduction will be considered here. Geographic segregation (even though it prevents interbreeding) will not be considered.

☞ Reproductive barriers prevent interbreeding between closely related species.

What is the difference between prezygotic and postzygotic isolating mechanisms?

The various reproductive barriers which isolate the gene pools of species are classified as either prezygotic or postzygotic depending on whether they function before or after the formation of zygotes. (See Campbell, Table 22.1)

☞ Prezygotic barriers impede mating between species or hinder fertilization of the ova should members of different species attempt to mate.

☞ In the event fertilization does occur, postzygotic barriers prevent the hybrid zygote from developing into a viable, fertile adult.

A. **Prezygotic Barriers** (p. 460–461)

What are five prezygotic isolating mechanisms? What is an example of each?

1. **Habitat Isolation** (p. 460)

Two species living in different habitats within the same area may encounter each other rarely if at all, even though they are not technically geographically isolated.

☞ For example, two species of garter snakes (*Thamnophis*) occur in the same areas but for intrinsic reasons, one species lives mainly in water and the other is mainly terrestrial.

☞ Since these two species live primarily in separate habitats, they seldom come into contact as they are ecologically isolated.

2. **Temporal Isolation** (p. 460)

Two species that breed at different times of the day, seasons, or years cannot mix their gametes.

☞ For example, brown trout and rainbow trout cohabit the same streams, but brown trout breed in the fall and rainbow trout breed in the spring.

☞ Since they breed at different times of the year, their gametes have no opportunity to contact each other and reproductive isolation is maintained.

3. **Behavioral Isolation** (p. 460)

Special signals and elaborate behavior to attract mates, which are unique to each species, are probably the most important reproductive barriers among closely related species.

☞ Male fireflies of different species signal to females of the same species by blinking their lights in a characteristic pattern; females discriminate among the different signals and respond only to flashes of their own species by flashing back and attracting the males.

Many animals recognize mates by sensing pheromones (distinctive chemical signals).

☞ Female Gypsy moths attract males by emitting a volatile compound to which the olfactory organs of male Gypsy moths are specifically tuned: when a male detects this pheromone, it follows the scent to the female.

☞ Males of other moth species do not recognize this chemical as a sexual attractant.

Other factors may also act as behavioral isolating mechanisms:
- Eastern and western meadowlarks are almost identical in morphology and habitat, and their ranges overlap in the central United States.
- They retain their species integrity partly because of the difference in their songs, which enables them to recognize potential mates as members of their own kind.

4. Mechanical Isolation (p. 460)

Mechanical isolation

Anatomical incompatibility may prevent sperm transfer when closely related species attempt to mate.
- For example, male dragonflies use a pair of special appendages to clasp females during copulation; when a male tries to mount a female of a different species, he is unsuccessful because his clasping appendages do not fit the female's form well enough to grip securely.
- In plants that are pollinated by insects or other animals, the floral anatomy is often adapted to a specific pollinator that transfers pollen only among plants of the same species.

5. Gametic Isolation (p. 460–461)

Gametic isolation

Gametes of different species that meet rarely fuse to form a zygote.
- For animals that utilize internal fertilization, the sperm of one species may not be able to survive the internal environment of the female reproductive tract of a different species.
- Cross-specific fertilization is also uncommon for animals that utilize external fertilization due to a lack of gamete recognition.

Gamete recognition may be based on the presence of specific molecules on the coats around the egg which adhere only to complementary molecules on sperm cells of the same species.
- Similar mechanisms of molecular recognition enables a flower to discriminate between pollen of the same species and pollen of different species.

B. Postzygotic Barriers (p. 461)

What are three postzygotic isolating mechanisms? What is an example of each?

When prezygotic barriers are crossed and a hybrid zygote forms, one of several postzygotic barriers may prevent development of a viable, fertile hybrid.

1. Hybrid Inviability (p. 461)

Hybrid inviability

Genetic incompatibility between the two species may abort development of the hybrid at some embryonic stage.
- For example, several species of frogs in the genus *Rana* live in the same regions and habitats.
- They occasionally hybridize but the hybrids generally do not complete development, and those that do are frail and soon die.

Hybrid sterility

2. Hybrid Sterility (p. 461)

Why are many hybrids sterile?

If two species mate and produce hybrid offspring that are viable, reproductive isolation is intact if the hybrids are sterile because genes cannot flow from one species' gene pool to the other.
- One cause of this barrier is that if chromosomes of the two parent species differ in number or structure, meiosis cannot produce normal gametes in the hybrid.
- The most familiar case is the mule which is produced by crossing a donkey and a horse; very rarely are mules able to backbreed with either parent species.

3. Hybrid Breakdown (p. 461)

Hybrid breakdown

In your own words, how does hybrid breakdown maintain separate species even if gene flow occurs?

When some species cross-mate, the first generation hybrids are viable and fertile, but when these hybrids mate with one another or with either parent species, offspring of the next generation are feeble or sterile.
- For example, different cotton species can produce fertile hybrids, breakdown occurs in the next generation when progeny of the hybrids die in their seeds or grow into weak defective plants.

C. Introgression (p. 461–462)

How does introgression affect the gene pool of a species?

Introgression = The transplantation of alleles between species.
- Introgression occurs when alleles occasionally seep through all reproductive barriers and pass between the gene pools of closely related species when fertile hybrids mate successfully with one of the parent species.

Introgression

For example, corn (*Zea mays*) contains some alleles traceable to the closely related wild grass teosinte (*Zea mexicana*).
- Introgressions occur when the two species hybridize and a small number of the hybrids manage to cross with corn plants.
- The transplant of alleles increases the genetic variation that can be exploited by breeders attempting to produce new corn varieties by artificial selection.
- Occasional hybridization does not erase the boundary between corn and teosinte as only a very rare introgression occurs, the isolation of the two gene pools is not seriously breached and the two species remain distinct.

III. THE BIOGEOGRAPHY OF SPECIATION (p. 462–467)

Reproductive barriers form boundaries around species, and the evolution of these barriers is the key biological event in the origin of new species.
- An essential episode in the origin of a species occurs when the gene pool of a population is separated from other populations of the parent species.

☞ This genetically isolated splinter group can then follow its own evolutionary course as changes in allele frequencies caused by selection, genetic drift, and mutations occur undiluted by gene flow from other populations.

Speciation episodes can be classified into two modes based on the geographical relationship of a new species to its ancestral species: allopatric speciation and sympatric speciation.

A. **Allopatric Speciation** (p. 462–465)

In your own words, what characterizes the allopatric speciation model? What are the roles of intraspecific variation and geographical isolation in this model?

Allopatric speciation

This type of speciation occurs when the initial block to gene flow is a geographical barrier that physically isolates the population.

Geological processes can cause once sympatric populations (had overlapping ranges) to become allopatric (having separate ranges).
  ☞ Such occurrences include emergence of mountain ranges, movement of glaciers, formation of land bridges, subsidence of large lakes.
  ☞ Also small populations may become geographically isolated when individuals from the parent population travel to a new location.

The extent of development of a geographical barrier necessary to isolate two populations depends on the ability of the organisms to disperse due to the mobility of animals or the dispersibility of spores, pollen and seeds of plants.
  ☞ For example, the Grand Canyon is an impassable barrier to small rodents but is easily crossed by birds.

An Example of Allopatric Speciation:

50,000 years ago, during an ice age, the Death Valley region of California and Nevada had a rainy climate and a system of interconnecting lakes and rivers.
  ☞ 10,000 years ago a drying trend began and by 4,000 years ago, the region had become a desert.
  ☞ Presently, isolated springs in deep clefts between rocky walls are the only remnants of the lake and river networks
  ☞ Living in many of these isolated springs are small pupfishes (*Cyprinodon* spp.).
  ☞ Each inhabited spring contains its own species of pupfish which is adapted to that pool and found nowhere else in the world.
  ☞ The various pupfish species probably descended from a single ancestral species whose range was fragmented when the region became arid, thus isolating several small populations that diverged in their evolution as they adapted to their spring's environment.

Conditions Favoring Allopatric Speciation:

When populations become allopatric, speciation can potentially occur as the isolated gene pools accumulate differences by microevolution that may cause the populations to diverge in phenotype.

☞ A small isolated population is more likely to change substantially enough to become a new species than is a large isolated population.

Why are peripheral isolates susceptible to speciation if geographic barriers arise?

The geographic isolation of a small population usually occurs at the fringe of the parent population's range. This *peripheral isolate* is a good candidate for speciation for three reasons:

1. The gene pool of the peripheral isolate probably differs from that of the parent population initially, since fringe inhabiters usually represent the extremes of any genotypic and phenotypic clines in an original sympatric population. With a small peripheral isolate, there will be a founder effect with chance resulting in a gene pool that is not representative of the gene pool of the parental population.

2. Genetic drift will continue to change the gene pool of the small peripheral isolate until a large population is formed. New mutations or combinations of alleles that are neutral in adaptive value may become fixed in the population by chance alone, causing phenotypic divergence from the parent population.

3. Since the peripheral isolate inhabits a frontier with a somewhat different environment, it will probably be exposed to different selection pressures than those encountered by the parental population. Thus, evolution caused by selection is likely to take a different direction in the peripheral isolate than in the parental population.

These three factors can cause a peripheral isolate to diverge from the parental population, as long as the gene pools are isolated.

Due to the severity of a fringe environment, most peripheral isolates do not survive long enough to speciate.

Although most peripheral isolates become extinct, evolutionary biologists agree that a small population can accumulate enough genetic change to become a new morphospecies in only hundreds to thousands of generations.

Adaptive Radiation on Island Chains

What is the adaptive radiation model? According to this model, how might it be possible for there to be many closely-related sympatric species, even if geographic isolation is necessary for them to evolve?

Allopatric speciation has occurred on island chains where founding populations, which have strayed or become passively dispersed from their ancestral populations, have evolved in isolation.

Adaptive radiation = Emergence of numerous species from a common ancestor introduced to an environment presenting a diversity of new opportunities and problems.

Examples of adaptive radiation are the endemic species of the Galapagos Islands which descended from small populations which floated, flew, or were blown from South America to the islands. Darwin's finches can be used to illustrate a model for such adaptive radiation on island chains.

- ☞ A single dispersal event may have seeded one island with a peripheral isolate of the ancestral finch which diverged as it underwent allopatric speciation.

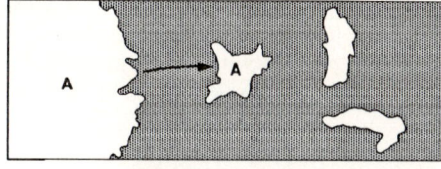

- ☞ A few individuals of this new species may have reached neighboring islands, forming new peripheral isolates which also speciated.

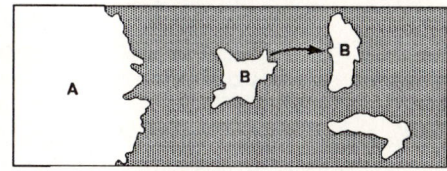

- ☞ After diverging on the island it invaded, a new species could recolonize the island from which its founding population emigrated and coexist with the ancestral species or form still another species.

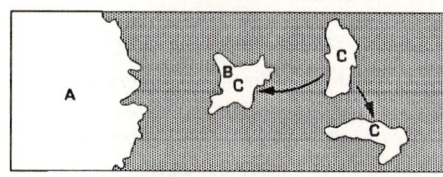

- ☞ Multiple invasions of islands could eventually lead to coexistence of several species on each island since the islands are distant enough from each other to permit geographic isolation, but near enough for occasional dispersal.

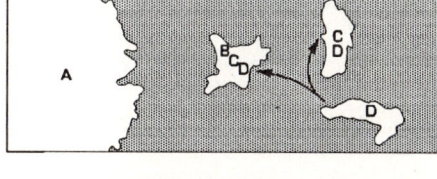

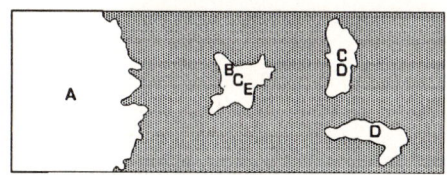

Thirteen species of finches have evolved on the Galapagos Archipelago from a single ancestral species.
- ☞ Each island presently has multiple species with as many as ten on some islands.
- ☞ In contrast, the more isolated Cocos Island has only one finch species. It is derived from an ancestral species that immigrated to this remote island which is so isolated that additional island hopping is unlikely.
- ☞ Similar evolutionary events have occurred on the Hawaiian Archipelago, which provide a great amount of physical diversity for evolutionary change.

Islands close enough to a mainland to allow free movement from the island to the mainland are not characterized by endemic species, since there is no long-term reproductive isolation (e.g. Florida Keys).

344   The Origin of Species

Polyploidy

Sympatric speciation

Autopolyploid

Tetraploid

Oenothera

Allopolyploid

B.   **Sympatric Speciation** (p. 465–467)

What is sympatric speciation? How can polyploidy cause reproductive isolation?

Sympatric speciation = A mode of speciation characterized by new species forming within the range of parent populations; reproductive isolation evolves without geographical isolation.
- ☞ Can occur quickly (in one generation) if a genetic change results in a reproductive barrier between the mutants and the parent population.
- ☞ Sympatric speciation may apply to a substantial proportion of plant species but its importance to animal evolution is less certain.
- ☞ Some plant species originate from improper cell division that results in extra sets of chromosomes. (See Campbell, Figure 22.12)

What is the difference between autopolyploidy and allopolyploidy?

Autopolyploid = An organism that results from a single species that doubles its chromosome number to the tetraploid state.
- ☞ Tetraploids can self-fertilize or mate with other tetraploids.
- ☞ The mutants cannot interbreed with the diploids of the parent population because hybrids would be inviable or sterile.
- ☞ Thus, an instantaneous special genetic event produces a postzygotic barrier which isolates the gene pool of the mutant in just one generation.
- ☞ Sympatric speciation by autopolyploidy was first discovered by Hugo De Vries in the early 20th century while working with *Oenothera*, the evening primrose.

Allopolyploid = A polyploid hybrid resulting from contributions of two different species.
- ☞ More common than autopolyploidy.
- ☞ Potential evolution of an allopolyploid begins when two different species interbreed and a hybrid is produced. (See Campbell, Figure 22.12b)
- ☞ Such interspecific hybrids are usually sterile, because the haploid set of chromosomes from one species cannot pair during meiosis with the haploid set of chromosomes from the second species.
- ☞ These sterile hybrids may actually be more vigorous than the parent species and propagate asexually.

At least two mechanisms can transform these sterile hybrids into fertile polyploids:
1. During the history of the hybrid clone, mitotic nondisjunction in the reproductive tissue may double the chromosome number.
   - ☞ The hybrid clone will then be able to produce gametes since each chromosome will have a homologue to synapse with during meiosis.
   - ☞ Gametes from this fertile tetraploid could unite and produce a new species of interbreeding individuals, reproductively isolated from both parent species.

2. Meiotic nondisjunction in one species produces an unreduced (diploid) gamete.

- ☞ This abnormal gamete fuses with a normal haploid gamete of a second species and produces a triploid hybrid.
- ☞ The triploid hybrid will be sterile, but may propagate asexually.
- ☞ During the history of this sterile triploid clone, meiotic nondisjunction again produces an unreduced gamete (triploid).
- ☞ Combination of this triploid gamete with a normal haploid gamete from the second parent species would result in a fertile hybrid with homologous pairs of chromosomes.
- ☞ This allopolyploid would have a chromosome number equal to the sum of the chromosome numbers of the two ancestral species (as in 1 above).

Speciation of polyploids (especially allopolyploids) has been very important in plant evolution.
- ☞ Some allopolyploids are very vigorous because they contain the best qualities of the two parent species.
- ☞ The accidents required to produce these new plant species (interspecific hybridization coupled with nondisjunction) have occurred often enough that between 25% and 50% of all plant species are polyploids.
- ☞ *Triticum aestivum*, bread wheat, is a 42 chromosome allopolyploid that is believed to have originated about 8000 years ago as a hybrid of a 28 chromosome cultivated wheat and a 14 chromosome wild grass.
- ☞ Other important polyploid species include oats, cotton, potatoes, and tobacco.

Plant geneticists are presently inducing these genetic accidents to produce new polyploids which will combine high yield and disease resistance.

IV. GENETIC MECHANISMS OF SPECIATION (p. 467–469)

What are two modes of speciation that are based on genetic mechanisms of change?

Another method of classifying modes of speciation is based on genetic mechanisms rather than biogeographical factors. This alternative method groups speciation mechanisms into two categories: speciation by divergence and speciation by peak shifts.

Divergence

A. Speciation by Divergence (p. 467–468)

Why must reproductive barriers develop if divergent speciation is to occur?

Two populations which adapt to different environments accumulate differences in the frequencies of genotypes and phenotypes.

- During this gradual adaptive divergence of the two gene pools, reproductive barriers may evolve between the two populations.
- Evolution of reproductive barriers would differentiate these populations into two species.

What may happen when sympatry is restored between previously allopatric populations?

It is possible to test some closely related allopatric populations to determine if speciation has occurred by attempting to hybridize representatives of the two populations in the laboratory.

- Some previously allopatric populations have come back into contact naturally at a border between their geographical ranges.
- Of many possibilities, two extremes may occur when sympatry is naturally restored:

    1. The two populations may hybridize freely and the gene pools mix when together, indicating that evolutionary divergence did not result in reproductive isolation and speciation.

    2. Evolutionary divergence of the two populations during their allopatric period results in reproductive barriers that maintain the two populations as separate species when they become sympatric again.

There are many intermediate cases between these two extremes. Limited hybridization between reunited populations map produce hybrids less fit than offspring from parents belonging to the same population.

Semispecies

Semispecies = Populations that are not totally reproductively isolated.

What are two hypotheses for how intrinsic isolating mechanisms may evolve? How is the role of natural selection viewed differently between the two hypotheses?

One hypothesis supports the idea that meeting of semispecies should reinforce prezygotic reproductive barriers between populations.

- Any individuals with prezygotic barriers to mating outside their population will usually leave a greater number of fertile offspring than individuals reproductively isolated from other populations by only postzygotic barriers.
- If true, this hypothesis holds that a speciation episode, which is not quite completed during the geographical isolation, will be completed when the two populations come back into contact and natural selection favors those individuals least likely to mate outside of their population.
- Postzygotic barriers must have evolved before prezygotic barriers for this type of reinforcement of reproductive isolation to be a step in the speciation process.

☞ Some geneticists question the importance of reinforcement to speciation. Prezygotic barriers probably evolve between populations more often during geographical isolation than as a consequence of reinforcement after reunion of the populations.

A key point in evolution by divergence is that reproductive barriers can arise without being favored directly by natural selection.
- ☞ Divergence of two populations is due to their adaptation to separate environments, with reproductive isolation being a secondary development.
- ☞ Postzygotic barriers may be pleiotropic effects of interspecific differences in those genes that control development. For example, hybrids may be inviable if both sets of genes for rRNA synthesis are not active (e.g. hybrids between *D. melanogaster* and *D. simulans*).
- ☞ Gradual genetic divergence of two populations may also result in the evolution of prezygotic barriers. For instance, an ecological barrier to inbreeding may secondarily result from the adaptation of an insect population to a new host plant different from the original population's host.

In some isolated populations, reproductive isolation has evolved more directly from sexual selection.
- ☞ For example, in *Drosophila heteroneura*, the male's wide head enhances reproductive success with females of the same species. It reduces the probability that a male *D. heteroneura* will mate with females of other species.
- ☞ Sexual selection, in this case, probably evolved as an adaptation for enhanced reproductive success. A secondary consequence is that it prevents interbreeding with other *Drosophila* species.
- ☞ Since reproductive barriers usually evolve when populations are allopatric, they do not function directly to isolate the gene pools of populations. For this reason, the emphasis on reproductive isolating mechanisms is one criticism of the "biological species concept".
- ☞ An alternative approach is the *recognition concept of species*.   _Recognition concept of species_

The recognition concept of species assumes that the reproductive adaptations of a species consists of a set of characteristics that maximize successful mating with members of the same population.
- ☞ Reproductive isolation from other populations would thus be a side effect.
- ☞ The recognition concept thus focuses on characteristics which are actually subject to natural selection in isolated populations that are in the process of speciation.

B. **Speciation by Peak Shifts** (p. 468–469)

In your own words, what is Sewall Wright's evolutionary metaphore, the adaptive landscape? According to this model, how does a population reach an alternate peak?

The concept of the "adaptive landscape" was proposed by Sewell Wright in the 1930's.   _Adaptive landscape_
Sewell Wright

Adaptive peaks

- The landscape includes many *adaptive peaks* separated by valleys.
- Each adaptive peak is an equilibrium state in which allelic frequencies maximize the average fitness of the population's members.
- The landscape could include many adaptive peaks, even under stable environmental conditions, but natural selection tends to maintain the population at a single peak.

For a population to reach an alternate peak by a gene pool change, it must pass through a valley where genetic combinations are of low average fitness.
- Thus, if a slight change in allele frequencies at one or more loci pushes a population from its adaptive peak, natural selection will tend to push that population back to its original adaptive peak.

Environmental changes redefine the landscape, making new adaptive peaks possible.
- For a population to survive the new conditions, it must reach a new adaptive peak through microevolution of its gene pool.

What causes peak shifts? What is the role of nonadaptive change in this speciation process? What is the role of adaptive evolution?

Peak shift

Geneticists use the term *peak shift* for speciation caused by non-adaptive changes in the genetic system.
- Peak shifts may be caused by founder effect and bottlenecks.
- Genetic drift can remove a small population from its original adaptive peak and if the gene pool is sufficiently destabilized, new adaptive peaks may be reached.
- The population that survives will be pushed to new adaptive peaks by natural selection in succeeding generations.
- Thus adaptive evolution plays a major role in peak shift, but genetic drift is the mechanism that makes the shift possible.
- Peak shifts can also occur in bottlenecked populations, even if there is no environmental change, due to the many adaptive peaks possible under the same environmental conditions.

With founder effect, the splinter population is affected not only by genetic drift moving the gene pool randomly over the adaptive landscape, but also by new environmental conditions which form a new set of adaptive peaks.
- The combination of genetic drift followed by natural selection under new environmental conditions may be responsible for the rapid radiation of island species (e.g. Hawaiian *Drosophila*).

Plant speciation by polyploidy depends on peak shift.
- Fertile polyploids which are reproductively isolated from their ancestral species, also represent new genetic combinations which originated by random processes instead of natural selection.
- The new genetic combinations redefine the adaptive landscape and make new adaptive peaks possible.
- The polyploid species survive only if natural selection produces combinations of alleles that move the population to one of the new adaptive peaks.

C. **How Much Genetic Change Is Required For Speciation?** (*p. 469*)

No generalizations can be made about genetic distance between closely related species.
- ☞ Reproductive barriers may result from changes in many loci or only a few.
- ☞ Two species of Hawaiian *Drosophila* differ at only one locus which determines head shape, an important factor in mate recognition. The phenotypic effect of different alleles at this locus is multiplied by epistasis involving at least ten other loci.
- ☞ Changes in one gene in a coadapted gene complex can substantially impact the development of an organism.

These examples indicate that massive genetic changes involving many loci are not necessary for speciation to occur.

V. **GRADUAL AND PUNCTUATED INTERPRETATIONS OF SPECIATION** (*p. 469–471*)

What are some points of agreement and disagreement between the two schools of thought about the tempo of speciation (gradualism vs. punctuated equilibrium)?

Traditional evolutionary trees diagram the descent of species from ancestral forms as branches that gradually diverge with each new species evolving continuously over long spans of time. (See Campbell, Figure 22.16a)
- ☞ The theory behind such a tree is the extrapolation of microevolutionary processes (allele frequency changes in the gene pool) to the divergence of species.
- ☞ Big changes thus occur due to the accumulation of many small changes.

Paleontologists rarely find gradual transitions of fossil forms but often observe species appearing as new forms suddenly in the rock layers.
- ☞ These species persist virtually unchanged and then disappear as suddenly as they appeared.
- ☞ Even Darwin, who believed species from a common ancestral stock evolve differences gradually, was perplexed by the lack of transitional forms in the fossil record.

Advocates of *punctuated equilibrium* have redrawn the evolutionary tree to represent fossil evidence for evolution occurring in spurts of relatively rapid change instead of gradual divergence. (See Campbell, Figure 22.16b)
- ☞ This model was proposed by Niles Eldredge and Stephen Jay Gould in 1972.
- ☞ It depicts species undergoing most of their morphological modification as they first separate from the parent species then showing little change as they produce new species.
- ☞ In this theory gradual change is replaced with long periods of stasis punctuated with episodes of speciation.
- ☞ The origin of new polyploid plants through genome changes is one mechanism of sudden speciation.
- ☞ Allopatric speciation of a splinter population separated from its parent population by geographical barriers may also be rapid. For a population facing new environmental conditions, genetic drift and selection can cause significant change in only a few hundred or thousand generations.

Margin notes: Gradualism; Punctuated equilibrium; Niles Eldredge, Stephen Jay Gould

A few thousand generations is considered rapid in reference to the geologic time scale.
- ☞ The fossil record indicates that successful species survive for a few million years on average.
- ☞ For example, if a species survives for five million years and most of its morphological changes occur in the first 50,000 years, then the speciation episode occurred in just 1% of its lifetime.
- ☞ With this time scale, a species will appear suddenly in rocks of a certain age, linger relatively unchanged for millions of years, then become extinct.
- ☞ While forming, the species may have gradually accumulated modifications, but with reference to its overall history, its formation was sudden.

An evolutionary spurt preceding a longer period of morphological stasis would explain why paleontologists find so few transitional fossils.
- ☞ It also explains why contemporary species are seen as discrete entities since species spend only a small fraction of their time becoming unique.
- ☞ Since a few populations are observed in the process of speciation, there will be some ambiguous cases where distinction between closely related species is difficult. (See Campbell, Figure 22.5)

Because "sudden" can refer to thousands of years on the geological time scale, differing opinions of punctuationalists and gradualists about the rate of speciation may be more a function of time perspective than conceptual difference. There is clear disagreement, however, over how much a species changes after its origin.
- ☞ In a species adapted to an environment that stays the same, natural selection would counter changes in the gene pool. Once selection during speciation produces new complexes of coadapted genes, mutations are likely to impose disharmony on the genome and disrupt the development of the organism.
- ☞ Stabilizing selection would thus hold a population at one adaptive peak to produce long periods of stasis.

Some gradualists feel that stasis is an illusion since many species may continue to change in undetectable ways after they have diverged from the parent population.
- ☞ Changes in internal anatomy may go unnoticed by paleontologists as fossils only show external anatomy and skeletons.
- ☞ Population geneticists also point out that many microevolutionary changes occur at the molecular level without affecting morphology.

Some biologists even contest the long periods of morphological stasis in the history of species. Peter Sheldon has analyzed approximately 15,000 trilobite fossils which represent an unusually complete record.
- ☞ Paleontologists have arranged the fossils into several evolutionary lineages and classified the youngest and oldest as different species based on morphological differences.
- ☞ Based on his extensive study, Sheldon concludes that specific species boundaries cannot be drawn among these trilobite fossils. He found a gradual change in the morphological characteristics used to name the new species.
- ☞ Sheldon suggests that additional extensive studies of fossil morphology where specific lineages are preserved should be carried out to assess the relative importance of gradual and punctuated tempos in the origin of new species.

23   TRACING PHYLOGENY: MACROEVOLUTION, THE FOSSIL RECORD, AND SYSTEMATICS

CHAPTER OUTLINE

What is macroevolution? How is it different from microevolution?

Macroevolution = The origin of taxonomic groups higher than the species level.

☞ Macroevolution is concerned with major events in the history of life and encompasses the origin of new designs such as: feathers and wings of birds, upright posture of humans, increasing brain size in mammals, diversification of certain groups of organisms following evolutionary breakthroughs, and extinction.

☞ Macroevolution extends the study of evolution to include global environmental and biological change.

Is macroevolutionary change primarily a cumulative product of microevolution working gradually over time, or is it the product of mechanisms other than the gradual modification of populations by natural selection?

I. THE RECORD OF THE ROCKS (*p. 475–478*)

Why is the fossil record important to the study of macroevolution?

Biologists reconstruct evolutionary history by studying the succession of organisms in the fossil record.

☞ Paleobiologists are paleontologists who specialize in animal fossils.
☞ Paleobotanists are paleontologists who specialize in plant fossils.

    A.   How Fossils Form (*p. 475*)

What are fossils, and how do they form?

Fossil = Any preserved remnant or impression left by an organism that lived in the past.

Some fossils, found as thin films pressed between layers of sandstone or shale, retain organic material.

☞ Paleobotanists have found leaves millions of years old that are still green with chlorophyll and preserved well enough that their organic composition and ultrastructure could be analyzed. (See Campbell, Figure 23.2a)

☞ Rarely, an entire organism has been fossilized, which could only have happened if the individual was buried in a medium that prevented bacteria and fungi from decomposing the body. (See Campbell, Figure 23.2f)

Macroevolution

Fossil

Fossils usually form from mineral rich hard parts of organisms (bones, teeth, shells of invertebrates) since organic substances usually decay rapidly.

☞ Paleontologists usually find parts of skulls, bone fragments, or teeth, although nearly complete skeletons of dinosaurs and other forms have been found.

☞ Many of the parts found have been hardened by *petrification*, which occurs when minerals dissolved in groundwater seep into the tissues of dead organisms and replace organic matter.

Other fossils found by paleontologists are replicas, cast from molds left when corpses were covered by mud or sand. (See Campbell, Figure 23.2e)

B. Limitations of the Fossil Record (p. 475–477)

Why is the fossil record incomplete?

A fossil represents a sequence of improbable events: an organism had to die in the right place and at the proper time for burial conditions favoring fossilization; the rock layer containing the fossil had to escape geologic events (erosion, pressure, extreme heat) which would have distorted or destroyed the rock; the fossil had to be exposed and not destroyed; and someone who knew what they were doing had to find the fossil.

☞ A large fraction of species that have lived probably left no fossils.
☞ Most fossils that were formed have probably been destroyed.
☞ Only a small number of existing fossils have been discovered.
☞ Consequently, the fossil record is comprised primarily of species that lived a long time, were abundant and widespread, and had shells or hard skeletons.

The fossil record provides the outline of macroevolution, but the evolutionary relationships between modern organisms must be studied to provide the details.

C. The Fossil Record and the Geological Time Scale (p. 477–478)

Several methods are used to determine the age of fossils which makes them useful in studies of macroevolution.

How are strata of sedimentary rocks formed?

Sedimentary rocks are the richest sources of fossils.

☞ These rocks form from deposits of sand and silt which have weathered or eroded from the land and are carried by rivers to seas and swamps.
☞ New deposits pile on and compress older sediments below into rock.
☞ Sand is compressed into sandstone and mud into shale.
☞ Aquatic organisms (and terrestrial ones swept into the water) settle into the sediments when they die, and some are preserved as fossils.

Sedimentation may occur when the sea-level changes or lakes and swamps dry and refill.

☞ The rate of sedimentation and the types of particles that sediment vary with time when a region is submerged.
☞ These different periods of sedimentation result in the formation or rock layers called *strata*.

Tracing Phylogeny: Macroevolution, the Fossil Record, and Systematics

1. **Relative Dating** (p. 477)

    Younger strata are superimposed on top of older ones.
    - ☞ The succession of fossil species chronicles macroevolution, since fossils in each layer represent organisms present at the time of sedimentation.

What are index fossils? How are they used to correlate the relative ages of strata?

Strata from different locations can often be correlated by the presence of similar fossils, known as *index fossils*.
- ☞ The shells of widespread marine organisms are the best index fossils for correlating strata from different areas.
- ☞ Gaps in the sequence may appear in any particular area because of the area being above sea level (which prevents sedimentation) or by subsequent erosion.

Geologists have formulated a sequence of geological periods by comparing many different sites. (See Campbell, Table 23.1)
- ☞ These periods are grouped into four eras with boundaries between the eras marking major transitions in the lifeforms fossilized in the rocks.
- ☞ Periods within each era are subdivided into shorter intervals called *epochs*.

This record of the rocks presents a chronicle of the <u>relative</u> ages of the fossils; showing the order in which groups of species present in the sequence of rock strata evolved.

2. **Absolute Dating** (p. 477–478)

What is the difference between relative dating and absolute dating? What is the most common method for absolute dating of rocks and fossils?

Absolute dating is not errorless, but does give the age in years, not in relative terms (e.g. before, after).

The most common method for determining the age of rocks and fossils on an absolute time scale is *radioactive dating*.
- ☞ Fossils contain isotopes of elements that accumulated in the living organisms.
- ☞ Since each radioactive isotope has a half-life, it can be used to date fossils by comparing the ratio of certain isotopes (e.g. ^{14}C and ^{12}C) in a living organism to the ratio of the same isotopes in the fossil.
- ☞ The half-life of an isotope is not affected by temperature, pressure or other environmental variables.
- ☞ Carbon-14 has a half-life of 5600 years, meaning that one-half of the carbon-14 in a specimen will be gone in 5600 years; half of the remaining carbon-14 would disappear from the specimen in the next 5600 years; this would continue until all of the carbon-14 had disappeared.
- ☞ Thus a sample beginning with 8 g of carbon-14 would have 4 g left after 5600 years and 2 g after 11,200 years.

Relative dating

Index fossils

Absolute dating

Radioactive dating

Half-life

Carbon-14

354 Tracing Phylogeny: Macroevolution, the Fossil Record, and Systematics

- Carbon-14 is useful in dating fossils less than 50,000 years old due to its relatively short half-life.
- Paleobiologists use other radioactive isotopes with longer half-lives to date older fossils. For example, potassium-40 has a half-life of 1.3 billion years and is reliable for dating rocks (and fossils within those rocks) hundreds of millions of years old.
- An error of + or − 10% is present with radioactive dating.

Other than radioactive isotopes, what else may be used to date fossils?

Other "clocks" may also be used to date fossils.
- Amino acids can have either left-handed (L-form) or right-handed (D-form) symmetry.
- Living organisms only synthesize L-form amino acids to incorporate into proteins.
- After an organism dies, L-form amino acids are slowly converted to D-form.
- The ratio of L-form to D-form amino acids can be measured in fossils.
- Knowing the rate of chemical conversion allows this ratio to be used in determining how long the organism has been dead.
- This clock is most reliable in environments where the climate has not changed significantly since the conversion is temperature sensitive.

The dating of rocks and fossils they contain has enabled researchers to determine the geological periods. (See Campbell, Table 23.1)

II. MECHANISMS OF MACROEVOLUTION (*p. 478–489*)

Evolutionary biologists interested in macroevolution try to determine the processes involved in large-scale evolutionary changes that can be traced through paleobiology and taxonomy.

   A. The Origins of Evolutionary Novelties (*p. 480–482*)

Higher taxa such as families and classes are defined by evolutionary novelties (e.g. birds have feathers; humans have more upright posture and larger brains than chimpanzees).

One mechanism by which these new designs evolve may be the gradual refinement of existing structures for new functions.

      1. Preadaptation (*p. 480*)

What is preadaptation, and how can it result in macroevolutionary change?

Preadaptation

Preadaptation is a term applied to a structure that evolved in one context and became co-opted for another function.
- Natural selection can improve an existing structure in context of its current utility, but cannot anticipate the future.

- For example, the honeycombed bones and feathers of birds did not evolve as adaptations for flight, but must have been beneficial to the bipedal reptilian ancestors of birds and later, through modification, become functional for flying.
- Preadaptation cannot be proven, but provides an explanation for how novel designs can arise gradually through a series of intermediate stages, each having some function in the organism.

2. **Development and Macroevolution** (p. 480–482)

The evolution of complex structures (e.g. wings) requires such large modifications that changes at many gene loci are probably involved.
- In other cases, relatively few changes in the genome can cause major modifications in morphology (e.g. humans vs chimpanzees).
- Thus, slight genetic divergence can become magnified into major differences.

How can modification of regulatory genes result in macroevolutionary change?

In animal development, a system of regulatory genes coordinate activities of structural genes to guide the rate and pattern of development.
- A slight alteration of development becomes compounded in its effect on adult allometric growth (differences in relative rates of growth of various parts of the body) which helps to shape an organism.
- A slight change in these relative rates of growth will result in a substantial change in the adult. (See Campbell, Figure 23.3)
- Thus, altering the parameters of allometric growth is one way that relatively small genetic differences can have major morphological impact.

Allometric growth

Genetic changes that alter the timing of development can also produce novel organisms. A slight change in timing that retards the development of some organs in comparison to others produces a different kind of animal.

What is paedomorphosis? How may it have influenced human evolution?

Paedomorphosis (retention in an adult organism of juvenile features) may have contributed to human evolution.

Paedomorphosis

- Humans and chimpanzees are much more similar as fetuses than as adults.
- The fetal skulls are the same shape. (See Campbell, Figure 23.3)
- Different allometric growth patterns result in the adult human skull retaining some juvenile features.
- The human brain also continues to grow several years longer than the chimpanzee brain.
- Thus, the genetic changes responsible for humanness are not great, but have profound effects.

Since each regulatory gene may influence hundreds of structural genes, there is a potential for evolutionary novelties that define higher taxa to arise much faster than would occur by the accumulation of changes in only structural genes.

B. **The Difficulty of Interpreting Evolutionary Trends** (p. 482–484)

Using horse evolution as an example, why is it difficult to identify evolutionary trends from the fossil record?

In most cases, a correct single evolutionary progression cannot be produced from the fossil record.
- ☞ This is a result of divergence which may produce organisms that seem to fit a single progression, but actually represent related but divergent forms.
- ☞ Modern horses (*Equus*) are believed to have evolved from *Hyracotherium*.
- ☞ A single evolutionary progression would have to include: an increase in size, reduction of four toes to one, and modification of browsing teeth to grazing teeth.

Selection of certain fossils can produce an apparent single evolutionary progression of intermediate organisms between *Hyracotherium* and modern horses (*Equus*). (See Campbell, Figure 23.6a)
- ☞ This is a misrepresentation as a more complete phylogeny shows that *Equus* is the only survivor of a much more complicated evolutionary tree. (See Campbell, Figure 23.6b)
- ☞ Examination of the complete fossil sequence also shows that the transition from *Hyracotherium* to *Equus* was not a smooth gradation but involved a number of transitional steps which included a series of speciation episodes and several adaptive radiations.

Evolution has produced many genuine trends.
- ☞ Punctuated equilibrium seems to have produced a trend toward larger size in the titanotheres.
- ☞ The titanotheres were about as large as elephants and the fossil record shows that they evolved from a mouse-sized ancestral mammal.
- ☞ Although each species in the various lineages of titanotheres remain the same size, there was a succession of progressively larger species.

Branching evolution can produce a trend even if some new species counter the trend.
- ☞ There was an overall trend in reptilian evolution toward large size during the Mesozoic era which eventually produced the dinosaurs.
- ☞ This trend was sustained even though some new species were smaller than their parental species.

Even with an equal large-to-larger and large-to-smaller ratio, the trend would be maintained if the larger groups speciated at a greater rate than smaller groups or lasted longer before becoming extinct.

Tracing Phylogeny: Macroevolution, the Fossil Record, and Systematics

According to Steven Stanley, what role does species selection or differential speciation play in macroevolution?

- ☞ Steven Stanley originated this view of macroevolution which holds that species are analogous to individuals with speciation being their birth and extinction their death.
- ☞ According to the Stanley model, an evolutionary trend is produced by *species selection* which is analogous to the production of a trend within a population by natural selection.
- ☞ The species that live longest and generate the greatest number of species determine the direction of major evolutionary trends.
- ☞ Differential speciation thus may play a role in macroevolution similar to the role of differential reproduction in microevolution.

Are evolutionary trends predetermined?

No intrinsic drive toward a preordained state of being is indicated by the presence of an evolutionary trend.
- ☞ Evolution is a response to interactions between organisms and their current environments.
- ☞ An evolutionary trend may cease or reverse itself under changing conditions. For example, the conditions of the Mesozoic era favored giant reptiles, but by the end of that era, smaller species prevailed.

C. Continental Drift and the Biogeography of Macroevolution
(p. 485–486)

Macroevolution has dimension in space as well as time.
- ☞ Biogeography was a major influence on Darwin and Wallace in developing their views on evolution.
- ☞ Drifting of continents is the major geographical factor correlated with the spatial distribution of life. (See Campbell, Figure 23.9)

What is continental drift? What are its geographical consequences?

Continental drift results from the movement of great plates of crust and upper mantle that float on the Earth's molten core.
- ☞ The relative positions of two land masses to each other changes unless they are embedded on the same plate.
- ☞ North America and Europe are drifting apart at a rate of 2 cm per year.
- ☞ Where two plates meet (boundaries), many important geological phenomena occur: mountain building, volcanism, and earthquakes.
- ☞ The formation of volcanic islands (e.g. the Galapagos) opens new environments for founders and adaptive radiation.

How has continental drift influenced the spatial distribution of life? How did the formation of Pangaea and its subsequent breakup influence macroevolution?

Plate movements continually rearrange geography, however, two occurrences had important impacts on life: the formation of Pangaea and the subsequent breakup of Pangaea.

Steven Stanley

Species selection

Continental drift

358 Tracing Phylogeny: Macroevolution, the Fossil Record, and Systematics

Pangaea

☞ At the end of the Paleozoic era (250 million years ago), plate movements brought all land masses together into a super-continent called *Pangaea*.
☞ Species evolving in isolation where brought together and competition increased.
☞ Total shoreline was reduced and the ocean basins became deeper (draining much of the remaining shallow coastal seas).
☞ Terrestrial organisms were affected as continental interior habitats (and their harsher environments) increased in size.
☞ Marine species (which inhabit primarily the shallow coastal areas) were greatly affected by reduction of habitat.
☞ Changes in ocean currents would have affected both terrestrial and marine organisms.
☞ Overall diversity was thus impacted by extinctions and increased opportunities for surviving species.

During the early Mesozoic era (about 180 million years ago) Pangaea began to breakup due to continuing continental drift.
☞ This isolated the fauna and flora occupying different plates.
☞ The biogeographical realms were formed and divergence of organisms in the different realms continued.
☞ The separation of Pangaea also resulted in the geographical distribution of many groups such as marsupial and placental mammals.

D. **Punctuations in the History of Biological Diversity** (p. 486–489)

The evolution of modern life has included long, relatively quiescent periods punctuated by briefer intervals of more extensive turnover in species composition.
☞ These intervals of extensive turnover included explosive adaptive radiations of major taxa as well as mass extinctions.

1. **Examples of Major Adaptive Radiations** (p. 486–487)

What is an adaptive zone? How can radiation into new adaptive zones result in macroevolutionary change? What are some examples of past radiations?

The evolution of some novel characteristic opened the way to new *adaptive zones* allowing many taxa to diversify greatly early in their history.

Adaptive zones

☞ Evolution of wings allowed insects to enter an adaptive zone with abundant new food sources and adaptive radiation resulted in hundreds of thousands of variations on the basic insect body plan.

A large increase in the diversity of sea animals occurred at the boundary between the Precambrian era and the Paleozoic era.
☞ Precambrian rock contains the oldest animals (700 million years old) which were shell-less invertebrates that differed from their successors found in Paleozoic rock.
☞ Nearly all the extant animal phyla and many extinct phyla evolved during the first 10 to 20 million years of the Cambrian (early Paleozoic era).

- This episode of great diversification was keyed in part to the origin of shells and skeletons in a few key taxa.
- Shells and skeletons opened a new adaptive zone by making many new complex body designs possible and altering the basis of predator-prey relationships.

An empty adaptive zone can be exploited only if the appropriate evolutionary novelties arise. Conversely, an evolutionary novelty cannot enable organisms to exploit adaptive zones that are occupied or that do not exist.
- Flying insects existed for 100 million years before the appearance of flying reptiles and birds that fed on them.
- Mammals existed 75 million years before their first large adaptive radiation.
- The major adaptive radiation of mammals in the early Cenozoic era may have resulted from the ecological void created with the extinction of the dinosaurs.
- Mass extinctions have often opened adaptive zones and allowed new adaptive radiations.

2. **Examples of Mass Extinctions** (*p. 487–489*)

How can mass extinctions occur?

Extinctions may be caused by habitat destruction or by unfavorable environmental changes.
- Many very well adapted marine species would become extinct if the ocean's temperature fell only a few degrees.
- Even with stable physical factors, changes in biological factors may cause extinctions.
- Since many species coexist in each community, an evolutionary change in one species will probably impact the other species.
- The evolution of shells by some Cambrian animals may have contributed to the extinction of some shell-less forms.

Extinction is inevitable in a changing world.
- The average rate of extinction has been between 2.0 and 4.6 families (each family may include many species) per million years.

What were the major biological consequences of the Permian and Cretaceous extinctions?

There have been periods of global environmental changes which greatly disrupted life and resulted in mass extinctions.
- During these periods, the rate of extinction has escalated to as high 19.3 families per million years.
- Two (of approximately a dozen or so) mass extinction episodes have been studied extensively by paleobiologists.
- They are recognized primarily from the decimation of hard-bodied animals of shallow seas which have the most complete fossil record.

Extinctions

The Permian extinctions (the boundary between the Paleozoic and Mesozoic eras) eliminated over 90% of the species of marine animals about 250 million years ago.
- ☞ Terrestrial life was probably also affected greatly.
- ☞ This occurred about the time Pangaea was formed by the merging of continents which disturbed many habitats and altered the climate.

The Cretaceous extinction (the boundary between the Mesozoic and Cenozoic eras) occurred about 65 million years ago.
- ☞ More than 50% of the marine species and many terrestrial plants and animals (including dinosaurs) were eliminated.
- ☞ During this time the climate was cooling and many shallow seas receded from continental lowlands.
- ☞ Increased volcanic activity during this time may have contributed to the cooling by releasing materials into the atmosphere and blocking the sunlight.

Evidence also indicates that an asteroid or comet struck the Earth while the Cretaceous extinctions were in progress.
- ☞ Iridium, an element rare on earth but common in meteorites, is found in large quantities in the clay layer separating Mesozoic and Cenozoic sediments. (See Campbell, Figure 23.13)
- ☞ Walter and Luis Alvarez (and colleagues), after studying this clay layer, proposed that it is fallout from a huge cloud of dust ejected into the atmosphere when an asteroid hit Earth.
- ☞ This cloud would have both blocked the sunlight and severely disturbed the climate for several months.
- ☞ Although the asteroid hit the earth during this time, it may not have caused the mass extinction of this period.
- ☞ Many paleobiologists point out that while the extinctions were rapid (on a geological time scale), they were not abrupt and that climatic changes due to continental drift and other processes are sufficient to cause the mass extinctions.

How do mass extinctions affect biological diversity?

Mass extinctions, whatever the cause, profoundly affect biological diversity.
- ☞ Not only are many species eliminated, but those that survive are able to undergo new adaptive radiations into the vacated adaptive zones and produce new diversity.

III. SYSTEMATICS: TRACING PHYLOGENY (p. 489–496)

<u>Phylogeny</u> = The evolutionary history of a species or group of related species.
- ☞ Phylogeny is usually diagrammed as phylogenetic trees that trace inferred evolutionary relationships.

<u>Systematics</u> = The branch of biology concerned with the diversity of life and the reconstruction of phylogenetic histories.

Taxonomy = Identification and classification of species; is a component of systematics.

What is the difference between systematics and taxonomy?

   A.   Taxonomy (p. 489–491)

What contribution did Carolus Linnaeus make to biology?

The taxonomic system used today was developed by Linnaeus in the eighteenth century.
- The two main features of this system are the assignment of a *binomial* (two part Latin name) to each species and a filing system for grouping species into a hierarchy of increasingly general categories.
- The first word of the binomial is the *genus* (pl. genera); the second word is the *specific epithet* of the species.
- The scientific name of a species combines the genus and specific epithet.
- Each genus can include many species of similar organisms. For example, *Felis cattus* is the domestic cat; *Felis lynx* is the lynx.
- Use of the scientific name defines the organism referred to and removes ambiguity.

What are the major taxonomic categories from the most to least inclusive?

Hierarchical classification formalizes the grouping of organisms.
- Binomial nomenclature is the first step in grouping: similar species are grouped in the same genus.
- The system then progresses into broader categories: similar genera are grouped in the same family, families are grouped into orders, orders are grouped into classes, classes are grouped into phyla, and phyla are grouped into kingdoms.
- The more closely related two species are, the more levels they share: (See Campbell, Appendix Two)

Category	Domestic Cat	Bobcat	Lion	Dog
specific epithet	cattus	rufus	leo	familiaris
genus	Felis	Felis	Panthera	Canis
family	Felidae	Felidae	Felidae	Canidae
order	Carnivora	Carnivora	Carnivora	Carnivora
class	Mammalia	Mammalia	Mammalia	Mammalia
phylum	Chordata	Chordata	Chordata	Chordata
kingdom	Animalia	Animalia	Animalia	Animalia

What are the main objectives of taxonomy? What is the difference between a taxon and a category?

The two main objectives of taxonomy are to sort out and identify separate species and to order species into the broader taxonomic categories.
- In sorting, closely related organisms are assigned to separate species (with the proper binomial) and described using the diagnostic characteristics which distinguish the species from one another.

Taxon

- In categorizing, the species are grouped into broader categories from genera to kingdoms.
- In some cases, intermediate categories (i.e. subclasses; between orders and classes) are also used.
- The named taxonomic unit at any level is called a *taxon* (pl. taxa).
- Rules of nomenclature have been established: the genus name and specific epithet are italicized, all taxa from the genus level and higher are capitalized.

Only species exist in nature as biologically cohesive units, bonded by interbreeding and bounded by reproductive isolation from all other species.
- Species can be distinguished objectively by using the characteristics unique to the organisms.
- Grouping species into higher taxa is subjective involving judgement by the taxonomists, and disagreement exists among taxonomists in many cases with regards to groupings in the higher taxa.

What is the goal of systematics? What characterizes monophyletic, polyphyletic and paraphyletic groupings? What is an example of each?

The goal of systematics is to have classification reflect the evolutionary affinities of species.
- Groups subordinate to other groups in the taxonomic hierarchy should represent finer and finer branching of phylogenetic trees. (See Campbell, Figure 23.14)

Monophyletic taxon

- A *monophyletic taxon* is one where a single ancestor gave rise to all species in that taxon and to no species placed in any other taxon. For example, Family Ursidae evolved from a common ancestor. (Taxon 1)

Polyphyletic taxon

- A *polyphyletic taxon* is one whose members are derived from two or more ancestral forms not common to all members. For example, Kingdom Plantae includes both vascular plants and mosses which evolved from different algal ancestors. (Taxon 2)

Paraphyletic taxon

- A *paraphyletic taxon* is one that excludes species that share a common ancestor that gave rise to the species included in the taxon. For example, lass Reptilia excludes the Class Aves although a reptilian ancestor common to all reptiles is shared. (Taxon 3)

B. **Sorting Homology from Analogy** (*p. 491–492*)

When constructing a phylogeny, why is it important to distinguish between homologous and analogous character traits? What is the difference between homologous and analogous structures?

Taxonomists classify species into higher taxa based on the extent of similarities in morphology and other characteristics.

<u>Homology</u> = Likeness attributed to shared ancestry.   *Homology*
- The forelimbs of mammals are homologous, they share a similarity in the skeletal support that has a genealogical basis.
- Homology must be distinguished from analogy in evolutionary trees.

<u>Analogy</u> = Similarities due to convergent evolution, not common ancestry.   *Analogy*

<u>Convergent evolution</u> = Acquisition of similar characteristics in species from different evolutionary branches due to sharing similar ecological roles with natural selection shaping analogous adaptations.   *Convergent evolution*

The distinction between homology and analogy is some times relative: the wings of birds and bats are modifications of the vertebrate forelimb, thus the appendages are homologous; as wings they are analogous since they evolved independently from the forelimbs of different flightless ancestors.

Homology must be sorted from analogy to reconstruct phylogenetic trees on the basis of homologous similarities.
- Generally, the greater the amount of homology, the more closely related the species.
- Adaptation and convergence often obscure homologies, although studies of embryonic development can expose homology that is not apparent in mature structures.
- Additionally, the more complex two similar structures are, the less likely it is that they have evolved independently.

C. **Molecular Systematics** (*p. 492–495*)

Molecular comparisons of proteins and DNA have become very useful in taxonomy.

Since the nucleotide sequences in DNA are inherited and they program the corresponding sequences of amino acids in proteins, examination of these macromolecules provide much information about evolutionary relationships.

1. **Protein Comparison** (*p. 492*)

What are the advantages of protein comparison as a taxonomic tool?

The primary structure of proteins is genetically programmed and a similarity in the amino acid sequence of two proteins from different species indicates that the genes for those proteins evolved from a common gene present in a shared ancestor.

Advantages of this taxonomic tool are:
- It is objective and quantitative.
- Additionally, it can be used to assess relationships between species so distantly related that no morphological similarities exist.

Cytochrome c is an ancient protein common to all aerobic organisms.
- The amino acid sequence has been determined for species ranging from bacteria to complex plants and animals.
- The sequence in cytochrome c is identical in chimpanzees and humans.
- The sequence differs at only one of the 104 positions in the rhesus monkeys.
- Chimpanzees, humans, and rhesus monkeys belong to the Order Primates. Comparing these sequences with nonprimate species shows greater differences (13 with the dog and 20 with the rattlesnake).
- Phylogenetic trees based on cytochrome c are consistent with evidence from comparative anatomy and the fossil record.

2. DNA Comparison (p. 492–494)

What are three techniques used for DNA comparison in molecular systematics? What information does each provide?

Comparing the genes or genomes of two species is the most direct measure of common inheritance from shared ancestors.

Three methods can be used for DNA comparison: *DNA-DNA hybridization, restriction mapping*, and *DNA sequencing*.

DNA-DNA hybridization can compare whole genomes by measuring the degree of hydrogen bonding between single-stranded DNA obtained from two sources.
- DNA is extracted and the complimentary strands separated by heating.
- The single-stranded DNA from two species is mixed and cooled to allow double-stranded DNA reformation.
- The hybrid DNA is then reheated to separate the double strands with the temperature necessary to separate the hybrid DNA being indicative of the similarity in the DNA from the two species.
- The temperature correlation is based on the degree of bonding between the strands of the two species with more bonding occurring with greater similarity.
- The more extensive the pairing, the more heat is needed to separate the hybrid strand.
- Evolutionary trees constructed through DNA-DNA hybridization usually agree with those based on other methods, however, this technique is very beneficial in settling taxonomic debates that have not been finalized by other methods.

Tracing Phylogeny: Macroevolution, the Fossil Record, and Systematics 365

Restriction mapping provides precise information about the match-up in specific nucleotide sequences of the DNA.

☞ Restriction enzymes are used to obtain DNA fragments which can be separated by electrophoresis and compared to restriction fragments of other species.

☞ Two samples of DNA with similar maps for the locations of restriction sites will produce similar collections of fragments.

☞ The greater the divergence of two species from the common ancestor, the greater the differences in restriction sites and less similarity of the restriction fragments.

☞ This method works best when comparing small fragments of DNA.

☞ Mitochondrial DNA (mtDNA) is best suited for this type of comparison since it is smaller than nuclear DNA (produces smaller fragments) and mutates about ten times faster than nuclear DNA.

☞ The faster mutation rate of mtDNA allows it to be used to determine phylogenetic relationships between not only closely related species but also populations of the same species.

DNA sequencing determines the actual nucleotide sequence of a DNA segment.

☞ This is the most precise method of comparing DNA and requires DNA segments that have been cloned by recombinant DNA techniques.

☞ DNA sequencing and comparison shows exactly how much divergence there has been in the evolution of two genes derived from the same ancestral gene.

☞ Ribosomal RNA (rRNA) sequencing is a similar technique which can provide information about some of the earliest branching in phylogenetic relationships since genes for rRNA change very slowly.

☞ rRNA sequencing has been very useful in examining the relationships among bacteria.

3. **Molecular Clocks** (*p. 494–495*)

How can scientists use comparisons of homologous proteins and DNA to determine the time since divergence of two species? What characteristic makes the rate of protein change and the rate of DNA divergence useful as molecular clocks?

Proteins evolve at different rates, although each type of protein evolves at a relatively constant rate over time.

When comparing homologous proteins from taxa that are known to have diverged from common ancestors during certain periods in the past, the number of amino acid substitutions is proportional to the elapsed time since divergence.

☞ Homologous proteins of bats and dolphins are more similar than those of sharks and tuna.

☞ This is consistent with fossil evidence showing that tuna and sharks have been separated much longer than bats and dolphins.

Restriction mapping

DNA sequencing

Molecular clocks

Although protein comparisons are reliable, DNA comparisons may be even more promising since the difference in DNA between two taxa is more closely correlated with the time since divergence than is morphological differences.

Molecular clocks are calibrated by graphing the number of nucleotide or amino acid differences against the times for a series of evolutionary branch points known from the fossil record.
- ☞ The graph can then be used to determine the time of divergence between taxa for which no substantial fossil record is available.

The consistent rate of protein change and the rate of DNA divergence indicate that there exists a significant background of neutral mutations that gradually change the genome as a whole more than specific genetic changes associated with adaptation.
- ☞ *Neutral evolution* and molecular clocks are questioned by many evolutionary biologists, although modern systematists evaluate all available taxonomic evidence (including molecular information) before reconstructing phylogeny.

D. Schools of Taxonomy (p. 495–496)

The two significant features of a phylogenetic tree are the location of branch points along the tree and the degree of divergence between branches.
- ☞ The locations of branch points along the tree symbolize the relative times of origin for different taxa.
- ☞ The degree of divergence between branches represents how different two taxa have become since branching from a common ancestor.

What are the differences in approach that characterize the three schools of taxonomy: phenetics, cladistics, and classical evolutionary taxonomy?

The question of which property of phylogenetic trees should be most important in grouping species into taxa has divided taxonomy into three schools: *phenetics*, *cladistics*, and *classical evolutionary taxonomy*.

1. Phenetics (p. 495)

Phenetics makes no evolutionary assumptions and decides taxonomic affinities entirely on the basis of measurable similarities and differences.
- ☞ A comparison is made of as many characters (anatomical characteristics) as possible without attempting to sort homology from analogy.
- ☞ Data from these comparisons is used to produce *phenograms* (dichotomously branched trees).
- ☞ Pheneticists feel that the contribution of analogy to overall similarity will be overridden by the degree of homology if enough characters are compared.
- ☞ While supported by few taxonomists, phenetics has made important contributions to taxonomy.
- ☞ Critics of phenetics argue that overall phenotypic similarity is not a reliable index of phylogenetic proximity.

2. Cladistics (p. 495–496)

Cladistics classifies organisms according to the order in time that branches arise along a phylogenetic tree, without considering the degree of divergence.

- This produces a cladogram, a tree consisting of a series of dichotomous forks.
- Each branch point is defined by novel homologies unique to the various species on that branch.

What are plesiomorphic and apomorphic characters? How are they used to construct a cladogram?

- *Plesiomorphic* characters are primitive characteristics shared among all of the species of the cladogram and with the common ancestor.
- *Apomorphic* characters are derived characters which are homologies that evolved after a branch diverged from the phylogenetic tree.
- Each species in the cladogram is a mixture of plesiomorphic and apomorphic characters.
- The sharing of plesiomorphic characters indicates nothing about the pattern of evolutionary branching from a common ancestor.
- A major difficulty in cladistics is finding characters that are appropriate for each branch point. (See Campbell, Figure 23.19)

Some surprises are produced by cladistic analysis:

- The branch point between crocodiles and birds is more recent than between crocodiles and the lizards and snakes (a fact also supported by the fossil record) since crocodiles and birds share apomorphic characters not present in lizards and snakes.
- Using strict cladistic analysis, the Class Aves would be eliminated and the birds included in the Class Reptilia as a subclass or order.
- Birds are deemed superficially different because of the morphological changes associated with flight which have developed since their divergence from reptilian ancestors.

The extent of morphological divergence between evolutionary branches is ignored in cladistics, a fact often criticized by opponents of the strictly cladistic approach.

3. Classical Evolutionary Taxonomy (p. 496)

What criteria are used to construct phylogenies in classical evolutionary taxonomy?

Classical evolutionary taxonomy is an approach to taxonomy that attempts to balance the criteria of phenetics and cladistics by considering overall homology along with branching sequence.

- Predates both phenetics and cladistics, but now incorporates some ideas from both.
- In cases of taxonomic conflict, a subjective judgement is made about which type of information receives the highest priority.

Using crocodiles and birds as an example, how can the outcome of a phylogenetic analysis differ depending upon the criteria used?

In reference to crocodiles and birds, a classical taxonomist recognizes a closer genealogical relationship between crocodiles and birds than between crocodiles and lizards.
- ☞ However, classical taxonomists combine crocodiles and lizards in a taxon (Class Reptilia) that excludes birds because the ability to fly was an evolutionary advancement which allowed birds to enter a new adaptive zone.
- ☞ The resulting divergence of birds was so extensive that they are placed in a separate class (Aves).

Phenetics and cladistics would produce identical classifications if taxa that split at each branch point in phylogeny diverged in morphology at the same rate.
- ☞ Thus, the longer two taxa were separated, the greater the difference between them.
- ☞ This is not the case as birds show greater changes from the reptilian common ancestor than do the crocodiles.

Comparing proteins and DNA provide the best hope for taxonomic characters that diverge in time from a branch at a constant rate.
- ☞ However, classifications will probably continue to rely on morphological similarities and differences.

IV. IS A NEW EVOLUTIONARY SYNTHESIS NECESSARY? (p. 496–497)

No biological theory has produced as much debate and controversy as the Theory of Evolution.
- ☞ The debate has continued since Darwin published *The Origin of Species* 131 years ago and continues today.
- ☞ The closest thing to a consensus has been the modern synthesis which has dominated evolutionary theory for the past 50 years.
- ☞ This view of evolution is called a synthesis since the ideas presented were drawn from several disciplines (paleontology, biogeography, systematics, population genetics) and continues to incorporate discoveries from new fields such as molecular biology.

<u>Modern synthesis</u>

What is the modern synthesis? According to this conceptual framework, what is the relative importance of natural selection and gradualism to evolutionary change?

Darwin's view of life was reaffirmed by the modern synthesis and updated by applying principles of genetics.
- ☞ Its view is strictly uniformitarian in that large-scale evolutionary changes are gradual accumulations of many minute changes occurring over vast spans of time.
- ☞ Microevolution (changes in gene frequencies in populations) is extrapolated to explain most macroevolution.

The modern synthesis holds that natural selection is the major cause of evolution at all levels: populations adapt by natural selection, new species arise when isolated populations diverge as different adaptations evolve, and continued divergence due to natural selection differentiates the higher taxa.
- That mechanisms such as genetic drift and chromosomal mutations can cause rapid, nonadaptive evolution is recognized by the modern synthesis; but major emphasis of the synthesis is on gradualism and natural selection.

The continuing debate has been indicated earlier in this chapter.

In your own words, what is the modern debate over the mechanism of evolutionary change? What other mechanisms of change besides microevolution by natural selection have been proposed? Is it possible to integrate both the evolutionists' and punctuationalists' views of evolution into a new evolutionary synthesis?

Some evolutionists dissent from the view that evolution recorded in the fossil record can be explained by extrapolating the processes of microevolution.
- Many transitions in the fossil record are punctuational, not gradual.
- Gradualists hold that apparent abruptness is in part derived from the imperfection of the fossil record and in part is a semantic issue confused by the vastness of geological time (e.g. is a change occurring over 10,000 years sudden or gradual?).
- Punctuationalists counter that the imperfection of the fossil record is not enough to account for the rarity of transitional forms if speciation and the origin of higher taxa were primarily gradual extensions of microevolution.
- The debate also includes differing opinions about the degree to which microevolution compounded over time is sufficient to explain macroevolution.

A hierarchical theory that gives mechanisms other than microevolution by natural selection a larger role in macroevolution is favored by some evolutionists.
- In this theory, most new species begin as small populations isolated from their parent populations (geographical or genetic isolation).
- These isolates can evolve relatively rapidly, divergence from the parent population being due to genetic drift as much as to selection.
- Chance may cause speciation to begin before selection has produced new adaptations.
- Also, some major new adaptations may evolve with minimal genetic change if regulatory genes are involved.
- Chance also figures prominently in macroevolution. Continental drift and mass extinctions have probably had at least as much effect on the history of biological diversity as gradual adaptation caused by selection influencing gene pools at the population level.
- In a hierarchical theory of evolution, most evolutionary trends progress by species selection not by phyletic transition due to an accumulation of microevolutionary changes.
- Species selection being the differential survival and branching of separate species that change little after they come into existence.

All groups in this debate recognize that natural selection is the mechanism of adaptation and should be the centerpiece of evolutionary theory.

☞ Selection adapts a population to its environment with generation-to-generation changes in the gene pool.

☞ It is also natural selection that refines unique adaptations when a new species or higher taxon comes into existence.

☞ Although events leading to speciation and episodes of macroevolution may have little influence on adaptation, new species only persist long enough to be entered into the fossil record if they have adapted to their environment through natural selection.

The modern synthesis has never proposed that evolution is always smooth and gradual or that processes other than changes in gene pools due to selection are unimportant.

☞ The questions are not so much about the nature of evolutionary mechanisms as about their relative importance.

Debate about how life evolved is indicative that evolutionary biology is an active science and the debate will continue.

24   EARLY EARTH AND THE ORIGIN OF LIFE

CHAPTER OUTLINE

Extant organisms represent the evolution of a diverse assemblage of organisms. This diversity resulted from geologic and biologic changes that altered the environment.
- ☞ Geologic change: breakup of Pangaea.
- ☞ Biologic change: evolution of photosynthetic organisms that released oxygen.

How life actually began is speculative.
- ☞ Clues for a number of hypotheses are present in the molecules, metabolic capabilities and anatomical development of each species.
- ☞ These clues coupled with the fossil record have produced several theories about how life evolved on Earth.

I. THE ANTIQUITY OF LIFE (p. 505–506)

What evidence points to the antiquity of life?

Life appeared relatively early in the Earth's history.
- ☞ Fossils similar to spherical and filamentous prokaryotes have been recovered from stromatolites 3.5 billion years old in western Australia and southern Africa.
  - Stromatolites = Banded domes of sediment similar to the layered mats constructed by colonies of bacteria and cyanobacteria currently living in salty marshes.
- ☞ The fossils from western Australia appear to be of photosynthetic organisms which would indicate life evolved prior to the time of the fossils.
- ☞ Other fossils similar to the prokaryotes have been recovered from the Fig Tree Chert rock formation in southern Africa which date to 3.4 billion years.

Those early prokaryotic organisms probably originated within a few hundred million years after the cooling of Earth's crust.
- ☞ These early prokaryotes appeared at least 2 billion years before the oldest eukaryotic fossils.

II. THE ORIGIN OF LIFE (p. 506–512)

Life originated between 3.5 and 4.1 billion years ago. This is the timespan during which the Earth's crust began to solidify (4.1 billion) and bacteria advanced enough to build stromatolites were present (3.5 billion).

Stromatolites

The environment during this time was such that life could arise by spontaneous generation.

What evidence supports the hypothesis that chemical evolution resulting in life's origin occurred in four stages: a) abiotic synthesis of organic monomers, b) abiotic synthesis of polymers, c) formation of protobionts, and d) origin of genetic information?

One hypothesis on the appearance of the first living organisms is that these organisms were the products of a chemical evolution that occurred in four stages:

1. Abiotic synthesis and accumulation of small organic molecules (monomers = amino acids and nucleotides).
2. Joining of monomers into polymers (proteins and nucleic acids).
3. Formation of *protobionts* due to the aggregation of abiotically produced molecules which differed chemically from their surroundings.
4. Origin of heredity during or before protobiont appearance.

A. Abiotic Synthesis of Organic Monomers (p. 507–508)

What contributions did A.I. Oparin, J.B.S. Haldane, Stanley Miller and Harold Urey make towards developing a model for abiotic synthesis of organic molecules.

A.I. Oparin and J.B.S. Haldane (1920's) postulated that the reducing atmosphere and greater UV radiation on primitive Earth enhanced reactions joining simple molecules to produce the first organic molecules.
- Not possible today since O_2 attacks chemical bonds, removing electrons, and the ozone layer screens out most UV radiation.

What gases did the Earth's early atmosphere probably include? Was it an oxidizing or reducing atmosphere, and why is this important?

Stanley Miller and Harold Urey made an apparatus containing H_2O, H_2, CH_4 and NH_3 to create organic molecules.
- Now we know the atmosphere of early Earth probably included CO, CO_2, and N_2, and was less reducing than the Miller-Urey model and, thus, less favorable to formation of organic compounds.
- Various experiments have produced all 20 amino acids, ATP, some sugars, lipids and purine and pyrimidine bases of RNA and DNA.
- An important characteristic of the early atmosphere must have been the rarity of oxygen.

B. Abiotic Synthesis of Polymers (p. 508)

Polymers = Chains of similar building blocks or monomers.
- Are synthesized by dehydration reactions. For example:

$$\square\text{—H} + \text{OH—}\square \longrightarrow \square\text{—}\square + H_2O$$

How did Sidney Fox demonstrate that the abiotic synthesis of polymers is possible, even in the absence of enzymes?

- Proteinoids = Abiotically synthesized polypeptides.
  - First produced by Sidney Fox.
  - Can be made by dripping organic monomers onto hot sand, rock or clay.

Why might clay have been an important substratum for polymerization reactions required for life?

- Clay may have been an important substrate for abiotic synthesis of polymers since:
  1. Monomers bind to charged sites in clay.
  2. Metal ions could act as catalysts of dehydration reactions in clay.
  3. The many binding sites on clay could have brought many monomers close together and assisted in forming polymers.
- Pyrite may also have been important as it provides a charged surface and electrons freed during its formation could support bonding between molecules.

C. **The Formation of Protobionts** (p. 508–509)

What are protobionts? Why is it plausible that they may have preceded living cells?

- Protobionts = Aggregates of abiotically produced molecules able to maintain an internal environment different from their surroundings and exhibiting some life properties such as metabolism, excitability and self-replication (yet not able to precisely reproduce). Probable antecedents of first true cells.

What experimental evidence demonstrates that protobionts could have formed spontaneously from abiotically produced organic compounds?

- Evidence to Support This Hypothesis:
  - When mixed with cool water, proteinoids self-assemble into *microspheres* surrounded by a selectively permeable membrane.
  - *Liposomes* can form spontaneously when phospholipids form a bilayered membrane similar to those of living cells.
  - *Coacervates* (colloidal drops of polypeptides, nucleic acids and polysaccharides) self-assemble.

D. **The Origin of Genetic Information** (p. 509–511)

Today's cells transcribe DNA into RNA, which is then translated into proteins. This chain of command must have evolved from a simpler mechanism of heritable control.

Margin terms: Proteinoids; Sidney Fox; Protobionts; Microspheres; Liposomes; Coacervates

Early Earth and the Origin of Life

What evidence supports the hypothesis that RNA molecules were the first primitive genes?

One hypothesis proposes that before DNA, there existed a primitive mechanism for aligning amino acids along RNA molecules, which were the first genes.

Evidence to Support This Hypothesis:
- Short polymers of ribonucleotides that can base pair (5–10 bases without enzyme, up to 40 bases with zinc added as catalyst) have been produced abiotically in test tubes.
- RNA is autocatalytic, as indicated by *ribozymes* (RNA that acts as a catalyst to remove introns, or catalyze synthesis of mRNA, tRNA or rRNA).
- RNA folds uniquely depending on sequence (unlike DNA), thereby providing a mechanism for natural selection to act on different molecular shapes varying in stability and catalytic properties.

The next step may have been the formation of a membrane which concentrated favorable reactions benefitting unique molecular aggregates.

DNA may have replaced RNA as the genetic material since it is more stable.

E. Some Alternative Views (p. 511)

What are some alternative views for how several key steps in the origin of life could have occurred?

No one knows how life actually began on Earth. The chemical evolution described and supporting lab simulations indicate key steps that could have occurred.

Several alternatives have been proposed.
- *Panspermia* — Some organic compounds may have reached Earth by way of meteorites and comets. Organic compounds (e.g. amino acids) have been recovered from modern meteorites. These extraterrestrial organic compounds may have contributed to the pool of molecules which formed early life.
- Most researchers believe life first appeared in shallow water or moist sediments. Some now feel the first organisms developed on the sea floor due to the harsh conditions on the surface during that time. This position was strengthened in the 1970's by discovery of the deep sea vents. Hot water and minerals emitted from such vents may have provided the energy and chemicals necessary for early protobionts.
- Simpler hereditary systems may have preceded nucleic acid genes. Julius Rebek synthesized a simple organic molecule in 1991. The importance of this molecule was that it served as a template for self-replication. This discovery supported the idea held by some biologists that RNA strands are too complicated to be the first self-replicating molecules.

III. THE KINGDOMS OF LIFE (p. 512–513)

The two kingdom system (animals and plants) long prevailed, but was not suitable as biologists learned more about the structures and life histories of different organisms.

Describe the basis for the five-kingdom system proposed by Robert H. Whittaker and Lynn Margulis. What are the five kingdoms, and how are they distinguished from one another?

The five kingdom system proposed by Robert H. Whittaker and modified by Lynn Margulis is widely accepted today.

```
                    Living Organisms
                   /                \
              Prokaryotic         Eukaryotic
                  |              /          \
               MONERA    Simple multicells   Multicellular
                         or unicells        /           \
                              |        Autotrophic   Heterotrophic
                          PROTISTA          |         /         \
                                         PLANTAE  Absorptive   Ingestive
                                                  nutrition    nutrition
                                                      |            |
                                                    FUNGI       ANIMALIA
```

Plants, fungi, and animals are also distinguished by unique characteristics of structure and development.

Prokaryotes were the first to evolve between 4.1 and 3.5 billion years ago and predominated exclusively for at least 2 billion years.

25 PROKARYOTES AND THE ORIGINS OF METABOLIC DIVERSITY

CHAPTER OUTLINE

What characteristics distinguish the Kingdom Monera from the other four kingdoms?

Prokaryotes are commonly referred to as *bacteria*.
- A structurally and metabolically diverse group.
- The earliest living organisms having evolved about 3.5 billion years ago. Were the only forms of life for 2 billion years.

Cellular and molecular features unique to the Monera include:
- Prokaryotes are small and lack membrane-bound organelles.
- Most have cell walls but the composition and structure differ from those found in plants, fungi and protists.
- Prokaryotes differ from eukaryotes by having simpler genomes along with some differences in genetic replication, protein synthesis and recombination.

Prokaryotes dominate the biosphere and have a great impact on Earth.
- Prokaryotes are the most numerous organisms and can be found in all habitats.
- Very important to all other forms of life as decomposers and the key organisms in life-sustaining chemical cycles.
- A small percentage cause disease.
- Many form symbiotic relationships with other prokaryotes and eukaryotes.

I. **PROKARYOTIC FORM AND FUNCTION** (p. 516–522)

   A. **The Morphology of Prokaryotes** (p. 516)

Prokaryotes may be unicellular, colonial, or have simple multicellular form with a division of labor between specialized cells.
- Most have diameters of 1–5μm (compared to eukaryotic diameters of 10–100μm).

What are the three most common shapes of bacteria?
- Cells have a diversity of shapes, the most common being spherical (*cocci*), rod-shaped (*bacilli*), and spiral (*spirilla*).

   B. **The Cell Surface** (p. 516–518)

What are the primary functions of the cell wall in prokaryotes?

Prokaryotes' external cell walls:
- Maintain the cell shape.

Kingdom Monera

Prokaryotes, Bacteria

Cocci, Bacilli, Spirilla

- Protect the cell.
- Prevent the cell from bursting in a hypoosmotic environment.
- Are made of *peptidoglycan*.

What is the composition of prokaryotic cell walls?

Peptidoglycan = Modified sugar polymers cross-linked by short polypeptides.
- Exact composition varies among species.
- Some antibiotics work by preventing formation of the cross-links in peptidoglycan, thus preventing the formation of a functional cell wall.

How can structural differences and staining properties of the cell wall be used to classify bacteria into two groups? Why are disease-causing gram-negative bacteria species generally more pathogenic than disease-causing gram-positive bacteria?

Gram stain = A stain used to distinguish two groups of bacteria by virtue of a structural difference in their cell walls.

1. Gram-positive bacteria.
   - Have simple cell walls with large amounts of peptidoglycan.
   - Stain blue.

2. Gram-negative bacteria.
   - Have more complex cell walls with smaller amounts of peptidoglycan.
   - An outer lipopolysaccharide-containing membrane covers the cell wall.
   - Stain pink.
   - These cells are more often disease-causing than Gram-positive bacteria.
   - Lipopolysaccharides are often toxic and the outer membrane helps protect these bacteria from host defense systems.

Capsule = A gelatinous secretion of some prokaryotes which provides cells with additional protection, helps them adhere to hosts and helps form aggregates.

Pili = Surface appendages used for adherence to a host (in the case of a pathogen), or for transferring DNA when bacteria conjugate.

C. **Motility of Prokaryotes** (*p. 518–519*)

What are three mechanisms motile bacteria use to move?

Motile bacteria use one of three mechanisms to move:
1. Flagella.

How do prokaryotic flagella work? Why are they considered not to be homologous to eukaryotic flagella?

- Filaments, composed of chains of the protein *flagellin*, are attached to another protein that forms a hook which is inserted into the basal apparatus.

Peptidoglycan

Gram stain

Gram-positive

Gram-negative

Capsule

Pili

Flagella

Flagellin

Spirochetes

Taxis
Phototactic, Chemotactic
Magnetotactic

Cyanobacteria

- They are distributed on the cell surface or concentrated at one or both ends of the cell.
- Their rotation is powered by the diffusion of H^+ into the cell. The H^+ gradient is maintained by an ATP-driven proton pump.
- Prokaryotic flagella differ from eukaryotic flagella in that they are:
  a. Unique in structure and function. Prokaryotic flagella lack the "9 + 2" microtubular structure and rotate rather than whip back and forth like eukaryotic flagella.
  b. Not covered by plasma membrane.
  c. One-tenth the width of eukaryotic flagella.

2. Axial filaments.
   - Characteristic of spirochetes, spiral-shaped bacteria.
   - Several filaments spiral around the cell inside the cell wall.
   - Similar to prokaryotic flagella in structure, axial filaments are attached to basal motors at either end of the cell. Filaments attached at opposite ends move relative to each other, rotating the cell like a corkscrew.

3. Gliding.
   - Some bacteria move by gliding through a secreted slime.

Taxis = Movement to or away from a stimulus. The stimulus can be light (phototactic), a chemical (chemotactic), or a magnetic field (magnetotactic).

During taxis (directed movement), bacteria move by *running* and *tumbling* movements, enabled by rotation of flagella either counterclockwise or clockwise respectively. This causes flagella to move coordinately about each other (for a run), or to separate and randomize movements (for a tumble).

D. Internal Membranous Organization (*p. 519*)

Prokaryotes lack the diverse internal membranes characteristic of eukaryotes.

If prokaryotes lack a system of internal membranes and membrane-bound organelles, where do photosynthesis and cellular respiration take place in prokaryotic cells?

Some prokaryotes, however, do have specialized membranes, formed by invaginations of the plasma membranes.
- Infoldings of plasma membrane function in cellular respiration of aerobic bacteria.
- Cyanobacteria have thylakoid membranes that contain chlorophyll and that function in photosynthesis.

E. The Prokaryotic Genome (*p. 519–520*)

How does the organization of the prokaryotic genome differ from that in eukaryotic cells?

The prokaryotic genome has only 1/1000 as much DNA as the eukaryotic genome.

Genophore = The bacterial chromosome, usually one double-stranded, circular DNA molecule.
- ☞ This DNA is concentrated in the nucleoid region, and is not surrounded by a membrane; therefore, there is no true nucleus.
- ☞ Has very little protein associated with the DNA.

Many bacteria also have *plasmids*.

Plasmids = Smaller rings of DNA having supplemental (usually not essential) genes for functions such as antibiotic resistance or metabolism of unusual nutrients.
- ☞ Replicate independently of the genophore.
- ☞ Transferred between partners during conjugation.

F. **Growth, Reproduction, and Gene Exchange** (p. 520–521)

How do monerans reproduce?

Neither mitosis nor meiosis occur in the Monera.
- ☞ Reproduction is asexual by *binary fission*.
- ☞ DNA synthesis is almost continuous.

Growth in the numbers of cells is geometric in an environment with unlimited resources.
- ☞ Generation time is usually 1–3 hours, although some can be 20 minutes in optimal environments.
- ☞ At high concentrations of cells, growth slows due to accumulation of toxic wastes, lack of nourishment, etc.
- ☞ Optimal growth requirements vary depending upon the species.

How are endospores formed? Why are endospore-forming bacteria so important to the food-canning industry?

Endospores = Resistant cells formed by some bacteria.
- ☞ Contains one copy of the cell's chromosome surrounded by a thick wall.
- ☞ When endospores form, the original cell replicates its chromosome and surrounds one copy with a durable wall. The original surrounding cell disintegrates, releasing the resistant endospore.
- ☞ Can survive adverse environmental conditions and toxins after the original cell disintegrates.
- ☞ Since some endospores can survive boiling water for a short time, home canners and the food canning industry must take special precautions to kill endospores of dangerous bacteria.
- ☞ May remain dormant for many years until proper environmental conditions return.

If bacterial reproduction is asexual, how does genetic recombination occur in bacteria?

Although meiosis and syngamy do not occur in prokaryotes, genetic recombination can take place through three mechanisms:

Transformation = The process by which external DNA is incorporated by bacterial cells.

Conjugation

Transduction

Conjugation = The direct transfer of genes from one bacterium to another.

Transduction = The transfer of genes between bacteria by viruses.

G. Metabolic Diversity (*p. 521–522*)

Metabolic diversity is greater in Monera than in all other kingdoms combined.

What are the four possible modes of bacterial nutrition?

Bacteria can be separated into four major groups depending on how they obtain energy and carbon:

Photoautotrophs

Photoautotrophs = Use light energy to synthesize organic compounds from CO_2.

Photoheterotrophs

Photoheterotrophs = must obtain organic carbon but use light to generate ATP.

Chemoautotrophs

Chemoautotrophs = Need only CO_2 as a carbon source and obtain energy by oxidizing inorganic compounds such as H_2S, NH_3 and Fe^{2+}.

Chemoheterotrophs

Chemoheterotrophs = Must obtain organic molecules for energy and as a source of carbon. Most bacteria are in this group.

What distinguishes a saprobe from a parasite?

The chemoheterotrophs can be broken down into subgroups: *saprobes* and *parasites*.

Saprobes
- Saprobes are decomposers that absorb nutrients from dead organic matter.

Parasites
- Parasites are bacteria that absorb nutrients from body fluids of living hosts.

The chemoheterotrophs are a very diverse group, some have very strict requirements while others are extremely versatile. Almost any molecule can serve as food for some species (e.g. some degrade petroleum and are used to clean oil spills). Those compounds that cannot be used are considered *non-biodegradable* (e.g. some plastics).

Non-biodegradable

How are bacteria categorized by the effect that oxygen has on growth?

Bacteria differ in their response to oxygen.

Obligate aerobes

Obligate aerobes = Bacteria needing O_2 for cellular respiration.

Facultative anaerobes

Facultative anaerobes = Those that use O_2 when present, but in its absence can grow using fermentation.

Obligate anaerobes = Those that are poisoned by oxygen.

Obligate anaerobes

What role do prokaryotes play in the cycling of nitrogen through ecosystems?

Prokaryotes are extremely important to the cycling of nitrogen through ecosystems.

Nitrogen fixation
- Nitrogen fixation ($N_2 \rightarrow NH_3$) is unique to certain prokaryotes (cyanobacteria) and is the only mechanism that makes atmospheric nitrogen available to organisms for incorporation into organic compounds.

- Some chemoautotrophic bacteria (*Nitrosomonas*) convert $NH_3 \rightarrow NO_2^-$.
- Facultative anaerobes, such as *Pseudomonas*, denitrify NO_2^- or NO_3^- to atmospheric N_2.

II. THE DIVERSITY OF PROKARYOTES (p. 522–524)

How has molecular systematics been used in developing a moneran classification?

Due to ancient diversification and obscure fossil records, prokaryotic taxonomy has, until recently, been a classification of convenience in identification rather than a reflection of evolutionary relationships.
- Molecular comparisons of amino acid sequences of homologous proteins and base sequences of RNA and DNA are now used by systematists to determine evolutionary relationships.
- The use of molecular systematics (especially ribosomal RNA comparisons) have shown that prokaryotes split into at least two divergent lineages very early: *archaebacteria* and *eubacteria*.

A. Archaebacteria (p. 523–524)

What unique characteristics distinguish the archaebacteria from the eubacteria?

Some characteristics of archaebacteria include:
- Cell walls lack peptidoglycan.
- Plasma membranes have a unique lipid composition.
- RNA polymerase and ribosomal protein are more like those of eukaryotes than of eubacteria.

What are the three main groups of archaebacteria? What features are characteristic of each group?

Three Main Groups of Archaebacteria:

1. Methanogens
   - Use H_2 to reduce CO_2 to CH_4.
   - Strict anaerobes.
   - Some species are important decomposers in marshes and swamps (form marsh gas).
   - Other species are important symbionts in termites and herbivores.

2. Extreme halophiles
   - Live in waters of extreme salinity (15–20%).
   - They have the pigment bacteriorhodopsin in their plasma membrane which absorbs light to pump H^+ ions out of the cell.
   - This pigment is also responsible for the pink color of the colonies.

3. Thermoacidophiles
   - Live in habitats of 60–80°C and pH 2–4.

B. Eubacteria (p. 524)

What are the major groups of eubacteria? What features, including mode of nutrition, characterize each group? What is a representative example of each?

The major groups of eubacteria include a very diverse assemblage of organisms.

1. Actinomycetes
   - Colonial with branching hyphae.
   - Spores are produced by fragmentation of the end of hyphae.
   - Most live in organic litter of soil.
   - 40 genera. Example: *Streptomyces*.

2. Chemoautotrophic bacteria
   - Oxidize inorganic substances (e.g. NH_3, H_2S) for energy and use carbon from CO_2 to synthesize organic molecules.
   - Obligate aerobes.
   - Common in aerated soils.
   - 25 genera. Example: *Nitrobacter*.

3. Cyanobacteria
   - Photoautotrophs; may be solitary, multicellular, or colonial.
   - Photosynthesis is similar to that found in plants; chlorophyll *a* and two photosystems.
   - Chlorophylls and electron-transport systems embedded in thylakoid membranes.
   - *Phycobilins* (accessory pigments) arranged in complexes on surface of thylakoids.
   - Cell walls usually thick and gelatinous.
   - Motility by gliding.
   - Inhabit freshwater primarily; some species are marine, in moist soils or associated with lichens.
   - *Heterocysts* in some filamentous genera; these specialized cells contain the nitrogen-fixing enzyme complex.
   - 32 genera. Example: *Anabaena*.

4. Endospore-forming bacteria
   - Endospores (dehydrated cells with thick walls) form internally as a survival mechanism under harsh conditions.
   - Gram-positive, flagellated rods.
   - Some are aerobes, some obligate anaerobes.
   - 6 genera. Example: *Clostridium*.

5. Enteric bacteria
   - Gram-negative, facultative anaerobes.
   - Various metabolic capabilities in the group with some using NO_3^- as the electron donor in anaerobic respiration.
   - Found in the intestinal tract of animals.
   - Most are non-pathogenic, some are pathogenic.
   - 34 genera. Example: *Escherichia*.

6. Mycoplasmas

How are mycoplasmas unique from other prokaryotes?

- Smallest of all cells with 100–200 nm diameters.
- Extracellular forms and the only prokaryotes without cell walls.
- Saprobes and animal pathogens.
- 6 genera. Example: *Mycoplasma*.

7. Myxobacteria
- Chemoheterotrophs found in soil.
- Motility by gliding.
- Form a fruiting body through cells aggregating during unfavorable conditions.
- Spores released from the fruiting body when favorable conditions return form new colonies.
- 8 genera. Example: *Myxococcus*.

8. Nitrogen-fixing aerobic bacteria
- Both free-living and mutualistic species found in this group.
- Mutualistic species may be found in root nodules of legumes where they are important for proper plant nutrition.
- 5 genera. Example: *Rhizobium*.

9. Phototrophic anaerobic bacteria
- Most are strict anaerobes.
- Photosynthesis reduces NADP$^+$ with electrons extracted from such molecules as H$_2$S, <u>not</u> from water. Thus, oxygen is <u>not</u> released by photosynthesis in this group.
- Includes distantly related groups such as the purple sulfur bacteria and green sulfur bacteria.
- Inhabit pond, lake, and ocean sediments.
- 27 genera. Example: *Chromatium*.

10. Pseudomonads
- Typically rod-shaped, gram-negative cells with a flagellum at one end.
- Chemoheterotrophs capable of metabolizing unusual nutrients.
- 5 genera. Example: *Pseudomonas* which has species in nearly all aquatic and soil habitats.

11. Rickettsias and Chlamydias
- Obligate intracellular parasites of animals.
- Reduced gram-negative cell wall and metabolism varying with host cell infected.
- Rickettsias utilize both arthropods and mammals in their life cycles.
- Chlamydias cause diseases which are passed directly between vertebrate hosts.
- 15 genera. Example: *Chlamydia*.

Mycoplasmas

Myxobacteria

Purple sulfur bacteria

Green sulfur bacteria

Pseudomonads

Rickettsias, Chlamydias

Treponema

12. Spirochetes
    - ☞ Shallowly helical, very thin cells up to 0.25 nm in length.
    - ☞ Unique internal flagellar filaments produce a corkscrew-like movement.
    - ☞ Free-living saprobes and parasitic species.
    - ☞ 7 genera. Example: *Treponema*.

III. **THE ORIGINS OF METABOLIC DIVERSITY** (p. 524–528)

Based on current evidence, what are four plausible scenarios for the evolution of metabolic diversity?

Prokaryotes evolved all forms of nutrition and most metabolic pathways eons before eukaryotes arose.

A. **The Origin of Glycolysis** (p. 524)

First prokaryotes were probably chemoheterotrophs that absorbed free organic compounds generated by abiotic synthesis.

The universal role of ATP implies that prokaryotes used that molecule for energy very early in evolution.
- ☞ As ATP supplies were depleted, natural selection favored those prokaryotes that would regenerate ATP from ADP, leading to step by step evolution of glycolysis and other catabolic pathways.

Glycolysis is the only metabolic pathway common to all modern organisms and does not require O_2 (which was not abundant on early Earth).

B. **The Origin of Electron Transport Chains and Chemiosmosis** (p. 524–525)

Chemiosmotic ATP synthesis probably evolved in early prokaryotes as it is a common mechanism in all five kingdoms.

Early prokaryotes may have used the transmembrane pumps to help regulate their internal pH by expelling hydrogen ions. Energy (ATP) would have been necessary to drive these pumps.

ATP may have been saved by the first electron transport chains by coupling oxidation of organic acids to the transport of H^+ out of the cell.

Some bacteria may have evolved electron transport chains so efficient that more H^+ was extruded than was necessary for pH regulation. These cells could then utilize the influx of H^+ to reverse the proton pump and generate ATP.
- ☞ Some modern bacteria use this form of energy metabolism (= anaerobic respiration).

C. **The Origin of Photosynthesis** (p. 525)

As the supply of free ATP and abiotically produced organic molecules was depleted, natural selection may have favored organisms that could make their own organic molecules from inorganic precursors.

Light absorbing pigments in the earliest prokaryotes may have provided protection to the cells by absorbing excess light energy, especially ultraviolet, that could be harmful.
- ☞ These energized pigments may have then been coupled with electron transport systems to power ATP synthesis.
- ☞ Bacteriorhodopsin, the light-energy capturing pigment in the membrane of extreme halophiles, uses light energy to pump H^+ out of the cell to produce a gradient of hydrogen ions. This gradient provides the power for chemiosmotic production of ATP.

Components of electron transport chains that functioned in anaerobic respiration in other prokaryotes may have been co-opted to also provide reducing power. For example, H_2S could be used as a source of electrons and hydrogen for fixing CO_2.
- ☞ The modern purple and green sulfur bacteria are believed the most similar to early prokaryotes with regards to obtaining nutrition.

Some prokaryotes eventually arose that could use H_2O as the electron source. Thus evolved cyanobacteria, which released oxygen.

D. Cyanobacteria, the Oxygen Revolution, and the Origins of Cellular Respiration (p. 525–528)

Cyanobacteria evolved at least 2.5 billion years ago.

Oxygen released by photosynthesis may have first reacted with dissolved iron ions to precipitate as iron oxide (supported by geological evidence of deposits), preventing accumulation of free O_2.
- ☞ Precipitation of iron oxide would have eventually depleted the supply of dissolved iron and O_2 would have accumulated in the seas.
- ☞ As seas became saturated with O_2, the gas was released to the atmosphere.
- ☞ As O_2 accumulated, many species became extinct while others survived in anaerobic environments (including some archaebacteria) and others evolved with antioxidant mechanisms.
- ☞ Aerobic respiration may have originated as a modification of electron transport chains used in photosynthesis. The purple non-sulfur bacteria are photoheterotrophs which still use a hybrid electron transport system between a photosynthetic and respiratory system.
- ☞ Other bacteria lineages reverted to chemoheterotrophic nutrition with electron transport chains adapted only to aerobic respiration.

Thus, all major forms of nutrition evolved among prokaryotes before the first eukaryotes arose.

IV. THE IMPORTANCE OF PROKARYOTES (p. 528–530)

A. Prokaryotes and Chemical Cycles (p. 528)

Why does all life on earth depend upon the metabolic diversity of prokaryotes?

Prokaryotes are critical links in the recycling of chemical elements between the biological and physical components of ecosystems.

Decomposers = Bacteria that decompose dead organisms and waste of live organisms to return elements such as carbon and nitrogen to the environment in inorganic forms needed for reassimilation by other organisms, many of which are also bacteria.

Autotrophic bacteria = Bacteria that fix CO_2, thus supporting food chains.

Cyanobacteria supplement plants in restoring oxygen to the atmosphere as well as fixing nitrogen into nitrogenous compounds used by other organisms.

Other prokaryotes also support cycling of nitrogen, sulfur, iron and hydrogen.

B. Symbiotic Bacteria (*p. 528–529*)

Most prokaryotes form associations with other organisms; usually with other bacterial species possessing complementary metabolisms.

Symbiosis = Ecological relationships between organisms of different species that are in direct contact.
- Usually the smaller organisms, the *symbiont*, lives within or on the larger *host*.

What are three categories of symbiotic relationships? What characterizes each type of symbiosis?

Three Categories of Symbiosis:

Mutualism = Symbiosis in which both partners benefit.
- For example, nitrogen-fixing bacteria in root nodules of certain plants fix nitrogen to be used by the plant, which in turn furnishes sugar and other nutrients to the bacteria.

Commensalism = Symbiosis in which the symbiont benefits while neither helping nor harming the host.

Parasitism = Symbiosis in which the symbiont (*parasite*) benefits at the expense of the host.

Symbiosis is believed to have played a major role not only in the evolution of prokaryotes, but also in the evolution of early protistans.

C. Bacteria and Disease (*p. 529–530*)

About 1/2 of human disease is caused by bacteria. To cause a disease, the bacteria must invade the host, evade or resist the host's internal defenses long enough to grow, and harm the host.

Some pathogens are *opportunistic*.

Opportunistic = Normal inhabitants of the body that become pathogenic only when defenses are weakened by other factors such as poor nutrition or other infections.
- For example, *Streptococcus pneumonia*.

What are Koch's postulates that are used as guidelines to substantiate a specific pathogen as the cause of a disease?

Robert Koch was the first to determine a direct connection between specific bacteria and certain diseases.

Koch's postulates = Four criteria to substantiate a specific pathogen as the cause for a disease are:
1. Find the same pathogen in each diseased individual.
2. Isolate the pathogen from a diseased subject and grow it in a pure culture.
3. Use cultured pathogen to induce the disease in experimental animals.
4. Isolate the same pathogen in the diseased experimental animal.

Some pathogens cause disease by growth and invasion of tissues but others cause disease by production of a *toxin*.

What are the differences between exotoxins and endotoxins? What types of symptoms do they each induce?

Two Major Types of Toxin:

Exotoxins = Proteins secreted by bacterial cells.
- Among the most potent poisons known.
- To cause disease, organisms itself does not have to be present: the toxin is enough.
- Elicits specific symptoms.
- For example, botulism toxin.

Endotoxins = Toxic component of outer membranes of some Gram-negative bacteria.
- All induce the general symptoms of fever and aches.
- Examples are *Salmonella typhi* (typhoid fever) and *Salmonella* induced food poisoning.

D. Putting Bacteria to Work (p. 530)

How do humans exploit the metabolic diversity of prokaryotes for scientific and commercial purposes? How is Streptomyces used commercially?

Humans use bacteria for a diversity of purposes.
- More than half of the antibiotics used to treat bacterial diseases come from cultures of various species of *Streptomyces* maintained by pharmaceutical companies.
- As simple models of life to learn about metabolism and molecular biology. (*E. coli* is the best understood of all organisms.)
- To digest organic wastes at sewage treatment plants.
- Some species of pseudomonads are used to decompose pesticides and other synthetic compounds.
- To produce products that we can purify such as chemicals like acetone, vitamins and antibiotics, and insulin. Genetic engineering has expanded these uses.
- To convert milk into yogurt and cheese.

26

PROTISTS AND THE ORIGIN OF EUKARYOTES

CHAPTER OUTLINE

I. CHARACTERISTICS OF PROTISTS (*p. 534*)

What characteristics distinguish protists from organisms in the other four kingdoms?

Protists are eukaryotic organisms having true nuclei, membrane-enclosed organelles, "9+2" flagella and cilia. Mitosis and meiosis, occur in most species.
- ☞ Most are unicellular, some are colonial, and some are multicellular with tissues arranged in simple body plans.

They are found in almost all moist environments.
- ☞ Free-living species are found in the seas, freshwater systems, and moist terrestrial habitats.
- ☞ Symbiotic species are found in the body fluids, tissues, and cells of the host.

Almost all are aerobic, using mitochondria for cellular respiration.
- ☞ Anaerobic forms lack mitochondria and live in anaerobic environments or have mutualistic respiring bacteria.
- ☞ May be autotrophic, heterotrophic, or mixotrophic.

Almost all have flagella or cilia (not homologous to prokaryotic flagella) at some time in life cycle.
- ☞ Cilia and flagella are extensions of the cytoplasm.
- ☞ Cilia and flagella have the same basic structure; they differ in that cilia are shorter and more numerous.

Cell division is varied.
- ☞ Unique mitotic divisions in many groups.
- ☞ All can reproduce asexually; some can reproduce sexually.
- ☞ Some form resistant cysts when stressed.

II. THE ORIGIN OF EUKARYOTES (*p. 534–536*)

A. The Antiquity of Eukaryotes (*p. 535*)

Precambrian *acritarchs*, dated 1.5 billion years old, are accepted as the oldest eukaryotic fossils known. They are similar in size and appearance to contemporary algal cysts.
- ☞ Eukaryotes evolved after O_2 accumulated in the primitive atmosphere.

Protists

Acritarchs

B. Models of Eukaryotic Origins (p. 535–536)

What are the current hypothetical models for eukaryotic origins?

The *autogenous model* proposes that eukaryotic cells evolved by specialization of internal membranes derived from prokaryotic plasma membranes. (See Campbell, Figure 26.2a)

☞ Most endomembranous structures are believed to have differentiated from invaginations of the plasma membrane.

☞ Double-membrane organelles (mitochondria and chloroplasts) may have evolved by secondary invagination or more complex membrane folding.

Autogenous model

The *endosymbiotic model* proposes that certain prokaryotic species, called *endosymbionts* lived within larger prokaryotes. (See Campbell, Figure 26.2b)

☞ Focuses mainly on mitochondria and chloroplasts.

☞ Chloroplasts are thought to have descended from endosymbiotic photosynthesizing prokaryotes living in larger cells.

☞ Mitochondria are postulated to be descendants of prokaryotic aerobic heterotrophs that may have been parasites or undigested prey of larger prokaryotes. The association progressed from parasitism or predation to mutualism.

Endosymbiotic model

Endosymbionts

What evidence do proponents of the endosymbiotic hypothesis use to support their case?

Evidence for the endosymbiont model is that mitochondria and chloroplasts:

☞ Are appropriate size to be descendants of eubacteria.

☞ Have inner membranes containing several enzymes and transport systems similar to those on prokaryotic plasma membranes.

☞ Replicate by splitting, as in prokaryotes.

☞ DNA is circular and not associated with histones or other proteins, as in prokaryotes.

☞ Contain their own components for DNA transcription and translation into proteins.

☞ Have ribosomes similar to prokaryotic ribosomes.

☞ Molecular systematics lends evidence to support this theory.

☞ Many extant organisms are involved in endosymbiotic relationships.

Are the autogenous and endosymbiotic models mutually exclusive?

Debate about which model of eukaryotic origin is most accurate continues. Evidence for the polyphyletic origin of eukaryotes has also been presented which may indicate both models may be at least partially accurate.

III. BOUNDARIES OF THE KINGDOM PROTISTA (p. 536–537)

In 1969, Robert H. Whittaker popularized the five kingdom taxonomic system and placed only unicellular eukaryotes in Protista.

Robert H. Whittaker

Why do modern biologists recommend expanding the original boundaries of the Kingdom Protista?

Now some multicellular algae and funguslike organisms, previously classified with the plants and fungi, are placed in Kingdom Protista because they are believed to be more closely related to unicellular forms than to true plants or fungi.
- ☞ The taxonomy of protists is still in a state of flux.

It is convenient to identify protozoa (animal-like protists), algae (plantlike protists) and funguslike protists. This <u>informal</u> grouping does <u>not</u> reflect evolutionary relationships.

What is meant by the statement that the Kingdom Protista is a polyphyletic group?

IV. PROTOZOA (p. 537–542)

The term *protozoa* refers to a very diverse group of heterotrophic protists. The subdivision of this group into different phyla is based on how they feed and move.

What are the six major protozoan phyla? What modes of locomotion, feeding and reproduction characterize each one? Which phyla contain parasitic species? What environments are the free-living species likely to inhabit?

A. Rhizopoda (p. 537–538)

How do amoebas move?

Rhizopoda (= rootlike feet) include the amoebas and their relatives.
- ☞ Simplest of protists.
- ☞ All unicellular.
- ☞ No flagellated stages in life cycle.
- ☞ *Pseudopodia* form as cellular extensions and function in feeding and movement. (See Campbell, Figure 26.3)
- ☞ All reproduction is by asexual mechanisms: no meiosis or sexual reproduction occurs.
- ☞ During mitosis, spindle fibers form, but typical stages of mitosis are not apparent in most species.
- ☞ Rhizopods inhabit freshwater, marine and soil habitats.
- ☞ Most are free-living, although some are parasitic.

B. Actinopoda (p. 538)

Actinopoda (= ray feet) possess *axopodia*.

<u>Axopodia</u> = Projections reinforced by bundles of microtubules thinly covered by cytoplasm.
- ☞ Used to increase surface area which helps the organisms float and function in feeding.

Most actinopods are planktonic.

The two main groups of Actinopods are the *heliozoans* and the *radiolarians*.
- *Heliozoans* live primarily in fresh water. (See Campbell, Figure 26.4a)
- *Radiolarians* are primarily marine and have delicate shells, usually made of silica. (See Campbell, Figure 26.4b)

C. Foraminifera (p. 539)

Forams have porous, multi-chambered, calcium carbonate shells. (See Campbell, Figure 26.5)
- Most live in the sand or attach to algae and rocks; some are planktonic.
- Are exclusively marine.
- Have cytoplasmic strands that extend through the shell's pores and function in swimming, feeding and shell formation.
- Foram shells are an important component of sediments and sedimentary rocks.

D. Apicomplexa (p. 539)

Why is the Phylum Apicomplexa so named?

All species of the phylum Apicomplexa are parasites of animals.
- The infectious cells produced in the life cycle are called *sporozoites*.
- The apex of sporozoites has organelles for penetrating host cells and tissues; it is for these apical organelles that the phylum is named.
- Life cycles are intricate with both sexual and asexual reproduction occurring; often requiring two or more different host species. (See Campbell, Figure 26.6)

What is the organism that causes malaria? What is its life history? (See Campbell, Figure 26.6)

For example, several species of *Plasmodium* cause malaria. This is a potentially fatal and relatively common tropical disease resulting in at least one million deaths each year.

E. Zoomastigophora (p. 539–540)

All species within the phylum Zoomastigophora are heterotrophs that absorb organic molecules or phagocytize prey.
- Use whip-like flagella to move.
- Most are singular, some are colonial.
- Many are free-living although a large number are symbiotic.

What is the organism that causes African sleeping sickness? How is it spread and why is it difficult to control?

Species of *Trypanosoma* cause African sleeping sickness. Spread by the bite of the tsetse fly, this parasite disease is difficult to control.
- There is a large natural reservoir for the disease since many tropical African mammals harbor the parasite.
- Trypanosomes can evade the host's immune system by rearranging genes coding for the molecular composition of coat proteins.

Heliozoans, Radiolarians

Foraminifera

Apicomplexa

Sporozoites

Plasmodium

Zoomastigophora

Trypanosoma

F. Ciliophora (p. 541–542)

Species within the phylum Ciliophora use cilia to move and feed.

- ☞ Cilia may be dispersed over surface, or clustered in fewer rows or tufts.
- ☞ Some move on leg-like *cirri* (many cilia bonded together).
- ☞ Others have rows of tightly packed cilia that function together as locomotor membranes (for example, *Stentor*).
- ☞ In some, cilia are associated with *trichocysts*, bulblike organelles that discharge sticky proteinaceous threads, the function of which is still unknown.

What is the function of contractile vacuoles in freshwater ciliates?

Most ciliates exist as solitary cells in fresh water.

- ☞ Freshwater ciliates, such as *Paramecium*, constantly take in water by osmosis from the hypoosmotic environment.
- ☞ *Contractile vacuoles* periodically expel excess water through the plasma membrane. (See Campbell, Figure 26.8c)

How do macronuclei and micronuclei differ in appearance and function?

Ciliates possess two types of nuclei: one large *macronucleus* and usually several small *micronuclei*.

Characteristics of the macronucleus:

- ☞ Large, has over 50 copies of the genome.
- ☞ Genes are packaged in a large number of small units, each with hundreds of copies of just a few genes.
- ☞ Controls everyday functions of the cell by synthesizing RNA and is necessary for asexual reproduction during binary fission.
- ☞ Does not undergo mitosis.

Characteristics of the micronucleus:

- ☞ Small, may number from 1 to 80 micronuclei, depending on the species.
- ☞ Does not function in growth, maintenance or asexual reproduction.
- ☞ Functions in *conjugation* a sexual process which produces genetic variation. (See Campbell, Figure 26.9)

What happens to the micronuclei and macronucleus during conjugation in Paramecium caudatum*? How does this process create genetic variation?*

V. ALGAE (p. 542–551)

What characteristics have been used to determine phylogenetic relationships among algae? Using these characteristics, how would you distinguish among the algal divisions?

Algal protists are aquatic organisms which all have chlorophyll *a* (like cyanobacteria and plants). Phylogenetic relationships among algal divisions have been determined by differences in:

- ☞ Accessory pigments such as carotenoids, xanthophylls, phycobillins, and other forms of chlorophyll.

☞ Chloroplast structure.
☞ Cell wall chemistry.
☞ Number, type and position of flagella.
☞ Food storage product.

A. Dinoflagellata (p. 543–544)

Dinoflagellates are components of phytoplankton and they provide the foundation of most marine food chains.

☞ Most are unicellular, some are colonial.
☞ Some live as photosynthetic symbionts of the cnidarians that build coral reefs.
☞ May cause *red tides* by explosive growth (bloom). These dinoflagellates produce a toxin that is concentrated by invertebrates, including shellfish.
☞ Some lack chloroplasts and live as parasites; a few carnivorous species are known.
☞ Have brownish plastids containing chlorophyll *a*, chlorophyll *c* and a mix of carotenoids, including *peridinin* (found only in this phylum).
☞ Food is stored as starch.
☞ Cell surface is reinforced by cellulose plates with flagella in perpendicular grooves, creating its whirling movement and resulting in a characteristic shape. (See Campbell, Figure 26.10)
☞ Chromosomes lack histones and are always condensed.
☞ Has no mitotic stages.
☞ Kinetochores are attached to the nuclear envelope and chromosomes distributed to daughter cells by the splitting of the nucleus.

B. Chrysophyta (p. 544)

The phylum Chrysophyta includes the golden algae. (See Campbell, Figure 26.11)

☞ Live among freshwater plankton; most are colonial.
☞ Have flagellated cells with both flagella attached near one end of the cell.
☞ Store carbohydrates in the form of *laminarin*, a polysaccharide.
☞ Plastids have chlorophyll *a*, chlorophyll *c*, yellow and brown carotenoids, and xanthophyll.
☞ Survive environmental stress by forming resistant cysts.

C. Bacillariophyta (p. 545)

The phylum Bacillariophyta includes the diatoms.

☞ Mostly unicellular organisms with overlapping glasslike walls of hydrated silica in an organic matrix. (See Campbell, Figure 26.12)
☞ Have the same photosynthetic pigments as in Chrysophyta.
☞ Components of freshwater and marine plankton.
☞ Store food in a form of oil, which also makes cells buoyant.
☞ Many have a gliding movement produced by chemical secretions.
☞ Usually reproduce asexually; sexual stages are rare.
☞ Some produce resistant cysts.

394 Protists and the Origin of Eukaryotes

Euglenophyta

Paramylon

Chlorophyta

Lichen

Volvox

Ulva

Isogamy

Anisogamy

Oogamy

D. Euglenophyta (p. 545)

Euglenophytes are common inhabitants of freshwater systems. (See Campbell, Figure 26.13)
- Have chlorophyll *a*, chlorophyll *b* (also in true plants), carotenoids, and xanthophyll.
- Possess flexible internal protein plates; no cell wall present.
- Store food as the polysaccharide *paramylon*.
- Have 1–3 apical flagella which pull the cell through the water.
- Are heterotrophic in the absence of light.

E. Chlorophyta (p. 545–548)

Chlorophyta (green algae):
- Mostly freshwater, but some are marine.
- Have plant-like chloroplasts and are believed to be ancestors to the plant kingdom.
- Some unicellular types live as plankton, inhabit damp soil, coat snow surfaces, or are symbionts.
- When living symbiotically with fungi = *lichen*.
- Colonial forms are often filamentous ("pond scum").
- Multicellular forms may have large, complex structures resembling true plants.

What are three possible evolutionary trends that led to multicellularity in the Chlorophyta?

Evolutionary trends that probably produced colonial and multicellular forms from flagellated unicellular ancestors are:
- Formation of colonies of individual cells (as seen in *Volvox*).
- Repeated division of nuclei with no cytoplasmic division (as in *Bryopsis*).
- Formation of multicellular thalli (as in *Ulva*).

Most chlorophytes have complex life histories involving sexual and asexual reproductive stages. (See Campbell, Figure 26.15)
- Nearly all reproduce sexually via biflagellated gametes.

What type of gametes characterize isogamy, anisogamy and oogamy?

- The morphology of the gametes produced varies with three possible types of fertilization occurring in this group:
  1. <u>Isogamy</u> = The male and female gametes are morphologically indistinguishable.
  2. <u>Anisogamy</u> = The male and female gametes differ in size or morphology.
  3. <u>Oogamy</u> = A flagellated male gamete (sperm) fertilizes a nonmotile female gamete (egg).

What is the difference between a sporophyte and a gametophyte? In an alternation of generations, where does meiosis occur, and what are the products of meiosis? What generation produces gametes and by what process are they produced?

Some multicellular green algae combine sexual reproduction with *alternation of generations*.

Alternation of generations = Alternation between a multicellular haploid (*gametophyte*) and diploid (*sporophyte*) generation. (See Campbell, Figure 26.16)
- ☞ Haploid gametophytes produce gametes (n) by mitosis.
- ☞ Diploid sporophytes produce spores (n) by meiosis.

Alternation of generations

Gametophyte, Sporophyte

Referring to Campbell, Figures 26.15, 26.16 and 26.19, how do the life cycles of Chlamydomonas, Ulva and Laminaria differ? What stages are haploid? What stages are diploid?

F. **Evolutionary Adaptations of Seaweed** (*p. 548–549*)

Seaweeds are large, multicellular marine algae which are found in the intertidal and subtidal zones of coastal waters.
- ☞ A diverse group of algae including members of the Chlorophyta (green algae), Phaeophyta (brown algae) and Rhodophyta (red algae).

The habitat of seaweeds, particularly the intertidal zone, poses several challenges to the survival of these organisms.
- ☞ Movement of the water due to wave action and winds produces a physically active habitat.
- ☞ Tidal rhythms result in the seaweeds being alternately covered by seawater and exposed to direct sunlight and the drying conditions of the air.

What structural and biochemical adaptations of seaweeds enhance their survival in intertidal and subtidal zones of coastal waters?

Seaweeds have evolved several unique structural and biochemical adaptations to survive the conditions of their habitats.

Structural adaptations found in seaweeds are a result of their complex multicellular anatomy. Some forms have differentiated tissues and organs analogous to those of plants.
- ☞ The body of a seaweed is called a *thallus*. It is plantlike in appearance but has no true roots, stems, or leaves.
- ☞ A thallus consists of a rootlike *holdfast* (maintains position), a stemlike *stipe* (supports the blades), and leaflike *blades* (large surfaces for photosynthesis).
- ☞ Floats which help suspend blades near the water surface are present in some brown algae.

Thallus

Holdfast

Stipe, Blades

Biochemical adaptations in some red and brown algae reinforce the anatomical adaptations and enhance survival.
- ☞ Cellulose cell walls also contain gel-forming polysaccharides (algin in brown algae; carageenan in red algae) which cushion the thalli against wave action and prevent desication during low tide.
- ☞ Some red algae retard grazing by marine invertebrates by incorporating large amounts of calcium carbonate into their cell walls.

Seaweeds are used by humans in a variety of ways: food, nutrient supplements, additives, lubricants, and microbiological culture media.

G. Phaeophyta (p. 549–550)

Phaeophyta (brown algae):
- Largest and most complex protists.
- All are multicellular and most are marine inhabitants.
- Have chlorophyll *a*, chlorophyll *c*, and the carotenoid pigment *fucoxanthin*.
- Store carbohydrate food reserves in the form of laminarin.
- Cell walls made of cellulose and *algin*.
- A variety of life cycles in the group, with most having alternation of generations. In some the two generations are *isomorphic*, while in others they are *heteromorphic*. (See Campbell, Figure 26.19)

H. Rhodophyta (p. 550–551)

Rhodophyta (red algae) are primarily warm, tropical marine inhabitants, although some are found in fresh water and soil.
- Contain chlorophyll *a*, carotenoids, phycobilins, and chlorophyll *d* in some.
- Red color of plastids due to the accessory pigment, phycoerythrin, a phycobili-protein.
- Carbohydrate food reserves stored as floridean starch (similar to glycogen).
- Cell walls are cellulose with agar and carageenan.
- Color of the thallus may vary (even in a single species) with depth as pigmentation changes to optimize photosynthesis.
- Most Rhodophytes are multicellular.
- Most thalli are filamentous, branched and lacy.
- Some *coralline algae* have hard calcareous cell walls.

All Rhodophytes reproduce sexually.
- Alternation of generations is common.
- Have no flagellated stages, unlike other algal protists.

I. Evolutionary Relationships Among Algae (p. 551)

According to the endosymbiotic model, what are the possible origins of chloroplasts? What evidence supports this?

The endosymbiotic model suggests chloroplasts originated at least three separate times from different prokaryotic endosymbionts.
- Examination of chloroplast structure and photosynthetic pigments provides the three lineages of algal protists: a red line (Rhodophyta), a green line (Chlorophyta), and a brown line (Phaeophyta).

Red lineage
- Chloroplast thylakoid arrangement and pigment composition closely resemble that of cyanobacteria.
- Red algae may have descended from prokaryotes possessing endosymbiotic cyanobacteria.

Green lineage
- ☞ Chloroplasts are similar to those of *Prochlorothrix*, a photosynthetic prokaryote.
- ☞ Chlorophyll *a* and chlorophyll *b* are found in both.
- ☞ *Prochlorothrix* is one of the few prokaryotes with chlorophyll *b*.
- ☞ Green algae may have evolved from a symbiotic relationship involving a prokaryote similar to *Prochlorothrix*.

Prochlorothrix

Brown lineage
- ☞ Thylakoids of this lineage usually occur in stacks of three.
- ☞ Chlorophyll *c* and certain xanthophylls are present as accessory pigments.
- ☞ No probable prokaryotic ancestor to these chloroplasts has yet been identified.

VI. FUNGUSLIKE PROTISTS (p. 551–554)

How do the life cycles of plasmodial and cellular slime molds differ?

The slime molds' similarity to fungi is probably the result of convergence.

A. Myxomycota (p. 552–553)

The phylum Myxomycota consists of the plasmodial slime molds.
- ☞ Are all heterotrophs although many are pigmented.

Myxomycota

Plasmodial slime molds

Plasmodium = Feeding stage of life cycle consisting of an amoeboid, coenocytic (multi-nucleated cytoplasm undivided by membranes) mass.

Plasmodium

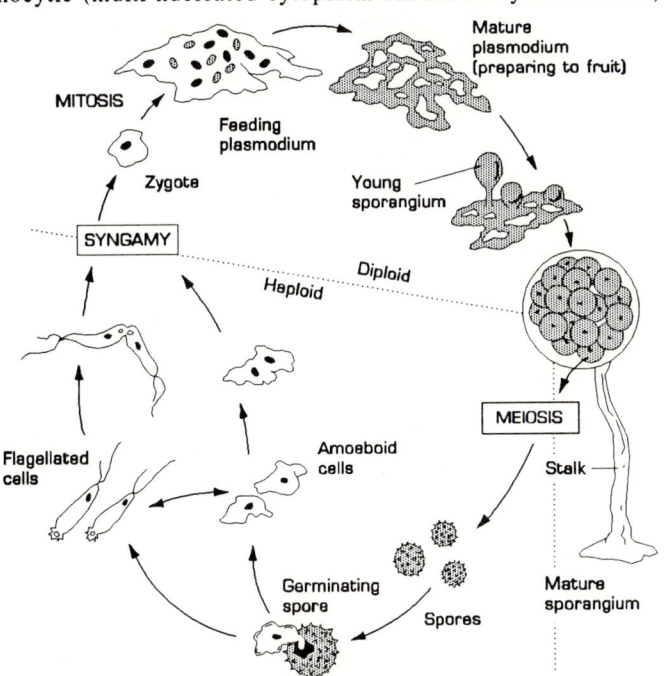

- ☞ In most species, the nuclei of plasmodia are diploid and exhibit synchronous mitotic divisions.
- ☞ Engulfs food by phagocytosis.

- Lives in moist soil and leaf mulch.
- When stressed, the plasmodium ceases growth and forms sexually reproductive structures called fruiting bodies, or *sporangia*.

B. **Acrasiomycota** (p. 553)

The phylum Acrasiomycota includes the cellular slime molds.

- Feeding stage of life cycle consists of individual, solitary haploid cells.
- When the food supply is depleted, cells aggregate to form a mass similar to those of myxomycota but cells remain separate (not coenocytic).
- Fruiting bodies function in asexual reproduction (unlike plasmodial slime molds).
- Only one genus has flagellated stages.

C. **Oomycota** (p. 553–554)

The phylum Oomycota includes water molds, white rusts, and downy mildews.

- Have coenocytic hyphae that are not homologous to fungal hyphae.

Why are oomycetes considered to be protists rather than true fungi?

- Cell walls are made of cellulose, rather than chitin found in true fungi.
- Diploid condition in the life cycle prevails in most species.
- Biflagellated cells are present in the life cycles, while fungi lack flagellated cells.

In water molds:
- ☞ A large egg is fertilized by a smaller sperm cell to form a resistant zygote. (See Campbell, Figure 26.23)
- ☞ Usually are saprophytes in fresh water and are important decomposers, some are parasitic.

What are some examples of oomycetes? Why are they economically important?

White rusts and downy mildews:
- ☞ Usually are parasitic on terrestrial plants.
- ☞ Disperse by windblown spores, but may also form flagellated zoospores.
- ☞ Some of the most important plant pathogens.

White rusts, Downy mildews

D. **Chytridiomycota** (p. 554)

Chytridiomycota

How does Chytridiomycota resemble Fungi? What is their probable relationship to the fungal kingdom?

Most are microscopic.
- ☞ Unicellular, coenocytic forms are the simplest funguslike protists.
- ☞ More complex species possess branched, coenocytic hyphae.
- ☞ Usually inhabit aquatic systems or moist soils.
- ☞ Many are saprobes while some are parasitic in algae, pollen grains, and some insects.
- ☞ Produce motile spores and flagellated gametes.
- ☞ Some members of the Chytridiomycota are similar to the true fungi; sharing the branched hyphal structure and having cell walls composed of chitin.
- ☞ Possibly the ancestral group to the true fungi.

VII. THE ORIGINS OF MULTICELLULARITY (p. 554–556)

What is the most widely accepted hypothesis for the evolution of multicellularity?

Eukaryotes were more complex than prokaryotes, allowing for greater morphological variation.

Earliest "multicellular" forms may have been colonies of interconnected cells.
- ☞ Evolution of multicellularity from colonial aggregates involved cellular specialization and division of labor.
- ☞ Earliest such specialization probably involved locomotion and reproduction.
- ☞ As cells became more interdependent, some lost their flagella and performed other functions.
- ☞ Further division of labor may have separated sex cells from somatic cells.
- ☞ Multicellular forms more complex than filamentous algae appeared approximately 700 million years ago.
- ☞ Seaweeds and other complex algae were abundant during the Cambrian period, about 570 million years ago.

27 PLANTS AND THE COLONIZATION OF LAND

CHAPTER OUTLINE

Plants appeared on land about 425 million years ago, and the evolutionary history of the plant kingdom reflects increasing adaptation to the terrestrial environment.

I. **AN INTRODUCTION TO THE PLANT KINGDOM** (p. 559–561)

   A. **General Characteristics of Plants** (p. 559–560)

   What distinguishes plants from organisms in the other four kingdoms? What characteristics do plants share with their green algal ancestors?

   Plants are multicellular eukaryotes that are photosynthetic autotrophs. They share the following characteristics with their green algal ancestors:
   - Chloroplasts with the photosynthetic pigments: chlorophyll *a*, chlorophyll *b*, and carotenoids.
   - Cell walls containing cellulose.
   - Food reserve is starch that is stored in plastids.

   What are some adaptations, unique to the plant kingdom, that are advantageous for a terrestrial existence?

   As plants adapted to terrestrial life, they evolved complex bodies with cell specialization for different functions.
   - Aerial plant parts are coated with a waxy *cuticle* that helps prevent desiccation.
   - Though gas exchange cannot occur across the waxy cuticle, CO_2 and O_2 can diffuse between the leaf's interior and the surrounding air through *stomata*, microscopic pores on the leaf's surface.

   B. **A Generalized View of Plant Reproduction and Life Cycles** (p. 560)

   Why was a new mode of reproduction necessary for plants to survive on land?

   With the move from an aquatic to terrestrial environment, a new mode of reproduction was necessary to solve two problems:
   1. *Gametes must be dispersed in a nonaquatic environment.* Plants produce gametes within *gametangia*, organs with protective jackets of sterile (nonreproductive) cells that prevent gametes from drying out. The egg is fertilized within the female organ.
   2. *Embryos must be protected against desiccation.* The zygote develops into an embryo that is retained for awhile within the female gametangia's jacket of protective cells.

What is meant by an "alternation of generations"? What is the difference between a sporophyte and a gametophyte? In an alternation of generations, where does meiosis occur, and what are the products of meiosis? What generation produces gametes and by what process are they produced?

NOTE: It is extremely important that you understand the generalized life cycle in plants, because it will be used as a frame of reference to compare life cycles among the plant divisions. Learning this now, will be well worth the effort.

Most plants reproduce sexually, and most are also capable of asexual propagation. All plants have life cycles with an *alternation of generations*.

- A haploid *gametophyte* generation produces and alternates with a diploid *sporophyte* generation. The sporophyte, in turn, produces gametophytes.
- The life cycles are *heteromorphic*; that is, sporophytes and gametophytes differ in morphology.
- The sporophyte is larger and more noticeable than the gametophyte in all plants but mosses and their relatives.

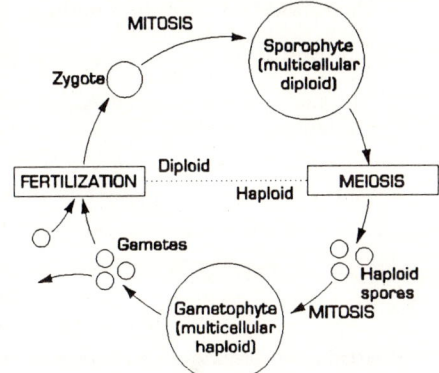

A comparison of life cycles among plant divisions is instructive because:
- It points to an important trend in plant evolution: reduction of the haploid gametophyte generation and dominance of the diploid sporophyte.
- Certain life cycle features are adaptations to a terrestrial environment; for example the replacement of flagellated sperm by pollen.

C. Some Highlights of Plant Evolution (p. 560–561)

What major periods of plant evolution opened new adaptive zones on land?

There are four major periods of plant evolution that opened new adaptive zones on land:

1. Origin of plants from aquatic ancestors (probably green algae) in the Silurian (425 million years ago).
   - Cuticle and jacketed gametangia evolved that protected gametes and embryos.
   - *Vascular tissue* evolved with conducting cells that transport water and nutrients throughout the plant.
2. Diversification of seedless vascular plants, such as ferns, during the early Devonian about 400 mya.

Alternation of generations

Gametophyte

Sporophyte

Heteromorphic

Silurian

Vascular tissue

Devonian

Seed

Gymnosperms

Cretaceous

Angiosperms

Division

3. Origin of the seed near the end of the Devonian about 360 mya.

   Seed = Plant embryo packaged with a store of food within a resistant coat.
   - ☞ Early seed plants bore seeds as naked structures and evolved into gymnosperms including conifers.
   - ☞ Conifers and ferns coexisted in the landscape for more than 200 million years.

4. Emergence of flowering plants during the early Cretaceous, about 130 mya.
   - ☞ Unlike gymnosperms, flowering plants bear seeds within the flower's protective ovaries.
   - ☞ Most contemporary plants are flowering plants or *angiosperms*.

D. **Classification of Plants** (*p. 561*)

What is the difference between the categories division and phylum?

The major taxonomic category of plants is the *division*; it is comparable to phylum, the highest category in the animal kingdom.

Using the classification scheme presented in the text, what are the ten plant divisions? What is the common name for plants in each division? How are they categorized into nonvascular, vascular seedless and vascular seed plants?

A CLASSIFICATION OF PLANTS

	Common Name	Approximate Number of Extant Species
Nonvascular Plants		
Division Bryophyta	Mosses	10,000
Division Hepatophyta	Liverworts	6,500
Division Anthocerophyta	Hornworts	100
Vascular Seedless Plants		
Division Psilophyta	Whiskferns	10 to 13
Division Lycophyta	Club mosses	1,000
Division Sphenophyta	Horsetails	15
Division Pterophyta	Ferns	12,000
Vascular Seed Plants		
Division Coniferophyta	Conifers	550
Division Cycadophyta	Cycads	100
Division Ginkgophyta	Ginkgo	1
Division Gnetophyta	Gnetae	70
Division Anthophyta	Flowering plants	235,000

II. THE MOVE ONTO LAND (p. 562–566)

A. A Case for Green Algae as the Ancestors of Plants (p. 562)

What evidence can be used to defend the position that plants evolved from green algae?

Plants probably evolved from green algae. Chlorophyta shares the following characteristics with plants:
1. Both have Chlorophyll a, Chlorophyll b and beta-carotene.
2. Chloroplasts of green algae have thylakoid membranes stacked into grana like those of plants.
3. Cell walls of both are made of cellulose.
4. Both store carbohydrates as starch.
5. In both, vesicles derived from Golgi bodies form a cell plate that divides the cytoplasm during cytokinesis.

Filamentous chlorophytes were probably the first green algae to colonize land.
- During the Silurian, land was subjected to flooding and draining; thus, natural selection would favor algae that could survive when not submerged.
- Cuticles and jacketed reproductive organs evolved, adaptations that made life above the water line possible.
- The terrestrial frontier offered unfiltered sunlight, rich soil, and the absence of herbivores.

Bryophytes and vascular plants probably diverged before the origin of vascular tissue characteristic of all other plants.

B. Bryophytes: Divisions Bryophyta, Hepatophyta, and Anthocerophyta (p. 562–563)

Bryophyta

What two adaptations made the bryophytes' move onto land possible?

Bryophytes have two adaptations that made the move onto land possible:
- A waxy cuticle that prevents desiccation.
- Gametangia that protect developing gametes.
   a. *Antheridium*, or male gametangium, produces flagellated sperm cells.
   b. *Archegonium*, or female gametangium, produces a single egg; fertilization occurs within the archegonium, and the zygote develops into an embryo within the protective jacket of the female organ.

Antheridium

Archegonium

How are bryophytes still tied to water?

Bryophytes are not totally free from their ancestral aquatic habitat.
- They need water to reproduce. Their flagellated sperm cells must swim from the antheridium to the archegonium to fertilize the egg.
- Most have no vascular tissue to carry water from the soil to aerial plant parts; they imbibe water and distribute it throughout the plant by the relatively slow processes of diffusion, capillary action, and cytoplasmic streaming.

Bryophytes lack woody tissue and cannot support tall plants on land; they may sprawl horizontally as mats, but always have a low profile.

What are the three divisions of bryophytes? What characterizes each division?

1. **Mosses (Division Bryophyta)** (p. 563)

    A tight pack of many moss plants forms a spongy mat that can absorb and retain water.
    - Each plant grips the substratum with *rhizoids*, elongate cells or cellular filaments.
    - Photosynthesis occurs mostly in the small stemlike and leaflike structures found in upper parts of the plant; these structures are not homologous with stems and leaves in vascular plants.

Referring to Campbell, Figure 27.5, what distinguishes the life cycle of bryophytes from the life cycles of vascular plants? Where do meiosis, spore production, gamete production and fertilization occur? What generations are free-living?

There is an alternation of haploid and diploid generations in the moss life cycle. (See Campbell, Figure 27.5)
- The sporophyte (2n) produces haploid spores by meiosis in a *sporangium*; the spores divide by mitosis to form new gametophytes.
- Contrary to the life cycles of vascular plants, the haploid gametophyte is the dominant generation in mosses and other bryophytes. Sporophytes are generally smaller and depend on the gametophyte for water and nutrients.

2. **Liverworts (Division Hepatophyta)** (p. 563)

    Less conspicuous than mosses, their bodies are sometimes lobed.
    - Their life cycle is similar to mosses. Their sporangia have *elaters*, coil-shaped cells that spring out of the capsule and disperse spores.
    - Can also reproduce asexually from *gemmae* (small bundles of cells that can bounce out of cups on the surface of the gametophyte when hit by rainwater).

3. **Hornworts (Division Anthocerophyta)** (p. 563)

    Hornworts:
    - Resemble liverworts, but sporophytes are horn-shaped, elongated capsules that grow from the mat-like gametophyte.
    - Their photosynthetic cells have only one large chloroplast, unlike the many smaller ones of other plants.

C. **Terrestrial Adaptations of Vascular Plants** (p. 563–566)

How are conditions in an aquatic environment different from those faced by plants in a terrestrial environment? What sporophyte and life cycle modifications in vascular plants contributed to their success on land?

The following terrestrial adaptations evolved in vascular plants:

1. Subterranean *roots* that absorb water and minerals from the soil, and an aerial *shoot* system of stems and leaves to make food. In a terrestrial environment, the resources of water and light are spatially segregated.

2. *Lignin*, a material embedded in the cellulose matrix of plant cell walls, provides support necessary in a terrestrial environment with a nonsupportive medium of air. This is not needed in aquatic habitats since plants are essentially weightless there.

3. A continuous vascular system transports material from roots to shoots and vice versa.

    Xylem = Dead, tube-shaped cells that carry water and minerals from roots to aerial parts of the plant. The lignified walls of xylem also function in support.
    Phloem = Living cells that distribute sugars, amino acids and other nutrients throughout the plant.

4. Pollen and seeds evolved in some lineages.

5. Increased dominance of the diploid sporophyte.

D. The Earliest Vascular Plants (p. 566)

Oldest fossilized vascular plant is *Cooksonia* (late Silurian). Other early vascular plants include:
- *Zosterophyllum*, an early Devonian ancestor of *Lycophyta*, that had clusters of lateral sporangia near stem tips.
- *Rhynia* or similar plants were probably in the lineage that led to all other divisions of extant plants. They had single, terminal sporangia.

III. SEEDLESS VASCULAR PLANTS (p. 566–570)

What are the four existing divisions of seedless vascular plants? What characterizes each division? What are some examples?

The earliest vascular plants were seedless. Today there are four divisions of seedless vascular plants.

A. Division Psilophyta (p. 566–567)

Characteristics of Psilophyta include:
- True roots and leaves are absent, subterranean rhizomes are covered with hair-like rhizoids, and shoots have scales which lack vascular tissue.
- Dichotomous branching.
- The gametophytes are subterranean and lack chlorophyll, depending on symbiotic soil fungi for food.
- Flagellated sperm swim through the soil from antheridia to the archegonium of the gametophyte.
- The sporophyte emerges from the gametophyte, which then dies.
- Has only two genera: *Psilotum* (whiskferns) and *Tmesipteris*.

B. Division Lycophyta (p. 567)

What is the difference between homosporous and heterosporous? What are the consequences of heterospory?

The Division Lycophyta include the club mosses and ground pines.
- Survived through the Devonian period and dominated land during the Carboniferous Period (340–280 million years ago).
- Some are temperate, low-growing plants with rhizomes and true leaves.
- Many tropical plants (e.g. some species of *Lycopodium*) are *epiphytes* (plants that use another organism as a substratum but are not parasites). The sporangia of *Lycopodium* are borne on *sporophylls* (leaves specialized for reproduction that in some are clustered at branch tips into club-shaped *strobili*, and hence the name club moss).
- Spores develop into inconspicuous gametophytes. The non-photosynthetic gametophytes are nurtured by symbiotic fungi.
- Most are *homosporous* (making only one type of spore which develops into a bisexual gametophyte).
- Genus *Selaginella* is *heterosporous*, having *megaspores* which develop into gametophytes bearing archegonia, and *microspores* which develop into gametophytes with antheridia. The gametophytes are unisexual, either male or female.

What are the differences among spore, sporophyte, sporophyll and sporangium?

C. Division Sphenophyta (p. 567)

The division Sphenophyta includes the horsetails; it survived through the Devonian and reached its zenith during the Carboniferous period.

Equisetum is the only extant genus.
- Live in damp locations and have flagellated sperm.
- Homosporous.
- Sporophyte generation is conspicuous.
- Have photosynthetic, free-living gametophytes (not dependent on the sporophyte for food).

D. Division Pterophyta (p. 568)

Appearing in the Devonian, ferns radiated into diverse species that coexisted with tree lycopods and horsetails in the great Carboniferous forests.
- Most diverse in the tropics, ferns are the most well represented seedless plants in modern floras; there are more than 12,000 existing species of ferns.

What characterizes the sporophyte of ferns?

Fern leaves are generally much larger than those of lycopods and probably evolved in a different way.
- Lycopods have *microphylls*, small leaves that probably evolved as emergences from the stem that contained a single strand of vascular tissue.

Plants and the Colonization of Land 407

☞ Ferns have *megaphylls*, leaves with a branched system of veins. Megaphylls probably evolved from webbing formed between separate branches growing close together.

Most ferns have *fronds*, compound leaves that are divided into several leaflets.
  ☞ The emerging frond is coiled into a *fiddlehead* that unfurls as it grows.
  ☞ Leaves may sprout directly from a prostrate stem (bracken and sword ferns) or from upright stems many meters tall (tropical tree ferns).

Referring to Campbell, Figure 27.14, what major differences distinguish the fern life cycle from the moss life cycle? Are ferns homosporous or heterosporous? Which is the dominant generation?

Ferns are *homosporous* and the conspicuous leafy fern plant is the sporophyte. (See Campbell, Figure 27.14)
  ☞ Specialized sporophylls bear sporangia on their undersides; many ferns have sporangia arranged in clusters called *sori* and are equipped with springlike devices that catapult spores into the air, where they can be blown by the wind far from their origin.
  ☞ The spore is the dispersal stage.
  ☞ The free-living gametophyte is small and fragile, requiring a moist habitat.
  ☞ Water is necessary for fertilization, since the flagellated sperm cells must swim from the antheridium to the archegonium, where fertilization takes place.
  ☞ The sporophyte embryo develops protected within the archegonium.

E. The Coal Forests (*p. 569–570*)

How is coal formed? During which geological period were the most extensive coal beds produced?

During the Carboniferous period, much of the land was covered in shallow seas and swamps.
  ☞ Organic rubble of the plants above accumulated as peat.
  ☞ When later covered by the sea and sediments, heat and pressure transformed the peat into coal.

IV. TERRESTRIAL ADAPTATIONS OF SEED PLANTS (*p. 570–571*)

What terrestrial adaptations contributed to the success of seed plants?

Three life cycle modifications contributed to seed plant success:

1. Gametophytes became reduced and retained in the moist reproductive tissue of the sporophyte generation (not independent).
2. Pollination evolved, so plants were no longer tied to water for fertilization.

Megaphylls

Fronds

Fiddlehead

Sori

Carboniferous

408 Plants and the Colonization of Land

3. The evolution of the seed.
   - Zygote develops into an embryo packaged with a food supply within a protective seed coat.
   - Seeds replace spores as main means of dispersal.

V. GYMNOSPERMS (p. 571–573)

What characteristic distinguishes gymnosperms from the flowering plants? What are the four divisions of gymnosperms? What characterizes each division?

Gymnosperms appear in the fossil record much earlier than flowering plants, and they:
- Lack enclosed chambers in which seeds develop.
- Are grouped into four divisions: Cycadophyta, Ginkgophyta, Gnetophyta and Coniferophyta.

A. Division Coniferophyta (p. 571–572)

Division Coniferophyta is the largest division of gymnosperms:
- Most are large trees and are evergreens.
- Include pines, firs, spruces, larches, yews, junipers, cedars, cypresses, and redwoods.
- Includes some of the tallest (redwoods and some eucalyptus); largest (giant sequoias); and oldest (bristle cone pine) living organisms.
- Most lumber and paper pulp is from conifer wood.

Describe how the needle-shaped leaves of pines and firs are adapted to dry conditions?

Needle-shaped conifer leaves are adapted to dry conditions.
- Thick cuticle covers the leaf.
- Stomata are in pits, reducing water loss.
- Despite the shape, needles are megaphylls, as are leaves of all seed plants.

The Life History of a Pine

What characterizes the life cycle of a pine? How does it differ from moss and fern life cycles?

The life cycle of pine, a representative conifer, is characterized by the following:

The multicellular sporophyte is the most conspicuous stage.
- The pine tree is a sporophyte, with its sporangia located on cones.

The multicellular gametophyte generation is reduced and develops from haploid spores that are retained within sporangia.
- The male gametophyte is the pollen grain; note that there is *no antheridium*.
- The female gametophyte consists of multicellular nutritive tissue and an *archegonium* that develops within an *ovule*.

Notes (margin): Cycadophyta, Ginkgophyta; Gnetophyta, Coniferophyta; Ovule

What is the consequence of the pine's heterosporous life cycle? How do ovulate and pollen cones differ?

Conifer life cycles are *heterosporous*; male and female gametophytes develop from different types of spores produced by separate cones.

☞ Trees of most pine species bear both pollen cones and ovulate cones, which develop on different branches.

☞ Pollen cones have *microsporangia*; cells in these sporangia undergo meiosis producing haploid *microspores*, small spores that develop into pollen grains — the male gametophytes.

☞ Ovulate cones have *megasporangia*; cells in these sporangia undergo meiosis producing large *megaspores* that develop into the female gametophyte. Each ovule initially includes a sporangium (nucellus) enclosed in protective integuments with a single opening, the *micropyle*.

In your own words, describe the life history of a pine. Which structures are part of the gametophyte generation and which are part of the sporophyte generation?

Conifer

Cones

Pollen cones

Ovulate cones

It takes nearly three years to complete the pine life cycle, which progresses through a complicated series of events to produce mature seeds.
- ☞ Windblown pollen falls onto the ovulate cone and is drawn into the ovule through the micropyle.
- ☞ The pollen grain germinates in the ovule, forming a pollen tube that begins to digest its way through the nucellus.
- ☞ A megaspore mother cell in the nucellus undergoes *meiosis* producing four haploid megaspores, one of which will survive; it divides repeatedly by *mitosis* producing the immature female gametophyte.
- ☞ Two or three archegonia, each with an egg, then develop within the multicellular gametophyte.
- ☞ More than a year after pollination, the eggs are ready to be fertilized; two sperm cells have developed and the pollen tube has grown through the nucellus to the female gametophyte.
- ☞ Fertilization occurs when one of the sperm nuclei unites with the egg nucleus. All eggs in an ovule may be fertilized, but usually only one zygote develops into an embryo.
- ☞ The pine embryo, or new sporophyte, has a rudimentary root and several embryonic leaves. It is embedded in the female gametophyte, which nourishes the embryo until it is capable of photosynthesis. The ovule has developed into a pine seed, which consists of an embryo (2n), its food source (n), and a surrounding seed coat (2n) derived from the parent tree.
- ☞ Scales of the ovulate cone separate, and the winged seeds are carried by the wind to new locations. Note, that with the seed plants, the seed has replaced the spore as the mode of dispersal.
- ☞ A seed that lands in a habitable place germinates, its embryo emerging as a pine seedling.

What is the difference between pollination and fertilization?

What are the major structures in a pine seed? Which of these structures are old sporophyte, gametophyte and new sporophyte?

    B.   **The History of Gymnosperms** (*p. 572*)

Gymnosperms descended from Devonian progymnosperms.
- ☞ Adaptive radiation during the Carboniferous and Permian periods led to today's divisions.
- ☞ During the Permian, Earth became warmer and drier; therefore, lycopods, horsetails and ferns (previously dominant) were largely replaced by conifers and their relatives, the cycads.
- ☞ This large change marks the end of the Paleozoic era and the beginning of the Mesozoic era.

VI.   **ANGIOSPERMS** (*p. 574–579*)

What are the two classes of the division Anthophyta? What are some common examples of each?

Flowering plants are the most widespread and diverse.

Plants and the Colonization of Land **411**

- There is only one division, Anthophyta, with two classes, Monocotyledones (monocots) and Dicotyledones (dicots).
- Most use insects and animals for transferring pollen, and, therefore, are less dependent on wind and have less random pollination.

What are some refinements in vascular tissue that developed during angiosperm evolution?

Vascular tissue also became more refined during angiosperm evolution.

Conifers have *tracheids*, water-conducting cells that are:
- An early type of xylem cell.
- Elongate, tapered cells that function in both mechanical support and water movement up the plant.

Most angiosperms also have *vessel elements* that are:
- Shorter, wider cells than the more primitive tracheids.
- Arranged end to end forming continuous tubes.
- Compared to tracheids, vessel elements are more specialized for conducting water, but less specialized for support.

Angiosperm xylem is reinforced by other cell types called *fibers*, which are:
- Specialized for support with a thick lignified wall.
- Evolved in conifers. (Note that conifer xylem contains both fibers and tracheids, but not vessel elements.)

    A.   The Flower (*p. 575*)

    Flower = The reproductive structure of an angiosperm which is a compressed shoot with four whorls of modified leaves.

Using a diagram, where are the following floral structures located: sepals, petals, stamens, carpels, stigma, and ovary? What are their respective functions?

Parts of the Flower:

Sepals — Sterile, enclose the bud.

Petals — Sterile, aid in attracting pollinators.

Stamen — Produces the pollen.

Carpel — Evolved from a seed-bearing leaf that became rolled into a tube.

Stigma — Part of the carpel which is a sticky structure that receives the pollen.

Ovary — Part of the carpel that protects the *ovules*, which develop into seeds after fertilization.

Anthophyta
Monocotyledones, Dicotyledones

Tracheids

Vessel elements

Fibers

Flower

Sepals

Petals

Stamen

Carpel

Stigma

Ovary, Ovules

What are four commonly recognized evolutionary trends in floral structure?

Four Evolutionary Trends in Various Angiosperm Lineages:
1. The number of floral parts have become reduced.
2. Floral parts have become fused.
3. Symmetry has changed from radial to bilateral.
4. The ovary has dropped below the petals and sepals, where the ovules are better protected.

B. The Fruit (*p. 575*)

What is a fruit? What fruit modifications help disperse seeds?

Fruit = A ripened ovary that protects dormant seeds and aids in their dispersal; some fruits (like apples) incorporate other floral parts along with the ovary.

Aggregate fruits = Several ovaries that are part of the same flower (e.g. raspberry).

Multiple fruit = One that develops from several separate flowers (e.g. pineapple).

Modifications of fruits that help disperse seeds include:
- Seeds are within fruits that act as kites or propellers to aid in wind dispersal.
- Fruits are modified as burrs that cling to animal fur.
- Fruits are edible and seeds pass through the digestive tract of herbivores unharmed, dispersing seeds miles away.

C. Life Cycle of an Angiosperm (*p. 576–577*)

In your own words, describe the generalized life cycle of an angiosperm. Which structures are haploid? How does the angiosperm life cycle differ from the pine life cycle?

Life cycles of angiosperms are heterosporous and the two types of sporangia are found in the flower:
- Microsporangia in anthers produce microspores that form male gametophytes.
- Megasporangia in ovules produce megaspores that develop into female gametophytes.

Immature male gametophytes:
- Are pollen grains, which develop within the anthers of stamens.
- Each pollen grain has two haploid nuclei that will participate in *double fertilization* characteristic of angiosperms.

Female gametophytes:
- Do not produce an archegonium.
- Are located within an ovule.
- Consist of only a few cells — an embryo sac with eight haploid nuclei in seven cells (a large central cell has two haploid nuclei).
- One of the cells is the egg.

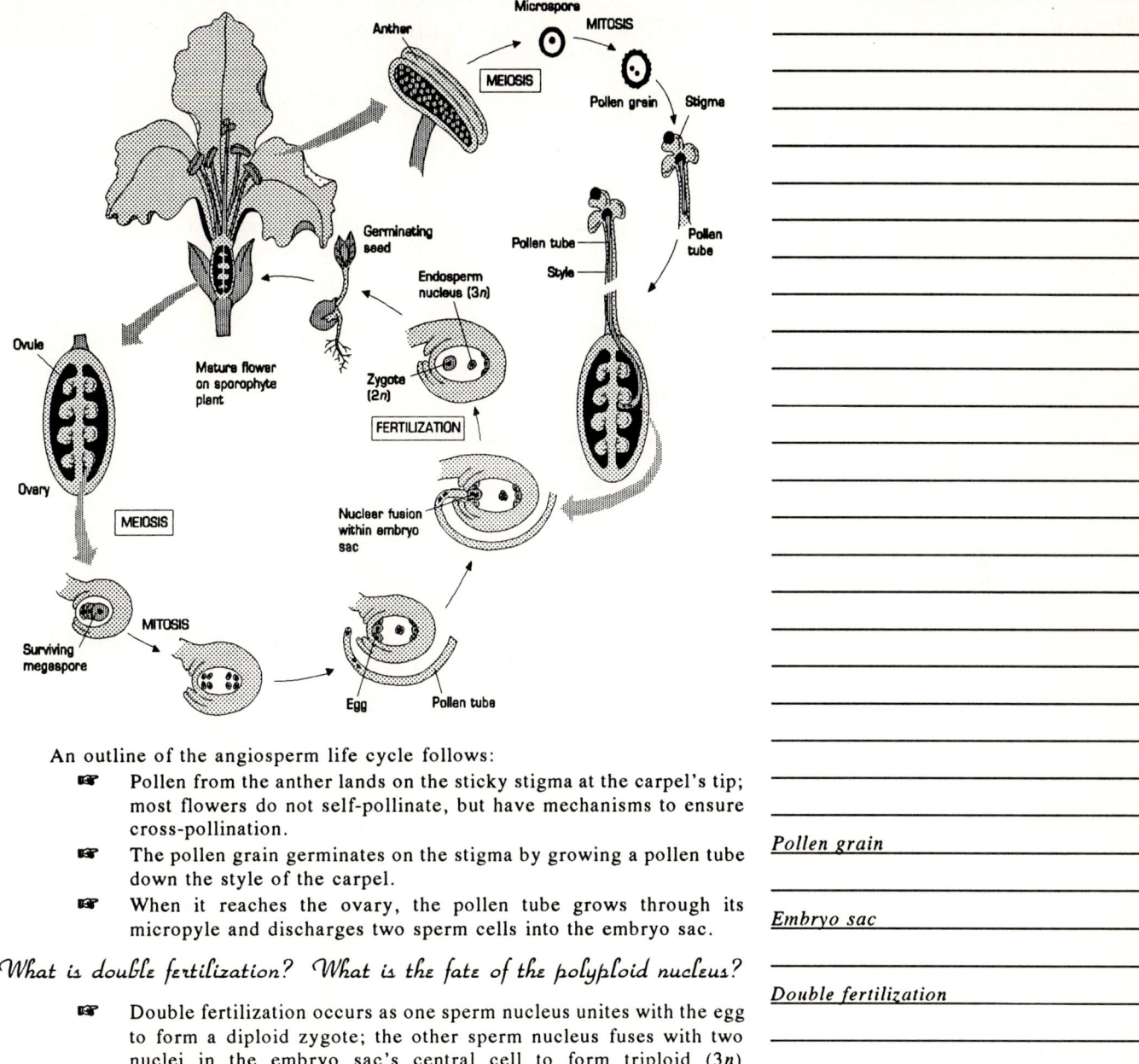

An outline of the angiosperm life cycle follows:
- ☞ Pollen from the anther lands on the sticky stigma at the carpel's tip; most flowers do not self-pollinate, but have mechanisms to ensure cross-pollination.
- ☞ The pollen grain germinates on the stigma by growing a pollen tube down the style of the carpel.
- ☞ When it reaches the ovary, the pollen tube grows through its micropyle and discharges two sperm cells into the embryo sac.

What is double fertilization? What is the fate of the polyploid nucleus?
- ☞ Double fertilization occurs as one sperm nucleus unites with the egg to form a diploid zygote; the other sperm nucleus fuses with two nuclei in the embryo sac's central cell to form triploid (3*n*) endosperm.
- ☞ After double fertilization, the ovule matures into a seed.

What are the major structures of an angiosperm seed? How does it differ from that of a pine?

The seed is a mature ovule, consisting of:
1. *Embryo.* The zygote develops into an embryo with a rudimentary root and one (in monocots) or two (in dicots) cotyledons or seed leaves.

414   Plants and the Colonization of Land

Endosperm

Seed coat

Cotyledons

2. *Endosperm.* The triploid nucleus in the embryo sac divides repeatedly forming triploid endosperm, rich in starch and other food reserves.
3. *Seed coat.* This is derived from the integuments (outer layers of the ovules).

Monocots and dicots use endosperm differently.
- ☞ Monocot seeds store most food in the endosperm;
- ☞ Dicots generally restock most of the nutrients in the developing cotyledons.

In a suitable environment the seed coat ruptures and the embryo emerges as a seedling, using the food stored in the endosperm and cotyledons.

D. **The Rise of Angiosperms** (p. 577)

Why do paleobotanists have difficulty piecing together the origin of angiosperms? What are some current theories about how flowering plants may have evolved?

Angiosperms showed a relatively sudden appearance in the fossil record with no clear transitional links to ancestors.
- ☞ Earliest fossils are early Cretaceous (approximately 120 million years ago).
- ☞ By the end of the Cretaceous (65 million years ago), angiosperms became dominant, as they are today.

There are two theories about their sudden appearance:
1. Artifact of an imperfect fossil record: angiosperms originated where fossilization was unlikely.
2. *Punctuated equilibrium*: angiosperms did evolve and radiate relatively abruptly.

Angiosperms perhaps evolved from seed ferns, an extinct group of unspecialized gymnosperms.

E. **Relationships Between Angiosperms and Animals** (p. 577–579)

How may animals have influenced the evolution of terrestrial plants, and vice versa?

Coevolution

Coevolution between angiosperms and their pollinators led to diversity of flowers. For example, many pollinators are specific for a flower.
- ☞ The pollinator gets a monopoly on a food source.
- ☞ The flower's pollen is sure to go to a flower of the same species.

As a fruit ripens, it becomes high in sugar, fragrance and colored to attract animals <u>only</u> when seeds are ready for dispersal.

Flowering plants provide nearly all our foods and many products.
- ☞ Endosperm of grain seeds is the main food source for most people.
- ☞ People affect plant evolution by breeding in agriculture and horticulture.
- ☞ Plants are also used for building material, clothing, fiber, drugs, perfume and decoration.

28   FUNGI

CHAPTER OUTLINE

I.   **CHARACTERISTICS OF FUNGI** (p. 583–585)

What characteristics distinguish fungi from organisms in the other four kingdoms?

Fungi are eukaryotes; nearly all are multicellular (although yeasts are unicellular). Their nutrition, structural organization, growth and reproduction distinguish them from organisms in the other four kingdoms.

    A.   **Nutrition and Habitats** (p. 583–584)

How do fungi acquire their nutrients?

        Fungi are heterotrophs that acquire nutrients by *absorption*.
- They secrete hydrolytic enzymes and acids to decompose complex molecules into simpler ones that can be absorbed.
- Specialized into three main types:
  1. *Saprophytes*, which absorb nutrients from dead organic material (decomposers).   _Saprophytes_
  2. *Parasitic fungi*, which absorb nutrients from body fluids of living hosts. Some are pathogenic.   _Parasitic_
  3. *Mutualistic fungi*, which absorb nutrients from a host but reciprocate to benefit the host.   _Mutualistic_

    B.   **Structure** (p. 584–585)

What is the basic body plant of a fungus? What is the difference between septate and aseptate (coenocytic) hyphae?

        The basic structural unit of a fungal vegetative body (*mycelium*) is the *hypha*.

        Hyphae = Filaments forming the fungal body.
- The network is known collectively as a *mycelium*.   _Hyphae, Mycelium_
- The filamentous mycelial structure provides an enormous surface area for absorbing nutrients.
- Cell walls are made of chitin.
- May be divided into cells by cross-walls or *septa* (septate hyphae).   _Septa_
- *Coenocytic fungi* have aseptate hyphae (a continuous cytoplasmic mass with hundreds of nuclei).   _Coenocytic fungi_
- Parasitic fungi usually have some modified hyphae called *haustoria*, nutrient absorbing hyphae that penetrate host tissue but remain outside the host cell membrane.   _Haustoria_

True fungi have no flagellated stages in their life cycle. This characteristic is partially why the Chytridiomycota and Oomycota have been moved to the Protista.

C. Growth and Reproduction (*p. 585*)

How do non-motile fungi seek new food sources? How do they disperse to new locations?

Mycelial growth is adapted to the absorptive mode of nutrition.
- Mycelia grow in length, not girth, which maximizes the surface area for absorption.
- Mycelia grow rapidly, as much as a kilometer of hyphae each day. Fast growth can occur because cytoplasmic streaming carries molecules synthesized by the mycelium to the growing hyphal tips.
- Since fungi are nonmotile, they cannot search for food or mates. Instead, they grow in hyphal length to reach new food sources and territory.

Fungal chromosomes and nuclei are relatively small, and the nuclei divide differently from most other eukaryotes.
- During mitosis, the nuclear envelope remains intact from prophase to anaphase; the spindle is inside the nuclear envelope.
- After anaphase, the nuclear envelope pinches in two, and the spindle disappears.

Fungi reproduce by releasing spores that are:
- Usually unicellular, haploid, and of various shapes and sizes.
- Produced either sexually (by meiosis) or asexually (by mitosis). In favorable conditions, fungi generally produce enormous numbers of spores asexually. For many fungi (not all), sexual reproduction occurs only as a contingency for stressful environmental conditions.
- The agent of dispersal responsible for the wide geographic distribution of fungi. Carried by wind or water, spores germinate if they land in a moist place with an appropriate substratum.

Except for transient diploid stages in sexual life cycles, fungal hyphae and spores are haploid. Some mycelia may, however, be genetically heterogeneous resulting from fusion of hyphae with different nuclei.
- The different nuclei may stay in separate parts of the same mycelium.
- Alternatively, the different nuclei may mingle and even exchange genes in a process similar to crossing over.

How does the sexual cycle in fungi differ from other eukaryotic organisms?

During the sexual cycle in fungi, *syngamy* occurs in two stages that are separated in time.

<u>Syngamy</u> = The sexual union of haploid cells from two individuals. In fungi, syngamy occurs in two stages:
1. *Plasmogamy*, the fusion of cytoplasm.
2. *Karyogamy*, the fusion of nuclei.

What are some advantages to the dikaryotic state?

After plasmogamy, haploid nuclei from each parent pair up but do not fuse, forming a *dikaryon*.

☞ Nuclear pairs in dikaryons may exist and divide synchronously for months or years.
☞ The dikaryotic condition has some advantages of diploidy; one haploid genome may compensate for harmful mutations in the other nucleus.
☞ Eventually, the haploid nuclei fuse forming a diploid cell that immediately undergoes meiosis.

II. THE DIVERSITY OF FUNGI (p. 582–590)

What characteristics are used to distinguish among the fungal divisions: Zygomycota, Ascomycota, Basidiomycota and Deuteromycota? How does asexual and sexual reproduction differ among these divisions? For what structures are the divisions named?

A. Division Zygomycota (p. 586–587)

Fungi in the division Zygomycota are characterized by the presence of dikaryotic *zygosporangia*, resistant structures formed during sexual reproduction.

☞ Zygomycetes are mostly terrestrial and live in soil or on decaying organic material.
☞ Some form *mycorrhizae*, mutualistic associations with plant roots.
☞ Zygomycete hyphae are coenocytic; septa are found only in reproductive cells.

Refer to Campbell, Figure 28.4 for the life cycle of the zygomycete, *Rhizopus stolonifer*, a common bread mold.

☞ The mycelium consists of horizontal hyphae that spread out and penetrate the food source.

Under favorable environmental conditions, *Rhizopus* reproduces asexually:

☞ *Sporangia* develop at the tips of upright hyphae.
☞ Mitosis produces hundreds of haploid spores that are dispersed through air.
☞ If they land in a moist, favorable environment, spores germinate into new mycelia.

In unfavorable conditions, *Rhizopus* begins its sexual cycle of reproduction:

☞ In unfavorable conditions, mycelia of opposite mating types (+ and −) form *gametangia* that contain several haploid nuclei walled off by a septum.
☞ *Plasmogamy* of the + and − gametangia occurs, and the haploid nuclei pair up forming a dikaryotic *zygosporangium* that is metabolically inactive and resistant to desiccation and freezing.
☞ When conditions become favorable, *karyogamy* occurs between paired nuclei; the resulting diploid nuclei immediately undergo meiosis producing genetically diverse haploid spores.

Dikaryon

Zygomycota

Rhizopus

Sporangia

418 *Fungi*

- *Ascomycota*
- *Conidia*
- *Ascospores*
- *Ascocarps*

☞ The zygosporangium germinates a sporangium that releases the genetically recombined haploid spores.
☞ If they land in a moist, favorable environment, spores germinate into new mycelia.

Even though air currents are not a very precise way to disperse spores, *Rhizopus* releases so many that enough land in hospitable places. Some zygomycetes, however, can actually aim their spores.

☞ For example, *Pilobolus*, a fungus that decomposes animal dung, bends sporangium-bearing hyphae toward light, where grass is likely to be growing.
☞ The sporangium is shot out of the hypha, dispersing spores away from the dung and onto surrounding grass. If an herbivore eats the grass and consumes the spores, the asexual life cycle is completed when the animal disperses the spores in its feces.

B. Division Ascomycota (p. 587–589)

Ascomycetes include unicellular yeasts and complex multicellular cup fungi.

☞ Hyphae are septate.
☞ In asexual reproduction, the tips of specialized hyphae form *conidia*, which are chains of haploid, asexual spores that are usually wind dispersed.
☞ In sexual reproduction, haploid mycelia of opposite mating strains fuse. One acts as "female" and produces an *ascogonium* which receives haploid nuclei from the antheridium of the "male."

The ascogonium grows hyphae with dikaryotic cells. Syngamy is delayed.

↓

In terminal cells of dikaryotic hyphae — syngamy occurs.

↓

Meiosis forms 4 haploid nuclei which undergo mitotic division to yield eight haploid nuclei.

↓

The nuclei form walls and become *ascospores* within an *ascus*, the sac of sexually produced spores. Multiple asci may form an *ascocarp*.

What is the difference between conidia and ascospores? What process produces each?

How can ascomycetes be useful to geneticists studying genetic recombination?

Ascocarps = Fruiting structures of many asci packed together.
☞ The ascospores of each ascus are lined up in a row in the order in which they formed from a single zygote, allowing geneticists to study genetic recombination.

- Unicellular yeasts appear dissimilar, but produce the equivalent of an ascus during sexual reproduction and bud during asexual reproduction in a manner similar to the formation of conidia. Thus, they are classified as ascomycetes.
- Includes important decomposers and both mutualistic and parasitic symbionts.
- Many live symbiotically with algae as lichens.

C. **Division Basidiomycota** (p. 589–590)

Basidiomycota

The division Basidiomycota, or club fungi, includes mushrooms, shelf fungi, puffballs, and stinkhorns. Basidiomycetes:
- Are named for a transient diploid stage called the *basidium*, a club-shaped spore-producing structure.

Basidium

- Are important decomposers of wood and other plant material. Saprobic basidiomycetes can decompose the complex polymer lignin, an abundant component of wood.
- Include mycorrhiza-forming mutualists and plant parasites. Many shelf fungi are tree parasites that function later as saprobes after the trees die.
- Include mushroom-forming fungi, only a few of which are strictly parasitic. About half are saprobic and the other half form mycorrhizae.
- Include the rusts and smuts, which are plant parasites.

What characterizes the basidiomycete life cycle? In your own words describe the life cycle of a mushroom-forming basidiomycete. Where do plasmogamy and karyogamy occur?

Basidiomycete life cycles are characterized by a long-lived *dikaryotic* mycelium that reproduces sexually by producing fruiting bodies called *basidiocarps*. (See Campbell, Figure 28.8)

Basidiocarps

- Basidiocarps house many *basidia* that produce sexual spores.

Refer to Campbell, Figure 28.8 for the life cycle of a mushroom-forming basidiomycete.
- Haploid basidiospores grow into short-lived haploid mycelia. Under certain environmental conditions, *plasmogamy* occurs between two haploid mycelia of opposite mating types (+ and −).
- The resulting dikaryotic mycelium grows; depending upon the species, it may form mycorrhizae with trees. Certain environmental cues stimulate the mycelium to produce mushrooms (basidiocarps). A "fairy ring" is an expanding ring of living mycelium that produces mushrooms above it; it slowly increases in diameter, about 30cm per year.
- The mushroom cap supports and protects a large surface area of gills; *karyogamy* in the terminal dikaryotic cells lining the gills produces diploid *basidia*.
- Each basidium immediately undergoes meiosis producing four haploid basidiospores. When mature, these spores drop from the cap and are dispersed by wind.

Asexual reproduction occurs less often than in ascomycetes, but also results in conidia formation.

III. SOME UNIQUE LIFESTYLES OF FUNGI (p. 590–593)

    A. Molds (p. 591)

Mold = A rapidly growing, asexually reproducing fungus.
- May be saprobes or parasites on a great variety of substrates.
- Molds only include asexual stages; they may be zygomycetes, ascomycetes, basidiomycetes or fungi with no known sexual stage.

Why are fungi classified as Deuteromycota?

Since molds are classified by their sexual stages (zygosporangium, ascogonium or basidium), molds with no known sexual stage cannot be classified as zygomycetes, ascomycetes or basidiomycetes.
- Molds with no known sexual stages are classified as *Deuteromycota* or *imperfect fungi*.
- Imperfect fungi reproduce asexually by producing conidia on specialized hyphae called *conidiphores*.
- Deuteromycetes are sources of antibiotics. For example, *Penicillium* produces penicillin.
- Other commercial uses of imperfect fungi include flavoring for cheeses such as blue cheese, Brie, Camembert and Roquefort; fermenting food products such as soybeans; and providing pharmaceuticals such as cyclosporine.
- Some deuteromycetes are predatory soil fungi that kill small animals such as soil nematodes.

Many are probably related to Ascomycetes, suggested by their asexual formation of conidia.

    B. Yeasts (p. 591–592)

What is a yeast? How do yeasts reproduce?

Yeasts are unicellular fungi that inhabit liquid or moist habitats; some can alternate between mycelium or yeast, depending on the amount of liquid in the environment.

Yeasts reproduce:
- Asexually by simple cell division or by budding off from a parent cell. Some are classified as Deuteromycota, if no sexual stages are known.
- Sexually by forming asci (Ascomycota) or basidia (Basidiomycota).

How are yeasts beneficial to humans? How are they harmful?

Though humans have used yeasts to raise bread and ferment alcoholic beverages for thousands of years, only recently have they been separated into pure culture for more controlled human use.
- *Saccharomyces cerevisiae* is the most important of all domesticated fungi. Highly active metabolically, this ascomycete is available as baker's and brewer's yeast.
- In an aerobic environment, baker's yeast respires, releasing small bubbles of carbon dioxide that leaven dough; cultured anaerobically, *Saccharomyces* ferments sugars to alcohol.
- Researchers use *Saccharomyces* to study eukaryotic molecular genetics because it is easy to culture and manipulate.

Some yeasts cause problems for humans.
- ☞ *Rhodotorula*, a pink yeast, grows on shower curtains and other moist surfaces.
- ☞ *Candida*, a normal inhabitant of moist human tissues, can become pathogenic when there is a change in pH or other environmental factor; or when someone's immune system is compromised.

C. Lichens (*p. 592–593*)

What is the basic anatomy of a lichen? How do lichens reproduce?

<u>Lichen</u> = Highly integrated symbiotic association of algal cells (usually filamentous green algae or blue-green algae) with fungal hyphae (usually ascomycetes).
- ☞ Can reproduce asexually as units from fragments or by dispersing airborne starters called *soredia*.
- ☞ Can reproduce independently, either sexually or asexually.
- ☞ The algae provides fungus with food.

What is the basis for the debate whether symbiosis in lichens is parasitic or mutualistic?

There are two theories as to the role of the fungi:
1. They benefit the algae.
   - ☞ May provide algae with water, minerals and protection.
   - ☞ Argument for mutualism is that lichen can survive in habitats that are inhospitable to either organism alone.
2. The symbiosis is more of a controlled parasitism.
   - ☞ The fungus actually kills some algae, though not as fast as the algae replenishes itself.

Lichens are named for their fungal component.
- ☞ They are informally categorized as *foliose* (leafy), *fruticose* (shrubby) or *crustose* (crusty).

Why are lichens ecologically important?

Lichens are important pioneers, breaking down rock and allowing for colonization by other plants.
- ☞ Some can tolerate severe cold.
- ☞ Photosynthesis occurs when lichen water content is 65–90%.

Lichens are sensitive to air pollution due to their mode of mineral uptake.

D. Mycorrhizae (*p. 593*)

How is the mutualistic relationship in mycorrhizae beneficial to both the fungus and the plant? Why are mycorrhizae important to natural ecosystems and agriculture?

Mycorrhizae are specific, mutualistic associations of plant roots and fungi.
- ☞ The fungi increase the absorptive surface of roots and exchanges soil minerals.

<u>Lichen</u>

<u>Soredia</u>

<u>Mycorrhizae</u>

- Mycorrhizae are seen in 90% of trees (fungus is usually a basidiomycete) and the majority of small vascular plants.
- Are necessary for optimal plant growth. (See Campbell, Chapter 33)

IV. ECOLOGICAL IMPORTANCE OF FUNGI (p. 593–595)

Why are fungi ecologically and commercially important?

A. Fungi as Decomposers (p. 593–594)

Fungi and bacteria are the principal decomposers that reduce complex polysaccharides and proteins to simple organic compounds that plants can assimilate as raw material for photosynthesis.

B. Fungi as Spoilers (p. 594)

Fungi decompose food, wood and even certain plastics.

C. Pathogenic Fungi (p. 594)

Many fungi are pathogenic (e.g. athletes foot, ringworm and yeast infections).

Plants are particularly susceptible. For example, Dutch elm disease, caused by an ascomycete, drastically changed the landscape of northeastern U.S.

Ergots = Purple structure on rye caused by an ascomycete.
- Causes gangrene, hallucinations and burning sensations (St. Anthony's fire).
- Produces lysergic acid, from which LSD is made.

Toxins from fungi may be used in weak doses for medical purposes such as treating high blood pressure.

D. Edible Fungi (p. 594–595)

Fungi are consumed as food by a variety of animals, including humans.
- In the U.S., mushroom (basidiomycete) consumption is usually restricted to one species of *Agaricus*, which is cultivated commercially on compost in the dark.
- In many other countries, however, people eat a variety of cultivated and wild mushrooms.
- Truffles prized by gourmets are underground ascocarps of mycelia that are mycorrhizal on tree roots. The fruiting bodies (ascocarps) release strong odors that attract mammals and insects — consumers that excavate the truffles and disperse their spores.
- Since it is difficult for novices to distinguish between poisonous and edible mushrooms, only qualified experts at identification should collect wild mushrooms for eating.

Ergots

V. THE EVOLUTION OF FUNGI (*p. 595*)

What is a plausible scenario for fungal phylogeny? What are two possible ancestors of Zygomycota?

The origin of fungi is unknown.
- Red algae or conjugating green algae ⟶ Zygomycota ⟶ Ascomycota ⟶ Basidiomycota is one possible evolutionary scenario.
- The oldest undisputed fossils are 450–500 million years old.
- All major groups of fungi evolved by the end of the Carboniferous period (approximately 300 million years ago).
- Plants and fungi moved from water to land together.

29 INVERTEBRATES AND THE ORIGIN OF ANIMAL DIVERSITY

CHAPTER OUTLINE

I. **DEFINING "ANIMALS"** (*p. 598–599*)

Over one million species of animals are living today; 95% of these are invertebrates.

A. Features Shared by Most Animals

What distinguishes animals from other organisms?

The features that are shared by most animals include:
- Multicellular, eukaryotic organisms.
- Heterotrophic via ingestion.

  Ingestion = Eating other organisms or decomposing organic matter (*detritus*). This mode of nutrition distinguishes animals from the plants and fungi.

- Carbohydrate reserves are stored as glycogen.
- No cell walls are present; animals do have intercellular junctions present: desmosomes, gap junctions, and tight junctions.
- Highly differentiated body cells which are organized into tissues, organs and organ systems for such specialized functions as digestion, internal transport, gas exchange, movement, coordination, excretion, and reproduction.
- Nervous tissue (impulse conduction) and muscle tissue (movement) are unique to animals.
- Reproduction is typically sexual with flagellated sperm fertilizing nonmotile eggs to form diploid zygotes. A diploid stage dominates the life cycle.
- The zygote undergoes a series of mitotic divisions known as *cleavage* which produces a *blastula* in most animals.
- *Gastrulation* occurs after the blastula has formed; during this process, the embryonic forms of adult body tissues are produced.

Development in some animals is direct to maturation while the life cycles of others include larva which undergo *metamorphosis* into a sexually mature adults.

The seas contain the greatest diversity of animal phyla, although many groups are also represented in fresh water and terrestrial habitats.
- Each of these habitats present special problems to the animals living there.

Ingestion

Glycogen

Cleavage, Blastula
Gastrulation

Metamorphosis

Phylogeny

Radiata

Bilateria

Cephalization

II. **CLUES TO ANIMAL PHYLOGENY** (*p. 599–602*)

Phylogeny = The evolutionary history of a species or group of related species. Based primarily on clues from comparative anatomy and embryology.

  A. **Major Branches of the Animal Kingdom** (*p. 599–600*)

  1. Subkingdom Parazoa
     - Contains only the phylum Porifera.
     - Unique development and simple anatomy separates this phylum from the other animals.

What is the difference between radial symmetry and bilateral symmetry?

  2. Subkingdom Eumetazoa (all other animal phyla)

  Branch Radiata:
  - Have radial symmetry.
  - These animals have an oral (top) and aboral (bottom) side, but no front, back, left, or right sides.

  Branch Bilateria:
  - Have bilateral symmetry.
  - Bilaterally symmetrical animals have dorsal (top), ventral (bottom), anterior (head), posterior (tail), left and right body surfaces.
  - These animals exhibit *cephalization* (an evolutionary trend toward concentration of sensory structures at the anterior end).

Care must be taken when assigning an animal to an evolutionary line. Some, such as the phylum Echinodermata, show a secondary radial symmetry which evolved as an adaptation to their lifestyle.

  B. **Development and Body Plan** (*p. 600–601*)

  1. **Germ Layers**

What is the difference between diploblastic and triploblastic animals?

The early embryo of all eumetazoans undergoes *gastrulation*. Concentric germ layers develop which form the various tissues and organs as development continues.
- All bilaterally symmetrical animals develop three germ layers and are termed *triploblastic*.
- Some eumetazoans develop only two germ layers and are termed *diploblastic* (e.g. phylum Cnidaria).

Three Germ Layers of an Early Embryo:

Ectoderm
- Covers the surface of the embryo.
- Forms the outer covering of the animal and the central nervous system in some phyla.

Endoderm
- Innermost germ layer which lines the *archenteron* (primitive gut).
- Forms the lining of the digestive tract and outpocketings give rise to the liver and lungs of vertebrates.

Mesoderm
- Located between the ectoderm and endoderm in triploblastic animals.
- Forms the muscles and most organs located between the digestive tract and outer covering of the animal.

2. Body Cavity

What is an acoelomate animal? A pseudocoelomate animal? A coelomate animal?

Acoelomate = No body cavity is present between the digestive tract and the outer body wall.
- This area is filled with cells.
- The phylum Platyhelminthes is acoelomate.

Pseudocoelomate = A fluid-filled body cavity separates the digestive tract and the outer body wall. (See Campbell, Figure 29.5)
- This cavity (the *pseudocoelom*) is not completely lined with mesoderm.
- The phylum Nematoda (and some others) are pseudocoelomate.

Coelomate = A fluid-filled body cavity completely lined with mesoderm (the *coelom*) separates the digestive tract from the outer body wall.
- Mesenteries connect the inner and outer mesoderm layers and suspend the internal organs in the coelom.
- The Phylum Annelida (and others) are coelomates.

Invertebrates and the Origin of Animal Diversity

Functions of a Fluid-Filled Body Cavity:
- Fluid cushions the organs to prevent injury.
- Internal organs can grow and move independently of the outer body wall.
- Serves as a hydrostatic skeleton in soft bodied coelomates such as earthworms.

C. **Protostome-Deuterostome Dichotomy** (p. 601–602)

What distinguishes a protostome from a deuterostome?

The coelomate phyla can be divided into two distinct evolutionary lines, *protostomes* and *deuterostomes*, which are distinguished by differences in development.

Developmental differences between protostomes and deuterostomes include: cleavage patterns, fate of the blastopore, and coelom formation.

1. **Cleavage Patterns** (p. 602)

What is the difference between spiral cleavage and radial cleavage? Between determinate cleavage and indeterminate cleavage?

Protostomes undergo spiral cleavage and determinate cleavage during their development.

Spiral cleavage = Cleavage in which the planes of cell division are diagonal to the vertical axis of the embryo.

Determinate cleavage = Cleavage in which the developmental fate of each embryonic cell is established very early. A cell isolated from the four-cell stage of an embryo will not develop fully.

Deuterostomes undergo radial cleavage and indeterminate cleavage during their development.

Radial cleavage = Cleavage during which the cleavage planes are either parallel or perpendicular to the vertical axis of the embryo.

Indeterminate cleavage = Cleavage in which each early embryonic cell retains the capacity to develop into a complete embryo if isolated from other cells. This type of cleavage in the human zygote results in identical twins.

2. **Fate of the Blastopore** (p. 602)

Blastopore = The first opening of the archenteron which forms during gastrulation.
- In Protostomes, the blastopore forms the mouth.
- In Deuterostomes, the blastopore forms the anus.

3. Coelom Formation (p. 602)

What is the difference between schizocoelous coelom formation and enterocoelous coelom formation?

- Schizocoelous = This refers to the manner in which the coelom develops. As the archenteron forms, the coelom begins as splits within the solid mesodermal mass.
  - ☞ Coelom formation found in Protostomes.

- Enterocoelous = Coelom development in which the mesoderm arises as lateral outpocketings of the archenteron with hollows that become the coelomic cavities.
  - ☞ Coelom formation found in Deuterostomes.

SUMMARY OF PROTOSTOME-DEUTEROSTOME DICHOTOMY	
PROTOSTOMES	DEUTEROSTOMES
Spiral plane of cleavage	Radial plane of cleavage
Determinate cleavage	Indeterminate cleavage
Blastopore forms the mouth in adults	Blastopore forms the anus in adults
Schizocoelous coelom formation	Enterocoelous coelom formation

III. PARAZOA (p. 603–604)

A. Phylum Porifera

Porifera (sponges) are the only member of the Subkingdom Parazoa due to their unique development and simple anatomy.

- ☞ Lack organs.
- ☞ No true tissues, only two layers of loosely associated unspecialized cells.
- ☞ No nerves or muscles, but individual cells detect and react to environmental changes.
- ☞ Size ranges from 1 cm to 2 m.
- ☞ All are suspension-feeders (= filter-feeders).

Possibly evolved from colonial choanoflagellates.

Approximately 9,000 species, mostly marine with only about 100 in fresh water.

How do the different parts of a sponge function in its lifestyle?

Parts of the Sponge:

Spongocoel = Central cavity of sponge.

Porocyte = Cells which form pores; possess a hollow channel through the center which extends from the outer surface (incurrent pore) to spongocoel.

Epidermis = Single layer of flattened cells which forms outer surface of the sponge.

Choanocyte = Collar cell, majority of cells which line the spongocoel; possess a flagellum which is ringed by a collar of finger-like projections. Flagellar movement moves water and food particles which are trapped on the collar and later phagocytized.

Mesohyl = The gelatinous layer located between the two layers of the sponge body wall (epidermis and choanocytes).

Amoebocyte = Wandering, pseudopod bearing cells in the mesohyl; function in food uptake from choanocytes, food digestion, nutrient distribution to other cells, formation of skeletal fibers, gamete formation.

Osculum = Larger excurrent opening of the spongocoel.

Spicule = Sharp, calcium carbonate or silica structures in the mesohyl which form the skeletal fibers of many sponges.

Spongin = Flexible, proteinaceous skeletal fibers in the mesohyl of some sponges.

Sponges are hermaphroditic but usually cross-fertilize.
- Eggs and sperm form in the mesohyl from differentiated amoebocytes or choanocytes.
- Eggs remain in the mesohyl.
- Sperm are released into excurrent flow of spongocoel and are then drawn in with incurrent flow of second sponge.
- Sperm penetrate into mesohyl and fertilization occurs.
- The zygote develops into a flagellated larva which is released into the spongocoel and escapes with the excurrent water through the osculum.
- Surviving larvae settle on the substratum and develop. In most cases the larva turns inside-out during metamorphosis, moving the flagellated cells to the inside.

Sponges possess extensive regeneration abilities for repair and asexual reproduction.

IV. RADIATA (p. 604–608)

Eumetazoa with radial symmetry.

A. Phylum Cnidaria (p. 604–606)

There are more than 10,000 species in the phylum Cnidaria, most of which are marine. The phylum contains hydras, jellyfish, sea anemones and corals.

What characteristics would you look for to determine if an animal was a member of the phylum Cnidaria?

Some characteristics of Cnidarians include: (See Campbell, Figure 29.9)
- ☞ Radial symmetry.
- ☞ Diploblastic.
- ☞ Simple, sac-like body.
- ☞ Gastrovascular cavity present: a central digestive cavity with only one opening (functions as mouth and anus).

What is the difference between a polyp and a medusa?

There are two body plan variations for Cnidarians: sessile *polyp* and motile, floating *medusa*:

Polyp = Cylindrical forms which adhere to the substratum by the aboral end of the body stalk and wave their tentacles to contact prey.

Medusa = flattened, oral opening down, disc-shaped form. Moves freely in water by passive drifting and weak bell contractions. Tentacles dangle from the oral surface which points downward.

Some species of cnidarians exist only as polyps, some only as medusae, and others are dimorphic (both polyp and medusa stages in their life cycles).

Cnidarians are carnivorous.
- ☞ Tentacles around the mouth/anus capture prey animals and push them through the mouth/anus into the gastrovascular cavity.
- ☞ Digestion begins in the gastrovascular cavity with the undigested remains being expelled through the mouth/anus.

Cnidocytes = Specialized cells of the epidermis that function in capture of prey and defense. They are arranged along the tentacles and contain nematocysts (stinging capsules). (See Campbell, Figure 29.10)

Muscles and nerves occur in their simplest forms in the Phylum Cnidaria.
- ☞ Epidermal and gastrodermal cells have bundles of microfilaments arranged into contractile fibers.
- ☞ The gastrovascular cavity, when filled with water, acts as a hydrostatic skeleton against which the contractile fibers can work to change the animal's shape.

A simple nerve net coordinates movement; no brain is present.

Hydrozoa

Scyphozoa

Anthozoa

Ctenophora

How are the three classes of Cnidaria adapted to their different habitats?

1. **Class Hydrozoa**

   Most alternate polyp and medusa forms in the life cycle although the polyp is the dominate stage. Some are colonial (e.g. *Obelia*, Campbell, Figure 29.12) while others are solitary (e.g. *Hydra*).

   Hydra is unique in that only the polyp stage is present. They usually reproduce asexually by budding; however, in unfavorable conditions they reproduce sexually. In this case a resistant zygote is formed and remains dormant until environmental conditions improve.

2. **Class Scyphozoa**

   The planktonic medusa (jellyfish) is the most prominent stage of the life cycle.

   Coastal species usually pass through a small polyp stage during the life cycle while open ocean species have eliminated the polyp entirely.

3. **Class Anthozoa**

   This class contains sea anemones and corals.

   Species occur only as polyps.

   Coral animals may be solitary or colonial and secrete external skeletons of calcium carbonate. Each polyp generation builds on the skeletal remains of earlier generations. In this way, coral reefs are formed. Coral is the rock-like external skeletons.

B. **Phylum Ctenophora** (p. 606–608)

The Phylum Ctenophora contains the comb jellies. There are approximately 100 species which are all marine.

What characteristics distinguish members of the phylum Ctenophora from other animals?

Some characteristics of Ctenophorans include:
- A resemblance to the medusa of Cnidarians.
- Transparent body, 1–10 cm in diameter.
- Possess eight rows of cilia which are used for locomotion.
- One pair of long retractable tentacles are present; cnidocytes have been found in only one species.
- A sensory organ containing calcareous particles is present. The particles settle to the low point of the organ which then acts as an orientation cue. Nerves extending from the sensory organ to the combs of cilia coordinate movement.

V. **ACOELOMATES** (p. 608–610)

   A. **Phylum Platyhelminthes** (p. 608–609)

What are the distinguishing characteristics of the phylum Platyhelminthes?

Characteristics of the Phylum Platyhelminthes include:
- Triploblastic.
- Acoelomate.
- Bilaterally symmetrical with dorsoventrally flattened bodies.
- Contain a gastrovascular cavity.

What anatomical advancements are present in the phylum Platyhelminthes but not in the phylum Cnidaria?

The phylum Platyhelminthes differs from the phylum Cnidaria in that they possess several distinct organs and organ systems and are triploblastic (develop from three-layered embryos: ectoderm, endoderm and mesoderm).

There are more than 20,000 species of Platyhelminthes which are divided into four classes:
- Class Turbellaria.
- Classes Trematoda and Monogenea.
- Class Cestoda.

What structural adaptations or life cycle specializations permit members of the four classes of Platyhelminthes to exist in their different habitats?

1. **Class Turbellaria** (p. 608–609)

    Mostly free-living, marine species; a few species are found in freshwater and moist terrestrial habitats.

    Planarians are familiar and common freshwater forms. (See Campbell, Figure 29.15)
    - Carnivorous, feed on small animals and carrion.
    - Move by using cilia on the ventral dermis to glide along a film of mucus. Muscular contractions produce undulations which allow some to swim.
    - On the head are a pair of eyespots which detect light and a pair of lateral auricles that are olfactory sensors.
    - Possess a rudimentary brain which is capable of simple learning.
    - Reproduce either asexually or sexually. Asexual is by regeneration: mid-body constriction separates the parent into two halves, each of which regenerates the missing portion. Sexual is by cross-fertilization of these hermaphroditic forms.
    - No specialized organs for gas exchange or circulation. Gas exchange is by diffusion (flattened body form places all cells close to water). Highly ramified gastrovascular cavity distributes food throughout the body.
    - Flame cell excretory apparatus present which functions primarily to maintain osmotic balance of the animal. Nitrogenous waste (ammonia) diffuses directly from cells to the water.

2. **Classes Trematoda and Monogenea** (p. 609)

    All members of these two classes are parasitic.

Platyhelminthes

Gastrovascular cavity

Turbellaria

Flame cell

Trematoda, Monogenea

Flukes are members of the class Trematoda.
- Suckers are usually present for attaching to host internal organs.
- Primary organ system is the reproductive system; majority are hermaphroditic.
- Life cycles include alternations of sexual and asexual stages with asexual development taking place in an intermediate host. Larvae produced by asexual development infect the final hosts where maturation and sexual reproduction occurs. (See Campbell, Figure 29.16)
- *Schistosoma* spp. (blood flukes) infect 200 million people worldwide.

Members of the class Monogenea are mostly external parasites of fish.
- Structures with large and small hooks are used for attaching to the host animal.
- All are hermaphroditic and reproduce sexually.

3. **Class Cestoda** (*p. 609*)

How does the anatomy of a tapeworm differ from other platyhelminths?

Adult tapeworms parasitize the digestive system of vertebrates.
- Possess a scolex (head) which may be armed with suckers and/or hooks that help maintain position by attaching to the intestinal lining.
- Posterior to the scolex is a long ribbon of units called *proglottids*. A proglottid is filled with reproductive organs.
- No digestive system is present.

The life cycle of a tapeworm includes an intermediate host.
- Mature proglottids filled with eggs are released from the posterior end of the worm and pass from the body with the feces.
- Eggs are eaten by an intermediate host and a larva develops, usually in muscle tissue. The final host becomes infected when it eats an intermediate host containing larvae.
- Humans can become infected by eating poorly cooked beef or pork containing larvae.

B. **Phylum Nemertea** (*p. 609–610*)

There are approximately 900 species, mostly marine with a few in fresh water and damp soil.

What characteristics distinguish members of the phylum Nemertea from other animals?

The Phylum Nemertea contains the proboscis worms. (See Campbell, Figure 29.17)
- Size from 1 mm to more than 30 m.
- Some active swimmers, others burrow in sand.
- Possess a long, retractable hollow tube (proboscis) which is used to probe the environment, capture prey, and as defense against predators.

Invertebrates and the Origin of Animal Diversity 433

- Excretory, sensory and nervous systems are similar to planarians.
- Acoelomate.

Some biologists believe proboscis worms evolved from flatworms. What evolutionary advancements in the digestive and circulatory systems are present in the Nemertea but not in the Platyhelminthes?

Differences Between Nemertea and Platyhelminthes:
1. Nemertea possess a simple circulatory system which consists of vessels through which blood flows. Some species have red blood cells containing a form of hemoglobin which transports oxygen. No heart is present, but body muscle contractions move the blood through vessels.
2. Nemertea possess a complete digestive system with two openings, a mouth and an anus.

VI. PSEUDOCOELOMATES (p. 610–611)

The pseudocoelomate body plan probably arose independently several times. They are thought to be more closely related to protostomes than deuterostomes, but more study of their true association is required.

A. Phylum Rotifera (p. 610)

What features of the phylum Rotifera distinguish this group from the other pseudocoelomates?

There are approximately 1,800 species of rotifers.
- Small, mainly freshwater organisms, although some are marine and others are found in damp soil.
- Size range from 0.05–2.0 mm
- Pseudocoelomate.
- Complete digestive system.
- Bear a crown of cilia that draws a vortex of water into the mouth.
- Posterior to the mouth, a jaw-like organ grinds the microscopic food organisms suspended in the water.

What is parthenogenesis? What is the advantage gained by rotifers having two ways in which they can reproduce?

Reproduction in rotifers may be by parthenogenesis or sexual in unfavorable conditions.
- In parthenogenesis, unfertilized eggs develops into females.
- Some species consist only of females.
- Other species produce two types of eggs, one that develop into females, the other into degenerate males.
- Males produce sperm that fertilize eggs which develop into resistant zygotes that survive desiccation.
- When conditions improve, the zygotes break dormancy and develop into a new female generation that reproduces by parthenogenesis until unfavorable conditions return.

Rotifera

Complete digestive system

Parthenogenesis

Rotifers have no regeneration or repair abilities.

Rotifers contain a certain and consistent number of cells as adults. Zygotes undergo a specific number of divisions and the adult contains a fixed number of cells.

B. Phylum Nematoda (*p. 610–611*)

What distinguishing characteristics separate the phylum Nematoda from other pseudocoelomates?

There are more than 80,000 species of roundworms.
- Cylindrical with tapered ends.
- Very numerous in both species and individuals.
- Found in fresh water, marine, moist soil, tissues of plants, and tissues and body fluids of animals.
- Size ranges from 1.0 mm to more than 1 m.
- Complete digestive tract.
- Tough transparent cuticle.
- Longitudinal muscles which provide whip-like movement.
- Dioecious with females larger than males.
- Sexual reproduction only, with internal fertilization.
- Female may produce 100,000 or more resistant eggs per day.

Why are nematodes important?

Free-living forms are important in decomposition and nutrient cycling.

Plant parasitic forms are important agricultural pests.

Animal parasitic forms can be hazardous to health (*Trichinella spiralis* in humans via poorly cooked pork).

Like rotifers, nematodes have a fixed number of cells as adults.

VII. PROTOSTOMES (SCHIZOCOELOMATES) (*p. 611–626*)

A. Phylum Mollusca (*p. 611–615*)

What are the distinguishing characteristics of the phylum Mollusca?

More than 50,000 species of snails, slugs, oysters, clams, octopuses, and squids.

Molluscs are mainly marine, though some inhabit fresh water and many snails and slugs are terrestrial.

Molluscs are soft-bodied, but most are protected by a hard calcium carbonate shell.
- Squids and octopuses have reduced, internalized shells or no shell.

What are the functions of the three primary body parts (foot, visceral mass, and mantle) of a mollusc?

Body consists of three primary parts: muscular *foot* for locomotion, a *visceral mass* containing most of the internal organs, and a *mantle*, a heavy fold of tissue that surrounds the visceral mass and secretes the shell.

A radula is present in many and functions as a rasping tongue to scrap food from surfaces.

Some species are monoecious while most are dioecious.
- ☞ Gonads are located in the visceral mass.

Some zoologists believe the mollusks evolved from ancestral annelids (although true segmentation is absent) while others believe that mollusks arose independently of annelids from flat-worm ancestors.

What modifications in basic body plan are seen in the four classes of molluscs (Polyplacophora, Gastropoda, Bivalvia and Cephalopoda) which permit them to inhabit their specific habitats?

1. **Class Polyplacophora** (p. 612)

   The Class Polyplacophora contains the chitons.
   - ☞ Marine species.
   - ☞ Oval with shell divided into eight dorsal plates. (See Campbell, Figure 29.21)
   - ☞ Cling to rocks along the shore at low tide using the foot as a suction cup to grip the rock.
   - ☞ Moves slowly on its broad, muscular foot when submerged.

2. **Class Gastropoda** (p. 612–613)

   The Class Gastropoda contains the snails and slugs.
   - ☞ Largest molluscan class with more than 40,000 species.
   - ☞ Mostly marine but many are freshwater or terrestrial.
   - ☞ Body protected by a shell (absent in slugs and nudibranchs) which may be conical or flattened.
   - ☞ Many species have distinct heads with eyes at the tips of tentacles.
   - ☞ Movement results from a rippling motion along the elongated foot.

Foot

Visceral mass, Mantle

Radula

Polyplacophora

Gastropoda

Bivalvia

Mantle cavity

Incurrent siphon

Excurrent siphon

Cephalopoda

- Torsion during embryonic development is a distinctive characteristic: uneven growth in the visceral mass causes the visceral mass to rotate 180 degrees, placing the anus above the head in adults.
- Most gastropods are herbivorous, using the radula to graze on plant material; several groups are predatory and possess modified radulae.
- Most aquatic gastropods exchange gases via gills; terrestrial forms have lost the gills and utilize a vascularized mantle cavity for gas exchange.

3. **Class Bivalvia** (*p. 613–614*)

The Class Bivalvia contains the clams, oysters, mussels and scallops.

Possess a shell divided into two halves. (See Campbell, Figure 29.25)
- The shell halves are hinged at the mid-dorsal line and are drawn together by two adductor muscles to protect the animal.
- Bivalves may extend the foot for motility or anchorage when the shell is open.
- The mantle cavity (between shells) contains gills which function in gas exchange and feeding.
- Most are filter feeders that trap small food particles in the mucus coating of the gills and then use cilia to move the particles to the mouth.
- Water enters the mantle cavity through an incurrent siphon, passes over the gills, and then exits through an excurrent siphon.
- No radula is present.

Bivalves lead sedentary lives. They use the foot as an anchor in sand or mud. Sessile mussels secrete threads that anchor them to rocks, docks or other hard surfaces.

Scallops can propel themselves along the sea floor by flapping their shells.

4. **Class Cephalopoda** (*p. 614–615*)

Why are the cephalopods considered the most advanced members of the phylum Mollusca?

The Class Cephalopoda contains the squids and octopuses.

Cephalopods are agile carnivores.
- Use beak-like jaws to crush prey.
- Mouth is at the base of the foot which is drawn out into several long tentacles.

A mantle covers the visceral mass, but the shell is either reduced and internal (squids) or totally absent (octopuses). The chambered nautilus is the only shelled cephalopod alive today.

Squids swim backwards in open water by drawing water into the mantle cavity, and then firing a jetstream of water through the excurrent siphon which points anteriorly. Directional changes can be made by pointing the siphon in different directions.
- ☞ Octopuses usually don't swim in open water, but move along the sea floor in search of food.

Cephalopods are the only mollusks with a *closed circulatory system* in which the blood is always contained in vessels.

Closed circulatory system

Cephalopods have well developed nervous systems with complex brains capable of learning. They also have well developed sense organs.

Most squid are less than 75 cm long but the giant squid may reach 17 m and weigh 2 tons.

Ancestors were probably shelled, carnivorous forms — the *ammonites*.
- ☞ These cephalopods were the dominant invertebrate predators in the oceans until they became extinct.

Ammonites

B. Phylum Annelida (p. 615–617)

There are more than 15,000 species of annelids. They have segmented bodies and range in size from less than 1 mm to 3 m. There are marine, freshwater and terrestrial (in damp soil) annelids.

Annelida

What characteristics distinguish the phylum Annelida from other animals? What two evolutionary advancements are exhibited by members of this phylum?

Annelids have a coelom partitioned by septa. The digestive tract, longitudinal blood vessels, and nerve cords penetrate the septa and extend the length of the animal. (See Campbell, Figure 29.27)

The complete digestive system is divided into several specialized parts: pharynx, esophagus, crop, gizzard, and intestine. Each part plays a specific role in the digestive process.

Pharynx, Esophagus
Crop, Gizzard

Annelids have a closed circulatory system.
- ☞ Hemoglobin is present in blood cells.
- ☞ Dorsal and ventral longitudinal vessels are connected by segmental pairs of vessels.
- ☞ Five pairs of hearts circle the esophagus.
- ☞ Gas exchange is across the moist skin.

An excretory system of paired metanephridia is found in each segment; each metanephridium has a nephrostome (ciliated funnel) and exits the body through an exterior pore.

Metanephridium, Nephrostome

Nervous system composed of a pair of cerebral ganglia above and anterior to the pharynx; a nerve ring around the pharynx connects these ganglia to a subpharyngeal ganglion, from which a pair of fused nerve cords run posteriorly; along these ventral nerve cords are fused segmental ganglia.

Margin notes:
- Cocoon, Clitellum
- Fragmentation
- Oligochaeta
- Casts
- Polychaeta
- Parapodia
- Setae

Hermaphroditic but cross-fertilize during sexual reproduction; eggs are deposited in a protective cocoon produced by the *clitellum*.
- ☞ Asexual reproduction occurs by fragmentation followed by regeneration.

How does a fluid-filled septate coelom help an annelid move?

Movement involves coordinating longitudinal and circular muscles in each segment with the fluid-filled coelom functioning as a hydrostatic skeleton.
- ☞ Circular muscle contraction results in each segment becoming thinner and elongating; longitudinal muscle contraction results in the segment becoming shorter and thicker.
- ☞ Waves of alternating contractions pass down the body.
- ☞ Most aquatic annelids are bottom-dwellers that burrow, although some swim in pursuit of food.

What distinguishing characteristics separate the three classes of annelids (Oligochaeta, Polychaeta, and Hirudinea)?

1. **Class Oligochaeta** (p. 616)

The Class Oligochaeta contains earthworms and a variety of aquatic species.

Earthworms ingest soil, extract nutrients in the digestive system and deposit undigested material (mixed with mucus from the digestive tract) as *casts* through the anus.
- ☞ Important to farmers as they till the soil and castings improve soil texture.

Charles Darwin estimated that 1 acre of British farmland had about 50,000 earthworms that produced 18 tons of castings per year.

2. **Class Polychaeta** (p. 616–617)

The Class Polychaeta contains mostly marine species. (See Campbell, Figure 29.28a and b)

A few drift and swim in the plankton, some crawl along the sea floor, and many live in tubes they construct by mixing sand and shell bits with mucus.
- ☞ Tube-dwellers include the fanworms that feed by trapping suspended food particles in their feathery filters which are extended from the tubes.

Each segment has a pair of *parapodia*: parapodia are highly vascularized paddle-like structures that function in gas exchange and locomotion.
- ☞ Traction for locomotion is proved by several *setae* present on each parapodium.

Invertebrates and the Origin of Animal Diversity 439

3. **Class Hirudinea** (*p. 617*)

The Class Hirudinea contains the leeches.
- ☛ Majority are freshwater but some are terrestrial in moist vegetation.
- ☛ Many are carnivorous and feed on small invertebrates while some attach temporarily to feed on blood.
- ☛ Size ranges from 1–30 cm in length.

What is hirudin and how does it function in the feeding of leeches?

Some blood-feeding forms have a pair of blade-like jaws that slit the host's skin while others secrete enzymes that digest a hole in the skin.
- ☛ An anesthetic is secreted by the leech to prevent detection of the incision by the host.
- ☛ Leeches also secrete *hirudin* which prevents blood coagulation during feeding.
- ☛ Leeches may ingest up to 10 times their weight in blood at a single meal and may not feed again for several months.

Leeches are currently used to treat bruised tissues and for stimulating circulation of blood to fingers and toes reattached after being severed in accidents.

The presence of a true coelom and segmentation are two important evolutionary advances present in the annelids.
- ☛ The coelom serves as a hydrostatic skeleton, permits development of complex organ systems, protects internal structures, and permits the internal organs to function separately from the body wall muscles.
- ☛ Segmentation provided for the specialization of different body regions.

C. **Phylum Arthropoda** (*p. 617–626*)

The Phylum Arthropoda is the largest phylum of animals with approximately 1,000,000 described species.

Arthropods are the most successful phylum based on species diversity, distribution, and numbers of individuals.

What characteristics distinguish the phylum Arthropoda from other animals?

1. **Characteristics of Arthropods** (*p. 617–618*)

The success and great diversity of arthropods is related to their segmentation, jointed appendages, and hard exoskeleton.

The segmentation in this group is much more advanced than that found in annelids. In the arthropods, different segments of the body and their associated appendages have become specialized to perform specialized functions.

Jointed appendages are modified for walking, feeding, sensory reception, copulation and defense.

Hirudinea

Hirudin

Arthropoda

Cuticle

Chitin

Hemolymph

Tracheal system, Book-lungs

What are the advantages and disadvantages of jointed appendages? Of a hardened exoskeleton?

> The arthropod body is completely covered by the *cuticle*, an exoskeleton (external skeleton) constructed of layers of protein and chitin.
> - The cuticle is thin and flexible in some locations (joints) and thick and hard in others.
> - The exoskeleton provides protection and points of attachment for muscles that move the appendages.
> - The old exoskeleton must be shed for an arthropod to grow (molting) and a new one secreted.
>
> Well developed sense organs including eyes, olfactory receptors, and tactile receptors are present.
> - Extensive cephalization has occurred in this group.

What is the difference between a hemocoel and a coelom?

> An open circulatory system containing *hemolymph* is present.
> - Hemolymph leaves the heart through short arteries and passes into the sinuses (open spaces) which surround the tissues and organs; the hemolymph reenters the heart through pores equipped with valves.
> - The body sinuses comprise the hemocoel which is not part of the coelom.
> - The coelom is reduced in adult arthropods with the hemocoel being the main body cavity.
>
> Gas exchange structures are varied: feathery gills are found in aquatic species, tracheal system or book-lungs in terrestrial forms.

2. Arthropod Phylogeny and Classification (p. 618–626)

What evidence exists to suggest an evolutionary link between the Annelida and Arthropoda?

> Arthropods are segmented protostomes which probably evolved from annelids or a segmented protostome common ancestor.
> - Parapodia may have been forerunners of appendages.
>
> Early arthropods may have resembled onychophorans which have unjointed appendages; however, many fossils of jointed-legged animals resembling segmented worms support the evolutionary link between the Annelida and Arthropoda.

How many main evolutionary lines have been identified in the phylum Arthropoda? What characteristics distinguish the members of each line from each other?

> Four main evolutionary lines have been identified in the arthropods: Trilobitomorpha (extinct trilobites), Chelicerata, Uniramia and Crustacea.

Subphylum Trilobitomorpha

Early arthropods called *trilobites* were very numerous but became extinct approximately 280 million years ago. (See Campbell, Figure 29.31)
- Trilobites had extensive segmentation, but little appendage specialization. As evolution continued, the segments tended to fuse and appendages became specialized for a variety of functions.

Subphylum Chelicerata

Other early arthropods included chelicerates such as the eurypterids (sea scorpions) which were predaceous and up to 3 m in length; their appendages were more specialized than those of trilobites, with the most anterior ones being either pincers or fangs.
- Chelicerates are named for their feeding appendages, the *chelicerae*. Only one marine chelicerate remains, the horse-shoe crab; the others are terrestrial spiders, scorpions, ticks and mites of the Class Arachnida which lack antennae and have simple eyes.

Members of the Class Arachnida possess a cephalothorax with six pairs of appendages: chelicerae, pedipalps (used in feeding), and four pairs of walking legs. (See Campbell, Figure 29.34)
- Fang-like chelicerae, equipped with poison glands, are used to attack prey.
- Chelicerae and pedipalps masticate the prey while digestive juices are added to the tissues. This softens the food and the spider sucks up the liquid.

Gas exchange is by book-lungs (stacked plates in an internal chamber).

Spiders weave silken webs to capture prey.
- The proteinaceous silk is produced by abdominal glands and spun into fibers by spinnerets. The fibers harden on contact with air.
- Web production is apparently an inherited complex behavior.
- Silk fibers are also used for escape, egg covers, and wrapped around food presented to females during courtship.

Another major line of arthropod evolution gave rise to the subphyla Uniramia and Crustacea.
- These groups have jaw-like mandibles instead of claw-like chelicerae.
- Also distinguished by presence of sensory antennae and a pair of compound eyes.

Subphylum Uniramia

Includes millipedes, centipedes and insects: all possess one pair of antennae and unbranched (uniramous) appendages. Believed to have evolved on land.

Trilobites

Chelicerae

Arachnida
Cephalothorax, Pedipalps

Uniramia, Crustacea
Mandibles

Uniramous

☞ The movement onto land by uniramians (and chelicerates) was aided by the presence of the exoskeleton first evolved in marine forms.

What arthropod structure adaptive for an aquatic lifestyle was a preadaptation for living on land?

Preadaptation = A structure that evolves and functions in one environmental context but can perform additional functions when placed in some new environment.
☞ For example, the presence of a protective cuticle, which reduces water loss, helped the chelicerates and uniramians move onto land.
☞ A firm exoskeleton also provided support without the buoyancy of water.

Arthropods have been on land about 450 million years as shown by fossils.

What characteristics distinguish each of the following classes: Diplopoda, Chilopoda, and Insecta?

The class Diplopoda includes the millipedes.
☞ Worm-like with a large number of walking legs (2 pair per segment).
☞ Eat decaying leaves and other plant matter.
☞ Probably among the earliest land animals.

The class Chilopoda include the centipedes.
☞ Carnivorous.
☞ Head has one pair of antennae and three pairs of appendages modified as mouthparts (including mandibles).
☞ Each trunk segment has one pair of walking legs.
☞ Poison claws on the most anterior trunk segment are used to paralyze prey and for defense.

The class Insecta has a greater species diversity than all other forms of life combined. They inhabit terrestrial and freshwater environments, but only a few marine forms exist.

Entomology = The study of insects.

Oldest insect fossils are from the Devonian period.
☞ Diversity increased and flight evolved during the Carboniferous and Permian periods.
☞ A second major radiation of insects occurred during the Cretaceous which paralleled radiation of flowering plants.

Flight is the key to the success of insects, enabling them to escape predators, find food and mates, and disperse more easily than non-flying forms.
☞ One or two pairs of wings emerge from the dorsal side of the thorax in most.
☞ Wings are extensions of the cuticle and not modified appendages.

- Wings may have first evolved to help absorb heat, then developed further for gliding, and finally for flight.
- When in flight, muscles warp the shape of the entire cuticle covering the thorax. As the wings flap, they change angles allowing them to produce lift on both up and down strokes.

Dragonflies were among the first to fly and have two coordinated pairs of wings. Modifications are found in later evolved groups.
- Bees and wasps hook their wings together (act as one pair).
- Butterflies have overlapping anterior and posterior wings.
- Beetles have anterior wings modified to cover and protect the posterior (flying) wings.

There are three main regions to the anatomy of an insect, as illustrated by the grasshopper: head, thorax and abdomen. (See Campbell, Figure 29.37)

Head, Thorax, Abdomen

- Head segments are fused while segmentation is apparent on the thorax and abdomen.
- The head has: one pair of antennae, one pair of compound eyes, and several pairs of appendages modified as feeding mouthparts (for chewing or lapping, piercing, sucking — depending on the insect).
- The thorax has three pairs of walking legs and the wings.

Insects have several internal organ systems.
- Complete digestive system with specialized regions.
- Open circulatory system with hemolymph.
- Excretory organs are the *Malpighian tubules* which are outpocketings of the gut.
- Gas exchange is by a *tracheal system* which opens to the outside via spiracles that can open or close to regulate air and limit water loss.
- Nervous system is composed of a pair of ventral nerve cords (with several segmental ganglia) which meet in the head where the anterior ganglia are fused into a dorsal brain close to the sense organs.

Malpighian tubules

Tracheal system

Insects show complex behavior which is apparently inherited (e.g. social behavior of bees and ants).

Insects are dioecious and usually reproduce sexually with internal fertilization.
- In most, sperm are deposited directly into the female's vagina during copulation. Some males produce spermatophores which are picked up by the female.
- Inside the female, sperm are stored in the *spermatheca*.
- Most insects produce eggs although some flies are viviparous.
- Many insects mate only once in a lifetime with stored sperm capable of fertilizing many batches of eggs.
- During development, many insects undergo metamorphosis.

Spermatophores

Spermatheca

Viviparous

What is the difference between incomplete metamorphosis and complete metamorphosis?

<u>Incomplete metamorphosis</u> = Young resemble adults but are smaller and have different body proportions.
- For example, in grasshoppers a series of molts occur with each stage looking more like an adult until full size is reached.

<u>Complete metamorphosis</u> = Larval stages (e.g. maggot, grub, caterpillar) occur which are very different in appearance from adults.
- Larva eat and grow before becoming adults.
- Adults find mates and reproduce with the females laying eggs on the appropriate food source for the larval forms.

Insects impact humans in a number of ways including competing for food, serving as disease vectors, and pollinating many crops and orchards.

<u>Subphylum Crustacea</u>

What characteristics distinguish members of the subphylum Crustacea from other arthropods?

Includes the crustaceans which have two pairs of antennae and branched (biramous) appendages. Primarily aquatic and believed to have evolved in the ocean.

There are more than 40,000 species of crustaceans in marine and fresh waters.

The crustaceans have extensive appendage specialization.
- Two pairs of antennae, three or more pairs of mouthparts including mandibles, walking legs on the thorax, appendages present on the abdomen.
- Lost appendages can be regenerated.

Gas exchange may take place across thin areas of the cuticle (small forms) or by gills (large forms).

An open circulatory system is present with hemolymph.

Nitrogenous wastes are excreted by diffusion across thin areas of the cuticle.

Salt balance is regulated by a pair of specialized antennal or maxillary glands.

Most forms are dioecious and some males (e.g. lobsters) have a specialized pair of appendages to transfer sperm to the female's reproductive pore during copulation.
- Most aquatic crustaceans have a swimming larval stage.

Invertebrates and the Origin of Animal Diversity 445

> Decapods = Relatively large crustaceans that have a *carapace* (calcium carbonate hardened exoskeleton over the cephalothorax). Examples include crayfish in fresh water; lobsters, crabs and shrimp in marine.

> Isopods = Mostly small marine crustaceans but include terrestrial sow bugs and pill bugs. Terrestrial forms live in moist soil and damp areas.

> Copepods = Numerous small marine and freshwater planktonic crustaceans.

> Barnacles = Sessile crustaceans with parts of their cuticle hardened into shells.
> ☞ Barnacles feed by directing suspended particles toward the mouth with specialized appendages.

VIII. DEUTEROSTOMES (ENTEROCOELOMATES) (p. 626–629)

The deuterostomes, while a very diverse group, share characteristics which indicate their association: radial cleavage, enterocoelous coelom formation, and the blastopore forms the anus (with one exception).

A. The Lophophorate Animals (p. 626–627)

The Lophophorate animals contain three phyla: Phoronida, Bryozoa and Brachiopoda.

What is a lophophore and how does it function?

> These three phyla are grouped together due to presence of a *lophophore*. (See Campbell, Figure 29.40)

> Lophophore = Horseshoe-shaped or circular fold of the body wall bearing ciliated tentacles that surround the mouth at the anterior end of the animal.
> ☞ Cilia direct water toward the mouth between the tentacles which trap food particles for these suspension feeders.
> ☞ The presence of a lophophore in all three groups suggests a relationship among these phyla.

> The three phyla also possess a U-shaped digestive tract (the anus lies outside of the tentacles) and have no distinct head — both adaptations for a sessile existence.

Why is it difficult to assign the lophorate animals to either the protostomes or deuterostomes?

> Lophophorates are difficult to assign as protostomes or deuterostomes.
> ☞ Their embryonic development is not exactly like either protostomes or deuterostomes. In the Phoronida, the blastopore develops into the adult mouth.
> ☞ The lophophorate animals are closely related to deuterostomes, but the two groups branched from a common ancestor before all the deuterostome developmental characteristics evolved.

Phoronida

1. Phylum Phoronida

This phylum contains about 15 species of tube-dwelling marine worms.
- Length from 1 mm to 50 cm.
- Phoronids live buried in sand in chitinous tubes with the lophophore extended from the tube when feeding.

Bryozoa

2. Phylum Bryozoa

The Phylum Bryozoa contains the moss animals. There are 5,000 species which are mostly marine and widespread.

Bryozoans are small, colonial forms. (See Campbell, Figure 29.40a)
- In most the colony is enclosed within a hard exoskeleton and the lophophores are extended through pores when feeding.
- Some are important reef builders.

Brachiopoda

3. Phylum Brachiopoda

The Phylum Brachiopoda contains the lamp shells. There are approximately 330 extant species, all marine.
- More than 30,000 fossil species of the Paleozoic and Mesozoic have been identified.

The body of a brachiopod is enclosed by dorsal and ventral shell halves. (See Campbell, Figure 29.40b)
- Attach to the substratum by a stalk.
- Open the shell slightly to allow water to flow through the lophophore.

B. Phylum Echinodermata (p. 627–629)

There are 7,000 species of Echinoderms.

What distinguishing characteristics separate the phylum Echinodermata from other animals?

Echinodermata

Echinoderms are sessile or sedentary marine forms with radial symmetry as adults. (See Campbell, Figure 29.41)
- Internal and external parts radiate from the center, often as five spokes.
- A thin skin covers a hard calcareous plate-like exoskeleton.
- Most have bumps and spines which serve various functions.

A unique feature of Echinoderms is the *water vascular system*.

Water vascular system

Water vascular system = A network of hydraulic canals which branch into extensions called tube feet that function for locomotion, feeding and gas exchange.

Dioecious with sexual reproduction and external fertilization.
- Bilaterally symmetrical larvae metamorphose into radial adults.
- Early embryonic development exhibits the characteristics of deuterostomes.

What differences are found in the water vascular systems of the six classes of echinoderms (Asteroidea, Ophiuroidea, Echinoidea, Crinoidea, Holothuroidea and Concentricycloidea)? What other characteristics distinguish among these classes?

1. **Class Asteroidea**

    This class includes the sea stars which have five or more arms extending from a central disc. (See Campbell, Figure 29.42)

    Tube feet are found on the undersurface of the arms.
    - Tube feet are extended by fluid forced into each tube foot by contraction of the *ampulla* attached to each.
    - Suction cups at the end of each tube foot attach to the substratum and muscles in the tube foot wall contract and shorten the foot.

    Coordination of extension, attaching, contraction and release allow slow movement and attachment to prey.
    - Prey are obtained by attaching tube feet to the shells of clams and oysters.
    - The muscles of the mollusks fatigue and the shell is pulled open.
    - The sea stars evert their stomachs between the shell halves and secrete digestive juices onto the soft tissues of the mollusk.

    Sea stars have a strong ability to regenerate, although a single arm cannot regenerate an entire body. Fishermen chopping up sea stars may actually increase their numbers.

2. **Class Ophiuroidea**

    The Class Ophiuroidea contains the brittle stars.

    Brittle stars differ from sea stars in that brittle stars have:
    - Smaller central discs than sea stars.
    - Longer, more flexible arms than sea stars.
    - No suckers on their tube feet.
    - Locomotion by serpentine lashing of flexible arms.
    - Varying feeding mechanisms.

3. **Class Echinoidea**

    The Class Echinoidea contains the sea urchins and sand dollars.

    Echinoideans lack arms.
    - Five rows of tube feet present for slow movement.
    - Spines are pivoted to aid in locomotion.

    Echinoideans have a complex jaw-like structure present around the mouth which is used for feeding on seaweeds and other food.

    Sea urchins are spherical in shape while sand dollars are flattened in the oral-aboral axis.

Asteroidea

Tube feet

Ampulla

Ophiuroidea

Echinoidea

Crinoidea

4. Class Crinoidea

The Class Crinoidea contains the sea lilies.

Most sea lilies are sessile, attached to substratum by stalks.
- Arms circle the mouth (which points upward) and are used in suspension feeding. Motile sea lilies use their arms for locomotion as well as for feeding.

Conservative evolution — extant forms are very similar to fossilized forms from Ordovician period.

Holothuroidea

5. Class Holothuroidea

The Class Holothuroidea contains the sea cucumbers which have little resemblance to other echinoderms.
- Lack spines.
- Hard endoskeleton is reduced.
- Elongated in the oral-aboral axis.

Species in the Holothuroidea do possess five rows of tube feet, a part of the unique water vascular system.

Concentricycloidea

6. Class Concentricycloidea

This family contains the sea daisies which are small (less than 1 cm), disc-shaped marine animals.
- Live in deep water.
- Do not possess arms.
- Tube feet are located around the disc margin.
- Possess a rudimentary digestive system or an absorptive velum on the oral surface.
- Water vascular system consists of two concentric ring canals.

Chordata

C. Phylum Chordata (*p. 629*)

Diverged from common deuterostome ancestor with echinoderms at least 500 million years ago.

Contains three subphyla: Urochordata, Cephalochordata and Vertebrata.

IX. THE ORIGIN AND DIVERSIFICATION OF ANIMALS (*p. 629–631*)

What are the primary differences between the syncytial hypothesis and colonial hypothesis of animal origin?

There are two hypothesis about animal origins from unicellular ancestors: the syncytial hypothesis and the colonial hypothesis.

A. Syncytial Hypothesis

Syncytial hypothesis — *Syncytial hypothesis* = Animals arose from a multinucleate protoctist (perhaps a ciliate) that became subdivided by membranes into many cells.
- Early animals resembled a group of extant bilaterally symmetrical flatworms which lack digestive cavities (the Acoels).
- Acoels have a ciliated epidermis and a pharynx leading to a solid coenocytic mass that takes in food by phagocytosis and digests within vacuoles.

This hypothesis presents a problem when considering the Porifera and radiate phyla.

B. Colonial Hypothesis

Colonial hypothesis = Animals arose from ancestral forms which were heterotrophic colonial flagellates.
- ☞ These ancestral organisms would have been free-swimming, hollow spheres of heterotrophic cells with an anterior-posterior orientation and distinct reproductive and somatic cells.
- ☞ Extant choanoflagellates are believed to be similar to the ancestral forms. (See Campbell, Figure 29.43)

Several evolutionary changes would have been necessary before present phyla originated. At least two ideas on how the original multicellular animals developed have been proposed.
1. Cell layers may have developed as cells proliferated into the hollow center, resulting in a protoanimal having an outer layer of flagellated locomotor cells and an inner mass of digestive and reproductive cells. These hypothetical planuloids would have been a solid mass of cells similar to the planula larva of some cnidarians. (See Campbell, Figure 29.44a)
2. Cell layers formed by invagination of the hollow colonial flagellates would produce a two-layered protoanimal similar to the gastrula stage in animal development. (See Campbell, Figure 29.44b)

C. Protoanimals May Have Resembled *Trichoplax adhaerens*

Why is it difficult to resolve what the first animals looked like? Why are there discontinuities between Ediacaran fauna and Cambrian fauna?

This extant species is the simplest animal known. It consists of a ciliated epidermis covering a solid core of unspecialized cells.

All or none of these hypotheses may approach what occurred since no transitional fossils between unicellular animals and eumetazoans have been found.
- ☞ Oldest known fauna lived during the Ediacaran period (700–590 million years ago) as indicated by extensive study of fossils from the Ediacara Hills of Australia.
- ☞ Most of the organisms were soft-bodied forms superficially resembling cnidarian medusas, colonial cnidarians and annelids.

A second radiation during the early Paleozoic era (Cambrian period — 590 million years ago) produced a more diverse fauna which included representatives of nearly all modern phyla. Evolution of shells and hard skeletons opened new adaptive zones and altered predator-prey relations.
- ☞ Cambrian diversity is speculated to have resulted from environmental changes and increased nutrient content of the sea from expanded tropical shorelines due to the break up of the continents.

Some paleobiologists believe that the Ediacaran fauna indicates the Cambrian radiation is basically a diversification of body plans in the previous Precambrian worms and jellyfish.

30   THE VERTEBRATE GENEALOGY

CHAPTER OUTLINE

I. **PHYLUM CHORDATA** (p. 635–636)

The phylum Chordata includes three subphyla: Urochordata, Cephalochordata and Vertebrata.

    A. **Chordate Characteristics** (p. 635–636)

What four characteristics distinguish chordates from other animals? What are the functions of each?

Chordates are deuterostomes with four unique characteristics which appear at some time during the animal's life. These characteristics are: the notochord; a dorsal, hollow nerve cord; pharyngeal slits; and muscular postanal tail.

       1. **Notochord**

Notochord = A longitudinal, flexible rod located between the gut and nerve cord.
- Present in all chordate embryos.
- Composed of large, fluid-filled cells encased in a stiff, fibrous tissue.
- Extends through most of the length of the animal as a simple skeleton.

In some invertebrate chordates and primitive vertebrates it persists to support the adult.

In most vertebrates, a more complex, jointed skeleton develops and the notochord is retained in adults as the gelatinous material of the discs between the vertebrae.

       2. **Dorsal, Hollow Nerve Cord**

Develops in the embryo from a plate of dorsal ectoderm that rolls into a tube located dorsal to the notochord.
- Unique to chordates.
- The brain and spinal cord (central nervous system) develops from this nerve cord.
- Other animal phyla have solid, usually ventral nerve cords.

       3. **Pharyngeal Slits**

In nearly all chordate embryos, the lumen of the digestive tube opens to the outside through several pairs of slits located on the flanks of the pharynx.

☞ These pharyngeal gill slits probably functioned for filter-feeding in early chordates but became modified for gas exchange and other functions during vertebrate evolution.

4. **Muscular Postanal Tail**

Postanal tail

A tail extending beyond the anus, found in most chordates and contains skeletal elements and muscles.
  ☞ Provides much of the propulsive force in many aquatic species.

The digestive tract in most nonchordates extends nearly the whole length of the body.

II. CHORDATES WITHOUT BACKBONES (*p. 636–637*)

What distinguishing characteristics separate the subphylum Cephalochordata from the subphylum Urochordata?

A. Subphylum Cephalochordata (*p. 636–637*)

Cephalochordata

Known as *lancelets* due to bladelike shape. (See Campbell, Figure 30.3)

Notochord, dorsal nerve cord, numerous gill slits and postanal tail are prominent and persist.

Species in the Subphylum Cephalochordata are filter feeders.
  ☞ Marine: burrow tail first into sand with only the anterior exposed.
  ☞ Water is drawn into mouth by ciliary action and food is trapped on a mucous net secreted across the pharyngeal slits.
  ☞ Water exits through the slits and trapped food passes down the digestive tube.

Cephalochordates are feeble swimmers with fishlike motions.
  ☞ Frequently move to new locations
  ☞ Muscle segments are serially arranged in chevronlike rows and coordinated contraction flexes the notochord from side to side in a sinusoidal pattern.
  ☞ Muscle segments develop from blocks of mesoderm called *somites* that are arranged along each side of the notochord in the embryo.

Somites

B. Subphylum Urochordata (*p. 637*)

Urochordata

Species in the subphylum Urochordata include the tunicates.
  ☞ Entire animal is cloaked in a tunic made of a celluloselike carbohydrate called *tunicin*.

Tunicin

Most Urochordates are sessile marine animals which adhere to rocks, docks and boats.
  ☞ Some species are planktonic, while others are colonial.

The Urochordates are filter feeders. (See Campbell, Figure 30.4)
  ☞ Seawater enters through an incurrent siphon, passes through the slits of the pharynx into a chamber called the *atrium*, and exits via an excurrent siphon.

Atrium

  ☞ Food filtered from the water by a mucous net of the pharynx is moved by cilia into the intestine.
  ☞ The anus empties into the excurrent siphon.

When disturbed, tunicates eject a jet of water through the excurrent siphon (e.g. sea squirts).

Adult tunicates bear little resemblance to other chordates.
- Lack notochord, nerve cord and tail.
- Possess only pharyngeal slits.

Larval tunicates are free swimmers and possess all four chordate characteristics.
- Larva attach by the head on a surface and undergo metamorphosis to adult form.

III. THE ORIGIN OF VERTEBRATES (p. 637–638)

Vertebrates first appear in the fossil records in Cambrian rocks. Fossilized invertebrates resembling cephalochordates are found in Burgess Shale of British Columbia about 550 million years old, about 50 millon years older than known vertebrates.

Most zoologist feel the protovertebrates possessed all four chordate characteristics and were filter-feeders.
- May have resembled lancelets but were less specialized.
- More likely that protovertebrates were derived from a urochordatelike ancestor more similar to tunicate larvae. (See Campbell, Figure 30.4c)
- Major steps in protovertebrate evolution probably occurred early in the Cambrian period.

What is paedogenesis? What impact could paedogenesis have had on the evolution of vertebrates?

Paedogenesis = Precocious attainment of sexual maturity in a larva.
- Paedogenesis is believed to have had a major impact on vertebrate evolution.

Zoologists postulate that some early urochordatelike larval forms became sexually mature and reproduced before undergoing metamorphosis.
- If reproducing larvae were successful, natural selection may have reinforced the absence of metamorphosis and a vertebrae life cycle may have evolved.
- Some extant species of urochordates exist only as free-swimming larvalike forms.
- Eventually, some of the reproducing larvalike forms may have become more active with development of segmental muscles and stronger skeletal support in their tails, producing more powerful swimmers.

Actively foraging animals dealing with navigation in current and avoidance of predators need good sense organs.
- Natural selection may have resulted in the evolution of a protovertebrate with a distinct head equipped with a brain and acute sensory organs.

IV. VERTEBRATE CHARACTERISTICS (p. 638–640)

What specialized characteristics are found in the subphylum Vertebrata and how are they beneficial to the organisms?

Vertebrates retain chordate features while adding other specializations. (See Campbell, Table 30.1)
- ☞ Cephalization and a highly specialized brain to process information.
- ☞ Most possess a column of serially arranged skeletal units, vertebrae, that enclose the nerve cord. (See Campbell, Figure 30.1)
- ☞ The brain is enclosed in the skull.
- ☞ Skull plus vertebral column make up the axial skeleton. In many vertebrates, other axial elements (e.g. ribs and breastbone) are present.

Appendicular skeletons are composed of those skeletal elements supporting the two pairs of appendages (fins or limbs).

The vertebrate endoskeleton may be made of bone, cartilage or a combination of the two.
- ☞ The endoskeleton grows with the organism. Living cells that secrete the matrix are present.

Vertebrates have a closed circulatory system, composed of a heart (two to four chambers), arteries, capillaries and veins.
- ☞ Oxygen is transported by red blood cells containing hemoglobin.
- ☞ The blood is oxygenated as it flows through the skin or highly vascularized membranes lining gills or lungs.
- ☞ Waste products are removed from the blood as it passes through the kidneys which are compact excretory structures.

Most vertebrates are dioecious.
- ☞ Reproduction is nearly always sexual with either external or internal fertilization.
- ☞ Parthenogenesis occurs in some species of most classes.

Seven extant classes comprise the subphylum Vertebrata: Agnatha, Chondrichthyes, Osteichthyes, Amphibia, Reptilia, Aves, and Mammalia. The first three of these classes are fishes, while the remaining four are the tetrapod vertebrates.

V. CLASS AGNATHA (p. 640)

What characteristics distinguish the class Agnatha from other vertebrate classes? How do these characteristics adapt these fish to their lifestyle?

This class contains the jawless fishes and extinct, heavily armored ostracoderms.

Oldest fossilized vertebrates. Some found in Cambrian but most date to the Ordovician and Silurian (400 to 500 million years ago).
- ☞ Not sufficient fossil evidence to give an explanation of agnathan radiation in these periods.
- ☞ Early agnathans were small, less than 50 cm in length.
- ☞ Jawless, most lacked paired fins and were bottom-dwellers.
- ☞ Some were active and had paired fins.
- ☞ Were probably bottom- or suspension-feeders that trapped organic debris in their gill slits.
- ☞ Ostracoderms and most other agnathans declined and disappeared during the Devonian.

Cephalization

Axial skeleton

Agnatha

Ostracoderms

Extant forms include about 60 species of lampreys and hagfishes which lack paired appendages and external armor.

Lampreys are eel-shaped and feed by clamping their round mouths onto live fish.
- ☞ Once attached, they use a rasping tongue to penetrate the skin and feed on the prey's blood.
- ☞ Sea lampreys spend their larval development in freshwater streams and migrate to the sea or lakes as they mature.
- ☞ Larva are suspension-feeders that resemble lancelets.
- ☞ Some lamprey species feed only as larvae. Once they mature and reproduce, they die within a few days.

Hagfishes superficially resemble lampreys.
- ☞ Scavengers without rasping mouthparts.
- ☞ Some species will feed on sick or dead fish while others feed on marine worms.
- ☞ Lack a larval stage and are entirely marine.

VI. CLASS PLACODERMI (p. 641)

What distinguishes the class Placodermi from agnathans? What adaptations gave them a competitive advantage over agnathans and how?

The Placodermi is a group of armored fishes which replaced agnathans during the late Silurian and early Devonian.
- ☞ Most were less than 1 m in length, but some were up to 10 m long.
- ☞ Differed from agnathans in that they possessed paired fins and hinged jaws.
- ☞ Paired fins enhanced swimming ability and hinged jaws allowed more varied feeding habits including predation.

Hinged jaws evolved as modifications of the skeletal rods which previously supported the anterior pharyngeal gill slits.
- ☞ Remaining gill slits retained function as major gas exchange sites.
- ☞ Hinged jaws of vertebrates work in an up and down direction — those in arthropods work from side to side.

Placoderms and another group of jawed fishes, the acanthodians, radiated during the Devonian period (the Age of Fishes) and many new forms evolved in fresh and salt waters.
- ☞ Placoderms and acanthodians disappeared during the Carboniferous period.
- ☞ Ancestors of the placoderms and acanthodians also gave rise to early sharks and bony fishes of the Classes Chondrichthyes and Osteichthyes.

VII. CLASS CHONDRICHTHYES (p. 641–642)

This class contains about 750 extant species of cartilaginous fishes (e.g. sharks, skates, rays).

What distinguishing characteristics are found in members of the class Chondrichthyes? How do these characteristics adapt these fish to their lifestyle?

Species in the Class Chondrichthyes have flexible skeletons composed of cartilage, well-developed jaws and paired fins. (See Campbell, Figure 30.8)

Sharks have streamlined bodies and are swift swimmers.
- The tail provides propulsion.
- The dorsal fins serve as stabilizers.
- Pectoral and pelvic fins produce lift.

Sharks have some buoyancy due to large amounts of oil stored in liver, but must swim continuously to remain in water column.
- Continual swimming also produces water flow through mouth and over gills for gas exchange.
- Some sharks are known to rest on the sea floor and in caves; water is continuously pumped over the gills while resting.

Most sharks are carnivorous although the largest (whale shark) is a filter feeder.
- Prey may be swallowed whole or pieces may be torn from large prey.
- Teeth evolved as modified scales.

Digestive tract is proportionately shorter than in other vertebrates.
- A spiral valve, which increases surface area and slows food movement, is present in the intestine.

Sharks possess sharp vision (cannot distinguish color) and olfactory senses.
- Electric sensory regions that detect muscle contractions of prey are located on the head.
- A *lateral line system* is present along the flanks. It is composed of rows of microscopic organs sensitive to water pressure changes and detects vibrations.
- A pair of auditory organs also detect sound waves passing through the water.

Reproduce sexually with internal fertilization.
- A pair of claspers on the pelvic fins of males transfers sperm into the female reproductive tract.
- Some species are oviparous, some are ovoviviparous, and a few are viviparous.
- A cloaca (common chamber for reproductive, digestive and excretory systems) is present.

Rays are adapted to a bottom-dwelling life style.
- Dorsoventrally flattened bodies.
- Jaws are used to crush mollusks and crustaceans.
- Enlarged pectoral fins provide propulsion for swimming.
- Tail in many species is whiplike and, in some, bears venomous barbs.

Chondrichthyes

Cartilage

Spiral valve

Lateral line system

Cloaca

Osteichthyes

Operculum

VIII. CLASS OSTEICHTHYES (p. 643–644)

What distinguishing characteristics separate members of the class Osteichthyes from other fishes? What characteristics permitted this class of fishes to become so diversified?

This class contains the bony fishes.

Most numerous of vertebrate classes with more than 30,000 extant species.
- Abundant in marine and fresh waters.
- Range from 1 cm to 6 m in length.
- Skeleton is bony, reinforced with a matrix of calcium phosphate.
- Skin is covered with flattened bony scales.
- Skin glands produce mucus that reduces drag when swimming.
- Lateral line system is present as a row of tiny pits in the skin on both sides of the body.

Gas exchange occurs by drawing water over the four or five pairs of gills located in chambers covered by an *operculum*.
- Water is drawn into the mouth, through the pharynx and out between the gills by movement of the operculum and contraction of muscles within the gill chambers.
- Allows bony fishes to breath while stationary.

A swim bladder located dorsal to the digestive system provides buoyancy.
- Transfer of gases between blood and swim bladder varies bladder inflation and adjusts the density of the fish.

Bony fishes are very maneuverable swimmers with flexible fins providing better steering and propulsion than stiff fins in sharks.
- Fastest bony fish can swim to 80 km per hour in short bursts.
- Fusiform body shape common to all fast fishes and aquatic mammals. Body shape reduces drag produced by dense water (convergent evolution).

Most are oviparous and utilize external fertilization.
- Some are ovoviviparous or viviparous and utilize internal fertilization.
- Some display complex mating behavior.

Both cartilaginous and bony fishes diversified during the Devonian and Carboniferous periods.
- Sharks arose in the sea, bony fishes probably originated in fresh water.
- The swim bladder was modified from lungs of ancestral fishes which supplemented the gills for gas exchange in stagnant waters. (See Campbell, Figure 30.10)

By the end of the Devonian, three distinct subclasses of bony fishes had evolved: Antinopterygii (ray-finned fishes), Crossopterygii (lobe-finned fishes) and Dipnoi (lungfishes).

What characteristics distinguish the three subclasses (Actinopterygii, Crossopterygii, Dipnoi) of bony fishes?

A. Subclass Actinopterygii

Bony fish whose fins are supported mainly by flexible rays.

☞ Most familiar fishes.

Spread from fresh water to the seas and many returned to fresh water during evolution of the taxon.

Some bony fish (e.g. salmon and sea-run trout) reproduce in fresh water and mature in the sea.

B. Subclass Crossopterygii

Bony fish whose fins were fleshy, muscular and supported by extensions of the bony skeleton.

☞ Many were large, bottom-dwelling forms that used their paired fins to walk on the substratum.

Most forms had lungs and remained in fresh water.

☞ Lungs were used to aid their gills with gas exchange.
☞ Only one extant species, the coelocanth: it is marine and lungless.

Lobe-fin fishes of Devonian were numerous and important in vertebrate genealogy because they gave rise to amphibians. (See Campbell, Figure 30.12)

C. Subclass Dipnoi

Bony fish most of which have fleshy, muscular fins supported by extensions of the skeleton.

Remained in fresh water and used their lungs to aid the gills.

Three genera are extant in the Southern Hemisphere and live in stagnant ponds and swamps.

☞ Surface to gulp air into lungs connected to the pharynx.
☞ When ponds dry, lungfishes burrow in the mud and aestivate.

IX. CLASS AMPHIBIA (p. 645–648)

Amphibians were the first vertebrates to move onto land.

There are about 4,000 extant species of frogs, salamanders and caecilians.

A. Early Amphibians (p. 645–646)

What evidence supports the hypothesis that amphibians evolved from crossopterygians?

Early amphibians evolved from lobe-finned crossopterygians that adapted to environmental variations (drought and flooding) of the Devonian. (See Campbell, Figure 30.13)

- Skeletal structure of lobe-fins suggest that they could assist in movement on land.
- Fossil lobe-fins, such as *Eusthenopteron*, exhibited many anatomical similarities to early amphibians.
- Oldest amphibian fossils are from the late Devonian (350 million years ago).

Early amphibians were predators that ate insects and other invertebrates that had moved previously onto land.

Amphibians were the only vertebrates on land in the late Devonian and early Carboniferous.

Radiation of forms occurred during the early Carboniferous period.
- Some forms reached 4 m in length and some resembled reptiles.
- Amphibians began to decline during the late Carboniferous. At the beginning of the Triassic period (230 million years ago), most of the survivors resembled modern species.

B. Modern Amphibians (p. 646–648)

There are three extant orders of amphibians: Urodela (salamanders), Anura (frogs and toads) and Apoda (caecilians).

What characteristics can be used to distinguish among the three orders (Urodela, Anura, Apoda) of living amphibians?

The Order Urodela (salamanders) contains about 400 species.
- Some aquatic, some terrestrial.
- Terrestrial forms walk with a side-to-side bending of the body. Aquatic forms swim sinusoidally or walk along the bottom of streams or ponds.

The Order Anura (frogs and toads) contains about 3,500 species.
- No tail.
- Enlarged hindlegs provide better movement (hopping) than in urodeles.
- Capture prey by flicking the sticky tongue which is attached anteriorly.
- Predator avoidance aided by camouflage color patterns and distasteful or poisonous mucus secreted by skin glands.

The Order Apoda (caecilians) contains about 150 species.
- Legless and almost blind.
- Most burrow in moist tropical soils, some inhabit freshwater ponds and streams.

What changes occur in the life cycle of a frog that result in the change from an aquatic larval form to a semi-aquatic adult?

<u>The Frog Life Cycle</u>:

The frog exhibits change from aquatic larva to a terrestrial or semiaquatic predator. (See Campbell, Figure 30.15)

- The tadpole (larval stage) is usually aquatic, herbivorous and possesses internal gills, a lateral line system and a long, finned tail.
- The tadpole lacks legs and swims by undulating the tail.
- During metamorphosis, legs develop and the gills and lateral line system disappear.
- A young frog is tetrapod, has air-breathing lungs, a pair of external eardrums and a digestive system that can digest animal protein.

Many amphibians, including some frogs, do not have a tadpole stage.
- Some species in each order are strictly aquatic while others are strictly terrestrial.
- Urodeles and apodans have larva that more closely resemble adults and both larva and adults are carnivorous.
- Paedogenesis is common in some groups of urodeles.

What adaptations are found in semi-aquatic frogs which permit them to survive in water and on land?

Most amphibians maintain close ties with water and are most abundant in damp habitats.
- Terrestrial forms in arid habitats spend much of their time in burrows where humidity is high.

Gas exchange is primarily cutaneous and terrestrial forms must keep the skin moist.
- Lungs can aid in gas exchange although most are small and inefficient. Some forms lack lungs.
- Many species also exchange gases across moist surfaces of the mouth.

Amphibians are dioecious.
- Reproduce sexually usually with external fertilization in water (e.g. ponds, streams, temporary pools).
- In frogs, the male grasps the female and sperm are released as the female sheds her eggs. (See Campbell, Figure 30.15a)
- Eggs are unshelled and produced in large numbers by most species.

Some species exhibit parental behavior and produce small numbers of eggs.
- Males or females (species dependent) incubate eggs on their back, in the mouth or in the stomach.
- Some tropical species lay eggs in a moist foamy nest that prevents drying.
- Some species are ovoviviparous and a few are viviparous.

Exhibit complex and diverse social behavior especially during breeding season (e.g. vocalization by male anurans, migrations, navigation or chemical signaling).

X. CLASS REPTILIA (p. 648–651)

Reptilia is a diverse class with about 7,000 extant species and a wide array of extinct forms.
- Reptiles possess several adaptations to live on land.

What are the distinguishing characteristics of members of the class Reptilia? How do these characteristics serve as special adaptations to the terrestrial environment?

A. Reptilian Characteristics (p. 648)

Scales of reptiles contain the protein keratin which helps prevent dehydration.

Gas exchange via lungs. Many turtles also use moist cloacal surfaces.

Dioecious with sexual reproduction and internal fertilization.
- Most are oviparous and produce an *amniote egg* which contains the amniotic fluid and a shell which prevent desiccation and permitted movement into terrestrial life style.

Ectothermic = Use behavioral adaptations to absorb solar energy to regulate their body temperature.
- Due to ectothermy, reptiles can survive on a much lower caloric intake than mammals of comparable size.

Reptiles were abundant and diverse in the Mesozoic era.

B. The Age of Reptiles (p. 648–650)

1. **Origin and Early Evolutionary Radiation of Reptiles** (p. 648–649)

The oldest reptilian fossils are found in upper Carboniferous rock (300 million years old).
- Ancestors were probably Devonian amphibians.

Cotylosaurs, stem reptiles, were diverse by the end of the Carboniferous.
- Most were small, lizardlike insectivores and were ancestral stock for the various reptilian orders which arose in two waves of adaptive radiation. (See Campbell, Figure 30.16)

The first reptilian radiation was in the Permian period and gave rise to the *synapsids* and other groups.

Synapsids were terrestrial predators and gave rise to the *therapsid* lineage which were mammal-like reptiles.
- Therapsids were large, dog-sized predators from which mammals are believed to have evolved.

Some stem reptiles returned to the water and gave rise to the plesiosaurs and ichthyosaurs which became extinct.

Thecodonts also descended from the stem reptiles during the Permian and were ancestors to the dinosaurs, crocodilians and birds.

Both synapsids and thecodonts survived the Permian extinctions and were the dominant terrestrial vertebrates until well into the Triassic period.

2. **Dinosaurs and Their Relatives** (*p. 649–650*)

The second reptilian radiation began in the Triassic (200 million years ago) and several lineages evolved from the thecodonts.
- ☞ Two groups are important: the dinosaurs and pterosaurs (flying reptiles).

Dinosaurs varied in body shape, size and habitat.
- ☞ Some fossilized forms probably weighed nearly 100 tons.
- ☞ Evidence indicates that dinosaurs were agile, fast moving and *endothermic*.

What evidence indicates that dinosaurs were endothermic? How could dinosaurs have survived if they were not endothermic?

Endothermy = The ability to keep the body warm through metabolism.

Pterosaurs had wings formed from skin stretched from the body wall, along the forelimb to the tip of an elongate finger and supported by stiff fibers.
- ☞ Pterosaurs were active, endothermic flyers, not just gliders.

Skeptics of these descriptions feel that the Mesozoic climate was warm and consistent, and that basking may have been sufficient for maintaining body temperature.
- ☞ Low surface-to-volume ratios of large forms reduced fluctuations of body temperature vs air temperature; thus, dinosaurs may not have been endothermic.

3. **The Cretaceous Crisis** (*p. 650*)

What changes could have resulted in the extinction of dinosaurs during the Cretaceous Period?

In the Cretaceous (last period of the Mesozoic), the climate became cooler and more variable.

Mass extinctions occurred.
- ☞ Twenty-five percent of the families of marine invertebrates disappeared.
- ☞ Nearly all dinosaurs disappeared by the end of the Mesozoic (65 million years ago) over a period of five to ten million years.

Some scientists speculate that a change from gymnosperm dominated flora to angiosperm dominated flora during the Cretaceous contributed to the dinosaur decline since they may have been fixed on gymnosperms for food.

The few reptilian groups to survive gave rise to extant forms.

Thecodonts

Endothermy

C. Reptiles of Today (p. 650–651)

The largest and most diverse extant orders are the: Chelonia (turtles), Squamata (lizards and snakes), and Crocodilia (alligators and crocodiles). (See Campbell, Figure 30.17)

What characteristics can be used to distinguish between the three orders (Chelonia, Squamata, Crocodilia) of living reptiles?

1. Order Chelonia

Evolved from stem reptiles during the Mesozoic.
- Show little change.
- Protected from predators by a hard shell.
- All turtles, even aquatic species, lay their eggs on land.

2. Order Squamata

Lizards are the most numerous and diverse reptiles.
- Most are small.
- Many nest in crevices and decrease activity during cold periods.

Snakes descended from burrowing lizards.
- Limbless and most live above ground.
- Vestigial pelvic and limb bones present in primitive snakes (boas) are evidence of a limbed ancestor.

Snakes are carnivorous and have a number of adaptations for hunting prey.
- Chemical sensors.
- Sensitive to ground vibrations (although lacking eardrums).
- Sensitive heat-detecting organs present in pit vipers.
- Flicking tongue helps transmit odors toward olfactory organs on roof of mouth.
- Poisonous snakes inject a toxin through a pair of sharp, hollow teeth and loosely articulated jaws allow them to swallow large prey.

3. Order Crocodilia

Crocodilians are among the largest living reptiles. (See Campbell, Figure 30.17d)

Spend most of their time in the water, breathing air through upturned nostrils.

Confined to warm regions of Africa, China, Indonesia, India, Australia, South America and the southeastern United States.

Evolved from thecondonts and are the living reptiles most closely related to dinosaurs.

XI. CLASS AVES (p. 651–653)

The Class Aves contains the birds.

Evolved from reptiles during the Mesozoic era.
- ☞ Possess distinct reptilian characteristics such as the amniote egg and scales on the legs.

A. Characteristics of Birds (p. 651–652)

What are the distinguishing characteristics of the class Aves? What special adaptations are found which permit birds to move by flying?

Each part of the anatomy is modified in some way that enhances flight.
- ☞ The bones are honeycombed which provides strength while weighing little (e.g. skeleton of a frigate bird has a wingspan of more than 2 m but weighs only 4 oz).
- ☞ Some organ systems are reduced (only one ovary in females).
- ☞ Birds have no teeth (reduces weight) and food is ground in the gizzard.

Beak is made of keratin and evolution has produced many shapes in relation to the bird's diet.

Flying requires much energy production from active metabolism.
- ☞ Endothermic with insulation provided by feathers and a fat layer.
- ☞ Efficient circulatory system with a four-chambered heart that segregates oxygenated blood from unoxygenated blood.
- ☞ Efficient lungs with tubes connecting to elastic air sacs that help dissipate heat and trim density.

Acute vision and well-developed visual and coordinating areas of the brain aid in flying.

Complex behavior especially during breeding season when elaborate courtship rituals are performed.

Dioecious with sexual reproduction and internal fertilization.
- ☞ Sperm are transferred from the cloaca of the male to the cloaca of the female (males lack a penis) during copulation.
- ☞ Eggs are laid and must be kept warm through brooding by the female, male or both depending on the species.

Wings are airfoils, formed by the shape and arrangement of the feathers, that illustrate the same aerodynamic principles as airplane wings.
- ☞ Power is supplied to the wings by contraction of the large pectoral (breast) muscles which are anchored to a keel on the sternum (breastbone).
- ☞ Some birds have wings adapted for soaring (hawks) while others must beat their wings continuously to stay aloft (hummingbirds).

Feathers are made of keratin and are extremely light and strong. (See Campbell, Figure 30.20)
- ☞ Feathers evolved from the scales of reptiles and may have first functioned as insulation.
- ☞ Feathers also function to control air movements around the wing.

464 The Vertebrate Genealogy

B. **The Origin of Birds** (*p. 653*)

What evidence supports the hypothesis that birds evolved from reptiles?

Shared a common ancestor with *Archaeopteryx lithographica*.
- Fossils of *Archaeopteryx* recovered from limestone date to 150 million years.
- *Archaeopteryx* had clawed forelimbs, teeth, a long tail containing vertebrae and feathers.
- Not considered ancestor to modern birds, but a side branch of the avian lineage.
- Skeleton indicates a weak flyer; may have been a tree-dwelling glider.

A fossil discovered in Spain from rocks dating to 125 million years has a tail similar to modern birds and forelimb bones that would have allowed stronger, more active flight.

The discovery of a fossil, *Protoavis*, in Texas during 1986 has added additional controversy to the subject of bird origins.
- Shares several characteristics with dinosaurs and birds.
- Found in rock dated to 225 million years ago; 75 million years before *Archaeopteryx*.

Radical alteration of body form was necessary for evolution of flight, but flight provides many benefits.
- Allows aerial reconnaissance that enhances hunting and scavenging.
- Birds can exploit flying insects as an abundant, highly nutritious food resource.
- Escape mechanism from land-bound predators.
- Allows migration to utilize different food resources and seasonal breeding areas.

C. **Modern Birds** (*p. 653*)

There are about 8,600 extant species in 28 orders.

What structural differences are found between flightless birds (ratites) and flying birds (carinates)?

Most fly but several are flightless (ostrich, kiwi, and emu).
- Flightless birds are called *ratites* because the breastbone lacks a keel and large breast muscles used for flying are absent.
- Flying birds are referred to as *carinate* due to the presence of a sternal keel (carina) that supports the large breast muscles used in flying.
- Carinate birds exhibit a variety of feather colors, beak and foot shape, behavior and flight ability.

Almost 60% of extant species belong to one order of carinate birds (the *Passeriformes*, or *perching birds*) which includes the jays, swallows, sparrows, warblers and many others.
- Penguins are also carinate birds which do not fly but use powerful breast muscles in swimming.

The Vertebrate Genealogy 465

XII. CLASS MAMMALIA (p. 653–657)

There are about 4,500 species of extant mammals.

A. Mammalian Characteristics (p. 653–655)

What characteristics distinguish mammals from other vertebrates?

Species in the Class Mammalia have the following characteristics:
- Possess hair which is composed of keratin but is not believed to have evolved from reptilian scales. The hair provides insulation.
- Endothermic.
- Efficient respiratory system which utilizes the diaphragm for ventilation.
- Four-chambered heart which segregates oxygenated from unoxygenated blood.
- Mammary glands present which produce milk to nourish the young.
- Teeth are differentiated.

Mammals are dioecious with sexual reproduction and internal fertilization.
- Most are viviparous with the developing embryo receiving nutrients from the female across the *placenta*.
- A few are oviparous.

Mammals have large brains and are capable of learning.
- Parental care of long duration helps young learn from the parents.

B. Evolution of Mammals (p. 655)

What conditions contributed to the adaptive radiation of mammals which occurred during the Cenozoic?

Mammals evolved from therapsid ancestors during the Triassic period (220 million years ago).

Early mammals coexisted with dinosaurs throughout the Mesozoic.
- Mesozoic mammals were small, probably insectivorous and nocturnal.

With extinction of dinosaurs, new adaptive zones were available and the mammals underwent a massive radiation.
- Mammals continued to diversify during the Cenozoic.

There are three major groups of extant mammals: monotremes, marsupials and placental mammals.

C. Monotremes (p. 656)

Why are the monotremes considered the most primitive mammals?

The Monotremes include the platypuses and echidnas.

Monotremes are oviparous with a reptilianlike egg.
- Large amounts of yolk in the egg nourish developing embryos.

Mammalia

Placenta

Monotremes

466 *The Vertebrate Genealogy*

Monotremes possess hair and produce milk from specialized glands on the belly of females.
- After hatching, young suck milk from the fur of the mother who lacks nipples.

Mixture of ancestral reptilian and derived mammalian traits suggests that monotremes descended from an early branch of the mammalian lineage.
- Extant monotremes are found in Australia and New Guinea.

D. Marsupials (*p. 656*)

Marsupials

Marsupium

The Marsupials include opossums, kangaroos, koalas, and other mammals that complete their development in a *marsupium* (maternal pouch).

How does reproduction in marsupials differ from that of other mammals?

Marsupial eggs contain a moderate amount of yolk that nourishes the embryo during early development in the mother's reproductive tract.
- Young are born in an early stage of development and are small (about the size of a honeybee in kangaroos).
- The hindlegs are simple buds, but the forelimbs are strong enough for the young to climb from the female reproductive tract exit to the marsupium.
- In the marsupium, the young attaches to a teat and completes its development while nursing.

Convergent evolution in Australian marsupials has produced a diversity of forms which resemble placental counterparts in all ecological roles.
- Opossums are the only extant marsupials outside of the Australian region.
- South America had an extensive marsupial fauna during the Tertiary period as seen in the fossil record.

Why are marsupials so prevalent in Australia and rare in other parts of the world?

Plate tectonics and continental drift provide a mechanism which explains the distribution of fossil and modern marsupials.
- Fossil evidence indicates that marsupials probably originated in what is now North America and spread southward while the land masses were joined.
- The breakup of Pangaea produced two island continents: South America and Australia.
- With isolation, their marsupial faunas diversified away from the placental mammals that began adaptive radiation on the northern continents.
- Australia has remained isolated from other continents for about 65 million years, thus isolating its developing fauna.
- When North and South America joined at the isthmus of Panama, extensive migrations took place over the land bridge in both directions.
- The most important migrations occurred about 12 million years ago and again about 3 million years ago.

E. **Placental Mammals** (*p. 656-657*)

What advantage is found in the way placental mammals reproduce in comparison to other mammals? What other adaptations could have resulted in the diversity of living mammals?

Placental mammals have embryonic development completed within the uterus where the embryo is joined to the mother by the placenta. (See Campbell, Figure 30.24)

Adaptive radiation during the late Cretaceous and early Tertiary periods (about 70-45 million years ago) produced the orders of extant placental mammals. (See Campbell, Table 30.2)
- Fossil evidence indicates that placentals and marsupials diverged from a common ancestor about 80-100 million years ago; thus, they are more closely related than either is to the monotremes.

Most mammalogists favor a genealogy that recognizes at least four main evolutionary lines of placental mammals.

What comparisons and contrasts can be found among the four main evolutionary lines of placental mammals?

1. One lineage consists of the Orders Chiroptera (bats) and Insectivora (shrews) which resemble early mammals.
   - The modified forelimbs which serve a wings in bats probably evolved from insectivores that fed on flying insects.
   - Some bats feed on fruits while others bite mammals and lap the blood.
   - Most bats are nocturnal.

2. A second lineage consists of medium-sized herbivores.
   - Underwent a massive adaptive radiation during the Tertiary which led to such modern orders as the Lagomorpha (rabbits), Perissodactyla (odd-toed ungulates), Artiodactyla (even-toed ungulates), Sirenia (sea cows), Proboscidea (elephants) and Cetacea (whales, porpoises).
   - The largest whales are filter-feeders that trap large quantities of planktonic crustaceans in their *baleens*. Porpoises and some whales feed on fish and squid.
   - Cetaceans include the largest mammals. Cetaceans are very intelligent and social, using sound in communication and navigation.

3. The third evolutionary lineage produced the Order Carnivora which probably first appeared during the Cenozoic.
   - Includes the seals and their relatives which evolved from middle Cenozoic carnivores that became adapted for swimming.

4. The fourth lineage had the greatest adaptive radiation and produced the primate-rodent complex.
   - Includes the Orders Rodentia (rats, squirrels, beavers) and Primates (monkeys, apes, humans).

Placental mammals

Placenta

Adaptive radiation

Baleens

468 The Vertebrate Genealogy

XIII. THE HUMAN ANCESTRY (p. 657–666)

A. Evolutionary Trends in Primates (p. 657)

The first primates were small arboreal mammals. Dental structure suggests they descended from insectivores in the late Cretaceous.

Purgatorius unio, found in Montana, is considered to be the oldest primate.

What primate characteristics appear to be adaptations for living in trees?

Primates have been present for 65 million years (end of Mesozoic era) and are defined by characteristics shaped by natural selection for living in trees. These characteristics include:
- Limber shoulder joints which make it possible to *brachiate* (swing from one hold to the next).
- Dexterous hands for hanging on branches and manipulating food.
- Sensitive fingers with nails, not claws.
- Eyes are close together on the front of the face giving overlapping fields of vision for enhanced depth perception (necessary for brachiating).
- Excellent eye-hand coordination.
- Parental care with usually single births and long nurturing of offspring.

B. Modern Primates (p. 657–660)

Modern primates are divided into two suborders: prosimii (premonkeys) and Anthropoidea (monkeys, apes, humans).

Prosimians (lemurs, lorises, pottos, tarsiers) probably resemble early arboreal primates.

How can Old World monkeys be distinguished from New World monkeys? What occurence could ahve resulted in this divergence?

First anthropoids in the fossil record are monkeylike primates that probably evolved from a prosimian stock about 40 million years ago in Africa or Asia (South America and Africa had already separated).

New World monkeys and Old World monkeys have evolved along separate pathways for many millions of years. (See Campbell, Figure 30.27)
- Ancestors of New World monkeys may have reached South America by rafting from Africa or migration southward from North America. All New World monkeys are arboreal.
- Old World monkeys include arboreal and ground-dwelling forms.
- Most monkeys of both groups are diurnal and usually live in social bands.

There are four genera of apes: *Hylobates* (gibbons), *Pongo* (orangutans), *Gorilla* (gorillas) and *Pan* (chimpanzees). (See Campbell, Figure 30.28)
- Confined to the tropical regions of the Old World.
- Larger than monkeys (except the gibbons) with relatively long legs, short arms and no tails.
- Only gibbons and orangutans are primarily arboreal although all are capable of brachiation.

Brachiate

Prosimians

Social organization varies with the gorillas and chimpanzees being highly social.
- ☞ Apes have larger brains than monkeys and thus exhibit more adaptable behavior.

C. The Emergence of Humankind (p. 660–666)

<u>Paleoanthropology</u> = The study of human origins and evolution.

Fossil discoveries have removed many misconceptions about human evolution.

What is the significance of the three most prominent popular misconceptions about human evolution?

1. Some Popular Misconceptions (p. 660)

Our ancestors were chimpanzees or other modern apes. Humans and chimpanzees represent two divergent branches of the anthropoid lineage which evolved from a common, less specialized ancestor.

Human evolution represents a ladder with a series of steps leading directly from an ancestral anthropoid to <u>Homo sapiens</u>. This progression is usually shown as a line of fossil hominids becoming progressively more modern. Human evolution included many branches which led to dead ends with several different human species coexisting at times. Most evolutionary change occurred with the appearance of new hominid species, not phyletic change in an unbranched lineage. (See Campbell, Figure 30.29)

Various human characteristics like upright posture and an enlarged brain evolved in unison. Mosaic evolution occurred with different features evolving at different times. Some ancestral forms walked upright but had small brains.

Present understanding of our ancestry remains unclear even after dismissing some myths.

2. Early Anthropoids (p. 660–661)

What events could have led to the movement of early anthropoids out of the forests?

Oldest known fossils of apes are *Aegyptopithecus*, the "dawn ape," which was a cat-sized tree-dweller about 35 million years ago.

About 25 million years ago (during the Miocene epoch) descendants of the first apes diversified and spread to Eurasia.

About 20 million years ago, the Indian plate collided with Asia and the Himalayan range formed.
- ☞ Climate became drier and forests (African and Asian) contracted.

Some Miocene apes may have descended from the trees and begun living on the edge of the forests and foraging for food on the adjacent savanna.
- ☞ One of these African anthropoids was the human ancestor.

Margin notes: <u>Paleoanthropology</u>; <u>Aegyptopithecus</u>

Most anthropologists believe that humans and apes diverged from a common ancestor 4–5 million years ago.
- ☞ Evidence from the fossil record and DNA comparisons between humans and chimpanzees.

3. **Australopithecines: The First Humans** (p. 661–663)

What is the significance of the discovery of the fossil hominid Australopithecus africanus? Of the discovery of fossils of Australopithecus afarensis?

Australopithecus africanus was discovered by Raymond Dart in 1924.
- ☞ Additional fossils proved that *Australopithecus* was a hominid that walked fully erect and had humanlike teeth and hands.
- ☞ Brain was about one-third the size of modern humans.
- ☞ *Australopithecus* appeared about three million years ago and foraged on African savannas for nearly two million years.
- ☞ Two forms were present, one slender and one robust (some assign them to different species).

"Lucy" was discovered in 1974 in the Afar region of Ethiopia.
- ☞ Lucy is believed to have lived about 3 million years ago.
- ☞ Skeleton was 40% complete and small, about one meter tall with a head the size of a softball.
- ☞ Different enough to be placed in a different species, *Australopithecus afarensis*.
- ☞ Similar fossils have been discovered which date to about 4 million years.

Some anthropologists believe these older fossils also represent *A. afarensis*, others believe them to be separate species. This disagreement has resulted in two possible genealogies.
- ☞ Those anthropologists who believe the species are separate feel that *A. afarensis* was part of an australopithecine branch that had already diverged from the lineage leading to *Homo*.
- ☞ Others who believe Lucy and the similar fossils are a single species propose that *A. afarensis* was the common ancestor of two hominid branches, one leading to *A. africanus* and the other to *Homo*.

In either case, Lucy proves that upright posture evolved early in hominid history and this is supported by fossilized footprints more than 3.5 million years old.
- ☞ *Australopithecus* walked erect for more than a million years without a substantial enlargement of the brain. This posture may have freed the hands for other things.

Enlargement of the human brain is first evident in fossils dating to about 2 million years ago.
- ☞ Skulls with brain capacities of about 650 cubic centimeters have been found compared with the 500 cc capacity of *A. africanus*.
- ☞ Simple stone tools have been found at times with the larger-brained fossils.

☞ Some paleoanthropologists believe these advances warrant placing the larger-brained fossils in the genus *Homo* and naming them *Homo habilis*.
☞ Others feel that this is just a variant of *Australopithecus*.
☞ Regardless, it is clear from the fossils that after walking upright for more than a million years, hominids began to use their brains and hands to fashion tools.
☞ *Homo habilis* coexisted with the smaller-brained *Australopithecus* for nearly one million years.

According to one theory of human origins, *Australopithecus africanus* (and other australopithecines) and *Homo habilis* were two distinct lines of hominids. *Australopithecus africanus* was an evolutionary dead end with *Homo habilis* leading first to *Homo erectus* which later gave rise to *Homo sapiens*.

4. *Homo erectus* and Descendants (p. 663)

What is the association among Homo habilis, Homo erectus, and Homo sapiens?

Homo erectus was the first hominid to migrate out of Africa into Europe and Asia.
☞ Fossils known as Java Man and Peking Man are examples.

Homo erectus was taller and had a larger brain capacity than *Homo habilis*.
☞ *Homo erectus* fossils range in age from 1.6 million years to about 300,000 years.
☞ During that time the brain capacity increased to as large as 1200 cc.

To survive, *Homo erectus* lived in huts or caves, built fires, wore clothes of skins, and designed more elaborate stone tools than *Homo habilis*.
☞ *Homo erectus* was poorly equipped to live outside of the tropics but made up for the deficiencies with intelligence and social cooperation.
☞ Some descendants of *H. erectus* developed larger brain capacities and exhibited regional diversity.

The Neanderthals are the best known descendants of *H. erectus*.
☞ Neanderthals lived in Europe, the Middle East, and Asia from 130,000 to 35,000 years ago.
☞ They had heavier brows, less pronounced chins, and slightly larger brain capacities than modern man.
☞ Skilled tool makers who participated in burials and other rituals requiring abstract thought.

472 The Vertebrate Genealogy

Homo sapiens

Many paleoanthropologists group the African post-*Homo erectus* fossils with Neanderthals and other descendants from Asia and Australasia. They believe these fossils represent the earliest forms of *Homo sapiens*.

☞ Some post-*H. erectus* date to 300,000 years ago.

5. The Emergence of *Homo sapiens*: Out of Africa...But When?
(p. 663–664)

The debate over the origin of modern humans continues unabated with two widely divergent models currently being discussed. These are the *Multiregional Model* and the *Monogenesis Model*.

What evidence supports the Multiregional Model of human origins? The Monogenesis Model of human origins?

Multiregional Model

The *Multiregional Model* proposes: 1) Neanderthals and other post-*Homo erectus* hominids were ancestors to modern humans; and 2) modern humans evolved along the same lines in different parts of the world.

Years ago	AFRICAN	EUROPEAN	EAST ASIAN	AUSTRALASIAN
0				
100,000	Klasies	Neanderthal	Dali	Ngandong
300,000	Saldanha	Petralona	Zhoukoudian ("Beijing")	Sambungmachan
700,000	Olduvai	European	Lantian	Java

Homo erectus

☞ If this model is correct then the geographic diversity of humans originated more than a million years ago when *Homo erectus* spread from Africa to other continents.

☞ Supporters of the model believe the facial features and anatomical characteristics of each modern race resulted from the parallel, multiregional emergence of *Homo sapiens*.

☞ Supporters also feel that interbreeding among neighboring populations has provided opportunities for gene flow over the entire range and resulted in the genetic similarity of modern humans.

During the 1980's, some paleoanthropologists who interpreted the fossil record in a different way began to develop an alternative to the Multiregional Model. This alternative became known as the *Monogenesis Model*.

Monogenesis Model

The *Monogenesis Model* proposes: 1) *Homo erectus* was the ancestor to modern humans who evolved in Africa; and 2) modern humans dispersed from Africa, displacing the Neanderthals and other post-*H. erectus* hominids.

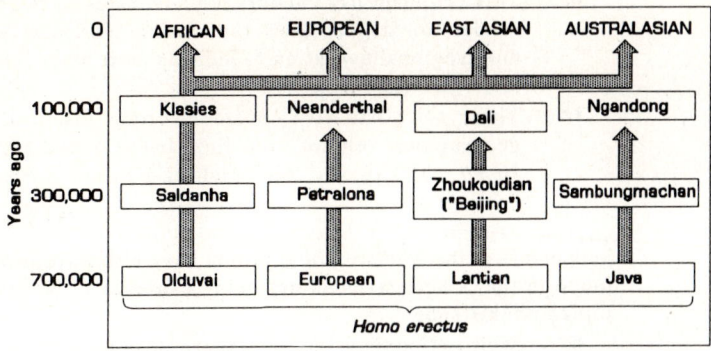

- If the Monogenesis Model is true, the recognized human races have developed from geographic diversification in the last 100,000 years.
- The focus of this model is on the relationship between Neanderthals and modern humans in Europe and the Middle East.
- The oldest fossils of modern *Homo sapiens* are about 100,000 years old. These were found in Africa and similar fossils have been recovered from caves in Israel.
- The fossils from Israel were found in caves near other caves containing Neanderthal-like fossils which date from 120,000−60,000 years ago — overlapping *H. sapiens* by about 40,000 years.
- Supporters of the Monogenesis Model interpret this information to mean that no interbreeding occurred during the time of coexistence since the two types of hominids persisted as distinct forms.
- This interpretation means the Neanderthals were not ancestors of modern humans since they coexisted and were probably evolutionary dead ends along with other dispersed post-*H. erectus* hominids.

Modern molecular techniques are being used to examine the question of human origins.
- In the late 1980's, a group of geneticists compared the mitochondrial DNA (mtDNA) from a multiracial group of more than 100 people from four different continents.
- The premise for this analysis is that the greater the differences in the mtDNA of two people, the longer the divergence from a common source-ancestral form.
- Analysis of the mtDNA comparisons resulted in the tracing of the source of all human mtDNA back to Africa with the divergence from that common source beginning about 200,000 years ago.
- These results appeared to support the Monogenesis Model in that the divergence from the common source was too late to represent dispersal of *Homo erectus*, but supported a later dispersal of modern humans.

Several researchers have challenged the interpretation of the mtDNA study, especially the methods used to construct evolutionary trees from this type of data and the reliability of mtDNA as a biological clock.

- ☞ This criticism has encouraged advocates of the Multiregional Model to argue that the fossil evidence supports the multiregional evolution of humans more strongly than African monogenesis.
- ☞ These scientists also consider certain fossils from different geographical regions to be links between that region's archaic *Homo sapiens* and the modern humans currently on that continent.

Supporters of the Multiregional Model have also interpreted the fossils found in the Israeli caves as not being evidence of Neanderthal-*Homo sapien* coexistence.
- ☞ Multiregionalists believe these fossils represent integrated fossils left by interbreeding populations of archaic *Homo sapiens* who immigrated to the region and gave rise to the modern humans of the area.

Debate continues about whether the Multiregional Model or the Monogenesis Model is more accurate.
- ☞ New evidence obtained by attempts to extract DNA from fossilized Neanderthal and other archaic *Homo sapien* skulls may shed new light on this question.

6. **Cultural Evolution: A New Force in the History of Life**
   (p. 664–666)

What is Cultural Evolution? What three stages are believed to have been involved?

Erect stance was a very radical anatomical change in our evolution and required major changes in the foot, pelvis and vertebral column.

Enlargement of the brain was a secondary alteration made possible by prolonging the growth period of the skull and its contents.

The brains of nonprimate mammalian fetuses grow rapidly, but growth slows and stops not long after birth.
- ☞ The brains of primates continue to grow after birth and the period is longer for a human than other primates.
- ☞ Parental care is lengthened due to this extended development which contributes to the child's learning.

Culture

The basis of culture (transmission of accumulated knowledge over generations) is transmitted by written and spoken language. Cultural evolution is thought to have occurred in three stages:

1. Nomads of the African grasslands made tools, organized communal activities and divided labors.

2. The development of agriculture in Eurasia and the Americas about 10,000–15,000 years ago encouraged permanent settlements.

3. The Industrial Revolution began in the eighteenth century.
   - ☞ Since then, new technology and the human population have escalated exponentially.

From the beginning to now, no significant biological change in humans has occurred.

Evolution of the human brain may have been anatomically simpler than acquiring an upright stance, but the consequences of cerebral growth have been enormous.

What impact has human cultural evolution had on other species?

Cultural evolution resulted in *Homo sapiens* becoming a species that could change the environment to meet its needs and not have to adapt to an environment through natural selection.
- Humans are the most numerous and widespread of large animals.

Cultural evolution outpaces biological evolution and we may be changing the world faster than many species can adapt.
- Rate of extinctions this century is 50 times greater than the average for the past 100,000 years.
- Overwhelming rate of extinction is due primarily to habitat destruction and chemical pollution, both functions of human cultural changes and overpopulation.
- Global temperature increase and alteration of world climates are a result of fossil fuel consumption.
- Destruction of tropical rain forests, which play a role in maintenance of atmospheric gas balance and moderating global weather, is startling.

The effect of *Homo sapiens* is the latest and may be the most devastating crisis in the history of life.

31 PLANT STRUCTURE AND GROWTH

CHAPTER OUTLINE

I. AN INTRODUCTION TO PLANT BIOLOGY: THE MAJOR THEMES (p. 674–676)

What important role do plants play in terrestrial ecosystems?

Plants form the foundation on which most terrestrial ecosystems are built. As the primary producers in most systems (through the process of photosynthesis), plants serve as the first link in the food chain which affects all species of animals in the system.

Plants were first studied by early humans who had to distinguish between edible and poisonous plants. These early humans later began to use plant products to make useful tools and other items.

Modern plant biology continues to center on how to use plants and plant products to benefit humans.

New methods and the discovery of unique, interesting experimental organisms have resulted in an ever expanding knowledge base in many areas of plant biology.
- ☞ Relating processes which occur at the molecular and cellular level to what is observed at the whole plant level.
- ☞ Correlating structure and function of plants.

II. BASIC MORPHOLOGY OF FLOWERING PLANTS: AN EVOLUTIONARY PERSPECTIVE (p. 676–681)

The two levels of plant architecture are morphology and anatomy.

<u>Plant morphology</u> = The study of the external structure of plants (e.g. arrangement of the parts of a flower).

<u>Plant anatomy</u> = The study of the internal structure of plants (e.g. arrangement of the cells and tissues in a leaf).

What characterizes angiosperms? How do monocots and dicots differ?

Angiosperms (flowering plants) are the most diverse and widespread of the plants with about 275,000 extant species.
- ☞ Characterized by flowers and fruits which are believed to have evolved to function in reproduction and seed dispersal.
- ☞ Divided into two taxonomic classes: *monocots* and *dicots* which possess either one or two seed leaves, respectively, in combination with other characteristics. (See Campbell, Figure 31.4)

How do root systems and shoot systems reflect the evolutionary history of plants as terrestrial organisms?

A plant can be divided into two basic systems, a subterranean *root system* and an aerial *shoot system* (stems, leaves, flowers). This two system arrangement reflects the evolutionary history of plants as terrestrial organisms. (See Campbell, Figure 31.5)

- Unlike the algal ancestors which are completely surrounded by nutrient rich water, terrestrial plants face a divided habitat: air is the source of CO_2 for photosynthesis and sunlight cannot penetrate into the soil; soil provides water and dissolved minerals to the plant.
- Each system depends on the other for survival of the whole plant.
- Roots depend on shoots for sugar and other organic nutrients.
- Shoots depend on roots for minerals, water and for support.
- *Xylem* conveys water and dissolved minerals to the shoots.
- *Phloem* conveys food from shoots to roots and other nonphotosynthetic parts, and from storage roots to actively growing shoots.

A. The Root System (p. 676–679)

Roots anchor plants, absorb and conduct water and minerals and store food.

What are the two major types of root systems? How do they differ?

There are two major types of root systems:

1. Taproot System
   - One large, vertical root (the *taproot*) produces many smaller secondary roots.
   - Seen in many dicots.
   - Provides firm anchorage.
   - Some taproots such as carrots, turnips, and sweet potatoes, are modified to store a large amount of reserve food.

2. Fibrous Root System
   - Mat of threadlike roots spread out below the soil surface.
   - Provides extensive exposure to soil water and minerals.
   - Roots are concentrated in the upper few centimeters of soil, preventing soil erosion.
   - Mostly seen in monocots.

Absorption of water is greatly increased by *root hairs*, which increase the surface area of the root. Roothairs are normally most numerous near the root tips.

Adventitious roots = Roots arising above ground from stems or leaves.
- Some, such as prop roots of corn, help support the plant stem.

B. The Shoot System (p. 679–681)

Shoot systems are comprised of vegetative shoots and floral shoots.
- Vegetative shoots consist of a stem and attached leaves; may be the main shoot or a branch.
- Floral shoots terminate in flowers.

1. **Stems** (*p. 679*)

   Morphology of a stem includes: (See Campbell, Figure 31.7)
   - ☞ *Nodes* are the points at which leaves are attached to stems.
   - ☞ *Internodes* are the stem segments between the nodes.
   - ☞ An *axillary bud* is an embryonic side shoot found in the angle formed by each leaf and the stem. Usually dormant.
   - ☞ A *terminal bud* is the bud on a shoot tip, usually has developing leaves and a compact series of nodes and internodes.

   What is the adaptive advantage of apical dominance?

   Growth of a shoot is usually concentrated at the *apex* of the shoot where the terminal bud is located. The presence of a terminal bud inhibits development of axillary buds, a condition called *apical dominance*.
   - ☞ Apical dominance appears to be an evolutionary adaptation to increase exposure of plant parts to light by concentrating resources on increasing plant height.

   Axillary buds begin to grow under certain conditions and after damage or removal of the terminal bud.
   - ☞ Some may develop into floral shoots.
   - ☞ Others develop into vegetative shoots with terminal buds, leaves, and axillary buds. This development results in branching which also increases exposure of plant parts to light.

   What are rhizomes, bulbs and stolons? Why are they considered to be modified stems rather than roots?

   Some plants have *modified stems* which many people mistake for roots. There are several types of modified stems, each of which performs a specific function. (See Campbell, Figure 31.8)
   - ☞ *Stolons* are horizontal stems growing along the surface of the ground (e.g. strawberry plants).
   - ☞ *Rhizomes* are horizontal stems growing underground (e.g. irises). Some end in enlarged tubers where food is stored (e.g. potatoes).
   - ☞ *Bulbs* are vertical, underground shoots with leaves modified for food storage (e.g. onions).

2. **Leaves** (*p. 679–681*)

   Leaves are the main photosynthetic organs of a plant.
   - ☞ Usually in the shape of a flattened blade.
   - ☞ *Petioles* join the leaf to the node of a stem. (Most monocots lack petioles; instead, the leaf base forms a sheath enclosing the stem.)
   - ☞ Monocot leaves have parallel major veins running the length of the blade.
   - ☞ Dicot leaves have a multi-branched network of major veins. Can be palmate or pinnate.
   - ☞ All leaves have numerous minor cross-veins.

What type of leaf characteristics do plant taxonomists use to classify plants?

Plant taxonomists use a variety of leaf characteristics to classify plants. (See Campbell, Figure 31.9)

Classification of leaf arrangement on stem:

Opposite = 2 leaves at each node 180° apart.

Alternate = Each node has one leaf and leaves at adjacent nodes point in opposite directions.

Whorled = A node has 3 or more leaves attached.

Simple leaf = One undivided blade.

Compound leaf = Divided into several leaflets.

Classification by leaf shape:
- ☞ Can be lanceolate, oval, cordate (heart-shaped) or triangular.

Classification by leaf margin:
- ☞ Can be entire (smooth), undulate, serrate or lobed.

How does leaf structure fit its function as the main photosynthetic organ of a plant? What modifications of leaf structure are adaptive for functions other than photosynthesis?

Some plants have leaves that have become adapted for functions other than photosynthesis. (See Campbell, Figure 31.10)
- ☞ Tendrils are modified leaflets that cling to supports.
- ☞ Spines of cacti function in protection.
- ☞ Many succulents have leaves modified for storing water.
- ☞ Some plants have brightly colored leaves that help attract pollinators to the flower.

III. BASIC PLANT ANATOMY: PLANT CELLS AND TISSUES (p. 681–686)

A. Types of Plant Cells (p. 682–685)

Each type of plant cell has structural adaptations that make it possible to perform that cell's function. Some are coupled with specific characteristics of the protoplast. (See Campbell, Figure 31.12)

How do parenchyma, collenchyma and sclerenchyma cells differ in structure? How are these differences a reflection of their different functions?

    1. Parenchyma Cells (p. 682–684)

Parenchyma cells are relatively unspecialized.
- ☞ Primary walls are thin and flexible.
- ☞ Lack secondary walls.
- ☞ The protoplast usually has a large central vacuole.
- ☞ Function in synthesizing and storing organic products.
- ☞ Some in stems and roots have colorless plastids that store starch.

Opposite

Alternate

Whorled, Simple leaf

Compound leaf

Parenchyma cells

☞ Most mature cells do not divide, but retain ability to divide and differentiate into other cell types under special conditions (e.g. repair and replacement after injury).

2. **Collenchyma Cells** (*p. 684*)

Collenchyma cells have protoplasts and usually lack secondary walls.
☞ Primary walls are unevenly thickened.
☞ Are usually grouped in strands or cylinders to support young parts of plants without restraining growth.
☞ Elongate as the young stems and leaves they support grow.

3. **Sclerenchyma Cells** (*p. 684*)

Sclerenchyma cells function in support.
☞ Have very rigid, thick secondary walls strengthened by lignin.
☞ So specialized for support that many lack protoplasts at functional maturity.
☞ At maturity, cannot elongate and may be dead, functioning only as support.
☞ Two forms of cells are *fibers* (long, slender, tapered cells occurring in bundles) and *sclerids* (shorter, irregularly-shaped cells).

4. **Tracheids and Vessel Elements: Water-Conducting Cells** (*p. 684–685*)

What are the functions of tracheids and vessel elements? How do their structures differ? Which is the most specialized for conducting water?

Xylem consists of two cell types, both with secondary walls and both dead at functional maturity.
☞ Secondary walls are deposited in spiral or ring patterns (which allows them to stretch) in parts of the plant that are still growing.

Tracheids are long, thing tapered cells having lignin-hardened secondary walls with *pits* (thinner regions where only primary walls are present). (See Campbell, Figure 31.13a)
☞ Water flows from cell to cell through pits.
☞ Also function in support.

Vessel elements are wider, shorter, thinner-walled, and less tapered. (See Campbell, Figure 31.13b and c)
☞ End walls are perforated for free flow of water through long chains of vessel elements called xylem vessels.
☞ More efficient as water conductors than tracheids.
☞ Evolved from tracheids.

5. **Sieve-Tube Members: Food-Conducting Cells** (*p. 685*)

What is the function of sieve-tube members? How does their structure differ in gymnosperms and angiosperms?

Sieve-tube members transport sucrose, other organic compounds, and some minerals.

- ☞ Chains of cells are collectively called *phloem*.
- ☞ Are alive at functional maturity.
- ☞ Protoplasts lack a nucleus, ribosomes and a distinct vacuole.

In ferns and gymnosperms, pores are distributed over the entire sieve cell while angiosperms possess sieve cells with pores concentrated at the ends.

Sieve plates = Porous end-walls between cells of angiosperms.
- ☞ Some cells have long strands of P protein whose function is unknown.
- ☞ Cells have the polysaccharide *callose* which is involved in formation of the sieve plates.
- ☞ A *companion cell* is connected to each sieve-tube member by many plasmodesmata; the companion's nucleus and ribosomes may also serve the sieve-tube member.

How does form fit function in water-conducting cells and sieve-tube members?

B. The Three Tissue Systems of a Plant (p. 685–686)

Plant cells are arranged into three tissue systems: dermal, vascular, and ground tissue.

What are the respective functions of the dermal, vascular and ground tissue systems?

Each of the three systems is continuous throughout the plant, although, specific characteristics and spatial relationships vary in different plant organs. (See Campbell, Figure 31.14)

1. Dermal Tissue System

Dermal tissue system or epidermis = Single layer of tightly packed cells covering the young parts of the plant.
- ☞ Functions in protection and has special characteristics consistent with the function of the organ it covers.
- ☞ For example, root hairs specialized for water and mineral absorption are extensions of epidermal cells near root tips.
- ☞ The waxy cuticle is secreted by epidermal cells of leaves and most stems.

2. Vascular Tissue System

Vascular tissue system = The xylem and phloem that functions in transport and support.

3. Ground Tissue System

Ground tissue system = Predominantly parenchyma that fills the space between dermal and vascular tissue systems.
- ☞ Has diverse functions including photosynthesis, storage and support.

IV. AN OVERVIEW OF PLANT GROWTH (p. 686–687)

How is growth in plants different from that in animals?

Indeterminate growth = Continued growth as long as the plant lives.

Most plants show indeterminate growth.
- In contrast, most animals cease growing after reaching a certain size (determinate growth).
- Certain plant organs, such as flower parts, show determinate growth.

How is indeterminate growth possible?

Indeterminate growth is made possible by *meristems* (perpetually embryonic tissues).
- Meristematic cells are unspecialized and divide to generate new cells near the growing point.
- New meristematic cells formed by division may remain in the region and produce new cells (*initials*) or be displaced and become incorporated and specialized into tissues (*derivatives*).

What is the difference between apical and lateral meristems? Primary and secondary growth?

- *Apical meristems*, located in root tips and shoot buds, supply cells for plants to grow in length.
- Primary growth is initiated by apical meristems and forms primary tissues organized into the 3 tissue systems.
- Secondary growth is the thickening of roots and shoots which occurs in woody plants due to development of *lateral meristems*.

Lateral meristems = Cylinders of dividing cells which extend the lengths of roots and shoots.
- Cell division in the lateral meristems produces secondary dermal tissues which are thicker and tougher than the epidermis it replaces.
- Also adds new layers of vascular tissues.

What are the differences among annual, biennial and perennial plants?

Some plants have finite life spans.
- Some are genetically determined.
- Some are environmentally determined.
- *Annuals* complete their life cycles in one year or growing season.
- *Biennials* typically have life spans of two years.
- *Perennials*, such as trees and some grasses, live many years.

V. A CLOSER LOOK AT PRIMARY GROWTH (p. 687–693)

The primary plant body with its three tissue systems is produced by primary growth.
- The youngest portions of woody plants and herbaceous plants are examples of primary plant bodies.
- Apical meristems are responsible for the primary growth of roots and shoots.

Plant Structure and Growth 483

A. **Primary Growth of Roots** (p. 687–689)

Where does primary growth of roots occur? What is the function of the root cap?

Root growth is concentrated near its tip and results in roots extending through the soil.
- ☞ The root tip is covered by a *root cap*, which protects the meristem and secretes a polysaccharide coating that lubricates the soil ahead of the growing root.

Root cap

What are the relative positions of the zones of cell division, cell elongation and cell differentiation in root tips? What occurs in each of these zones?

The root tip contains 3 zones of cells in successive stages of primary growth. Although described separately, these zones blend into a continuum. (See Campbell, Figure 31.16)

1. Zone of Cell Division
   - ☞ Located near the tip of the root; includes the apical meristem and its derivatives, primary meristems.
   - ☞ The apical meristem is centrally located; it produces the primary meristems and replaces cells of the root cap.
   - ☞ The primary meristems form as three concentric cylinders of cells (the protoderm, procambrium, ground meristem) which will produce the tissue systems of the roots.
   - ☞ The *quiescent center* is located near the center of the apical meristem. It is composed of resistant, slowly dividing cells which may serve as reserve replacement cells in case of damage to the meristem.

Zone of cell division

2. Zone of Cell Elongation
   - ☞ Region where cells elongate as much as 10 times their original length.
   - ☞ Elongation of cells in this region pushes the root tip through the soil.
   - ☞ Continued growth is sustained by the meristem's constant addition of new cells to the youngest end of the elongation zone.

Zone of cell elongation

3. Zone of Cell Differentiation
   - ☞ Farthest from the root tip.
   - ☞ Region where the new cells become specialized in structure and function and where the three tissue systems complete their differentiation.

Zone of cell differentiation

What are the three primary meristems of roots? Into what tissues do they develop?

The apical meristem produces three primary meristems, which in turn give rise to the three primary tissues of roots:

1. The protoderm is the outermost primary meristem.
   - ☞ Gives rise to the epidermis over which water and minerals must cross.
   - ☞ Root hairs are epidermal extensions which increase surface area.

Protoderm

2. The procambium forms a stele (central cylinder) where xylem and phloem develop.

Procambium

484   Plant Structure and Growth

Pith

Ground meristem

Cortex, Endodermis

Stele

Lateral roots

Pericycle

Leaf primordia

- ☞ In dicots, xylem radiates from the stele's center in 2 or more spokes, with phloem in between the spokes.
- ☞ In monocots, a central *pith* (core of parenchyma cells) is ringed by an alternating pattern of xylem and phloem.

3. The <u>ground meristem</u> is between the protoderm and procambium, and gives rise to the *ground tissue system*. The ground tissue:
   - ☞ Is mostly parenchyma and fills the cortex.
   - ☞ Stores food; the cell membranes are active in mineral uptake.
   - ☞ Fills the *cortex* (root area between the stele and epidermis).
   - ☞ Has *endodermis*, the single-cell tick, innermost layer of the cortex that forms the boundary between the cortex and the stele. [It selectively regulates passage of substances from soil to the vascular tissue of the stele.]

How do lateral roots form? What is the pericycle and what role does it play in lateral root formation?

Lateral roots may sprout from the outermost layer of the stele of a root.
- ☞ The *pericycle*, just inside the endodermis, is a layer of cells that may become meristematic and divide to form the lateral root.
- ☞ The lateral root maintains its vascular connection to the stele of the main root.

B. Primary Growth of Shoots (p. 689–693)

Shoot apical meristem is a dome-shaped mass of dividing cells at the tip of the terminal bud.
- ☞ Forms the primary meristems that differentiate into the 3 tissue systems.
- ☞ On the flanks of the apical dome are *leaf primordia* which form leaves.
- ☞ The meristem leaves meristematic cells at the base of the leaf primordia that develop into axillary buds.

How does primary shoot growth occur?

Most of the shoot elongation actually occurs due to growth of slightly older internodes below the shoot apex.
- ☞ Growth is a result of both cell division and elongation within the internode.
- ☞ *Intercalary meristems* are present at the base of each internode in grasses and some other plants. These tissues permit prolonged internode elongation along the length of the shoot.

In what way is the development of axillary buds different from the development of lateral roots? How is this a reflection of structural differences between roots and stems?

Axillary buds may form branches later in the life of the plant.
- ☞ Branches originate at the surface of the shoot and are connected to the vascular system which lies near the surface.
- ☞ This is a direct contrast to development of lateral roots which form deep in the pericycle of the root.

1. Primary Tissues of Stems (p. 691)

How is the vascular tissue of stems organized? How does its arrangement differ in dicots and monocots?

The vascular tissue of the stem is organized into strands of *vascular bundles*.
- ☞ Converge at the transition zone (shoot ⟶ root) to join the root stele.
- ☞ Each bundle is surrounded by ground tissue, including pith and cortex.

Vascular bundles

In dicots, bundles are arranged in a ring with pith inside and cortex outside.
- ☞ Xylem faces the pith, phloem faces the cortex.
- ☞ Pith and cortex are connected by *pith rays*.

Pith rays

In monocots, vascular bundles are scattered throughout the ground tissue resulting in no pith or cortex regions in the ground tissue.

The protoderm of the terminal bud gives rise to the epidermal portion of the dermal tissue system.

Ground tissue of stems is mostly parenchyma, strengthened in many plants by collenchyma located beneath the epidermis.

The protoderm of the terminal bud gives rise to the epidermal portion of the dermal tissue system.

2. Tissue Organization of Leaves (p. 691–692)

What is the function of the epidermis in leaves? Why is it important for aerial plant parts to have a cuticle and for roots not to have a cuticle?

Leaves are cloaked by *epidermis*.
- ☞ Protects against physical damage and pathogens.
- ☞ Waxy cuticle prevents water loss.

Epidermis

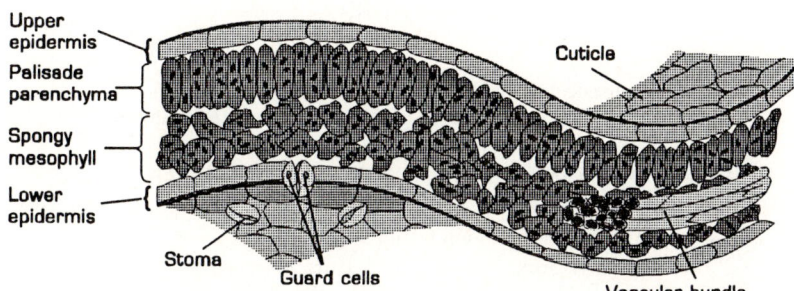

Upper epidermis | Palisade parenchyma | Spongy mesophyll | Lower epidermis | Stoma | Guard cells | Cuticle | Vascular bundle

- ☞ *Stomata* are pores flanked by *guard cells* which regulate gas exchange. Stomata also allow *transpiration* (water loss from plant by evaporation).
- ☞ Stomata are usually more numerous on the bottom surface of the leaf. Their location minimizes water loss.

Stomata, Guard cells

Transpiration

486 *Plant Structure and Growth*

Mesophyll

Palisade mesophyll

Spongy mesophyll

Leaf traces

What is the ground tissue of a leaf? How is it arranged in dicots? What is the functional value of that arrangement?

The ground tissue of a leaf is *mesophyll*.
☞ Consists mainly of parenchyma equipped with chloroplasts for photosynthesis.
☞ Dicots usually have two distinct layers:

Palisade mesophyll = Columnar cells of the upper half of leaf.

Spongy mesophyll = Irregularly-shaped cells surrounded by air spaces through which oxygen and CO_2 circulate. In lower half of leaf.

How is the arrangement of leaf vascular tissue important for its function?

The leaf vascular tissue is continuous with that of the stem through *leaf traces*.
☞ Continues in the petiole as a vein, which branches repeatedly throughout the mesophyll of the leaf blade.
☞ Functions also as a skeleton to support the shape of the leaf.

3. **Modular Shoot Construction and Phase Changes During Development** (*p. 692−693*)

Modular Shoot Construction:
☞ Shoots are constructed of a series of modules produced by the serial development of nodes and internodes.
☞ Each module consists of a stem, one or more leaves, and an axillary bud associated with each leaf.
☞ Elongation of the internode provides for primary growth of the plant.

Modules found in the body of a plant represent various ages of tissue. The age of a module is proportional to its distance from the apical meristem.
☞ The developmental phase of each module changes over time.
☞ There is a gradual change from a juvenile vegetative state to a mature vegetative state.
☞ In some plants, the mature shoot apex may undergo a second phase change to a flower-producing (reproductive) state.

VI. **A CLOSER LOOK AT SECONDARY GROWTH** (*p. 693−696*)

A. **Secondary Growth of Stems** (*p. 693−695*)

What are the consequences of secondary growth?

Secondary growth = Increase in girth produced by two <u>lateral</u> meristems: vascular cambium and cork cambium.
- ☞ Produces the secondary plant body which is comprised of the secondary tissues produced during growth in diameter.
- ☞ Occurs in all gymnosperms and most dicot angiosperms.
- ☞ Rare in monocots.

What sequence of events occurs during secondary growth of stems? What is the role of vascular cambium?

Vascular cambium forms when meristemic parenchyma cells develop between the primary xylem and primary phloem of each vascular bundle and in the pith rays between the bundles.
- ☞ Forms a continuous cylinder of dividing cells around the xylem and pith of the stem.
- ☞ *Ray initials* (cambium cells of pith rays) produce radial files of parenchyma cells (= *xylem rays*) which permit lateral transport of water and nutrients as well as storage of starch and other reserves.
- ☞ *Fusiform initials* (cambium cells in vascular bundles) produce new vascular tissues; secondary xylem to the inside of the vascular cambium; secondary phloem to the outside.
- ☞ Accumulated layers of secondary xylem produces wood; secondary phloem, and all tissues external to it, develop into bark.

What types of cells compose wood? How is wood formed?

Wood = Consists mostly of tracheids, vessel elements and fiber cells.
- ☞ Has rays of parenchyma cells radiating through secondary xylem to function in storage and lateral transport of water and solutes.
- ☞ Forms annual growth rings due to yearly activity: cambium dormancy, spring wood production and summer wood production.

The secondary phloem does not accumulate extensively: it develops into bark which eventually sloughs off the tree trunk.

What is the role of cork cambium in secondary growth? What is periderm? What is the function of cork? Lenticels?

Cork cambium is a cylinder of meristematic tissue that forms protective layers of the secondary plant body.
- ☞ First forms in the outer cortex of the stem.
- ☞ Inner cells remain meristematic, outer cells develop into cork cells which die.
- ☞ Dead cork cells protect the stem from damage and pathogens while reducing water loss.
- ☞ As continued secondary growth splits this, it is replaced by new cork cambium formed deeper in the cortex.
- ☞ When no cortex is left, it develops from parenchyma cells in the secondary phloem (only the youngest secondary phloem, internal to cork cambium, functions in sugar transport).

<u>Secondary growth</u>

<u>Vascular cambium</u>

<u>Ray initials</u>
<u>Xylem rays</u>

<u>Fusiform initials</u>

<u>Wood</u>

<u>Annual growth rings</u>

<u>Cork cambium</u>

488 *Plant Structure and Growth*

<u>Periderm</u>

<u>Suberin</u>

<u>Lenticels</u>

<u>Heartwood</u>

<u>Sapwood</u>

<u>Secondary xylem</u>

<u>Secondary phloem</u>

<u>Periderm</u> = The protective coat of the secondary plant body that replaces primary epidermis. It is made of cork cambium and layers of cork.
- ☞ Cork is produced by cork cambium (to its outside). The cells are impregnated with suberin, a waxy material, and are dead at functional maturity.

<u>Lenticels</u> = Spongy regions in the bark which permit gas exchange by living cells within the trunk.

Wood has two zones: (See Campbell, Figure 31.26)

1. *Heartwood*, the older layers of secondary xylem blocked with resins to become nonfunctional for water transport.
2. *Sapwood*, the younger secondary xylem, vascular cambium, young secondary phloem, and cork cambium. Conducts water and nutrients.

B. Secondary Growth of Roots (p. 695–696)

How do vascular cambium and cork cambium function in the secondary growth of roots?

Vascular cambium and cork cambium also function in secondary growth of roots.

The vascular cambium produces secondary xylem to its inside and secondary phloem to its outside.
- ☞ First located between xylem and phloem of the stele.
- ☞ As stele grows, cortex and epidermis split and are shed.

The cork cambium forms from the pericycle of the stele and produces the periderm, which becomes secondary dermal tissue.
- ☞ Periderm is impermeable to water.

Older roots become woody, and annual rings appear in the secondary xylem.

32   TRANSPORT IN PLANTS

CHAPTER OUTLINE

Land plants require transport system (unlike aquatic ancestors, photosynthetic organs have no direct access to water and minerals).

I. **BASIC TRANSPORT PROCESSES IN PLANTS: AN OVERVIEW** (p. 699–704)

What levels of transport exist in plants?

Three levels of transport occur in plants:
1. Uptake of water and solutes by individual cells.
2. Short-distance cell-to-cell transport at the level of tissues and organs.
3. Long-distance transport of sap in xylem and phloem at the whole-plant level.

    A. **Transport at the Cellular Level** (p. 699–704)

        1. **Passive and Active Transport of Solutes** (p. 700–701)

What controls the movement of solutes between a plant cell and the extracellular fluid?

The movement of solutes between a plant cell and the extracellular fluids is controlled by the selective permeability of the cell's plasma membrane. Solutes may move by passive or active transport.

What processes may move solutes across the cell's plasma membrane?

Passive transport occurs when a solute molecule diffuses across a membrane due to the presence of a concentration gradient.
- ☞ No direct expenditure of energy by the cell.
- ☞ Transport proteins embedded in the membrane may increase the speed at which solutes cross.

Transport proteins may facilitate diffusion by serving as carrier proteins or forming selective channels.
- ☞ Carrier proteins bind selectively to a solute molecule on one side of the membrane, undergo a conformational change, and release the solute molecule on the opposite side of the membrane.
- ☞ Selective channels are simply passageways by which selective molecules may enter and leave a cell; some gated selective channels are stimulated to open or close by environmental conditions.

Active transport occurs when a solute molecule is moved across a membrane against the concentration gradient by a cell which expends metabolic energy during the process.
- The *proton pump* is an active transporter important to plants.

What is a proton pump? What energy source drives this energy-requiring process? Why does a transmembrane proton gradient represent potential energy? How does the plant cell use this potential energy to transport solute molecules?

A proton pump hydrolyzes ATP and uses the energy to pump hydrogen ions (H^+) out of the cell.
- Produces a proton gradient with a higher concentration outside of the cell.
- Also produces a membrane potential since the inside of the plant cell is negative in relation to the outside of the cell.
- This membrane potential and the stored energy of the proton gradient are used by the plant to transport many different molecules.
- Potassium ions (K^+) are pulled into the cells because of the electrochemical gradient.
- Nitrate (NO_3^-) enters plant cells against the electrochemical gradient by *cotransporting* with hydrogen ions.

2. **Water Potential and Osmosis** (p. 701–703)

Osmosis results in the uptake or loss of water by the cell. Whether water moves into or out of a plant cell depends on which component, the cell or extracellular fluids, has the highest *water potential*.

What is water potential? How does solute concentration and pressure affect water potential? What equation describes this relationship?

Water potential (Ψ) = The physical property predicting the direction water will flow, governed by solute concentration and applied pressure.
- Water will always move across the membrane from the solution with the higher water potential to the one with lower water potential.
- Water potential is measured in units of pressure called *megapascals* (MPa); one MPa is equal to ten atmospheres of pressure.
- Pure water in an open container has a water potential of zero megapascals ($\Psi = 0$ MPa).
- Addition of solutes to water lowers the Ψ into the negative range.
- Increased pressure raises the Ψ into the positive range.
- A negative pressure, or *tension*, may also move water across a membrane; this *bulk flow* (movement of water due to pressure differences) is usually faster than the movement caused by different solute concentrations.

The effects of pressure and solute concentration on water potential are represented by:

$$\Psi = P - \pi$$

- ☞ P = physical pressure.
- ☞ π = osmotic pressure.
- ☞ A 0.1M solution has an osmotic pressure of 0.23 MPa; in an open container, physical pressure is zero; the water potential of this 0.1M solution would be −0.23 ($\Psi = 0 - 0.23 = -0.23$). Water would enter this solution due only to osmotic pressure.
- ☞ The addition of pressure to the solution could counter the affects of osmotic pressure by stopping net water movement (if P = 0.23) or by forcing water from the solution back into the pure water (if P ≥ 0.3).
- ☞ Similar changes would result if a negative pressure were applied to the pure water side of the membrane.

Plant cells will gain or lose water to intercellular fluids in relation to their water potential. (See Campbell, Figure 32.5)

If a flaccid cell is placed in a hyperosmotic environment, in what direction will there be a net water movement? How would a plant cell respond in this situation?

- ☞ A flaccid cell (P = 0) placed in a hyperosmotic solution will lose water by osmosis; the cell will plasmolyze (protoplast pulls away from the cell wall) in response.

Plasmolyze

If a flaccid cell is placed in a hypoosmotic environment, in what direction will there be net water movement? What happens to the plant cell when equilibrium is reached?

- ☞ A flaccid cell placed in a hypoosmotic solution will gain water by osmosis; the cell will swell and a *turgor pressure* develops; when pressure from the cell wall is equal to the osmotic pressure, an equilibrium is reached and no net water movement occurs (P = π; $\Psi = 0$).

Turgor pressure

3. **The Role of the Tonoplast** (p. 703–704)

Where is the tonoplast? How does it contribute to the proton gradient across the plasma membrane? How is the electrochemical gradient across the tonoplast related to its function?

Tonoplast = Membrane surrounding the large central vacuole found in plant cells.

Tonoplast

The tonoplast is important in regulating intracellular conditions.
- ☞ Transport proteins that control the movement of solutes between the cytosol and the vacuole are present in the tonoplast.

- Proton pumps in the tonoplast help the plasma membrane maintain a low H⁺ concentration in the cytosol by moving H⁺ *into* the vacuole; this also produces a tonoplast membrane potential.
- Several solutes are transported between the cytosol and vacuole due to this membrane potential and proton gradient.

B. **Short-Distance (Lateral) Transport at the Level of Tissues and Organs** (*p. 704*)

Why is short-distance transport called lateral transport?

This type of transport is usually along the radial axis of plant organs. This is why it is often referred to as *lateral transport*.

What three routes can absorbed water and minerals take, from the outer cells to the inner cells of the root?

Three routes are available for lateral transport in plant tissues and organs: (See Campbell, Figure 32.6)

1. Across the plasma membranes and cell walls. Solutes move from one cell to the next by repeatedly crossing plasma membranes and cell walls.

2. The *symplast* route. A symplast is the continuum of cytoplasm within a plant tissue formed by the plasmodesmata which pass through pores in the cell walls. Once water or a solute enters a cell by crossing a plasma membrane, the molecules can enter other cells by traveling through the plasmodesmata.

3. The *apoplast* route. An apoplast is the continuum between plant cells which is formed by the continuous matrix of cell walls. Water and solute molecules can move from one area of a root or other organ via the apoplast without entering a cell.

Water and solute molecules can move laterally in a plant organ by any one of these routes or by a combination through switching from one to another.

C. **Long-Distance Transport at the Whole-Plant Level** (*p. 704*)

Why is vascular tissue necessary for long-distance transport?

This type of transport is usually along the vertical axis of the plant (up and down) from the roots to the leaves and vice versa.

Vascular tissues are involved in this type of transport as diffusion would be too slow.
- Bulk flow (movement due to pressure differences) moves water and solutes through xylem vessels and sieve tubes.
- Transpiration reduces pressure in the leaf xylem; this creates a tension which pulls sap up through the xylem from the roots.
- Hydrostatic pressure develops at one end of the sieve tubes in the phloem; this forces the sap to the other end of the tube.

II. THE ABSORPTION OF WATER AND MINERALS BY ROOTS: A CLOSER LOOK (p. 704–706)

What path do water and minerals follow from outside the root to the shoot system?

Water and minerals:
- ☞ Enter through root epidermis, cross the cortex, pass into the stele, and flow upward in xylem.
- ☞ Most absorption occurs near root tips where the epidermis is permeable to water.
- ☞ Root hairs in this area increase the surface area available for absorption.
- ☞ Lateral transport of minerals and water through the root is usually by a combination of apoplastic and symplastic routes.

Cells cannot obtain a sufficient supply of mineral ions by diffusion alone: the soil solution is too dilute.
- ☞ Active transport of mineral ions involving carrier proteins permits root cells to accumulate essential minerals in very high concentrations.

How do water and minerals cross the root cortex from the epidermis to the stele? Which route involves selective mineral absorption?

Water and minerals cross the root cortex in two ways: through symplasts or the apoplasts.
- ☞ Only minerals using the symplastic route may move directly into the vascular tissues. They have been previously selected by a membrane.
- ☞ Minerals and water passing through apoplasts are blocked at the endodermis by a *Casparian strip* (a ring of suberin around each cell in the endodermis) and must enter an endodermal cell.

What is the function of the Casparian strip? Why is it necessary for selective absorption of minerals and water into the stele?

Water and mineral entrance into the stele must be through the cells of the endodermis.
- ☞ Casparian strips ensure that all substances entering the stele pass through at least one membrane, allowing only selected ions to pass into the stele. Also prevents stele contents from leaking back into the apoplast and out into the soil.

How are solutes transferred between the symplast and apoplast?
- ☞ Water and minerals enter the stele via symplast, but tracheids and xylem vessels are part of the apoplast.
- ☞ Endodermal and parenchymal cells selectively discharge minerals into the apoplast so they may enter the xylem. This action probably involves diffusion and active transport.
- ☞ Those minerals and water which move into the apoplast are free to enter the tracheids and xylem vessels.

Margin notes: Cortex; Endodermis; Casparian strip, Suberin; Tracheids, Xylem vessels

III. THE ASCENT OF XYLEM SAP: A CLOSER LOOK (p. 706–708)

Xylem sap flows upward at 15m per hour or faster.
- Xylem vessels are close to each leaf cell due to the branching of veins.

Water transported up from roots must replace that lost by *transpiration*.
- Transpiration is the evaporation of water from the aerial parts of a plant.
- The upward flow of xylem sap also provides nutrients (minerals) to the shoot system.

A. Pushing Xylem Sap: Root Pressure (p. 706)

How is root pressure created by some plants? How does it cause guttation?

When transpiration is low, ions pumped into the xylem decrease the stele's water potential and cause water flow into the stele. This uptake increases pressure which forces fluid up the xylem (= *root pressure*).
- Root pressure causes *guttation* (exudation of water droplets at leaf margins).
- The water droplets escape through specialized structures called *hydathodes* which relieve the pressure caused by more water entering the leaves than is lost by transpiration.

Root pressure is not the major mechanism driving the ascent of xylem sap.
- Cannot keep pace with transpiration.
- Can only force water up a few meters.

B. Pulling Xylem Sap: The Transpiration-Cohesion-Adhesion Mechanism (p. 706–708)

According to the transpiration-cohesion-adhesion theory, how can xylem sap be pulled upward in xylem vessels?

Transpirational pull:
- Gaseous water in damp intercellular leaf spaces diffuses into the drier atmosphere through stomata.
- The water lost is replaced by evaporation from mesophyll cells bordering the airspaces.
- The remaining water film, attracted by adhesion to the hydrophilic cell walls, retreats into the cell wall pores.
- Cohesion in this surface film of water resists an increase in the surface area of the film.
- The water film forms a meniscus due to the negative pressure caused by the adhesion and cohesion.
- This negative pressure pulls water from the xylem, through the mesophyll, toward the surface film on cells bordering the stomata.
- Water is moving through symplasts and apoplasts to a region of low water potential.
- Mechanism results in water from the xylem replacing water transpired through the stomata.

Why is a water potential gradient required for the passive flow of water through a plant, from soil?

    Cohesion and *adhesion* of water: <u>Cohesion, Adhesion</u>
- ☞ The transpirational pull on the xylem sap is transmitted to the soil solution. Cohesion of water due to H bonds allows for the pulling of water from the top of the plant without breaking the "chain."
- ☞ The adhesion of water (by H bonds) to the hydrophilic walls of xylem cells also helps pull against gravity.
- ☞ The small diameter of vessels and tracheids is important to the adhesion effect.
- ☞ The upward pull of sap causes tension (negative pressure) in xylem, which decreases water potential and allows passive flow of water from soil into stele.

    <u>Cavitation</u> = Formation of a water vapor pocket in xylem.   <u>Cavitation</u>
- ☞ Breaks the chain of water molecules and the pull is stopped.
- ☞ Vessels cannot function again unless refilled with water by root pressure. (This can only occur in small plants.)
- ☞ Pits between adjacent xylem vessels allow for detours around a cavitated area.
- ☞ Secondary growth also adds new xylem vessels each year.

  C. The Long-Distance Transport of Water in Plants by Solar-Powered Bulk Flow (*p. 708*)

Bulk flow is the movement of fluid due to pressure differences at opposite ends of a conduit.

How is the ascent of xylem sap solar-powered?

The upward movement of xylem sap results from the negative pressure caused by transpiration, which is due to the absorption of sunlight.
- ☞ Xylem vessels or chains of tracheids serve as the conduits in plants.
- ☞ Transpirational pull lowers the pressure at the upper (leaf) end of the conduit.
- ☞ Osmotic movement of water from cell to cell in roots and leaves are due to small gradients in water potential.
- ☞ In contrast, bulk flow through the xylem vessels depends only on pressure.

IV. THE CONTROL OF TRANSPIRATION: A CLOSER LOOK (*p. 708–712*)

Transpiration results in a tremendous water loss from the plant. This water is replaced by the upward movement of water through the xylem.

  A. The Photosynthesis-Transpiration Compromise (*p. 708–709*)

Large surface areas along a leaf's airspaces are needed for CO_2 intake for photosynthesis, but also results in greater surface area for evaporative water loss.
- ☞ Internal surface area of a leaf may be 10×–30× the external surface area.

Stomata are more concentrated on the bottom of leaves, away from the sun; this reduces evaporative loss.
- ☞ The waxy leaf cuticle prevents water loss from the rest of the leaf surface.

Why does the transpiration-to-photosynthesis ratio differ between C_3 and C_4 plants?

Transpiration:photosynthesis ratios measure efficiency of water use. This ratio is g H_2O lost/g CO_2 assimilated into organic material.
- ☞ Ratio of 600:1 is common in C_3 plants; 300:1 in C_4 plants.
- ☞ C_4 plants can assimilate CO_2 at greater rates than C_3 plants.

What are the benefits of transpiration?

Benefits of transpiration:
- ☞ Assists in mineral transfer from roots to shoots.
- ☞ Evaporative cooling reduces risk of leaf temperatures becoming too high for enzymes to function.

If transpiration exceeds delivery of water by xylem, plants wilt.
- ☞ Plants can adjust to reduce risk of wilting.
- ☞ Regulating the size of stomatal openings also reduces transpiration.

B. How Stomata Open and Close (p. 709–710)

How do guard cells control the stomatal aperture? How can this affect photosynthetic rate and transpiration?

Guard cells = Cells that flank stomata and control stomatal diameter by changing shape.
- ☞ When turgid, guard cells "buckle" due to microfibrils arranged in a radial manner and stomata open; when flaccid, guard cells sag and stomatal openings close.

How do K^+ fluxes across the guard cell membrane affect guard cell function?

The change in turgor pressure that regulates stomatal opening results from reversible uptake and loss of K^+ by guard cells. (See Campbell, Figure 32.9)
- ☞ Uptake of K^+ decreases guard cell water potential so H_2O is taken up, cells become turgid, and stoma opens. The tonoplast plays a role as most of the K^+ and water are stored in the vacuole.
- ☞ The increase in positive charge is countered by the uptake of chloride (Cl^-). H^+ ions are released from organic acids (which become negative) and are pumped out of the cell.
- ☞ Closing of the stomata results when K^+ exits the guard cells.
- ☞ Causes water loss and decreased turgor.

Guard cells

What three cues contribute to stomatal opening at dawn?

Stomata open at dawn because:
- Light induces guard cells to take up K⁺ due to activation of a blue-light receptor which activates proton pumps in the plasma membrane.
- Decrease of CO_2 in leaf air spaces due to photosynthesis in the mesophyll.
- An internal clock of guard cells will make them open even if kept in dark (a *circadian rhythm* approximates a 24 hour cycle).

Circadian rhythm

What environmental stresses can cause stomata to close during the daytime?

Guard cells may close stomata during the daytime if:
- There is a water deficiency resulting in flaccid guard cells.
- Production of abscisic acid (a hormone) by mesophyll cells signals guard cells to close.
- High temperature increases CO_2 in leaf air spaces due to increased respiration, closing guard cells.

   C.   **Evolutionary Adaptations that Reduce Transpiration** (*p. 710–712*)

How can xerophytes be adapted to arid climates?

Xerophytes are plants adapted to arid climates.
- Small, thick leaves (reduced surface area:volume, so less H_2O loss).
- A thick cuticle.
- Stomata are in depressions on the underside of leaves to protect from water loss due to drying winds.
- Some shed leaves in the driest time of the year.
- Cacti and others store water in stems during the wet season.

Xerophytes

How does crassulacean acid metabolism allow CAM plants to reduce the transpiration rate?

Plants of family Crassulaceae use CAM (crassulacean acid metabolism) to assimilate CO_2.
- CO_2 is assimilated at night into organic acids.
- The acids are broken down to transfer CO_2 in cells during the daytime and photosynthesis occurs.
- Therefore, stomata close during the day and the plant conserves water.

Crassulacean acid metabolism

V.   **TRANSPORT IN PHLOEM: A CLOSER LOOK** (*p. 712–715*)

Translocation = The transport of the products of photosynthesis by phloem to the rest of the plant.
- Sieve-tube members are the specialized cells of phloem that function in translocation.
- Phloem sap carries primarily sucrose, but also minerals, amino acids and hormones.

Translocation

A. Source-to-Sink Transport (p. 712–713)

What is source-to-sink transport? What determines the direction of phloem sap flow?

Phloem sap movement is not unidirectional; it moves through the sieve tubes from a source (production area) to a sink (use or storage area).

Source = Organ where sugar is produced by photosynthesis or by the breakdown of starch (usually leaves).

Sink = Organ that consumes or stores sugar (growing parts of plants, fruits, non-green stems and trunks, and others).

Sugar flows from source to sink.
- Source and sink depend on season. A tuber is the sink when stockpiling in the summer, but is the source in the spring.
- Minerals may also be transported to sinks.
- The sink is usually supplied by the nearest source.
- Direction of flow within a phloem element can change, depending on locations of the source and sink.

B. Phloem Loading and Unloading (p. 713–714)

How does phloem loading differ among plant species?

Sugar produced at a source must be loaded into sieve-tube members before it can be translocated to a sink.
- In some plant species, the sugar may move through the symplast from mesophyll cells to the sieve members.
- In other species, the sugar uses a combination of symplastic and apoplastic routes. (See Campbell, Figure 32.12)
- Some plants have *transfer cells*. These are modified companion cells which have numerous ingrowths of their walls. These structures increase the cells' surface area and enhances solute transfer between apoplast and symplast.

What is one explanation for how a proton pump can allow selective accumulation of sucrose in the symplast?

- In plants such as corn, sucrose is accumulated by active transport in sieve — tube members to $2\times - 3\times$ the concentration in mesophyll cells.
- Proton pumps power this transport by creating a H^+ gradient; the difference in H^+ concentration creates a form of stored energy.
- A membrane protein uses this energy to cotransport sucrose by coupling sugar transport to the diffusion of H^+ back into the cell.

Sucrose is unloaded at the sink end of sieve tubes.
- In some plants, sucrose is unloaded from the phloem by active transport.
- In other species diffusion moves the sucrose from the phloem into the cells of the sink.
- Both symplastic and apoplastic routes may be involved.

C. Pressure Flow (Bulk Flow) of Phloem Sap (*p. 714–715*)

Phloem sap flows up to 1 meter per hour, too fast for just diffusion or cytoplasmic streaming.

What causes phloem sap to flow from source to sink?

The flow is by a bulk flow (pressure-flow) mechanism. (See Campbell, Figure 32.13)
- ☞ Phloem loading causes high solute concentrations at the source end.
- ☞ Water potential decreases and water flows into tubes.
- ☞ Hydrostatic pressure is greatest at the source end of the tube.
- ☞ At the sink end, the potential is lower outside the tube due to the unloading of sugar, releasing hydrostatic pressure built up at the source end.
- ☞ The buildup of pressure at the source and the release of pressure at the sink causes source-to-sink flow.
- ☞ Xylem vessels recycle water from the sink to the source.

How have scientists tested this model of pressure-flow in phloem?

Aphids have been used to study flow in phloem. (See Campbell, Figure 32.14)
- ☞ The aphid stylet punctures phloem and the aphid is "force-fed" by the pressure.
- ☞ The aphid is removed from its stylet and flow is measured through the stylet.
- ☞ The flow has been shown to exert greater pressure and have a higher sugar concentration as you approach the source.

33 PLANT NUTRITION

CHAPTER OUTLINE

Plants must be nourished with CO_2, water and minerals; and require oxygen for cellular respiration. To obtain the required nutrients, plants have an enormous amount of surface area in contact with their environment: roots, shoots, leaves, branches, etc.

I. **NUTRITIONAL REQUIREMENTS OF PLANTS** (*p. 718−722*)

   A. **The Chemical Composition of Plants** (*p. 718−719*)

   Early ideas about plant nutrition:

   1. In the fourth century B.C., Aristotle thought soil was the substance for plant growth and that leaves only provided shade for fruit.

   2. In the 1600's, Jean-Baptiste von Helmont concluded plants grow mainly from the water it absorbs.
      - ☞ To discover if plants grew by absorbing soil, he planted a seed in a measured amount of soil, weighed the grown plant and measured how much soil was left. Soil did not decrease proportionately.

   3. In the 1700's, Stephen Hales postulated plants were nourished by air.

What is the chemical composition of plants including: a) percent of wet weight as water, b) percent of dry weight as organic substances, and c) percent of dry weight as inorganic minerals?

   Plants grow mainly by accumulating water in their cells central vacuoles.
   - ☞ Water is a nutrient since it supplies most of the hydrogen and some oxygen incorporated into organic compounds by photosynthesis.
   - ☞ 80−85% of a herbaceous plant is water.
   - ☞ More than 90% of the water absorbed is lost by transpiration.
   - ☞ The water retained functions as a solvent, it allows cell elongation and keeps cells turgid.

   By weight, the bulk of a plant's organic material is from assimilated CO_2.

   The composition of the dry weight of plants:
   - ☞ 95% is organic substances, mostly carbohydrates.
   - ☞ 5% is minerals, to some extent determined by soil composition.

Plant Nutrition 501

   B. **Essential Nutrients** (*p. 719–720*)

What is an essential nutrient? How is hydroponic culture used to determine which minerals are essential nutrients?

   An *essential nutrient* is one that is required for a plant to grow from a seed and complete its life cycle. (See Campbell, Table 33.1)
   ☞ Determined by hydroponic culture, in which roots are bathed in an aerated aqueous solution of known mineral content. If a mineral is omitted and plant growth is abnormal, it is essential.

What is the difference between a macronutrient and a micronutrient?

   Macronutrients = Elements required by plants in large amounts. (See Campbell, Table 33.1)

   Micronutrients = Elements required by plants in small amounts.
   ☞ These function primarily as cofactors of enzymatic reactions.
   ☞ Optimal concentrations vary for different plant species.

What are the nine macronutrients required by plants? What is their importance in normal plant structure and metabolism? (See Campbell, Table 33.1)

What are seven micronutrients required by plants? Why do plants only need minute quantities of these elements?

   C. **Mineral Deficiencies** (*p. 720–722*)

What determines the symptoms of a mineral deficiency?

   Symptoms of mineral deficiencies depend on:
   1. The role of the nutrient in the plant.
   2. Its mobility within the plant.
      ☞ Deficiencies of nutrients mobile in the plant appear in older organs first since some are preferentially shunted to growing parts.
      ☞ Deficiencies of immobile nutrients affect young parts of plant first because older tissues may have adequate reserves.

   Deficiencies of N, K, and P are the most common.
   ☞ Shortages of micronutrients are less common and often localized.
   ☞ Overdoses of some micronutrients can be toxic.

II. **SOIL** (*p. 722–725*)

Plants growing in an area are adapted to the texture and chemical composition of the soil.

Essential nutrient

Macronutrients

Micronutrients

A. Texture and Composition of Soils (p. 722–723)

How is soil formed?

Soil is produced by the weathering of solid rock. Living organisms may accelerate the process once they become established.

Topsoil = Mixture of decomposed rock of varying texture, living organisms, and humus.

Horizons = Distinct soil layers.

What determines the texture of topsoil? What are the types of soil particles from coarsest to smallest?

The texture of a topsoil depends on the particle size:
- ☞ Coarse sand has diameter of 0.2–2 mm.
- ☞ Sand is 20–200 μm.
- ☞ Silt is 2–20 μm.
- ☞ Clay is less than 2 μm.

What is the composition of loams? Why are they the most fertile soils?

Loams = Mixture of sand, silt and clay; are the most fertile soils because:
- ☞ Fine particles retain water and minerals.
- ☞ Coarse particles provide air spaces with oxygen for cellular respiration.

Soil contains many bacteria, fungi, algae, protists, insects, earthworms, nematodes and plant roots. These affect the soil composition. (For example, earthworms aerate soil; bacteria alter soil mineral composition).

How does humus contribute to the texture and composition of soil?

Humus = Decomposing organic material.
- ☞ Prevents clay from packing together.
- ☞ Builds a crumbly soil that retains water but is still porous for good root aeration.
- ☞ Acts as a reservoir of mineral nutrients.

Soil composition determines which plants may grow in it, and plant growth affects soil characteristics.

B. The Availability of Soil Water and Minerals (p. 723–724)

Why is it that plants cannot extract all of the water in soil?

Some water is bound so tightly to hydrophilic soil that it cannot be extracted by plants.

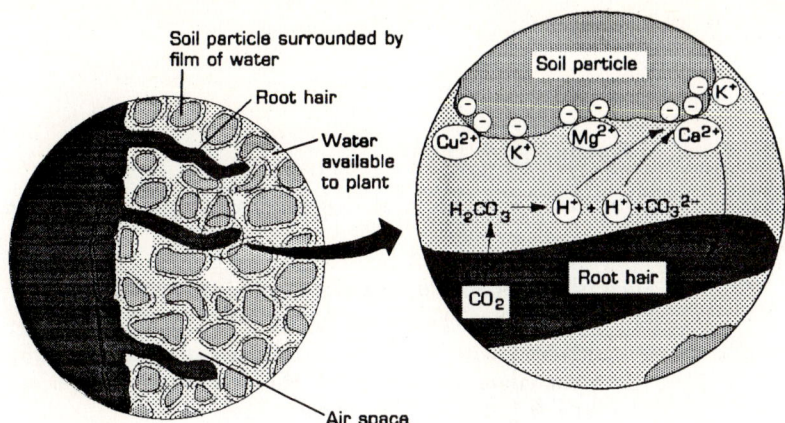

Water bound less tightly is generally available to the plant as a soil solution containing minerals. This solution is absorbed into the root hairs and passes via the apoplast to the endodermis.

How does the presence of clay in soil help prevent leaching of mineral cations?

Positively charged minerals (K^+, Ca^+, Mg^+) adhere by electrical attraction to negatively charged clay particles.
- ☞ Clay provides much surface area for binding.
- ☞ Prevents leaching of mineral nutrients.

What is cation exchange? Why is it necessary for plant nutrition? How can plants stimulate this process?

Cation exchange = H ions in soil displace positively charged mineral ions from clay, making them available to plants.
- ☞ Stimulated by roots which release acids to add H^+ to the soil solution.

Negatively charged minerals (NO_3^-, $H_2PO_4^-$, SO_4^-) are not tightly bound to soil particles.
- ☞ Tend to leach away more quickly.

Water and minerals are depleted from soil surrounding roots, but roots grow to bring the plant into contact with a new supply of nutrients.

    C.   Soil Management *(p. 724–725)*

Why is soil management necessary in agricultural systems, but not in natural ecosystems such as forests and grasslands?

Good soil management is necessary to maintain the fertility which may have taken centuries to develop through decomposition and accumulation of organic matter.

Agriculture is unnatural and depletes the mineral content of the soil, making soil less fertile. Crops also use more water than natural vegetation.

Cation exchange

Three important aspects of soil management are:
1. Fertilizers
2. Irrigation
3. Erosion prevention

1. Fertilizers (p. 724)

What are three mineral elements that are most commonly deficient in farm soils?

Fertilizers may be mined, chemically produced or organic.
- ☞ Usually enriched in N, P, K.
- ☞ Marked with 3 numbers corresponding to % nitrogen (as ammonium or nitrate), % phosphorus (as phosphoric acid), and % potassium (as potash).

What are the environmental consequences of overusing commercial fertilizers?

Organic fertilizers are manure, fishmeal, and compost.
- ☞ Release minerals more gradually than chemical fertilizers, thus, soil retains minerals longer. Excess minerals from chemical fertilizers may be leached from soil and may pollute streams and lakes.

How does soil pH determine the effectiveness of fertilizers and a plant's ability to absorb specific mineral nutrients?

To properly use fertilizers, one must consider the soil pH.
- ☞ Acidity affects cation exchange and the chemical form of the minerals.
- ☞ A change in soil pH may make one essential element more available while causing another to adhere so tightly to soil particles that it is unavailable.

2. Irrigation (p. 724–725)

Availability of water limits plant growth in many environments.

What problems result from farm irrigation in arid regions? What are several current approaches to solving these problems?

Problems of irrigation arid land:
- ☞ Huge drain of water resources.
- ☞ Can gradually make soil salty and infertile.

Solutions to these problems:
- ☞ Use of *drip irrigation*. Perforated pipes slowly drip water close to plant roots, which reduces water evaporation and drainage.
- ☞ Development of plant varieties that require less water or can tolerate more salinity.

Chemical fertilizers

Drip irrigation

3. **Erosion** (*p. 725*)

Wind and water erode away much of the topsoil each year.

What precautions can reduce wind and water erosion?

Measures to prevent these losses are:
- ☞ Rows of trees to divide fields act as windbreaks.
- ☞ Terracing hillsides helps prevent water erosion.
- ☞ Planting alfalfa and wheat provide good ground cover and protection.

Proper management makes soil a renewable resource which will remain fertile for many generations.

III. NITROGEN ASSIMILATION BY PLANTS (*p. 726–729*)

Plants require nitrogen to produce proteins, nucleic acids and other organic molecules.
- ☞ The nitrogen cannot be in gaseous form for plant use, but must be in the form of ammonium (NH_4^+) or nitrate (NO_3^-).

A. Nitrogen Fixation (*p. 726*)

What is nitrogen fixation? What is the overall equation that represents the conversion of gaseous nitrogen to ammonia?

<u>Nitrogen fixation</u> = The process of converting atmospheric nitrogen (gaseous state) to nitrogenous compounds that can be directly used by plants (nitrate or ammonia). (See Campbell, Figure 33.9)
- ☞ The process is catalyzed by the enzyme *nitrogenase*:

$$N_2 + 8e^- + 8H^+ + 16ATP \xrightarrow{\text{nitrogenase}} 2NH_3 + H_2 + 16ADP + 16 \; \textcircled{P}_i$$

- ☞ Some soil bacteria possess nitrogenase.
- ☞ Very energy consuming process, costing the bacteria at least 12 ATPs for each ammonia molecule synthesized.

What are the forms of nitrogen that plants can absorb? How are they used by plants?

- ☞ In the soil, ammonia is converted to the ammonium ion which plants can absorb:
$$NH_3 + H^+ \longrightarrow NH_4^+$$
- ☞ Plants acquire most of their nitrogen in the form of nitrate (NO_3^-), which is produced in soil by *nitrifying bacteria* that oxidize ammonium.
- ☞ Nitrogen absorbed by the plant is incorporated into organic compounds.
- ☞ Most plant species export nitrogen from the roots to the shoots (through xylem) in the form of organic compounds and amino acids.

What is the difference between nitrogen-fixing bacteria and nitrifying bacteria?

Nitrogen fixation

Nitrogenase

Nitrifying bacteria

Plant Nutrition

B. Symbiotic Nitrogen Fixation (p. 727–728)

Legumes (e.g. peas, beans, soybeans, peanuts, alfalfa, clover) have a built-in source of fixed nitrogen because they possess root *nodules*.

<u>Nodules</u> = Root swellings composed of plant cells that contain nitrogen-fixing bacteria of the genus *Rhizobium*. Each species of legume is associated with a particular species of *Rhizobium*.

Beginning with free-living rhizobial bacteria, how does a root nodule develop?

Nodules form as follows: (See Campbell, Figure 33.11)
- Roots secrete chemicals that attract nearby bacteria.
- Attracted bacteria emit chemicals which stimulate root hairs to elongate and curl to prepare for bacterial infection.
- Bacteria enter the root through an "infection thread" that carries them to the root cortex.
- Bacteria become enclosed in vesicles and assume a form called *bacteroids*.
- Bacteroids produce a chemical that induces the host's cells to divide and form a nodule.
- Nodules continue to grow and a connection with the xylem and phloem develops.

Why is the symbiosis between a legume and its nitrogen-fixing bacteria considered to be mutualistic?

This association is mutualistic; the bacteria supplies fixed nitrogen, and the plant supplies carbohydrates and other organic compounds.

What are two functions of leghemoglobin? Why is its synthesis evidence for coevolution?

<u>Leghemoglobin</u> = An iron-containing protein that binds oxygen.
- The plant and the bacteria each make a part of the molecule.
- Releases oxygen for the intense respiration needed to produce ATP for nitrogen fixation.
- Keeps the free oxygen concentration low in root nodules so that the oxygen cannot inhibit the function of nitrogenase.

Most of the ammonium produced is used by the nodules to make amino acids for export to the shoots and leaves.

What is the basis for crop rotation?

The basis for crop rotation is that, under favorable conditions, root nodules fix more nitrogen than the legume uses. The excess is secreted as ammonium into the soil.
- One year a nonlegume crop is planted, and the next year a legume is planted to restore the fixed nitrogen content of the soil.
- Legumes may be plowed under to further increase the fixed nitrogen content of the soil.

Some non-legumes host nitrogen-fixing symbionts.
- ☞ Alders and tropical grasses may host nitrogen-fixing actinomycetes.
- ☞ Rice farms culture a fern (*Azolla*), containing symbiotic nitrogen-fixing cyanobacteria, with the rice.

C. Improving the Protein Yield of Crops (*p. 728–729*)

A majority of the world's population depends mainly on plants for protein. Some food plants have a low protein content while others may be deficient in certain amino acids.

What are some agricultural research methods used to improve the quality and quantity of proteins in plant crops?

Ways in which to increase the protein content of plants are:

1. Plant breeding to create new varieties enriched in protein.
   - ☞ Unfortunately these "super" varieties require large quantities of nitrogen in the form of commercial fertilizer too expensive for many countries to afford.
2. Improving the productivity of symbiotic nitrogen fixation.
   - ☞ Mutant strains of *Rhizobium* have been isolated that continue to produce nitrogenase even after fixed nitrogen accumulates, thus releasing the excess into the soil.
   - ☞ *Rhizobium* varieties may be selected that fix N_2 at a lower cost in photosynthetic energy, yielding a higher total food content in the plant.
3. Genetic engineering.
   - ☞ May create varieties of *Rhizobium* that can infect nonlegumes.
   - ☞ May transfer the genes required for nitrogen-fixation directly into plant genomes, using bacterial plasmids as vectors. (See Campbell, Chapter 19)

IV. SOME NUTRITIONAL ADAPTATIONS OF PLANTS (*p. 729–731*)

What are some nutritional adaptations that have evolved among plants including parasitic plants, carnivorous plants and mycorrhizae?

Modifications to enhance nutrition have evolved among many plants.

A. Parasitic Plants (*p. 729–730*)

Some are photosynthetic but supplement nutrition by using *haustoria* (not homologous to those of parasitic fungi) to obtain xylem sap from its host plant (e.g. mistletoe).
- ☞ Some have ceased photosynthesis entirely, drawing all nutrients from the host plant (e.g. dodder).

Epiphytes grow on the surface of other plants, anchored by roots, but are not parasitic. Epiphytes nourish themselves from the water and minerals absorbed from rain.

Haustoria

Epiphytes

B. Carnivorous Plants (p. 730–731)

Carnivorous plants:
- Live in habitats with poor (usually nitrogen deficient) soil conditions.
- Are photosynthetic, but obtain some nitrogen and minerals by killing and digesting insects.
- Most insect traps evolved by modifications of leaves and usually are equipped with glands that secrete digestive juices.

C. Mycorrhizae (p. 731)

Mycorrhizae = Symbiotic associations (mutualistic) between plant roots and fungi. The fungus either forms a sheath around the root or penetrates root tissue.
- Helps the plant absorb water.
- Absorbs minerals and may secrete acid that increases mineral solubility and converts minerals to forms easily used by the plant.
- May help protect the plant against certain soil pathogens.
- The plant nourishes the fungus with photosynthetic products.

Almost all plants are capable of forming mycorrhizae if exposed to the proper species of fungi. Plants grow more vigorously when mycorrhizae are present.

Mycorrhizae may have permitted early plants to colonize land.
- Fossils indicate the earliest land plants possessed mycorrhizae.
- This mutualistic association may have allowed the early plants to obtain enough nutrients to survive colonization.

34

PLANT REPRODUCTION AND DEVELOPMENT

CHAPTER OUTLINE

Modifications in reproduction were key adaptations enabling plants to spread into a variety of terrestrial habitats.

I. **SEXUAL REPRODUCTION OF FLOWERING PLANTS** (*p. 734–744*)

   A. **The Angiosperm Life Cycle: An Overview** (*p. 735–736*)

In your own words, describe the angiosperm lifecycle. What are the gametophytes? (Refer to Figure 34.2 in the Campbell text.)

The angiosperm life cycle: (See Campbell, Figure 34.2)
- Includes an *alternation of generations*: an alternation between a multicellular haploid form (gametophyte) and diploid for (sporophyte) in the life cycle of a plant.
- The *sporophyte* is the recognizable "plant" most familiar to us. It produces haploid spores by meiosis in *sporangia*.
- Spores will undergo mitotic division and develop into the multicellular male and female *gametophytes*.
- Gametophytes produce gametes (sperm and egg) by mitosis. The gametes fuse to form a zygote which develops into a multicellular sporophyte.
- The sporophyte is dominant in the angiosperm life cycle with the gametophyte stages reduced to being totally dependent on the sporophyte.

Flowers are the reproductive structure of angiosperms.

What are the four floral parts in order from outside to inside of the flower?

Using a diagram of an idealized flower, where are the following structures: a) sepals; b) petals; c) stamen with filament and anther; and d) carpel with style, ovary, ovule and stigma? What are their respective functions?

<u>Alternation of generations</u>

<u>Sporophyte</u>

<u>Sporangia</u>

<u>Gametophytes</u>

- Evolved from compressed shoots with four whorls of modified leaves separated by very short internodes.
- The four sets of modified leaves are the: *sepals, petals, stamens,* and *carpels*.
- Stamens and carpels contain the sporangia.
- Male gametophytes develop in the stamen sporangia and female gametophytes develop in the carpel sporangia.

Stamens and carpels are the reproductive parts of a flower.
- They contain the sporangia where gametophytes develop.
- Female gametophytes develop in carpel sporangia as embryo sacs. This occurs inside the ovules which are at the base of the carpel and surrounded by ovaries.
- Male gametophytes develop in the stamen sporangia as pollen grains. These form at the stamen tips within chambers of the anthers.

Pollination occurs when wind- or animal-born pollen released from anthers lands on the stigma at the tip of a carpel.
- A pollen tube grows from the pollen grain, down the carpel, into the embryo sac.
- Sperm are discharged resulting in fertilization.
- The zygote will develop into an embryo; as the embryo grows, the ovule surrounding it develops into a seed.
- While seed formation is taking place, the entire ovary is developing into a *fruit* which will contain one or more seeds.

Seeds are dispersed from the source plant when fruits are moved about by the wind or animals.
- Seeds deposited in soil of the proper conditions (moisture, nutrients) will *germinate*.
- The embryo starts growing and develops into a new sporophyte.
- After flowers are produced by the sporophyte, a new generation of gametophytes develop and the life cycle continues.

B. More About Flowers (*p. 736–738*)

Flower = The angiosperm structure specialized for sexual reproduction.
- Several variations on the basic flower structure have evolved during the angiosperm history.

What is the difference between a complete and incomplete flower?

Complete flower = A flower with sepals, petals, stamens and carpels.

Plant Reproduction and Development   511

Incomplete flower = A flower that is missing one or more of the parts listed for a complete flower (e.g. most grasses do not have petals on their flowers).

How do perfect flowers differ from imperfect flowers?

Perfect flower = A flower having both stamens and carpels (may be incomplete by lacking either sepals or petals).

Imperfect flower = A flower that is either *staminate* (having stamens but no carpels) or *carpellate* (having carpels but no stamens) — a unisex flower.

What is the difference between monoecious and dioecious?

Monoecious = Plants having both staminate flowers and carpellate flowers on the same individual plant.

Dioecious = Plants having staminate flowers and carpellate flowers on separate individual plants of the species.

There are other variations in the overall flower structure. (Refer to Campbell, Figure 34.4.)

C. **Pollen Development: A Closer Look** (*p. 738*)

Pollen grain = The immature male gametophyte that develops within the anthers of stamens in an angiosperm.
- ☞ Are extremely durable. Their tough coats are resistent to biodegradation.
- ☞ Fossilized pollen has provided many important evolutionary clues.

How do pollen grains form in angiosperms? What will be the functions of the generative and tube cell nuclei?

- ☞ Formation of a pollen grain is as follows: (See Campbell, Figure 34.5a)

Within the sporangial chamber of an anther, diploid microsporocytes undergo meiosis to form four haploid microspores.
↓
The haploid microspore nucleus undergoes mitotic division to give rise to a *generative cell* and a *tube cell*.
↓
The wall of the microspore then thickens and becomes sculptured into a species-specific pattern.
↓
These two cells and the thickened wall are the pollen grain, an immature male gametophyte.

Incomplete flowers

Perfect flower

Imperfect flower

Monoecious

Dioecious

Pollen grain

Microspore

Generative cell

Tube cell

D. The Development of Ovules: A Closer Look (p. 738–739)

Ovules = Structures which form within the chambers of the plant ovary and contains the female gametophyte.

What is the female gametophyte in angiosperms? Where and how does it develop? What happens to each of its cells?

The female gametophyte is the *embryo sac*, and it develops as follows:

A megasporocyte in the sporangium of each ovule grows and goes through meiosis to form 4 haploid *megaspores* (only one usually survives).

↓

The remaining megaspore grows and its nucleus undergoes 3 mitotic divisions, forming 1 large cell with 8 haploid nuclei.

↓

Membranes partition this into a multicellular *embryo sac*.

Within the embryo sac:
- ☞ The egg cell is located at one end.
- ☞ 2 cells (*synergids*) flank the egg cell.
- ☞ At the opposite end are 3 antipodal cells.
- ☞ The other 2 nuclei (*polar nuclei*) share the cytoplasm of the large central cell.
- ☞ At the end containing the egg is the *micropyle* (an opening through the integuments surrounding the embryo sac).

E. Pollination and Fertilization (p. 739–740)

What is the difference between pollination and fertilization? How can pollen be transferred between flowers?

Pollination = The placement of pollen onto the stigma of a carpel.
- ☞ Some plants use wind to disperse pollen.
- ☞ Other plants have relationships with animals that transfer pollen directly between flowers.
- ☞ Some plants self-pollinate, but most have mechanisms that make self-pollination difficult or prevent self-pollination, ensuring genetic variety.

What are some mechanisms that prevent self-pollination? How does this contribute to genetic variation?

Most monoecious angiosperms have mechanisms to prevent self-pollination. These mechanisms thus contribute to genetic variation in the species by ensuring sperm and eggs are from different plants.
- ☞ The stamens and carpels mature at different times in some species.

☞ Structural arrangement of the flower in many species pollinated by animals reduces the chance that pollinators will transfer pollen from anthers to the stigma of the same flower.
☞ Other species are *self-incompatible*, a single-gene-based mechanism. If the pollen and stigma are homozygous for this gene (from the same flower) a biochemical block prevents the pollen grain from developing and fertilizing the egg.

What happens during double fertilization? What is the function of endosperm?

Angiosperms undergo *double fertilization*. (See Campbell, Figure 34.6)   *Double fertilization*
☞ After adhering to a stigma, the pollen grain germinates and extends a pollen tube between the cells of the style toward the ovary.
☞ The generative cell divides (mitosis) to form two sperm. (Pollen grain with tube enclosing two sperm = male gametophyte.)
☞ Directed by a chemical (usually calcium), the pollen tube enters through the micropyle and discharges its two sperm nuclei into the embryo sac.
☞ One sperm unites with the egg to form the zygote.
☞ The other sperm combines with the two polar nuclei to form a 3N nucleus in the large central cell of the embryo sac. This central cell will give rise to the *endosperm* which is a food storing tissue.   *Endosperm*

What happens after fertilization?

After fertilization is completed, each ovule will develop into a seed and the   *Seed*
ovary will develop into a fruit surrounding the seed(s).

F.   **The Seed** (*p. 740–741*)

Endosperm development begins before embryo development.
☞ The triploid nucleus divides to form a milky, multinucleate "supercell."
☞ This endosperm undergoes cytokinesis to form membranes and cell walls between the nuclei. Endosperm is rich in nutrients, which it provides to the developing embryo. In many dicots, food reserves of the endosperm are restocked in cotyledons, thus mature seeds have no endosperm.

How does a plant embryo develop from the first mitotic division to an embryonic plant with rudimentary organs?

In the development of the embryo: (See Campbell, Figure 34.7)
☞ The zygote's first mitotic division is transverse, creating a larger basal cell and a smaller terminal cell.
☞ The basal cell divides transversely to form the *suspensor*, which   *Suspensor*
anchors the embryo and transfers nutrients to it from the parent plant.
☞ The terminal cell divides to form a spherical *proembryo* attached to   *Proembryo*
the suspensor.
☞ Cotyledons appear as bumps on the proembryo and the embryo   *Cotyledons*
elongates.   *Protoderm*
☞ The embryo contains *protoderm*, *ground meristem* and *procambium*.   *Ground meristem, Procambium*

- Between the cotyledons is the apex of the embryonic shoot with its meristem.
- Where the suspensor attaches is the apex of the embryonic root with its meristem.
- After germination, the meristems at the root and shoot tips will sustain primary growth.
- Cells of the embryo have equivalent nuclei but differ in cytoplasmic composition and location. Position of an embryonic cell and its cytoplasmic contents helps direct development by providing cues to the genome of the nucleus.

Referring to Figure 34.7 in the Campbell text, where are the following structures of a seed: a) seed coat, b) embryo, c) hypocotyl, d) radicle, e) epicotyl, f) plumule, g) endosperm, h) cotyledons and i) shoot apex? What are their respective functions?

In mature seeds, the embryo is quiescent until germination.
- The seed dehydrates until its water content is only 5–15% by weight.
- The embryo is surrounded by endosperm, cotyledons, or both.
- The seed coat is formed from the integuments of the ovule.
- Below the cotyledons, the embryonic axis is termed the *hypocotyl*, which terminates in the *radicle*, or embryonic root.
- Above the cotyledons, the embryonic axis is termed the *epicotyl*, which terminates in the *plumule* (shoot tip with a pair of tiny leaves).

How do a monocot and dicot seed differ?

- The embryo of a monocot is enclosed in a sheath of *coleorhiza* covering the root and the *coleoptile*, covering the shoot. The *scutellum* (single cotyledon) has a large surface area and absorbs nutrients from the endosperm during germination.

G. **Development of the Fruit** (p. 741–742)

What is a fruit? How do fruits form?

A fruit is a ripened ovary.
- Pollination triggers hormonal changes that cause the ovary to grow.
- The wall thickens to become the *pericarp*.
- The transformation of a flower into a fruit parallels seed development and is called *fruit set*.
- In most plants, fruit does not develop without fertilization of ovules. (In *parthenocarpic* plants, fruit does develop without fertilization.)

How can fruits be classified? What is an example of each type?

Depending upon their origin, fruits can be classified as:
1. Simple fruits.
   - Fruit derived from a single ovary.
   - For example, cherry (fleshy) or soybean (dry).

☞ Mitosis resumes in the root and shoot apical meristems after germination.
☞ Enlargement of these newly produced cells results in most of the actual size increase.

Morphogenesis is the development of body shape and organization.
☞ Begins in the early divisions of the embryo to produce the cotyledons and rudimentary roots and shoots.
☞ Continues to shape the root and shoot systems as the plant grows.

Cellular differentiation is the divergence in structure and function of cells as they become specialized during the plant's development.
☞ Each cell of each organ is fixed in a certain location and performs a specific function (e.g. guard cells, xylem).
☞ Every organ of the plant body has a diversity of cells within its total structure.

1. **The Roles of Cell Division and Cell Expansion in Morphogenesis** (*p. 749–750*)

What is the consequence of cell expansion in morphogenesis?

Plant shape depends on the spatial orientations of cell divisions and cell expansions.
☞ Plant cells cannot move about as individuals within a developing organ due to their cell walls being cemented to those of neighboring cells.
☞ Since movement is eliminated, when the cell elongates, its growth is perpendicular to the orientation plane of division.

2. **The Cytoskeleton: Orienting the Plane of Cell Division** (*p. 750*)

How does the cytoskeleton orient the plane of division in plant cells?

During late interphase, the cytoskeleton of the cell becomes rearranged with the microtubules of the outer cytoplasm becoming concentrated into the *preprophase band*.
☞ The microtubules of the preprophase band disperse leaving behind an array of actin microfilaments.
☞ These microfilaments hold the nucleus in a fixed orientation until the spindle forms and then direct movement of the vesicles that produce the cell plate.
☞ The walls that develop at the end of cell division form along the plane established by the preprophase band.

3. **The Cytoskeleton: Determining the Direction of Cell Expansion** (*p. 750–752*)

How do plant cells grow?

Plant cells expand (elongate) when the cell wall yields to the turgor pressure of the cell.
☞ Cross-links between cellulose microfibrils in the cell wall are weakened by an acid secreted by the cell.
☞ The loosened wall permits uptake of water by the hyperosmotic cell; water uptake causes the cell to expand.

Morphogenesis

Cellular differentiation

Preprophase band

- Growth continues until the cross links become re-established firmly enough to offset the turgor pressure.
- About 90% of the cell's expansion is due to water uptake, although some cytoplasm is also produced by the cell.
- Most of the water entering the cell is stored in the large central vacuole which forms due to coalescence of small vacuoles as the cell grows.

Why do plant cells not expand equally in all directions when they grow?

Plant cells show very little increase in width as they elongate.
- The cellulose microfibrils in the innermost cell wall layers stretch very little; consequently, the cell expands in the direction perpendicular to the orientation of the microfibrils.
- Alignment of microfibrils in the wall mirrors the microtubule orientation found in the cortex. This is believed to result from microtubular control of the flow of cellulose-producing enzymes along the membrane.

B. Cellular Differentiation: The Basic Problem (p. 752)

What is the relationship between protein synthesis and the differentiation process?

The progressive development of specialized structures and functions in plant cells reflects the different types of proteins synthesized by different types of cells.

Xylem cells function in both transport within the plant and structural support.
- Cell walls are hardened by lignin which is produced by enzymes made by the cell.
- The final stage of differentiation includes the production of hydrolytic enzymes which destroy the protoplast. This leaves only the cell walls intact and permits the movement of xylem sap through the cells.

Guard cells regulate the size of the stomatal opening.
- Must have flexible walls, thus the enzymes that produce lignin are not produced.
- The protoplast remains intact and regulates ion exchange necessary to increase and decrease turgor.

If all cells in a plant have the same genome, what controls differentiation?

All cells in a plant possess a common genome. This has been proven by cloning whole plants from single somatic cells.
- All the genes necessary are present since these cells dedifferentiate in tissue cultures and then redifferentiate to produce the diversity of cells found in the plant.
- This ability indicates that differentiation is controlled by gene expression leading to the production of specific proteins.
- Different cell types (like xylem cells and guard cells) selectively express certain proteins at different times during their development; this results in the different developmental pathways that gives rise to the diverse cell types.

C. Pattern Formation and Positional Information (p. 752)

The organization in a plant can be seen in the characteristic pattern of cells in each tissue, the pattern of tissues in each organ, and the spatial organization of the organs on the plant.

Pattern formation = The development of specific structures in specific locations.
- ☞ Depends on *positional information*.

How does positional information influence pattern formation? How might morphogens provide positional information?

Positional information = Signals indicating a cell's location relative to other cells in an embryonic mass.
- ☞ Genes respond to these signals and their response effects the localized rates and planes of cell division and expansion.
- ☞ This signal detection continues in each cell as the organs develop and cells respond by differentiating into particular cell types.

Several hypotheses have been proposed as to how embryonic cells detect their positions. One hypothesis about positional information transmission is that it relies on gradients of chemicals called *morphogens*.
- ☞ A morphogen might diffuse from a shoot's apical meristem and the decreasing gradient farther from the source would indicate to cells their relative position from the tip.
- ☞ A second morphogen released from the outermost lateral cells would diffuse inward indicating the radial position to each cell.
- ☞ Each cell could thus determine its relative longitudinal and radial position from the gradients of these two morphogens.

D. Clonal Analysis of the Shoot Apex (p. 752–753)

What is clonal analysis? Using this technique, what have researchers learned about the role of positional information in differentiation?

Positional information is the basis for the processes involved in plant development: growth, morphogenesis, and differentiation. Plant developmental biologists have developed the technique of *clonal analysis* to study the relationships of these processes.
- ☞ Maps the cell lineages derived from each cell of the apical meristem, noting their position as the plant organs develop.
- ☞ Mapping is possible due to induced somatic mutations in each cell which can be used to distinguish that cell and its derived cells from neighboring cells.

This technique has been used to determine that the developmental fates of cells in the apex are somewhat predictable.
- ☞ Almost all cells developing from the outermost meristematic cells become part of the dermal tissues of leaves and stems.

It is not currently possible to predict what meristematic cells will develop into specific tissues and organs.
- ☞ The outermost cells usually divide on a plane perpendicular to the shoot surfaces.

Pattern formation

Positional information

Morphogens

Clonal analysis

- ☞ Random changes sometimes result in one of these cells dividing on a plane parallel to the surface; this indicates meristematic cells are not dedicated early in their development to forming specific tissues and organs.
- ☞ Consequently, it is the cell's final position in a developing organ which determines what type of cell it becomes, not the particular cell lineage to which it belongs.

E. Homeotic Mutations in Flower Development: In Search of the Genetic Basis of Pattern Formation (p. 753–754)

The shoot tip in flowering plants shifts from indeterminate growth to determinate growth when the flower is produced.
- ☞ The meristem is consumed during the formation of primordia for sepals, petals, stamens, and carpels.

What experimental evidence implicates organ-identity genes in the development of normal floral pattern?

Positional information is responsible for committing each primordium to develop into the organ of specific structure and function.
- ☞ Some of the genes that function in development of the floral pattern and are regulated by positional information have been identified.
- ☞ These *organ-identity genes* are responsible for development of normal floral patterns.
- ☞ Occasionally a normal floral pattern will not develop; for example, an extra whorl of sepals may develop instead of petals.
- ☞ These abnormal floral patterns are due to *homeotic mutations* in organ-identity genes which result in one type of body part developing where another part normally occurs.

Why are many plant developmental biologists using Arabidopsis thaliana as an experimental organism?

Arabidopsis thaliana is the experimental organism many plant biologists are now using to study plant development.
- ☞ Has a small genome that simplifies the search for specific genes.
- ☞ Several homeotic mutations affecting floral patterns have been identified.

What is one hypothesis for how positional information determines which organ-identity gene is expressed?

One hypothesis about how positional information influences a particular floral-organ primordium is based on the overall genetic basis of pattern formation.
- ☞ Organ-identity genes code for transcription factors that are regulatory proteins which help control expression of other genes.
- ☞ This control involves binding of the transcription factor to specific sites on the DNA, which affects transcription.
- ☞ It is believed that positional information determines which organ-identity gene is expressed, and the resulting transcription factor induces expression of these genes controlling development of specific organs.

Organ-identity genes

Homeotic mutations

35

CONTROL SYSTEMS IN PLANTS

CHAPTER OUTLINE

How can plants respond to their environment?

Plants respond to the environment.
- ☞ They can send signals between different parts of the plant.
- ☞ They track the time of day and the time of year.
- ☞ They sense and respond to gravity and direction of light, etc.
- ☞ They respond to environmental cues by adjusting their growth pattern and development.
- ☞ These control systems evolved as adaptations from interaction with the environments.

I. THE SEARCH FOR A PLANT HORMONE: A CASE STUDY IN THE SCIENTIFIC PROCESS (p. 757–758)

<u>Hormone</u> = A compound produced by one part of an organism that is translocated to other parts where it triggers a response in target cells and tissues.

<u>Phototropism</u> = Growth toward or away from light (e.g. growth of a shoot toward light).
- ☞ Results form differential growth of cells in opposite sides of a shoot or, in the case of a grass seedling, coleoptile.
- ☞ Cells on the darker side elongate faster.

What experimental evidence led to the discovery of a plant hormone? What were the individual contributions of Charles and Francis Darwin, Peter Boysen-Jensen, F.W. Kent and Kenneth Thimann?

Experiments on phototropism led to the discovery of a plant hormone.

1. Charles and Francis Darwin removed the tip of the coleoptile from a grass seedling (or covered it with an opaque cap) and it failed to grow toward light.
   - ☞ Concluded that the coleoptile tip was responsible for sensing light, and, since the curvature occurs some distance below the tip, the tip sends a signal to the elongating region.
2. Peter Boysen-Jensen separated the tip from the remainder of the coleoptile by a block of gelatin, preventing cellular contact but allowing chemical diffusion.
   - ☞ Seedlings behaved normally.
   - ☞ If an impenetrable barrier was substituted, no phototropic response occurred.
   - ☞ Demonstrated that the signal was a mobile substance.

Hormone

Phototropism

Charles and Francis Darwin

Peter Boysen-Jensen

524 Control Systems in Plants

F. W. Went

3. F.W. Went removed the coleoptile tip, placed it on an agar block, and then put the agar (without the tip) on decapitated coleoptiles kept in the dark.
    - ☞ If block was placed off-center, the plant curved away from the side with the block.
    - ☞ Concluded the agar block contained a chemical that diffused into it from the coleoptile tip, and that this chemical stimulated growth.
    - ☞ Went called this chemical an *auxin*.
4. Auxin was purified and characterized by Kenneth Thimann.

II. **FUNCTIONS OF PLANT HORMONES** (*p. 759–765*)

How may plant hormones control growth and development in a plant?

Hormones:
- ☞ Control plant growth and development by affecting division, elongation, and differentiation of cells.
- ☞ Effects depend on site of action, stage of plant growth and hormone concentration.
- ☞ The hormonal signal is amplified, perhaps by affecting gene expression, enzyme activity, or membrane properties.
- ☞ Reaction to hormones depends on hormonal balance (relative concentration of one hormone compared with others).

What are five classes of plant hormones? What are their major functions? Where are they produced in the plant?

So far, five classes of plant hormones have been identified. (See Campbell, Table 35.1)

1. Auxin (such as IAA)
2. Cytokinins (such as zeatin)
3. Gibberellins (such as GA_3)
4. Abscisic acid
5. Ethylene

  A. Auxin (*p. 759–761*)

Auxin

Auxin = A hormone that promotes elongation of young developing shoots or coleoptiles.
- ☞ Natural auxin is a compound named indoleacetic acid (IAA).

Indoleacetic acid

The apical meristem is a major site of auxin production.
- ☞ Stimulates cell growth only at concentrations between $10^{-8}-10^{-3}$ M.
- ☞ Moves from the apex down to the zone of cell elongation at a rate of 10 mm per hour.

What is a possible mechanism for polar transport of auxin?

Polar transport

Polar transport of auxin is unidirectional and requires metabolic energy. (See Campbell, Figure 35.5)
- ☞ IAA is actively transported down a stem by auxin carriers located on the basal ends of cells (carriers are absent on the apical ends).

- Movement of auxin is aided by the differences in pH between the acidic cell wall and the neutral cytoplasm.
- Energy is provided by a chemiosmotic mechanism.

According to the acid-growth hypothesis, how can auxin initiate cell elongation?

The acid-growth hypothesis states that cell elongation is due to stimulation of a proton pump which acidifies the cell wall.

- Acidification causes the crosslinks between the cellulose myofibrils of the cell walls to break.
- This loosens the wall, allowing turgor pressure to elongate the cell.

Along with functions listed in Campbell, Table 35.1, auxin also affects secondary growth by inducing vascular cambium cell division and differentiation of secondary xylem.

Why is 2,4-D widely used as a weed killer?

Auxins are used as herbicides. 2,4-D is a synthetic auxin which affects dicots selectively, allowing removal of broadleaf weeds from a lawn or grain field.

B. Cytokinins (p. 761–762)

What is a cytokinin? How does the ratio of cytokinin to auxin affect cell division and cell differentiation?

Cytokinins = Modified forms of adenine that stimulate cytokinesis.
- Move from the roots to target tissues by moving up in the xylem sap.
- Stimulate RNA and protein synthesis.

Cytokinins function in several areas of plant growth:
- Cell division and differentiation.
- Apical dominance.
- Anti-aging hormones.

Cytokinins, in conjunction with auxin, control cell division and differentiation.
- Stem parenchyma cells cultured without cytokinins grow very large and do not divide.
- Cytokinins added alone have no effect on cells grown in tissue culture.
- Equal concentrations of cytokinins and auxins stimulate cells to grow and divide, but they remain an undifferentiated callus.
- More cytokinin than auxin causes shoot buds to develop from the callus.
- More auxin than cytokinin causes roots to form.
- The new proteins produced by stimulation of RNA appear to be involved in cell division.

Margin notes: Acid-growth hypothesis; 2,4-D; Cytokinins

What is apical dominance? How does the check-and-balance control of lateral branching by auxins and cytokinins help the plant balance growth of the shoot and root systems?

Cytokinins and auxin are antagonistic and contribute to *apical dominance*, the terminal bud's ability to suppress axillary bud development.
- ☞ Auxin restrains axillary bud growth causing the shoot to lengthen.
- ☞ Cytokinins stimulate axillary bud growth.
- ☞ Auxin cannot suppress axillary bud growth once it has begun.
- ☞ Lower buds thus grow before higher ones since they are closer to the cytokinin source than the auxin source.
- ☞ Auxin stimulates growth of branch roots while cytokinins restrain it.
- ☞ This stimulation-inhibition action may help balance plant growth since an increase in the root system would signal the plant to produce more shoots.

Cytokinins can retard aging of some plant organs, perhaps by inhibiting protein breakdown, stimulating RNA and protein synthesis and mobilizing nutrients.
- ☞ May slow leaf deterioration on plants since detached leaves dipped in a cytokinin solution stay green longer.

C. Gibberellins (*p. 762–763*)

More than 70 different gibberellins, many naturally occurring, have been identified.

How does stem elongation and fruit growth depend upon a synergism between auxin and gibberellins?

Gibberellins are produced primarily in roots and young leaves.
- ☞ Stimulate growth in leaves and stems but show little effect on roots.
- ☞ Stimulate cell elongation and division in stems, possibly in conjunction with auxin.
- ☞ Gibberellins cause *bolting* (rapid growth of floral stems, which elevates flowers).

Fruit development is controlled by both gibberellins and auxin.
- ☞ In some plants both must be present for fruit set.
- ☞ Commercial application in spraying Thompson seedless grapes which grow large and farther apart after treatment.

What is the probable mechanism by which gibberellins trigger seed germination?

The release of gibberellins signals seeds to break dormancy and germinate.
- ☞ A high concentration of gibberellins is found in many seeds, especially in the embryo.
- ☞ Imbibed water appears to stimulate gibberellin release.
- ☞ Environmental cues may also cause gibberellin release in seeds which require special conditions to germinate.

☞ In cereal grains, gibberellins stimulate germination and support growth by stimulating synthesis of mRNA which codes for α-amylase. The α-amylase then digests the stored nutrients, making them available to the embryo and seedling.

In breaking both seed dormancy and bud dormancy, gibberellins act antagonistically with abscisic acid, which inhibits plant growth.

D. Abscisic Acid (ABA) (p. 764)

How does abscisic acid (ABA) help prepare a plant for winter?

Abscisic acid helps prepare plants for winter by suspending both primary and secondary growth.
- ☞ Directs leaf primordia to develop scales that protect dormant buds.
- ☞ Inhibits cell division in vascular cambium.

How does the ratio of ABA to gibberellin control seed dormancy? What effect does ABA have as a "stress hormone"?

The onset of seed dormancy is another time it is advantageous to suspend growth.
- ☞ The ratio of ABA:gibberellins determines whether seeds remain dormant or germinate in most cases.
- ☞ In other plants, seeds germinate when ABA is washed out of the seeds (desert plants) or degraded by some other stimulus such as sunlight.

ABA also acts as a stress hormone, closing stomata in times of water-stress.

E. Ethylene (p. 764–765)

What is the role of ethylene in plant senescence, fruit ripening and leaf abscission?

Ethylene = A gaseous hormone that diffuses through air spaces between plant cells. High auxin concentrations induce release of ethylene, which acts as a growth inhibitor.

Senescence (aging) is a natural process in plants that may occur at the cellular, organ, or whole plant level. Ethylene probably plays an important role at each level.
- ☞ Examples include: tracheid and cork cells which die before becoming fully functional; leaf fall in the autumn; withering of flowers; death of annuals after flowering.
- ☞ The best studied forms of senescence are fruit ripening and leaf abscission.

During fruit ripening, ethylene triggers senescence, and then the aging cells release more ethylene.
- ☞ The breakdown of cell walls and loss of chlorophyll are considered aging processes.
- ☞ The signal to ripen spreads from fruit to fruit since ethylene is a gas.

Abscisic acid

Stress hormone

Ethylene

Leaf abscission

What is leaf abscission? What are two environmental stimuli for leaf abscission?

Leaf abscission is an adaptation that prevents deciduous trees from desiccating during winter when roots cannot absorb water from the frozen ground.
- ☞ Before abscission, the leaf's essential elements are shunted to storage tissues in the stem from which they are recycled to new leaves in the spring.
- ☞ Environmental stimuli are shortening days and cooler temperatures.

When a leaf falls, the breakpoint is an abscission layer near the petiole base.
- ☞ Weak area since the small parenchyma cells have very thin walls and there are no fiber cells around the vascular tissue.
- ☞ Mechanics of abscission are controlled by a change in the balance of ethylene and auxin.
- ☞ Auxin decrease makes cells in the abscission layer more sensitive to ethylene. Cells then produce more ethylene which inhibits auxin production.
- ☞ Ethylene induces synthesis of enzymes that digest the polysaccharides in the cell walls, further weakening the abscission layer.
- ☞ Wind and weight cause the leaf to fall by causing a separation in the abscission layer.
- ☞ Even before the leaf falls, a layer of cork forms a protective scar on the twig's side of the abscission layer. The cork prevents pathogens from entering the plant.

III. PLANT MOVEMENTS (p. 766–769)

A. Tropisms (p. 766–767)

Tropisms

What is tropism? What are three primary stimuli that induce tropisms and a consequent change of body shape?

Tropisms = Growth responses that result in curvatures of whole plant organs toward or away from stimuli.
- ☞ Mechanism is a differential rate of cell elongation on opposite sides of the organ.

There are 3 primary stimuli which result in tropisms:
1. Phototropism is caused by light.
2. Gravitropism is a response to gravity.
3. Thigmotropism is a response to touch.

1. Phototropism (p. 766)

How does light cause a phototropic response?

Phototropism

Phototropism = Cells on the darker side of shoot elongate faster than cells on the bright side due to asymmetric distribution of auxins from the shoot tip.

Control Systems in Plants

- ☞ A photoreceptor sensitive to blue light is present in the shoot tip; this receptor is believed to be a yellow pigment related to riboflavin.
- ☞ Auxins move laterally across the tip from the bright to dark side by an unknown mechanism.

2. **Gravitropism** (*p. 766–767*)

<u>Gravitropism</u> = Orientation of a plant to a field of gravity.
- ☞ Roots display positive gravitropism, shoots display negative gravitropism.

How do plants apparently tell up from down? What is a plausible explanation for why roots display positive gravitropism?

The possible mechanism of gravitropism in roots:
- ☞ Specialized plastids containing dense starch grains (*statoliths*) aggregate in the low points of plant cells.
- ☞ Statoliths occur in certain root cap cells and, in the shoot, in parenchyma cells adjacent to vascular bundles.
- ☞ Aggregating statoliths trigger calcium redistribution which results in lateral transport of auxin in the root.
- ☞ The calcium and auxin are on the lower side of the elongation zone.
- ☞ Roots curve down because auxin <u>inhibits</u> root cell elongation at high concentration resulting in the cells on the upper side elongating faster than those on the lower side.
- ☞ The tropism continues until the root is growing straight down.

The mechanism in shoots is uncertain.

3. **Thigmotropism** (*p. 767*)

What is thigmotropism? Thigmomorphogenesis?

<u>Thigmotropism</u> = Directional growth in response to touch.
- ☞ Contact of tendrils stimulates a coiling response caused by differential growth of cells on opposite sides of the tendril.

<u>Thigmomorphogenesis</u> = Decrease in stem length with concomitant thickening caused by mechanical perturbation.
- ☞ Results from increased production of ethylene in response to chronic mechanical stimulation.

B. **Turgor Movements** (*p. 767–769*)

How can motor organs with pulvini cause rapid leaf movements and sleep movements?

<u>Turgor movements</u> = Reversible movements caused by changes in turgor pressure of specialized cells in response to stimuli.

Rapid leaf movements occur in plants such as *Mimosa*.
- ☞ When a compound leaf is touched, it collapses and folds together.

Gravitropism

Statoliths

Thigmotropism

Thigmomorphogenesis

Turgor movements

Rapid leaf movements

Control Systems in Plants

Pulvini

- ☞ Results from rapid loss of turgor within *pulvini* (special motor organs located in leaf joints).
- ☞ Motor cells lose potassium, which causes water loss by osmosis.

What is a plausible explanation for how a stimulus that causes rapid leaf movement can be transmitted through the plant?

- ☞ The stimulus and response travels wavelike through the plant at 1 cm/sec.
- ☞ This transmission is correlated with action potentials (electrical impulses) resembling those in animals but thousands of times slower.

Sleep movements

Sleep movements = Movement of leaves to a vertical position in evening and raising of leaves to a horizontal position in morning.
- ☞ Occurs in many legumes.
- ☞ Due to daily changes in turgor pressure of motor cells of pulvini.
- ☞ Migration of potassium ions from one side of the pulvinus to the other is the osmotic agent leading to reversible uptake and loss of water by motor cells.

IV. CIRCADIAN RHYTHMS AND THE BIOLOGICAL CLOCK (p. 769)

What is circadian rhythm? What happens when an organism is artificially maintained in a constant environment?

Circadian rhythm

Circadian rhythm = A physiological cycle with a frequency of about 24 hours that persists even when an organism is sheltered from environmental cues.
- ☞ Ubiquitous feature of all eukaryotes.

What are some common factors that entrain biological clocks?

Although the oscillator is endogenous, it is set to a period of precisely 24 hours by daily signals from the environment.
- ☞ The 24 hour cycle is synchronized with the Earth's light-dark cycle.
- ☞ When sheltered from environmental cues may deviate from 24 hours (called *free-running periods*) and can vary from 21 to 27 hours.

Free-running periods

V. PHOTOPERIODISM (p. 769–773)

What is photoperiodism?

Photoperiodism

Photoperiodism = A physiological response to day length.
- ☞ Seasonal events are important in plant life cycles.
- ☞ Plants detect the time of year by the *photoperiod* (relative lengths of night and day).

A. Photoperiods Control of Flowering (p. 769–771)

What are the differences among short-day plants, long-day plants and day-neutral plants? How do they depend upon critical night length?

W.W. Garner and H.A. Allard postulated that the amount of day length controls flowering and other responses to photoperiod.
- *Short-day plants* generally flower in late summer, fall and winter.
- *Long-day plants* generally flower in late spring and summer.
- *Day-neutral plants* are unaffected by photoperiod.

In the 1940's, it was discovered that a *critical night length*, not day length, actually controls flowering and other responses to photoperiod.
- If daytime period is broken by a brief exposure to darkness, there is no effect on flowering.
- If nighttime period is interrupted by brief exposure to light, photoperiodic responses are disrupted.
- Therefore, short-day plants flower if night is longer than a critical length and long-day plants need a night shorter than a critical length.

Some plants flower after a single exposure to the proper photoperiod.
- Some require several successive days of the proper photoperiod to bloom.
- Still others respond to photoperiod only if they have been previously exposed to another stimulus. For example, *vernalization* is a requirement for pretreatment with cold before flowering.

What evidence is there for the existence of a florigen?

Leaves detect the photoperiod while buds produce flowers.
- Only requires one leaf for a plant to detect photoperiod and floral buds develop.
- Most plant physiologists believe an as yet unidentified hormone called *florigen* is produced in the leaves and moves to the buds.

B. Phytochrome (*p. 771–773*)

Phytochrome = A protein containing a *chromophore* (light-absorbing component) responsible for a plant's response to photoperiod.
- Red light (λ of 660 nm) is most effective in interrupting night length.
- If an R flash (flash of red light) is followed by a far-red (FR) flash of light (λ of 730 nm), the plant perceives no interruption of night length. Only the wavelength of the last flash affects the plant's measurement of night length.

How can the interconversion of phytochrome act as a switching mechanism to help plants detect sunlight and trigger many plant responses to light?

Phytochrome alternates between two forms: P_r (red absorbing) and P_{fr} (far-red absorbing). The $P_r \rightleftarrows P_{fr}$ interconversion is a switching mechanism controlling various plant events.
- Plants synthesize P_r and, if kept in dark, it remains as P_r, but if the phytochrome is illuminated, some $P_r \longrightarrow P_{fr}$.
- P_{fr} triggers many plant responses to light.

532   Control Systems in Plants

> A shift in the $P_r \rightleftarrows P_{fr}$ equilibrium might cause changes (germination, increased growth) which would adjust the plant's growth and development in response to some environmental changes.

C. Role of the Biological Clock in Photoperiodism (p. 773)

In darkness, P_{fr} gradually reverts to P_r. This occurs every day after sunset.

Plants do not use the disappearance of P_{fr} to measure night length since:
> The conversion is complete within a few hours after sunset.
> Temperature affects the conversion rate, thus, it would not be reliable.

Night length is measured by the biological clock.
> Perhaps phytochrome synchronizes the clock to the environment.
> Clock measures night length very accurately (some short-day plants will not flower if night is even 1 minute shorter than the critical length).

VI. SIGNAL-TRANSDUCTION PATHWAYS IN PLANT CELLS (p. 773–775)

What is a signal-transduction pathway? What are the three main steps in signal transduction?

Signal-transduction pathway
Signal-transduction pathway = A mechanism linking a mechanical or chemical stimulus to a cellular response.

All hormones and environmental stimuli act on plant cells through signal-transduction pathways. Three steps are involved in each pathway: reception, transduction and induction. (See Campbell, Figure 35.18)

Reception
Reception is the detection of a hormone or environmental stimulus by the cell.
> May take various forms depending on the stimulus: absorption of a particular wavelength of light by a pigment within a cell; or the binding of a hormone to a specific protein receptor in the cell or on its membrane.

Target cells
> Reception of a hormone only occurs in *target cells* for that hormone. Target cells possess the specific protein receptor to which the hormone must bind; other cells do not possess the receptor.

Using the second messenger model as an example, how does a hormone elicit a response by the target cell? How does signal transduction amplify a response?

Transduction
Transduction in the pathway results in an amplification of the stimulus and its conversion into a chemical form which can activate the cell's responses.
> The hormone (first messenger) binds to a specific receptor and the hormone-receptor combination stimulates the *second messenger* (= a substance that increases in concentration within a cell stimulated by the first messenger).

Second messenger
> The receptor may be bound to the cell membrane and its activation result in a chemical change to the cell.

- ☞ Calcium ions appear to be important second messengers in many plant responses. Calcium ion concentration increases in the cell and the ions bind to the protein *calmodulin*.
- ☞ The calmodulin-calcium complex then activates other target molecules within the cell.
- ☞ Amplification of the signal results from a single first messenger molecule binding to its receptor giving rise to many second messengers which activate an even larger number of proteins and other molecules.

Induction is the pathway step in which the amplified signal induces the cell's specific response to the stimulus.
- ☞ Some responses occur rapidly: ABA stimulation of stomatal closing; auxin-induced acidification of cell walls during cell elongation.
- ☞ Other responses take longer, especially if they require changes in gene expression (thigmomorphogenesis).

36   AN INTRODUCTION TO ANIMAL STRUCTURE AND FUNCTION

CHAPTER OUTLINE

Common problems are faced by many animals: how to exchange gases, obtain nutrients, reproduce, etc.

By taking a comparative approach to animal form and function, you can gain an understanding how the common problems have been solved by the diverse group called "animals."

To understand how organisms function, two themes apply:

1. Correlation of structure and function.
2. Capacity of life to adjust to the environment in 2 temporal scales: over short term, by physiological responses; and over long term, by adaptation due to natural selection.

I. LEVELS OF STRUCTURAL ORGANIZATION (p. 782–787)

There is a structural hierarchy of life:
- Atoms ⟶ molecules ⟶ supramolecular structures ⟶ cell.
- The cell is the lowest level of organization that can live as an organism.

The hierarchy of multicellular organisms is: cell ⟶ tissues ⟶ organs ⟶ organ systems.

A. Animal Tissues (p. 783–787)

What is a tissue? Where are tissues in the hierarchy of structural organization?

Tissues = Groups of cells with common structure and function.
- Cells may be held together by a sticky coating or woven together in a fabric of extracellular fibers.

What are the four main categories of animal tissues?

There are four main categories of tissues: epithelial tissue, connective tissue, muscle tissue and nervous tissue.

1. Epithelial Tissue (p. 783–784)

What is the function of epithelial tissue? How is the structure of epithelium suited for this function?

Epithelial tissue covers the outside of the body and lines organs and body cavities.

Cells are closely joined. In some, cells are riveted by tight junctions. (See Campbell, Chapter 7)
- ☞ Function as barriers against mechanical injury, invading microbes, and fluid loss.
- ☞ Free surface is exposed to air or fluid. Cells at base are attached to a *basement membrane* which is a dense layer of extracellular material.

How is epithelial tissue categorized?

Epithelial tissue cells are categorized by the number of layers and shape of the free surface cells:
- ☞ *Simple epithelium* is one layer of cells.
- ☞ *Stratified epithelium* has multiple tiers of cells.
- ☞ *Pseudostratified epithelium* is one layer of cells that appear to be multiple since the cells vary in length.
- ☞ Cell shapes are: *cuboidal* (like dice), *columnar* (bricks on end), or *squamous* (like flat floor tile).
- ☞ A tissue may be described by a combination of terms such as stratified squamous epithelium.

<u>Simple, Stratified</u>

<u>Pseudostratified</u>

<u>Cuboidal, Columnar, Squamous</u>

In addition to being a protective barrier, what other function might epithelia perform?

Some epithelia are specialized for absorption or secretion of chemical solutions in addition to their protective role.
- ☞ Some epithelia are ciliated (e.g. the lining of the respiratory system).
- ☞ The *mucous membranes* lining the oral cavity and nasal passageways secrete mucus which moistens and lubricates the surfaces.
- ☞ The structure fits function. For example, simple squamous epithelium is leaky and is specialized for exchange of materials by diffusion. It is found in blood vessel linings and air sacs in the lungs.

   2.   **Connective Tissue** (*p. 784–786*)

<u>Connective tissue</u>

What is the function of connective tissue, and what distinguishes it from other tissues?

Functions to bind and support other tissues.

Has a sparse cell population scattered through an extracellular *matrix* (e.g. web of fibers in a homogenous ground substance).

What are the various types of connective tissue? What types of cells characterize each, and how does the cellular structure relate to the tissue's function?

There are several types of connective tissue: loose connective tissue, adipose tissue, fibrous connective tissue, cartilage, bone and blood.

Loose connective tissue

Collagenous fibers

Elastic fibers

Reticular fibers

Fibroblasts, Macrophages

Adipose tissue

Fibrous connective tissue

Tendons, Ligaments

Cartilage, Chondrin

Chondrocytes

Bone

Osteocytes

Haversian systems

Blood

Leukocytes

Loose connective tissue holds organs in place and attaches epithelia to underlying tissues. It consists of 3 types of fibers:
- *Collagenous fibers* are made of collagen in bundles of fibers of 3 collagen molecules each. Great tensile strength, resist stretching.
- *Elastic fibers* are long threads of the protein elastin. Elastic properties lend tissue a resilience to quickly return to the original shape.
- *Reticular fibers* are branched and form a tightly woven fabric joining connective tissue to adjacent tissues.

There are two types of cells in loose connective tissue:
- *Fibroblasts* secrete the proteins of the extracellular fibers.
- *Macrophages* are phagocytic amoeboid cells that function in immune defense of the body.

Adipose tissue is loose connective tissue specialized to store fat in adipose cells distributed throughout its matrix.
- Insulates the body and stores fuel molecules.
- Each adipose cell has one large fat droplet which can vary in size as fats are stored or utilized.

Fibrous connective tissue is dense due to a large number of collagenous fibers organized into parallel bundles for greater tensile strength.
- In tendons (attach muscles to bones) and ligaments (attach bones together at joints).

Cartilage is composed of collagenous fibers embedded in *chondrin* (a protein-carbohydrate complex secreted by chondrocytes).
- Chondrocytes = Cells confined to scattered spaces called *lacunae* in ground substance.
- Cartilage is strong yet somewhat flexible support material.
- Comprises the skeleton of all vertebrate embryos and some adult vertebrates such as sharks.

Bone is a mineralized connective tissue.
- Osteocytes deposit a matrix of collagen and calcium phosphate which hardens into the mineral hydroxyapatite.
- The combination of collagen and hydroxyapatite makes the bone harder than cartilage but not brittle.
- Bone consists of repeating *Haversian systems* (concentric layers of lamellae deposited around a central canal containing blood vessels and nerves).
- Osteocytes are located in spaces called *lacunae* surrounded by a hard matrix and are connected to each other by cell extensions called *canaliculi*.
- In long bones, only the outer area is hard and compact; the inner area is filled with spongy bone tissue called *marrow*.

Blood is a connective tissue with a liquid matrix of plasma and three types of blood cells:
- *Leukocytes* are white blood cells that function in immune defense.

An Introduction to Animal Structure and Function 537

- ☞ *Erythrocytes* are red blood cells that transport oxygen.
- ☞ *Platelets* function in blood clotting. (See Campbell, Chapters 38 and 39)
- ☞ Blood cells are made in red marrow near the ends of long bones.

3. **Muscle Tissue** (*p. 786–787*)

What is the function of muscle tissue, and what distinguishes it from other tissue types?

Muscle tissue consists of long, excitable cells capable of much contraction.
- ☞ In the cell cytoplasm are parallel bundles of the contractile proteins actin and myosin.
- ☞ Muscle is the most abundant tissue in most animals.

What are the three types of vertebrate muscle and how do they differ? How are these differences a reflection of function?

There are three types of muscle tissue in vertebrates: skeletal muscle, cardiac muscle and visceral muscle.

Skeletal muscle is responsible for voluntary movements.
- ☞ Attached to bones by tendons.
- ☞ Myofibrils are aligned to form a banded *striated* appearance.

Cardiac muscle forms the contractile wall of the heart.
- ☞ Striated but cardiac cells are branched.
- ☞ Ends of cells are joined by intercalated disks, which relay the impulse to contract from cell to cell.

Visceral muscle is smooth (unstriated) tissue in walls of internal organs.
- ☞ Spindle-shaped cells contract slowly but retain contracted condition a long time.
- ☞ Responsible for involuntary movements (e.g. churning of the stomach).

4. **Nervous Tissue** (*p. 787*)

Nervous tissue senses stimuli and transmits signals from one part of the animal to another.

What is the general structure of a neuron, and how does that structure fit its function?

Neuron = Nerve cell specialized to conduct an impulse.
- ☞ Consists of cell body and nerve processes (appendages) of two types: *dendrites* (conduct impulses to cell body), and *axons* (transmit impulses away from cell body). (See Campbell, Chapter 44)

Erythrocytes

Platelets

Muscle tissue

Skeletal muscle

Cardiac muscle

Visceral muscle

Nervous tissue

Neuron

Dendrites, Axons

538 *An Introduction to Animal Structure and Function*

B. Organs and Organ Systems (p. 787)

What is an organ? An organ system?

Tissues are organized into *organs* in all but the simplest animals.
- ☞ May be layered such as the dermis of a human.
- ☞ Many organs are suspended by sheets of connective tissue called *mesenteries*.

<u>Organ systems</u> = Several organs with separate functions that act in a coordinated manner (e.g. digestive, circulatory and respiratory systems).
- ☞ Systems are interdependent: an organism is a living whole greater than the sum of its parts.

II. SIZE, SHAPE AND THE EXTERNAL ENVIRONMENT (p. 787–789)

An animal's size, shape and symmetry affect how it interacts with the environment.
- ☞ To take in oxygen, all cells must be bathed in aqueous medium to permit diffusion across the plasma membrane.

What are several body shapes that maximize the external surface area in contact with the environment?

Some organisms have all cells in contact with the external environment:
- ☞ One-celled organisms (Protozoan).
- ☞ *Hydra* is two cells thick, with each layer contacting the external environment. (See Campbell, Figure 36.8)

How can animals with complex internal organization and small surface-to-volume ratio, have enough surface area to exchange materials with the environment?

More complex animals have internal surfaces specialized for exchanging materials with the environment.
- ☞ Extensive branching and folding provides these internal membranes with large surface areas.
- ☞ For example, lungs absorb oxygen and kidneys filter waste.
- ☞ Materials pass from these exchange surfaces to other cells through a circulatory system in vertebrates.
- ☞ Makes terrestrial life possible since only small tubes connect expansive epithelial surfaces to the external environment.

III. THE ANIMAL'S INTERNAL ENVIRONMENT (p. 789–791)

Where is interstitial fluid? What is its function?

<u>Interstitial fluid</u> = The internal environment in vertebrates.
- ☞ Fills spaces between cells.
- ☞ Exchanges nutrients and wastes with blood carried in capillaries.

Animals differ in the extent of control over internal environment (e.g. *Hydra* has little control, vertebrates have a lot).

Margin notes:
<u>Organs</u>
<u>Mesenteries</u>
<u>Organ systems</u>
<u>Interstitial fluid</u>

In your own words, what is homeostasis?

Homeostasis = "Steady state."
- French physiologist Claude Bernard first described the "constant internal milieu" in animals.
- Maintained by mechanisms which include three components: a receptor, a control center, and an effector.

What is the difference between negative and positive feedback? How do they help maintain homeostasis?

Homeostasis works via two mechanisms: (See Campbell, Figure 36.10a)

1. *Negative feedback* is analogous to a thermostat: the heater is turned on if the temperature goes below a set level, and the heater turns off if the temperature gets too high.
    - For example, if the human hypothalamus detects a high blood temperature, it sends nerve impulses to sweat glands, which increase sweat output and cause evaporative cooling.
    - When the body temperature returns to normal, no additional signals are sent.

2. *Positive feedback* occurs when a change in some variable causes a reaction which increases that change.
    - For example, during childbirth, the baby's head against the uterine opening stimulates contractions which cause greater pressure of the head against the uterine opening.

The homeostatic system permits regulated change of the internal environment to occur. These changes are usually short term and in response to specific conditions. This adaptive response is a major theme in organismal biology.

Homeostasis

Negative feedback

Positive feedback

37 ANIMAL NUTRITION

CHAPTER OUTLINE

I. FOOD TYPES AND FEEDING MECHANISMS (p. 794–795)

What are herbivores, carnivores and omnivores?

<u>Holotrophs</u> = Organisms that ingest other organisms.
- Herbivores eat plants, carnivores eat other animals, omnivores consume both.

What are the following feeding mechanisms: a) filter-feeding, b) substrate-feeding, c) deposit-feeding, and d) fluid feeding?

Suspension-feeders sift small food particles from the water. *Substrate-feeders* live on or in their food source. *Deposit-feeders*, such as earthworms, are a specific type of substrate-feeder that separates detritus from the soil ingested. *Fluid-feeders* suck nutrient-rich fluids from a living host.

Most animals are *bulk-feeders*, they feed by ingesting relatively large pieces of food.

II. DIGESTION: A COMPARATIVE INTRODUCTION (p. 795–799)

What is digestion? Why is it necessary?

<u>Digestion</u> = The process of breaking down food into small molecules the body can absorb.
- Enzymatically cleaves macromolecules into component monomers that can be used by the animal (polysaccharides and disaccharides to simple sugars; proteins to amino acids; fats to glycerol and fatty acids).

A. Enzymatic Hydrolysis (p. 796)

What is the role of hydrolysis in digestion?

Digestion breaks bonds in macromolecules by enzymatic addition of water (*hydrolysis*).
- Chemical digestion is usually preceded by mechanical fragmentation (e.g. chewing) that increases the surface area exposed to digestive juices.
- Occurs in a specialized compartment where the enzymes are contained so they don't damage the animal's own cells.
- After the nutrients are absorbed, undigested material is eliminated.

Animal Nutrition   541

B. **Intracellular Digestion in Food Vacuoles** (p. 796–797)

What is the difference between intracellular and extracellular digestion? Why must intracellular digestion be sequestered in a food vacuole? What are some examples of organisms which digest their food in vacuoles?

Intracellular digestion occurs in organelles where a single cell digests its food without digestive enzymes mixing with its cytoplasm.   *Intracellular digestion*
- Protozoa have food vacuoles that form around the food item and into which the hydrolytic enzymes are secreted. (See Campbell, Figure 37.6)
- Sponges also form food vacuoles whose contents are transferred to other cells. (See Campbell, Figure 37.7)

Extracellular digestion occurs within compartments that are continuous via passages with the outside of the body.   *Extracellular digestion*

C. **Digestion in Gastrovascular Cavities** (p. 797–798)

What is a gastrovascular cavity? Why are extracellular digestive cavities advantageous?

Gastrovascular cavities = Digestive sacs with a single opening.   *Gastrovascular cavities*
- Food enters the sac and the gastrodermis secretes digestive enzymes that break down the food.
- Food particles are taken into gastrodermal cells by phagocytosis and digestion continues intracellularly within food vacuoles.

Using Hydra as an example, how does a gastrovascular cavity function in both digestion and distribution of nutrients?

Digestion in *Hydra*. (See Campbell, Figure 37.8a)
- Food item is immobilized by stings from nematocysts on the tentacles.
- Tentacles then force prey through the mouth into the gastrovascular cavity.
- Specialized gastrodermal cells secrete digestive enzymes that fragments the soft tissues of the prey into tiny pieces.
- The small pieces are phagocytized by gastrodermal cells and surrounded by a food vacuole.
- Hydrolysis is completed by intracellular digestion.
- Undigested materials are expelled from the gastrovascular cavity.

What major animal phyla use gastrovascular cavities for digestion?

D. **Digestion in Alimentary Canals** (p. 798)

What advantages do complete digestive tracts have over gastrovascular cavities? What major animal phyla have alimentary tracts?

Alimentary canals = Digestive tubes running between two openings, the mouth (where food is initially ingested) and the anus (where undigested wastes are egested). (See Campbell, Figure 37.9)   *Alimentary canals*

Animal Nutrition

☞ Food moves along in one direction so that the tube can be organized into specialized regions that carry out digestion and absorption of nutrients in a stepwise fashion.
☞ The unidirectional passage of food and the specialization of function for different regions makes the alimentary canal more efficient.

III. THE MAMMALIAN DIGESTIVE SYSTEM (p. 799–807)

The digestive system in mammals includes not only the alimentary canal, but also accessory glands that secrete digestive juices into the canal through ducts.
☞ The digestive tract has a four-layered wall: the lumen is lined by a mucous membrane (mucosa), then a connective tissue layer followed by a layer of smooth muscle, and outermost is connective tissue attached to the membrane of the body cavity.

What is peristalsis? What is its role in the digestive tract?

☞ <u>Peristalsis</u> = Smooth muscle contractions that push food along the tract.
☞ *Sphincters* (modifications of the muscle layer into ringlike valves) occur at some junctions between compartments and regulate passage of materials through the system.
☞ Accessory organs are three pairs of salivary glands, the pancreas, the liver, and the gall bladder.

Refer to Figure 37.10 for orientation of the specialized compartments in the human system.

A. Oral Cavity (p. 799–800)

Physical and chemical digestion begins in the oral cavity.
☞ Physical breakdown of food results from chewing during which large pieces of food are broken down to smaller pieces; makes food easier to swallow and increases the surface area available for enzyme action.

How is salivation controlled? What are the functions of saliva? What is the role of salivary amylase in digestion?

Presence of food in the oral cavity causes salivary glands to deliver saliva, which contains mucin (protects the mouth from abrasion and lubricates food), buffers that neutralize acids, antibacterial agents, and *salivary amylase*, an enzyme that hydrolyzes starch to the disaccharide maltose or small polysaccharides.

The tongue manipulates food during chewing and forms food into a *bolus*, which is swallowed.

B. Pharynx (p. 800)

The pharynx serves as an intersection for both the digestive and respiratory systems.

<u>Peristalsis</u>

<u>Sphincters</u>

<u>Salivary glands, Pancreas</u>

<u>Liver, Gall bladder</u>

<u>Oral cavity</u>

<u>Salivary amylase</u>

<u>Bolus</u>

<u>Pharynx</u>

What is the sequence of events that occur as a result of the swallowing reflex?

When we swallow, the *epiglottis* moves to block the entrance of the windpipe, directing food through the *pharynx* and into the *esophagus*. (See Campbell, Figure 37.11)

    C. Esophagus (*p. 800–801*)

What is the function of the esophagus? How is peristalsis in the esophagus controlled?

The esophagus is a muscular tube which conducts food from the pharynx to the stomach.
- ☞ Peristalsis moves the bolus along the esophagus to the stomach.

The initial entrance of the bolus into the esophagus is voluntary (swallowing); once in, the peristalsis results from involuntary contraction of the smooth muscles.

    D. Stomach (*p. 801–802*)

The stomach is a large, saclike structure located just below the diaphragm on the left side of the abdominal wall.
- ☞ Has a very elastic wall containing folds called *rugae* that can expand to accommodate 2 liters of food. The storage capacity permits periodic feeding (meals).

What are the three types of secretory cells in the stomach epithelium? What substances do they secrete?

What is the normal pH of the stomach? What is the function of stomach acid?

What is the function of pepsin?

The epithelium that lines the lumen secretes *gastric juice*, which contains hydrochloric acid and pepsin.
- ☞ HCl makes gastric juice very acidic (pH 2) and functions to break up food, denature proteins in food, inactivate salivary amylase and kill most bacteria.
- ☞ Pepsin is synthesized and secreted as the *zymogen* pepsinogen.

  Zymogen = An inactive form of a protein digesting enzyme.

Why does the stomach normally not digest itself?

- ☞ Pepsinogen only mixes with the HCl in the lumen of the stomach since they are produced by different cells; the acid removes a short segment of amino acids from the pepsinogen which then becomes active pepsin due to exposure of its active site.
- ☞ Active pepsin activates other pepsinogen molecules and pepsin begins to cleave proteins into smaller polypeptides.
- ☞ A coating of mucus secreted by epithelial cells protects the stomach lining from digestion; however, the lining is still constantly eroded and replaced.

Margin notes: Epiglottis; Esophagus; Stomach; Rugae; Lumen, Gastric juice; Pepsin, Pepsinogen; Zymogen

Animal Nutrition

How are pepsin and acid secretion regulated? What are the roles of the hormones gastrin and enterogastrone?

- Gastric secretion is controlled by nerve impulses and the hormone *gastrin*, which is released into the blood and stimulates further secretion of gastric juice. About 3 liters of gastric juice are secreted each day.
- Stomach contents are mixed by the churning action of smooth muscles.

What is the function of the pyloric sphincter?

- The *pyloric sphincter* at the bottom of the stomach regulates the passage of *acid chyme* (the nutrient broth formed in the stomach) into the small intestine.

E. **Small Intestine** (p. 802–806)

The human small intestine is about 6 m in length and is the site of most enzymatic hydrolysis of food and absorption of nutrients.

What is the sequence of events that occurs in response to acid chyme entering the duodenum? What are the roles of: a) secretin, b) bicarbonate, c) cholecystokinin, d) bile, e) pancreatic enzymes, and f) enterogastrone?

Several other organs contribute to the digestion which occurs in the small intestine. (See Campbell, Figure 37.13)

- Entry of acid chyme into the *duodenum* stimulates the intestinal wall to release a hormone, *secretin*, that signals the pancreas to release bicarbonate to neutralize the acid.

What is the composition and function of bile? Where is it produced and stored?

- Another hormone released by cells of the duodenal wall, *cholecystokinin* (*CCK*) causes the gall bladder to release its store of bile salts (produced in the liver) that emulsify fats, allowing *lipase* to hydrolyze fat molecules.

What are the reactants and products for enzymatic digestion of carbohydrates, proteins and lipids? Where in the digestive system do these reactions occur? (Refer to Campbell, Table 37.1.)

- Pancreatic amylase hydrolyzes starch into the disaccharide maltose, which is hydrolyzed by <u>maltase</u> into glucose. Other *disaccharideases* split other disaccharides into monosaccharides. These enzymes are built into the membranes of the intestinal epithelium; consequently, the final breakdown of carbohydrates occurs where the products will be absorbed.

How are pancreatic zymogens for proteolytic enzymes activated in the duodenum? What is the role of the intestinal enzyme enterokinase?

- ☞ Protein digesting enzymes (trypsin, chymotrypsin, and carboxypeptidase) are secreted as inactive zymogens by the pancreas and are activated by an intestinal enzyme, *enterokinase*, in the lumen of the small intestine. The small intestine also secretes the protein-digesting enzymes *aminopeptidase* and *dipeptidases*.
- ☞ Since the protein-digesting enzymes from the pancreas and small intestine break bonds in specific areas of the polypeptide, protein digestion to amino acids is a combined effort from them all.
- ☞ *Nucleases* hydrolyze DNA and RNA in the food.
- ☞ The duodenum also releases *enterogastrone* that inhibits peristalsis in the stomach, thus slowing the entry of food into the small intestine.
- ☞ See Table 37.1 for a summary of digestion.

Where in the digestive tract does most nutrient absorption occur?

The remaining areas of the small intestine, the *jejunum* and *ileum*, are specialized for absorption of nutrients.

Why are the many folds, villi and microvilli important in the small intestine?

- ☞ Large folds in the walls are covered with projections called *villi*, which in turn have many microscopic *microvilli*; this greatly increases the surface area for absorption. (See Campbell, Figure 37.15)
- ☞ Penetrating the hollow core of each villus are capillaries and a tiny lymph vessel called a *lacteal*.

How are specific nutrients absorbed across the intestinal epithelium and across the capillary or lacteal wall? In each case, is the transport with or against the concentration gradient?

- ☞ Nutrients are absorbed across the 2 cell-thick epithelium and into the capillaries or lacteals. This may occur by diffusion or active transport.
- ☞ Amino acids, sugars, and lipoproteins (fat molecules bound to special proteins) enter the capillaries.

What happens to glycerol and fatty acids after they are absorbed into the intestinal epithelium? What happens to the chylomicrons and lipoproteins?

- ☞ Absorbed glycerol and fatty acids are recombined in epithelial cells to form fats; most are coated with proteins to form *chylomicrons* which enter the lacteals.

What is the function of the hepatic portal vein?

- ☞ Capillaries that drain nutrients away from the villi converge into the *hepatic portal vein*, which leads directly to the liver where various organic molecules are used, stored, or converted to a different form.

F. Large Intestine (p. 806-807)

Where in the digestive tract does most reabsorption of water occur?

The large intestine, or colon, connects to the small intestine at a T-shaped junction containing a sphincter; the blind end of the T is called the *cecum*.
- The *appendix* is a fingerlike extension composed of lymphoid tissue which extends from the cecum.
- The colon is about 1.5 m long and is in the shape of an inverted "U."
- The colon functions to complete reabsorption of water.
- Feces are moved along by peristalsis.

What is the main source of vitamin K in humans?
- Intestinal bacteria live on organic material in the feces and some produce vitamin K which is absorbed by the host.
- An abundance of salts is excreted by the colon lining.
- Feces are stored in the *rectum* and pass through the two sphincters (one involuntary, one voluntary) to the anus.

IV. EVOLUTIONARY ADAPTATIONS OF VERTEBRATE DIGESTIVE SYSTEMS (p. 807-809)

How are the following adaptations related to diet: a) variation in dentition, b) variation in length of the digestive tract, and c) fermentation chambers?

Examples of anatomical variation reflecting diet:
- Dentition varies between herbivores, carnivores and omnivores. Herbivores possess teeth with broad surfaces for grinding vegetation; carnivores have specialized teeth to kill prey; and omnivores possess both. (See Campbell, Figure 37.16)
- Length of the digestive system correlates with diet; herbivores and omnivores having relatively longer digestive tracts since vegetation is more difficult to digest than meat.
- Herbivorous mammals may also have special fermentation chambers where microbes digest cellulose to simple sugars and convert sugars to nutrients essential to the animal.
- *Ruminants* have evolved elaborate adaptations including a four-chambered stomach. (See Campbell, Figure 37.18)

V. NUTRITIONAL REQUIREMENTS (p. 809-815)

Why do animals need a healthful diet?

A healthful diet provides fuel for cellular respiration, supplies organic raw materials used to fabricate the animal's own molecules, and supplies *essential nutrients* which must be obtained in prefabricated form.

A. Food as Fuel (p. 809-811)

Monomers from any of the complex organic molecules can be used to produce energy, although those from carbohydrates and fats are used first.

Animal Nutrition 547

The energy content of food is measured in *kilocalories*; oxidation of a gram of fat liberates 9.5 kcal, twice that of a gram of carbohydrate or protein.

Kilocalories

What is basal metabolic rate (BMR)? How is it measured? What is the relationship between BMR and body size?

Basal metabolic rate (BMR) = The number of kilocalories a resting animal requires to fuel breathing, heartbeat, and in some, maintenance of body temperature.
- ☞ Birds and mammals have the highest BMR of vertebrates since they are endothermic.
- ☞ There is an inverse relationship between body size and metabolic rate, the smaller the animal, the higher the BMR.

Basal metabolic rate

More strenuous activities consume more calories. The liver and muscles store excess energy (calories) in the form of glycogen; further excess is stored in adipose tissue in the form of fat.
- ☞ When the diet is deficient in calories, glycogen stored in the liver and muscles is utilized first and fat is withdrawn from adipose tissues.

What are the effects of undernourishment or starvation?

- ☞ An *undernourished* person or animal is one whose diet is deficient in calories. If starvation persists, the body begins to breakdown its own proteins as a source of energy. The breakdown of the body's own proteins can cause muscles to atrophy and can result in the consumption of the brain's proteins.

Undernourished

B. Food for Fabrication (p. 811)

Heterotrophs cannot use inorganic materials to make organic molecules; they must obtain organic precursors for these molecules from the food they ingest.

Given a source of carbon and nitrogen, heterotrophs can fabricate a great variety of organic molecules by using enzymes to rearrange the molecular skeletons of precursors acquired from food.
- ☞ A single type of amino acid can supply the nitrogen necessary to build other amino acids.
- ☞ The liver is responsible for most of the conversion of nutrients from one type of organic molecules to another.

C. Essential Nutrients (p. 811–815)

What is the difference between malnourished and undernourished?

What are four classes of essential nutrients? What happens if they are deficient in the diet?

Essential nutrients = Chemicals an animal requires but cannot synthesize.
- ☞ Must be included in the diet.
- ☞ An animal is *malnourished* if their diet is missing one or more essential nutrients.
- ☞ Includes essential amino acids, essential fatty acids, vitamins, and minerals.

Essential nutrients

Malnourished

548 Animal Nutrition

Essential amino acids

Essential amino acids are those that must be obtained in the diet in a prefabricated form.
- Most animals can synthesize about half of the 20 kinds of amino acids needed to make proteins.
- Human adults can produce 12, leaving 8 as essential in the diet. (Human infants can only produce 11.)

Protein deficiency

Kwashiorkor

- *Protein deficiency* results when the diet lacks one or more essential amino acids. The syndrome known as *kwashiorkor* is a form of protein deficiency in Africa.
- The human body cannot store essential amino acids, thus a deficiency retards protein synthesis; is most frequent in individuals, who for economic or other reasons, have unbalanced diets.

Essential fatty acids

Essential fatty acids are those unsaturated fatty acids that cannot be produced by the body.
- An example in humans is linoleic acid which is required to produce some of the phospholipids necessary for membranes.
- Fatty acid deficiencies are rare as most diets include sufficient quantities.

What are the water-soluble and fat-soluble vitamins? How are they used by the body?

Vitamins

Vitamins are <u>organic</u> molecules required in the diet.
- Needed in much smaller quantities than amino acids or fatty acids.
- Most serve a catalytic function as coenzymes or parts of coenzymes.
- Vitamin deficiencies can cause very severe effects as shown in Table 37.2.
- Water-soluble vitamins are not stock-piled in the body tissues; amounts ingested in excess of body needs are excreted in the urine.
- Fat-soluble vitamins (vitamins A, D, E and K) can be held in the body; excess amounts are stored in body fat and may accumulate over time to toxic levels.
- If the body of an animal can synthesize a certain compound it is not a vitamin; consequently, a compound such as ascorbic acid is a vitamin for humans (vitamin C) and must be included in our diets, while it is not a vitamin in rabbits where the normal intestinal bacteria produce all that is needed.

What are the dietary sources, functions and effects of deficiency for the following required minerals in the human diet: calcium, phosphorus, sulfur, potassium, chlorine, sodium, magnesium and iron?

Minerals

Minerals are <u>inorganic</u> nutrients required in the diet.
- Needed in small quantities ranging from 1 mg to 2500 mg per day depending on the mineral.
- Some minerals serve structural and maintenance roles in the body (calcium, phosphorous) while others serve as part of enzymes (copper) or other molecules (iron).
- Refer to Table 37.3 for the mineral requirements of humans.

38

CIRCULATION AND GAS EXCHANGE

CHAPTER OUTLINE

The exchange of materials (whether nutrients, gases, or waste products) between an organism and its environment must take place across a moist cell membrane.
- ☞ The molecules must be dissolved in water in order to diffuse or be transported across the membrane.
- ☞ In protozoans, the entire external surface may be used for this exchange.
- ☞ The simple, multicellular animals (sponges, cnidarians) have body structures so that each cell is exposed to the surrounding waters.

Three-dimensional animals face the problem that some of their cells are isolated from their surroundings.
- ☞ These animals have specialized organs, where exchange with the environment occurs, coupled with special systems for the internal transport through body fluids to the cells.
- ☞ The association of specialized organs with an internal transport system not only reduces the distance over which molecules must diffuse to enter or leave a cell, but permits regulation of internal fluid composition.

I. **INTERNAL TRANSPORT IN INVERTEBRATES** (p. 819–820)

   A. **Gastrovascular Cavities** (p. 819)

What are the major animal phyla with gastrovascular cavities? Why do they not need a circulatory system?

Cnidarians have a body wall only 2 cells thick enclosing a central gastrovascular cavity.
- ☞ The gastrovascular cavity functions in digestion and distribution of nutrients.
- ☞ Gastrodermal cells lining the cavity have direct access to the nutrients produced by digestion, however, the structure of the body is such that nutrients only have a short distance to diffuse to the outer cell layer.

Planarians and other flatworms also have a gastrovascular cavity.
- ☞ The highly ramified structure of the cavity and the flattened body shape ensure all cells are exposed to cavity contents.

   B. **Open and Closed Circulatory Systems** (p. 819–820)

What is the difference between open and closed circulatory systems?

Open circulatory system = Hemolymph bathes internal organs directly while moving through sinuses. (See Campbell, Figure 38.3a)

Open circulatory system

Sinuses

550 Circulation and Gas Exchange

Using an arthropod as an example, how does hemolymph circulate?

- Circulation results from contractions of the dorsal vessel (heart) and body movements.
- Relaxation of the "heart" draws blood through the ostia (pores) into the vessel.

Closed circulatory system

Closed circulatory system = Blood is confined to vessels.

In Annelida, the closed circulatory system consists of two major vessels, the dorsal and ventral, that branch into smaller vessels which supply blood to organs. (See Campbell, Figure 38.3b)

How does blood circulate in an earthworm? How does it exchange materials with interstitial fluid?

- The dorsal vessel serves as the heart and pumps blood forward.
- Also has 5 pairs of vessels that loop around the digestive tract to connect the main vessels, and function as auxiliary hearts.
- Materials are exchanged with the interstitial fluids which come into direct contact with the cells.

II. CIRCULATION IN VERTEBRATES (p. 820–830)

What are the components of a vertebrate cardiovascular system?

The *cardiovascular system* consists of the heart, blood vessels and blood.

Heart, Atria
Ventricles

- The heart consists of one or two *atria*, chambers that receive blood, and one or two *ventricles*, chambers that pump blood out.

What is the difference between an artery and a vein?

Arteries
Capillaries

- *Arteries* carry blood away from the heart to organs, where they branch into smaller *arterioles*, that give rise to *capillaries* (the site of chemical exchange between blood and interstitial fluid).
- Capillaries rejoin ⟶ *venules* ⟶ *veins* that return blood to the heart.

Venules, Veins

A. Vertebrate Circulatory Schemes: An Evolutionary Perspective (p. 821–822)

What are the major differences among the circulatory schemes of fish, amphibians and mammals?

Fish have a 2-chambered heart with one atrium and one ventricle. (See Campbell, Figure 38.4a)

- Blood pumped from the ventricle ⟶ gills, where it is oxygenated and where CO_2 is disposed of ⟶ other organs ⟶ atrium of the heart ⟶ ventricle.

Capillary bed

- Blood flows through two capillary beds during each complete circuit; one in the gills, and a second in the organ systems. As blood flows through a capillary bed, blood pressure drops substantially.
- Blood flow to the tissues and back to the heart is aided by swimming motions.

Amphibians have a three-chambered heart with two atria and one ventricle. (See Campbell, Figure 38.4b)

What is the difference between pulmonary and systemic circuits? What is the function of each?

- Blood flows through a *pulmonary circuit* (to the lungs and skin) and a *systemic circuit* (to all other organs) in a scheme called *double circulation*.
- Blood flow pattern: ventricle → lungs and skin to become oxygenated → left atrium → ventricle → all other organs → right atrium.
- There is some mixing of oxygen-rich and oxygen-poor blood in the single ventricle although a ridge present int the ventricle diverts most of the oxygenated blood to the systemic circuit and most of the deoxygenated blood to the pulmonary circuit.

What is the advantage of double circulation over a single circuit?

Birds and mammals contain a four-chambered heart with two atria and two ventricles. (Please note that crocodiles also have a four-chambered heart.)
- Double circulation is similar to that of amphibians except that oxygenated and deoxygenated blood does not mix due to the presence of two ventricles.
- The complete separation of oxygenated and deoxygenated bloods increases the efficiency of oxygen delivery to the cells.

B.   The Heart (*p. 822–825*)

The wall of the heart consists mostly of cardiac muscle tissue.
- It is surrounded by a sac with a two-layered wall.
- Atria have thin walls and function as collection chambers for blood returning to the heart.
- Ventricles have thick, powerful walls that pump blood to the organs.

What path does blood take as it flows through the human heart? What structure does it pass through en route?

What is the difference between systole and diastole?

Heart cycle = Sequence of events during each heartbeat, lasting about 0.8 second.
- During *systole*, heart muscle contracts and the chambers pump blood.
- During *diastole*, the ventricles fill with blood.
- The first 0.1 seconds of systole involves atrial contraction forcing blood into the ventricles; the remaining 0.3 seconds is when the ventricles force blood into the arterial system.

What are the four heart valves? Where are they located? What is their respective function?

What are the events of the cardiac cycle? What causes the first and second heart sounds?

Margin notes:
Pulmonary circuit
Systemic circuit, Double circulation
Heart cycle
Systole
Diastole

552 Circulation and Gas Exchange

There are four valves in the heart which prevent backflow of blood during systole: (See Campbell, Figure 38.5b)

- Atrioventricular valves = Valves between each atrium and ventricle that keep blood from flowing back into the atria during ventricular contraction.
- Semilunar valves = Valves located where the aorta leaves the left ventricle and where the pulmonary artery leaves the right ventricle; prevents blood from flowing back into ventricles when they relax.

What is a heart murmur? What is its cause?

- A *heart murmur* is a defect in one or more of the valves that allows backflow of blood. Serious defects are usually corrected by surgical replacement of the valve.

What is a pulse? What is the relationship between size and pulse rate among different mammals?

Heart rate = Pulse; the number of heartbeats per minute.
- In humans, the average resting heartrate is 65–75 beats per minute; rates vary depending on the level of activity.
- There is an inverse relationship between animal body size and pulse; elephants have a rate of 25 beats per minute while some shrews have 600 beats per minute.

What is cardiac output? How is it affected by a change in heart rate or stroke volume?

Cardiac output = The volume of blood per minute that the left ventricle pumps into the systemic circuit; depends on heart rate and stroke volume.
- *Stroke volume* is the amount of blood pumped by the left ventricle each time it contracts. The average human stroke volume is about 75 ml per beat.
- The cardiac output of an average human is 5.25 L per minute (4875–5625 ml per minute)

What is myogenic? What are some unique properties of cardiac muscle which allow it to contract in a coordinated manner?

What is a pacemaker? Where are the two patches of nodal tissue in the human heart?

What is the origin and pathway of the action potential (cardiac impulse) in the normal human heart?

Cardiac muscle cells are myogenic (self-excitable) and can contract without input from the nervous system. The tempo of contraction is controlled by the *sinoatrial (SA) node*, a specialized region of the heart; the pacemaker.
- Located in the right atrium wall near the entrance of the anterior vena cava.
- Composed of nodal tissue which has characteristics of both muscle and nerve tissue.

Margin terms:
- Atrioventricular valves
- Semilunar valves
- Heart murmur
- Heart rate
- Cardiac output
- Stroke volume
- Myogenic
- Sinatrial node

Circulation and Gas Exchange

- ☞ Contraction of the nodal tissue initiates a wave of excitation that spreads rapidly from the SA node and causes the two atria to contract in unison.
- ☞ This wave of contraction will pass down the atria until it reaches the *atrioventricular (AV) node*; a second mass of nodal tissue located near the base of the wall separating the atria.

Atrioventricular node

Why is it important that the cardiac impulse be delayed at the AV node? What is the function and importance of the Purkinje fibers?

- ☞ The impulse is delayed at the AV node for 0.1 second to ensure the atria are completely empty before the ventricles contract.
- ☞ The impulse is then carried by a bundle branches to the tip of the ventricles; the impulse then travels through the Purkinje fibers upward through the ventricular walls.

Purkinje fibers

How can the pace of the SA node be modulated by sympathetic and parasympathetic nerves, changes in temperature, physical conditioning and exercise?

Although the SA node controls the rate of heartbeat, it is influenced by several factors:
- ☞ Two antagonistic sets of nerves influence the heart rate; one speeds up contractions in the SA node, the other slows contractions.
- ☞ Hormones influence the SA node with some like epinephrine increasing the rate.
- ☞ Others factors, including body temperature changes, also have an influence.

C. Blood Flow (p. 825–828)

How do structural differences between arteries and veins reflect the difference in their functions?

The walls of arteries and veins have 3 layers: an outer layer of connective tissue with elastic fibers, a middle layer of smooth muscle and elastic fibers, and an inner endothelium of simple squamous epithelium. (See Campbell, Figure 38.7b and 38.7c)

How does capillary structure differ from other vessels? How does this structure fit capillary function?

- ☞ Capillaries are comprised only of the endothelial lining which permits the exchange of chemicals with the interstitial fluids.

What is the law of continuity? Why is blood flow through capillaries slower than it is through arteries and veins?

There is a great difference in the speed at which blood flows through the various parts of the circulatory system.
- ☞ The velocity decreases in accordance with the *law of continuity* which states that a fluid will flow faster through narrow portions of a pipe than wider portions if the volume of flow remains constant.

Law of continuity

- An artery gives rise to so many capillaries that the total diameter of vessels is much greater in capillary beds than in the artery, thus blood flows more slowly in capillaries.
- Resistance to blood flow is greater in capillary beds since the blood contacts more epithelial surface area.
- Blood flows faster as it enters the veins since the cross-sectional area is decreased.

What is blood pressure and how is it measured?

How do peripheral resistance and cardiac output affect blood pressure?

Blood pressure = The hydrostatic force that blood exerts against a vessel wall.
- Greater in arteries than in veins and greatest during systole.
- *Peripheral resistance* results from impedance by arterioles; blood enters the arteries faster than it can leave. (Thus, there is a pressure even during diastole, driving blood into capillaries continuously.)
- Determined by cardiac output and degree of peripheral resistance. (Stress may trigger neural and hormonal responses that cause the smooth muscles of vessel walls to contract, constricting blood vessels and increasing resistance.)

How does blood return to the heart, even though it must travel from the lower extremities against gravity?

- In veins pressure is near zero; blood returns to the heart by the action of skeletal muscles flanking the veins. (Veins have valves that allow blood to flow only towards the heart.)
- Breathing also helps return blood to the heart since the pressure change in the thoracic cavity during inhalation cause the vena cava and large veins near the heart to expand and fill.

What is microcirculation? How is blood flow through capillary beds regulated?

Some blood streams directly from arterioles to venules through *thoroughfare channels*, which are always open.
- True capillaries branch from these channels.
- Passage of blood into a capillary is regulated by a sphincter at the capillary entrance.
- Distribution of blood to organs is controlled by dilation and constriction of arterioles and by the precapillary sphincters.
- All tissues and organs receive a sufficient supply of blood even though only 5–10% of the capillaries are carrying blood at any one time; the supply is adequate due to the vast number of capillaries present in each tissue.

D. Capillary Exchange (p. 828–829)

The exchange of materials between the blood and interstitial fluids that are in direct contact with the cells occurs across the thin walls of capillaries.
- The capillary wall is a single, "leaky" layer of flattened endothelial cells that overlap at their edges.

Margin notes: Blood pressure, Hydrostatic force; Peripheral resistance; Microcirculation; Thoroughfare channels

☞ Materials may cross these cells in vesicles (via endocytosis and exocytosis), by diffusion through the cell, or by bulk flow between the cells due to hydrostatic pressure.

How do osmotic pressure and hydrostatic pressure regulate fluid and solute exchange across capillaries?

Direction of fluid movement at any point along a capillary depends on the relative forces of hydrostatic pressure and osmotic pressure. (See Campbell, Figure 38.12)
☞ Fluid flows out of a capillary at the upstream end near an arteriole and into a capillary at the downstream end near a venule.
☞ About 99% of the fluid which leaves the blood at the arteriole end of the capillary, re-enters from the interstitial fluid at the venous end.

E. **The Lymphatic System** (p. 829–830)

What is the composition of lymph? How does the lymphatic system help the normal functioning of the circulatory system?

Capillary walls leak fluid and some blood proteins, which return to the blood via the *lymphatic system*. This is the 1% of fluid that does not re-enter the capillaries.
☞ The fluid, *lymph*, is similar in composition to interstitial fluid.
☞ The lymph enters the system by diffusing into lymph capillaries.
☞ The lymphatic system drains into the circulatory system at two locations near the shoulders.
☞ Lymph vessels have valves that prevent backflow and depend mainly on movement of skeletal muscles to squeeze fluid along (although rhythmic contractions of vessel walls also help circulate the lymph).

How does the lymphatic system help defend the body against infection?

☞ *Lymph nodes* filter the lymph and attack viruses and bacteria; this defense is carried out by specialized white blood cells that inhabit the lymph nodes.
☞ Lymph capillaries penetrate small intestine villi and absorb fats, thus transporting them from the digestive system to the circulatory system.

III. **MAMMALIAN BLOOD** (p. 830–832)

Why is vertebrate blood classified as connective tissue?

What are the components of blood? What are their respective functions?

Vertebrate blood is connective tissue with several cell types suspended in a liquid matrix called *plasma*.
☞ The average human has 4 to 6 liters of whole blood (plasma + formed elements).
☞ Formed elements are the cellular components of the blood and represent about 45% of the blood volume.

Lymphatic system

Lymph

Lymph capillaries

Lymph nodes

Plasma

556 Circulation and Gas Exchange

A. Plasma (p. 830)

Water accounts for 90% of plasma, which also contains electrolytes and plasma proteins.

Electrolytes

Electrolytes = Inorganic salts in the form of dissolved ions that help maintain osmotic balance between blood and interstitial fluid; some also help buffer the blood.
- ☞ Electrolyte balance in the blood is maintained by the kidneys.

Plasma proteins
Immunoglobins

Plasma proteins help buffer blood, help determine osmotic strength of blood, and contribute to its viscosity.
- ☞ Some escort lipids through blood, some are immunoglobins, some (*fibrinogens*) are clotting factors.

Clotting factors
- ☞ Serum is blood plasma that has had the clotting factors removed.

Plasma also contains substances in transit such as nutrients, metabolic wastes, respiratory gases, and hormones.

B. Blood Cells (p. 830-832)

Erythrocytes

Erythrocytes = Red blood cells; biconcave discs that function in transport of oxygen.
- ☞ Each cubic millimeter of human blood contains 5-6 million erythrocytes.
- ☞ Lack nuclei and mitochondria; generate ATP exclusively by anaerobic metabolism.

Hemoglobin
- ☞ Contains *hemoglobin*, an iron-containing protein that reversibly binds oxygen. About 250 million molecules of hemoglobin are in each erythrocyte.

What happens between formation of erythrocytes from stem cells to their destruction by phagocytes?

- ☞ Erythrocyte production is controlled by a negative-feedback mechanism; if tissues are not receiving enough oxygen, kidneys secrete the hormone *erythropoietin*, which stimulates production of erythrocytes from *pluripotent stem cells* in the bone marrow.

Erythropoietin, Pluripotent stem cells
- ☞ The average erythrocyte circulates in the blood for 3-4 months before being destroyed.

Leukocytes

Leukocytes = White blood cells that function in defense and immunity.

Basophils, Eosinophils
- ☞ There are 5 types of leukocytes: basophils, eosinophils, neutrophils, lymphocytes and monocytes.

Neutrophils, Lymphocytes, Monocytes
- ☞ Actually spend most time outside the circulatory system, patrolling through interstitial fluid; large numbers are found in the lymph nodes.
- ☞ Arise from stem cells in the bone marrow; some leave the marrow to mature in the spleen, thymus, tonsils, adenoids or lymph nodes.
- ☞ There are usually 5,000-10,000 leukocytes per cubic millimeter of blood although this number increases during an infection.

Platelets

Platelets originate as pinched-off cytoplasmic fragments of large cells in the bone marrow and function in blood clotting.

C. Blood Clotting (p. 832)

What is the sequence of events that occurs when blood clots? What prevents spontaneous clotting in the absence of injury?

A clot forms when platelets clump together to form a temporary plug and release clotting factors (some are also released from damaged cells) that initiate a complex reaction resulting in conversion of inactive *fibrinogen* to active *fibrin*. (See Campbell, Figure 38.15)

- Fibrin aggregates into threads that form the clot.
- Anticlotting factors normally prevent spontaneous clotting in the absence of injury.

Fibrinogen

Fibrin

IV. CARDIOVASCULAR DISEASE (p. 832–834)

Cardiovascular disease = Disease of the heart or blood vessels; cause more that 1/2 of all deaths in the U.S.

Cardiovascular disease

What is the difference between a thrombus and an embolus? Atherosclerosis and arteriosclerosis? Low-density lipoproteins (LDL's) and high-density lipoproteins (HDL's)?

Thrombus = A blood clot that blocks a key blood vessel.
- If it blocks the coronary arteries, a heart attack occurs.
- An *embolus* is a moving clot.
- If the thrombus or embolus blocks an artery in the brain, a stroke results.

Thrombus

Embolus

How does artherosclerosis affect the arteries?

Atherosclerosis = Plaques develop on the inner walls of arteries and narrow the bore of the vessels.
- Increase the risk of clot formation and heart attack.

Atherosclerosis, Plaques

Arteriosclerosis = A form of atherosclerosis in which plaques become hardened by calcium deposits.

Arteriosclerosis

Angina pectoris = Chest pains that occur when the heart receives insufficient oxygen due to the build up of plaques in the arteries.

Angina pectoris

Hypertension = High blood pressure; promotes atherosclerosis.

Hypertension

What factors have been correlated with an increased risk of cardiovascular disease?

Smoking, lack of exercise and a diet rich in animal fats correlate with increased risk of cardiovascular disease.

Abnormally high concentrations of *LDL's* (low density lipoproteins) in the blood correlate with atherosclerosis; *HDL's* (high density lipoproteins) actually reduce deposition of cholesterol in arterial plaques.

Low density lipoproteins

Hight density lipoproteins

V. GAS EXCHANGE IN ANIMALS (p. 835–846)

A. General Problems of Gas Exchange (p. 835)

Gas exchange = The traffic of O_2 and CO_2 between the animal and its environment.
- ☞ Supports cellular respiration by supplying O_2 and removing CO_2.

The *respiratory medium* (source of oxygen) for terrestrial animals is the air, while it is water for aquatic animals.
- ☞ Air is 21% oxygen while the amount of dissolved oxygen in water varies due to temperature, solute concentrations, and other factors.

Respiratory surface = Portion of the animal surface where gas exchange with the respiratory medium occurs.
- ☞ O_2 and CO_2 can only diffuse through membranes if they are first dissolved in the water that coats the respiratory surface.

B. Respiratory Organs: General Structure and Function (p. 835–836)

What are the general requirements for a respiratory surface? What variety of respiratory organs are adapted for this purpose?

Respiratory surfaces are generally thin, moist epithelium with a rich blood supply.
- ☞ Usually only a single cell layer separates the respiratory medium from the blood or capillaries.

Some animals with large surface area-to-volume ratios use their entire outer skin as a respiratory organ. Others have an extensively folded or branched localized region of body surface.
- ☞ External in aquatic animals (e.g. gills in fish).
- ☞ Invaginated into the body, opening to the atmosphere through narrow tubes, in terrestrial animals.

C. Gills: Respiratory Adaptations of Aquatic Animals (p. 836–838)

What are some respiratory adaptations of aquatic animals?

Gills are evaginations of the body surface specialized for gas exchange. (See Campbell, Figure 38.19)
- ☞ In some animals, gills are distributed over entire body; in others, are a localized region where the skin is finely dissected to form a feathery surface with a large area.
- ☞ Often sheltered by a protective covering.
- ☞ Must be efficient since although water keeps the respiratory surface wet, it has a lower oxygen concentration than air.

Ventilation = Any method of increasing the flow of the respiratory medium over the respiratory surface.
- ☞ Brings in a fresh supply of oxygen and removes CO_2.
- ☞ Due to the density and low oxygen concentration of water, fish must expend a large amount of energy to ventilate water.

Circulation and Gas Exchange 559

What is countercurrent exchange? Why is it more efficient than concurrent flow of water and blood?

Blood flows opposite to the direction in which water passes over gills, making efficient oxygen transfer to the blood possible by a process called *countercurrent exchange*. (See Campbell, Figure 38.20)

☞ Establishes a constant concentration gradient for oxygen between the blood and the water.

D. **Tracheae: Respiratory Adaptations of Insects** (*p. 838*)

What are the advantages and disadvantages of air as a respiratory medium? How are insect tracheal systems adapted for efficient gas exchange in a terrestrial environment?

Air has several advantages over water as a respiratory medium: a higher oxygen concentration; oxygen and carbon dioxide diffuse faster through air than water; respiratory surfaces do not have to be ventilated as thoroughly.

☞ The major disadvantage is that respiratory surfaces are continually desiccated.

Trachea are tiny air tubes that branch throughout the insect body; air enters the system through pores called *spiracles* and diffuses through the small branches which extend to the surface of nearly every cell.

☞ Some insects rely on diffusion for ventilation while others use rhythmic body movements.

☞ Cells are exposed directly to the respiratory medium so insects do not use their circulatory systems to transport O_2 and CO_2; a major reason why the open circulatory system works so well in insects.

E. **Lungs: Respiratory Adaptations of Terrestrial Vertebrates** (*p. 838–846*)

Lungs are heavily vascularized invaginations of the body surface that are restricted to one location.

☞ The circulatory system must transport oxygen from the lungs to the rest of the body.

☞ Various degrees of lung development are found in terrestrial vertebrates: frogs have simple balloonlike lungs with limited surface area; mammals have highly subdivided lungs with a large surface area.

In the human respiratory system, how does air move from air passageways to the alveolus? What structures must it pass through on the journey?

In mammals, the lungs are located in the thoracic cavity and consists of two layers held together by the surface tension of fluid between the layers. (See Campbell, Figure 38.22)

☞ Air entering the nostrils is filtered by hairs, warmed and humidified. It then travels through the pharynx, the larynx (which possesses vocal cords and functions as a voice box), and the cartilage-lined trachea that forks into two *bronchi* which further branch into finer bronchioles that dead-end in alveoli.

Countercurrent exchange

Trachea, Spiracles

Lungs

Pharynx, Larynx
Bronchi
Bronchioles, Alveoli

- Alveoli are lined with epithelium which serves as the respiratory surface.
- Oxygen dissolves in the moist film covering the epithelium and diffuses across to the capillaries covering each alveolus; carbon dioxide in the opposite direction by diffusion.

Vertebrates ventilate lungs by *breathing* (alternate inhalation and exhalation of air).
- Maintains a maximum oxygen concentration and minimum carbon dioxide concentration in the alveoli.

Frogs inhale by pushing air down the windpipe and exhale by elastic recoil of lungs and by compressing the lungs with the muscular body wall.
- Air is pulled into the mouth by lowering the floor region; this enlarges the oral cavity.
- The nostrils and mouth are closed and the floor of the mouth is raised, forcing air down the trachea.
- Oxygen is also absorbed across the lining of the mouth, which is ventilated by fluttering of the throat.

What is negative pressure breathing? How do respiratory movements in humans ventilate the lungs?

Mammals ventilate lungs by *negative pressure breathing* in two ways:

1. During vigorous exercise, rib muscles pull the ribs upwards which expands the rib cage, thus expanding the lungs since surface tension causes the lungs to follow.
   - Lung volume increases resulting in lower air pressure within alveoli, causing air to rush in.

What are the following lung volumes: a) tidal volume, b) vital capacity, and c) residual volume? What are the normal capacities for the human male?

2. When at rest, the *diaphragm* (thin sheet of muscle that forms the bottom wall of the thoracic cavity) contracts, enlarging the thoracic cavity.
   - Lowers pressure in lungs and causes inhalation.
   - *Tidal volume* is the volume of air an animal inhales and exhales with each breath. Averages about 500 ml in humans.
   - *Vital capacity* is the maximum air volume that can be inhaled and exhaled during forced breathing. Averages 4000–5000 ml in college-age males, a little less in females.
   - *Residual volume* is the amount of air that remains in the lungs even after forced exhalation.

Birds have complex ventilation:
- Besides lungs, they have 8 or 9 air sacs in their abdomen, neck and wings that serve to trim the density of the body and act as sinks for the heat dissipation by metabolism of flight muscles.
- Lungs have *parabronchi*; air flows through the entire system, lungs and air sacs, in only one direction regardless of whether it is inhaling or exhaling.

- ☞ The air sacs do not exchange gases, they serve as bellows to keep the air moving.
- ☞ The continuous flow of air through the porabronchi provides a constant supply of oxygen to the blood.

How is breathing controlled?

Breathing is an automatic action, we inhale when nerves in the *breathing center* of the medulla send impulses to the rib muscles or diaphragm, stimulating the muscles to contract.   *Breathing center*   *Medulla*
- ☞ This occurs about 10–14 times per minute and the degree of lung expansion is controlled by a negative-feedback mechanism involving stretch receptors in the lungs.
- ☞ The breathing center also monitors blood pH, which drops as blood CO_2 concentrations increase. When it senses an increased CO_2 level, the tempo and depth of breathing are increased.

How does oxygen move from the alveolus into the capillary? Why is a pressure gradient necessary?

Whether a gas enters or leaves the blood depends on the concentration of gases is measured as *partial pressure* (e.g. PO_2 = partial pressure of oxygen). Oxygen comprises about 21% of the atmosphere and carbon dioxide about 0.03%.   *Partial pressure*
- ☞ The partial pressure of a gas is the proportion of the total atmospheric pressure (760 mm) contributed by the gas; PO_2=160 mm (.21 × 760) and PCO_2=0.23 mm.
- ☞ A gas always diffuses from areas of high to those of low partial pressure.
- ☞ In the lungs, the blood exchanges gases with air in the alveoli and the PO_2 increases and PCO_2 is lowered.
- ☞ In systemic capillaries gradients of partial pressure favor diffusion of oxygen out of the blood and diffusion of CO_2 into it since cellular respiration rapidly depletes interstitial fluid of O_2 and adds CO_2. (See Campbell, Figure 38.25)

What is the difference between hemoglobin and hemocyanin?

In most animals, oxygen is carried by respiratory pigments in the blood since oxygen is not very soluble in water.   *Respiratory pigments*

1. In arthropods and mollusks, *hemocyanin* carries O_2.   *Hemocyanin*
   - ☞ The oxygen-binding component is copper.
   - ☞ Is always dissolved directly in plasma.

What is the structure of hemoglobin? What is the consequence of cooperative binding? How many oxygen molecules can a saturated hemoglobin molecule carry?

2. In vertebrates, *hemoglobin* is the oxygen-transporting pigment.   *Hemoglobin*
   - ☞ Consists of 4 subunits, each containing a heme group that bonds O_2; an iron atom is at the center of each heme group.
   - ☞ Binding of O_2 to one subunit induces a shape change that increases the affinity of the other 3 subunits for oxygen.

- The unloading of oxygen from one heme group also stimulates unloading from the other three.

What is the shape of the Hb-oxygen dissociation curve? How does the affinity of hemoglobin for oxygen change with oxygen concentration?

- The cooperative nature of this mechanism is evident in the dissociation curves depicted in Figure 38.26a.

What is the Bohr effect? How does the oxygen dissociation curve shift with changes in carbon dioxide concentration and with changes in pH?

What is the advantage of the Bohr shift?

- *Bohr shift* is the lowering of hemoglobins' affinity for oxygen upon a drop in pH. (See Campbell, Figure 38.26b) This occurs in active tissues due to the entrance of carbon dioxide into the blood.

Carbon dioxide is transported by the blood in three forms; dissolved CO_2 in the plasma (7%), bound to the amino groups of hemoglobin (23%), and as bicarbonate ions in the blood (70%).

How is carbon dioxide picked up at the tissues and deposited in the lungs? What is the role of carbonic anhydrase? In what form is most of the carbon dioxide transported?

Carbon dioxide diffuses into erythrocytes, where *carbonic anhydrase* catalyzes a reversible reaction wherein CO_2 converts into bicarbonate. (See Campbell, Figure 38.27)
- Carbonic acid forms first but quickly dissociates to bicarbonate and hydrogen ions.
- Bicarbonate then diffuses out of the erythrocyte and into the blood plasma.

How does hemoglobin act as a buffer?

- The hydrogen ions attach to hemoglobin and other proteins which results in only a slight change in the pH.
- The process is reversed in the lungs.

What are some respiratory adaptations in diving mammals?

Diving mammals such as seals, dolphins, and whales have special adaptations which allow them to make long, underwater dives.
- Weddell seals, which make dives to 200–500 m depths for 20 minutes or more, have been extensively studied.
- Can store large amounts of oxygen: 70% of the oxygen load is found in the blood (51% in humans) and they possess a higher myoglobin (an oxygen storing pigment) concentration in their muscles.
- The high oxygen load in the blood may be due to the higher blood volume in these seals (twice the amount of blood per kilogram of body weight as humans).

- ☞ They also have a very large spleen which probably contracts after a dive and forces more erythrocytes into the blood.
- ☞ The high concentration of myoglobin permits these seals to store about 25% of the oxygen in their muscles.

These adaptations provide diving mammals with a large oxygen reservoir at the beginning of a dive, but they also have several adaptations to conserve oxygen during the dive.

- ☞ A diving reflex slows the pulse and oxygen consumption declines as the cardiac output slows.
- ☞ Most of the blood is routed to the brain, spinal cord, eyes, adrenal glands, and placenta due to regulatory mechanisms that reduce blood flow to the muscles.
- ☞ The reduced blood flow to the muscles results in a depletion of the myoglobin oxygen reserves after about 20 minutes, the muscles then shift to fermentation for production of ATP.

39 THE BODY'S DEFENSES

CHAPTER OUTLINE

The vertebrate body possesses two mechanisms which protect it from potentially dangerous viruses, bacteria, other pathogens, and abnormal cells which could develop into cancer.

- One of these mechanisms is nonspecific, that is, it does not distinguish between infective agents.
- The second mechanism is specific in that it responds in a very specific manner (production of antibodies) to the particular type of infective agent.

I. NONSPECIFIC DEFENSE MECHANISMS (p. 850–854)

What is a nonspecific defense mechanism? What are the components involved with this form of defense?

Nonspecific defense mechanisms help prevent entry and spread of harmful microorganisms in an animal's body.

- An invading microbe must cross the external barrier formed by the skin and mucous membranes.
- If the external barrier is penetrated, the microbe encounters a second line of defense: interacting mechanisms of phagocytic white blood cells, antimicrobial proteins, and the inflammatory response.

A. The Skin and Mucous Membranes (p. 851)

How do the skin and mucous membranes function as defensive mechanisms? How are the physical barriers reinforced by chemical defenses?

The skin and mucous membranes act as a physical barrier preventing entry of pathogens, and as chemical barriers of anti-pathogen secretions.

- In humans, oil and sweat gland secretions acidify the skin (pH 3–5) which discourages microbial growth.
- The normal bacterial flora of the skin (adapted to the acidity) may release acids and other metabolic wastes to further inhibit pathogen growth.
- An enzyme (*lysozyme*) in perspiration, tears, and saliva attacks the cell walls of many bacteria and destroys other microbes.
- In the respiratory tract, nostril hairs filter inhaled particles and mucus traps microorganisms that are then swept out of the upper respiratory tract by cilia, thus preventing their entrance into the lungs.
- In the digestive tract, stomach acid kills many bacteria that enter with foods.

B. Phagocytic White Cells and Natural Killer Cells (p. 851–852)

Microbes that penetrate the skin encounter amoeboid white blood cells capable of phagocytosis or cell lysis.

What three types of phagocytic white cells function in nonspecific defense? What contribution does each make to the defense of the body? What are natural killer cells and how does their function differ from that of phagocytic white cells?

Neutrophils are cells that become phagocytic in infected tissue.
- Comprise 60–70% of total white cells.
- Attracted by chemical signals, they enter infected tissues by amoeboid movement; only live a few days as they destroy themselves when destroying pathogens.

Macrophages are large amoeboid cells that use pseudopodia to phagocytize microbes that are destroyed by digestive enzymes and reactive forms of oxygen within the cell.
- Most wander through interstitial fluid phagocytosing bacteria, viruses and cell debris.
- Some reside permanently in organs and connective tissues. They are fixed in place, but are located where they will have contact with infectious agents circulating in the blood and lymph.
- Develop from monocytes, which migrate from capillaries into interstitial fluid and represent about 5% of the total white cell count.

Eosinophils represent about 1.5% of the total white cell count but have limited phagocytic activity.
- Contain destructive enzymes in cytoplasmic granules which are discharged against the outer covering of the invading pathogen.
- Main contribution is defense against larger invaders such as parasitic worms.

Natural killer cells destroy the body's own infected cells, especially those harboring viruses.
- Also assault aberrant cells that could form tumors.
- Are not phagocytic, but attack the membrane, causing cell lysis.

C. Antimicrobial Proteins (p. 852)

A number of proteins function in nonspecific defense by either directly attacking microorganisms or impeding their reproduction.

What are complement proteins and interferons? How does each contribute to the nonspecific defense of the body?

The two most important, nonspecific protein groups are *complement* and the *interferons*.

Complement = A group of at least 20 proteins which interact with other defense mechanisms.

Phagocytosis

Neutrophils

Macrophages

Monocytes

Eosinophils

Natural killer cells

Complement

- These proteins interact in a series of steps which results in lysis of the invading microbes.
- Some components of the system function in chemotaxis as attractants to stimulate phagocyte movement into the infected site.

Interferons = Substances produced by virus-infected cells which help other cells resist infection by the virus.
- Secreted by infected cells as a nonspecific defense earlier than specific antibodies appear.
- Cannot save the infected cell, but their diffusion to neighboring cells stimulates production of proteins in those cells that inhibit viral replication.
- Not a virus-specific defense; interferon produced to infection by one strain of virus produces resistance in other viral infections.
- Most effective against short-term infections (colds and influenza).
- One type of interferon also activates phagocytes which enhances their ability to ingest and kill microorganisms.

D. **The Inflammatory Response** (p. 852–854)

What occurs in a localized inflammatory response? What triggers the response? What chemical signals initiate the response? How does it prevent the spread of infection to surrounding tissues?

A localized *inflammatory response* occurs when there is damage to a tissue due to physical injury or entry of microorganisms.
- Vasodilation of small vessels near the injury increases the blood supply to the area.
- The dilated vessels become more permeable allowing fluids to move into surrounding tissues.

Chemical signals are important for initiating an inflammatory response. (See Campbell, Figure 39.6)
- *Histamine* is released from injured circulating basophils and/or mast cells in the connective tissue.
- Released histamine causes localized vasodilation and the capillaries in the area become leakier.
- Prostaglandins are also released from white blood cells and damaged tissues. These and other substances promote blood flow to the injured area.
- Increased blood flow to the site of injury delivers clotting elements which help block the spread of pathogenic microbes and begin the repair process.

Migration of phagocytic cells into the injured area is also a result of increased blood flow and increased leakage from the capillaries. (See Campbell, Figure 39.6)
- Phagocytes are attracted to the damaged tissues by several chemical mediators including some complement proteins.
- Neutrophils arrive first, followed closely by monocytes which develop into macrophages.
- The neutrophils eliminate microorganisms and then die.

- ☞ Macrophages destroy pathogens and clean up the remains of damaged tissue cells and dead neutrophils.
- ☞ Dead cells and fluid leaked from the capillaries may accumulate as pus in the area before it is absorbed by the body.

How does a systemic inflammatory response differ from a localized response? What conditions can cause a systemic response?

More widespread (systemic) inflammatory responses may also occur in cases of severe infections (meningitis, appendicitis).
- ☞ The bone marrow may be stimulated to release more neutrophils by chemicals emitted by injured cells.
- ☞ There may also be a severalfold increase in the number of leukocytes within a few hours of response onset.
- ☞ A fever may develop in response to toxins produced by pathogens or due to *pyrogens* released by leukocytes.
- ☞ While a high fever is dangerous, moderate fevers inhibit the growth of microorganisms.
- ☞ Moderate fevers may facilitate phagocytosis and speed up tissue repairs.

II. THE IMMUNE SYSTEM AND SPECIFIC DEFENSE: SOME BASIC CONCEPTS (p. 854–860)

A. Basic Features of the Immune System (p. 854–855)

The *immune system* is the body's third line of defense and is specific in nature.
- ☞ Distinguished from nonspecific defenses by: specificity, diversity, self/nonself recognition, and memory.

Why is the immune system considered the body's third line of defense? What are the four basic features of the system? What is the relationship between an antigen and an antibody?

Specificity refers to this system's ability to recognize and eliminate particular microorganisms and foreign molecules.

Antigen = A foreign molecule that elicits an immune response.

Antibody = An antigen-binding immunoglobulin, produced by B cells, that functions as the effector in an immune response.
- ☞ Antigens may be molecules exhibited on the surface of, produced by, or released from bacteria, viruses, fungi, protozoans, parasitic worms, pollen, insect venom, transplanted organs, or worn-out cells.
- ☞ Each antigen has a unique molecular shape and stimulates production of an antibody that defends specifically against that particular antigen.
- ☞ The immune response is thus very specific and distinguishes between even closely related invaders.

Diversity refers to the immune system's ability to respond to numerous kinds of invaders which are recognized by their antigenic markers.

- Based on the wide variety of lymphocyte populations in the immune system.
- Each population of antibody-producing lymphocytes is stimulated by a specific antigen; the stimulated lymphocytes then synthesize and secrete the appropriate antibody.

Self/nonself recognition is the ability of the immune system to distinguish between the body's own molecules and foreign molecules.
- Failure of this system leads to autoimmune disorders which destroy the body's own tissues.

Memory refers to the immune system's ability to recognize previously encountered antigens and to react faster and more effectively to subsequent exposures.
- This acquired immunity has long been recognized as a resistance to some infections encountered earlier in life (e.g. chicken pox).

B. Active Versus Passive Acquired Immunity (p. 855)

How can you distinguish between active and passive immunity? What role does a vaccine play in each?

Active immunity is the immunity conferred by recovery from an infectious disease.
- Depends on the person's own immune system.
- May be acquired <u>naturally</u> from an infection to the body or <u>artificially</u> by vaccination.
- Vaccines may be inactivated bacterial toxins, killed microorganisms, or weakened living microorganisms; in all cases the organisms can no longer cause the disease but can act as antigens and stimulate an immune response.
- A person vaccinated against an infectious agent will show the same rapid, memory-based immunological response upon encountering the pathogen as someone who has had the disease.

Passive immunity is immunity which has been transferred from one individual to another.
- Natural occurrence when antibodies cross the placenta from a pregnant woman's system to her fetus.
- Passive immunity provides protection to newborns whose immune system is not fully operational.
- Some antibodies are transferred to nursing infants though the milk.
- Persists for only a few weeks or months after which the infant's own system defends its body.
- May also be transferred artificially from an animal or human already immune to the disease.
- Rabies is treated by injecting antibodies from people vaccinated against rabies; produces an immediate immunity important to quickly progressing infections.
- Artificial passive immunity is also of short duration.

C. The Immune System's Dual Defense: An Overview of Humoral and Cell-Mediated Responses (p. 855–856)

The body will mount either a humoral response or a cell-mediated response depending on the antigen which stimulates the system.

The Body's Defenses 569

How does humoral immunity differ from cell-mediated immunity? How are lymphocytes involved? What is the difference between B lymphocytes and T lymphocytes?

    Humoral immunity is the response activated against toxins, free bacteria, and viruses present in the body fluids.
- Antibodies to these types of antigens are synthesized by certain lymphocytes and then secreted as soluble proteins which circulate through the body in blood plasma and lymph.

    Cell-mediated immunity is the response to intracellular bacteria and viruses, fungi, protozoans, worms, transplanted tissues, and cancer cells.
- Depends on the direct action of certain types of lymphocytes rather than antibodies.

    Lymphocytes are responsible for both humoral and cell-mediate immunity; the different responses are due to the two main classes of lymphocytes in the body: B cells and T cells.
- B cells (B lymphocytes) form in the bone marrow and complete their maturation in the bone; these are responsible for the humoral immune response.
- T cells (T lymphocytes) also form in the bone marrow, then migrate to the thymus gland to mature; these cells are responsible for the cell-mediated immune response.
- Early B and T cells (as well as other lymphocytes) develop from pluripotent stem cells in the bone marrow and are very much alike, they only differentiate after reaching their site of maturation.

Where are mature B and T cells normally found in the body? What association is there between effector cells and B and T cells?

    Mature B cells and T cells are concentrated in the lymph nodes, spleen, and other lymphatic organs.
- These positions place lymphocytes where they are most likely to come into contact with antigens.
- *Antigen receptors* are present on the membranes of both types of cells.
- The antigen receptor on a B cell is a bound antibody molecule which will recognize a specific antigen.
- The T cell receptors are proteins embedded in the membrane which recognize specific antigens.

    Effector cells are the cells which actually defend the body during an immune response.
- Effector cells are a population of cells resulting from division of a lymphocyte which was activated by the binding of an antigen to its antigen receptor.
- Activated B cells give rise to *plasma cells* which secrete antibodies (humoral response) that eliminate the activating antigen.
- Activated T cells may produce two types of effector cells: *cytotoxic T cells* which destroy infected cells and cancer cells; and *helper T cells* which help stimulate both the humoral and cell-mediated responses.

Margin notes:
Humoral immunity
Cell-mediated immunity
Lymphocytes
B cells, T cells
Humoral immune response
Antigen receptors
T cell receptors
Effector cells
Plasma cells

The Body's Defenses

D. The Molecular Basis of Antigen-Antibody Specificity (p. 856–858)

What is an antigen? What is the structure of an antibody? How is the antibody structure associated with antigen-antibody specificity?

Antigens are usually proteins or large polysaccharides that make up a portion of the outer covering of pathogens or transplanted cells.

- May be components of the coats of viruses, capsules and cell walls of bacteria, or surfaces of other cell types.
- Molecules on the cell surface of transplanted tissues and organs or blood cells from other individuals are also recognized as foreign. (See Campbell, Figure 39.8)
- Antibodies recognize a localized region on the surface of an antigen (the *epitope*), not the entire antigen molecule.
- Several types of antibodies may be produced to a single bacterial cell since it may have different antigens on different areas and each bacterial antigen may possess more than one recognizable epitope.

Antibodies comprise a specific class of proteins called *immunoglobulins* (Igs). (See Campbell, Figure 39.9a)

- The structure of the immunoglobulin is associated with its function.
- Antibodies are Y-shaped molecules comprised of four polypeptide chains: two identical light chains and two identical heavy chains.
- All four chains have *constant* (C) regions that show little variation in amino acid sequence in antibodies that perform a particular type of defense.
- At the tips of the Y are found *variable* (V) regions of all four chains; the amino acid sequences in the variable region show extensive variation from antibody to antibody.
- The variable regions function as *antigen-binding sites* and their amino acid sequences result in specific shapes that fit and bind to specific antigen epitopes.
- The antigen-binding site is responsible for the antibody's ability to recognize its specific antigen epitope and the stem (constant) regions are responsible for the mechanism by which the antibody inactivates or destroys the antigenic invader.

What criteria can be used to distinguish among the five classes of immunoglobulins?

As mentioned earlier, there is little variation in the constant region among the antibodies. What variation is present has resulted in the immunoglobulins of mammals being divided into five classes. (See Table 39.1 for a summary.)

- *IgM*. Consists of five Y-shaped monomers arranged in a pentamer structure. Circulating antibodies which appear in response to an initial exposure to an antigen.
- *IgG*. A Y-shaped monomer. Most abundant circulating antibody; readily crosses blood vessels and enters tissue fluids; protects against bacteria, viruses, and toxins circulating in blood and lymph; triggers complement system action.

☞ *IgA*. A dimer consisting of two Y-shaped monomers. Produced primarily by cells abundant in mucous membranes; prevents attachment of bacteria and viruses to epithelial surfaces; also found in saliva, tears, perspiration, and colostrum.

☞ *IgD*. A Y-shaped monomer. Found primarily on external membranes of B cells; probably functions as an antigen-receptor which initiates differentiation of B cells.

☞ *IgE*. A Y-shaped monomer. Stem regions attach to receptors on mast cells and basophils; stimulates these cells to release histamine and other chemicals that cause allergic reactions when triggered by an antigen.

E. **Clonal Selection: The Cellular Basis of Immunological Specificity and Diversity** (*p. 858*)

What is clonal selection? How is clonal selection related to the body's ability to respond to a variety of antigens? How can a single antigen result in the production of several clones of effector T cells or plasma cells?

The ability of the immune system to respond to the wide variety of antigens which enter the body is based in the enormous diversity of antigen-specific lymphocytes present in the system.

☞ Each lymphocyte will recognize and respond to only one antigenic epitope.

☞ This specificity is determined during embryonic development before any antigens are encountered. (See Campbell, Figure 39.10)

☞ *Clonal selection* is the formation of a clone of effector cells specific for a particular antigen from an inactive lymphocyte which was stimulated by that antigen.

☞ When an antigen enters the body and binds to a lymphocyte, that lymphocyte is activated and begins to divide.

☞ The divisions produce a large number of identical effector cells (clones) which bind to the antigen that stimulated the response.

☞ If, for example, a B cell is activated, it will proliferate to produce a large number of plasma cells that will each secrete an antibody which functions as an antigen receptor for the specific antigen that activated the original B cell.

Each antigen thus activates a small number of the diverse group of lymphocytes. The activated cells proliferate to produce a clone of millions of effector cells which are specific for the original antigen.

☞ Because most antigens possess several active epitopes, several different clones of effector T cells or plasma cells may be produced in response to the same antigen.

F. **The Cellular Basis of Immunological Memory** (*p. 859–860*)

How does a primary immune response differ from a secondary immune response? What is the relationship to memory cells to these responses?

The *primary immune response* is the proliferation of lymphocytes to form clones of effector cells specific to an antigen during the body's <u>first</u> exposure to the antigen.

572 The Body's Defenses

- There is a 5 to 10 day lag period between exposure and maximum production of effector cells.
- The selected lymphocytes to the antigen are differentiating into effector T cells and plasma cells during the lag period.

Secondary immune response

A *secondary immune response* occurs when the body is exposed to a previously encountered antigen.
- The response is faster (3 to 5 days) and more prolonged than a primary response.
- The antibodies produced are also more effective at binding the antigen.

This ability to recognize a previously encountered antigen is known as immunological memory.

Memory cells

- Based on *memory cells* which are produced during clonal selection for effectors in a primary immune response.
- Memory cells are not active during the primary response and survive in the system for long periods. (Effector cells produced in the primary response are active and, thus, short-lived.)
- When the same antigen that caused a previous primary immune response again enters the body, the memory cells are activated and rapidly proliferate to form a new clone of effector cells and memory cells.
- These new clones of effector and memory cells are the secondary immune response.

G. The Development of Self-Tolerance (p. 860)

When does self-tolerance develop? Why is it important? What is the major histocompatibility complex?

Self-tolerance

Self-tolerance = The lack of a destructive immune response to the body's cells.
- Develops before birth when lymphocytes begin to mature in the embryo.
- Any lymphocytes with receptors for molecules present in the body at that time are destroyed; consequently, the body contains no lymphocytes with antigen receptors for its own molecules, only for foreign molecules.

Major histocompatibility complex

The *major histocompatibility complex* (MHC) is a groups of glycoproteins embedded in the plasma membranes of cells.
- Important "self-markers" coded for by a family of genes.
- There are at least 20 MHC genes and at least 50 alleles for each gene.
- The probability that two individuals will have matching MHC sets is virtually zero unless they are identical twins.
- There are two main classes of MHC molecules in the body: *Class I MHC* molecules are located on all nucleated cells of the body; *Class II MHC* molecules are found only on macrophages and B cells.

III. **THE HUMORAL RESPONSE: A CLOSER LOOK** (*p. 860–863*)

The humoral response occurs when an antigen binds to receptors on B cells which have specific receptors for the antigen epitopes.
- The B cells differentiate into a clone of plasma cells which begin to secrete antibodies.
- These antibodies are most effective against pathogens circulating in the blood or lymph.
- Memory cells are also produced and form the basis for secondary immune responses.

A. **The Activation of B Cells** (*p. 860–861*)

How is the humoral response stimulated? What steps are involved in B cell activation? How are helper T cells and class II MHC proteins involved?

The selective activation of a B cell by an antigen results in the formation of a clone of plasma cells and memory cells.
- The first step in the process is the binding of the antigen to specific antigen-receptors on the surface of the B cell.

The second step in B cell activation involves macrophages and helper T cells; this step ends with the production of plasma cells. (See Campbell, Figure 39.12)
- After macrophages phagocytize pathogens, pieces of the partially digested antigen molecules are displayed on the surface of the macrophage.
- These antigen molecules form complexes with class II MHC proteins embedded in the membrane and the macrophage now functions as an *antigen-presenting cell*.   *Antigen-presenting cell*
- A helper T cell with a receptor specific for the presented antigen binds to the self/nonself class II MHC protein-antigen complex.
- The T cell is activated by contact with the macrophage and proliferates to form a clone of helper T cells specific for the presented antigen.
- These helper T cells then stimulate B cells which have encountered the same antigen; recognition of these B cells also involves a class II MHC protein-antigen complex to which the receptor on the T cell binds.
- The T cell contact activates these B cells to form a clone of plasma cells.
- Each plasma cell (= effector cell) then secretes antibodies specific for the antigen.

What major difference is there between the interactions of helper T cells with macrophages and B cells? What are the differences in response to T-dependent antigens and T-independent antigens?

Both macrophages and B cells act as antigen-presenting cells in their interactions with helper T cells, but there is one major difference:

574 The Body's Defenses

- Each macrophage can display a number of different antigens depending on the type of pathogen phagocytized.
- B cells are specific and can bind to and display only one type of antigen.

Macrophages, which are nonspecific, can thus enhance specific defense by selectively activating helper T cells which in turn activate B cells specific for the antigen.
- Helper T cells are also antigen-specific and are activated only by macrophages presenting the proper class II MHC protein-antigen complex.

T-dependent antigens = Antigens that evoke the cooperative response involving macrophages, helper T cells, and B cells.
- These antigens cannot stimulate antibody production without T cell involvement.

T-independent antigens = Antigens that trigger humoral immune responses without macrophage or T cell involvement.
- Usually long chains of repeating units such as polysaccharides or protein subunits.
- B cells are stimulated directly by the antigen which probably binds simultaneously to several antigen receptors on the B cell surface.
- The antibody production (humoral response) is usually much weaker to that of T-dependent antigens.

Whether activated by T-dependent or T-independent antigens, a B cell gives rise to a clone of plasma cells.
- Each of these effector cells secretes up to 2,000 antibodies per second into the body fluids for its 4 to 5 day lifespan.
- The specific immunoglobulins help eliminate the foreign invader from the body.

B. How Antibodies Work (*p. 862*)

How are antibodies actually involved in the destruction of an antigen? How do the four effector mechanisms of the humoral response differ in mode of action?

Antibodies do not directly destroy an antigenic pathogen. The antibody binds to the antigen to form an antigen-antibody complex which tags the invader for destruction by one of several effector mechanisms. (See Campbell, Figure 39.13)
- *Neutralization* is the simplest mechanism. The antibody blocks viral attachment sites or coats a bacterial toxin, making them ineffective. Phagocytic cells eventually destroy the complex.
- *Agglutination* is another mechanism. Each antibody has two or more antigen-binding sites and can cross-link adjacent antigens. The cross-linking can result in clumps of a bacteria being held together by the antibodies, making it easier for phagocytes to engulf the mass.
- *Precipitation* is similar to agglutination but involves the cross-linking of soluble antigen molecules instead of cells. These immobile precipitates are easily engulfed by phagocytes.

☞ *Activation* of the complement system is another mechanism. Antibodies combine with complement proteins; this combination activates the complement which produces lesions in the foreign cell's membrane that results in cell lysis.

C. **Monoclonal Antibodies** (*p. 863*)

What are monoclonal antibodies? How are they produced? What current and potential medical uses do they possess?

Monoclonal antibodies = Defensive proteins produced by cells descended from a single cell; all antibodies produced by these cells are identical.

This technology permits the production of large quantities of antibodies quickly and at relatively little expense.
☞ Can be used in diagnostic labs to detect pathogenic microbes in clinical samples.
☞ Form the basis for over-the-counter pregnancy tests.
☞ Are used as therapeutic agents.
☞ Show promise in treating cancer when combined with a toxin that would destroy the cancer cell.

Monoclonal antibodies are produced by *hybridoma* cells.
☞ Hybridoma cells are hybrid cells resulting from the fusion of certain cancer cells (*myelomas*) with normal antibody-producing plasma cells.
☞ The cancer cell is used since it can be cultured indefinitely (other cells can only be cultured for a few generations).
☞ Plasma cells expressing the desired antibody are mixed with myeloma cells and some of the cell two cell types fuse.
☞ The hybridoma cells are isolated and cultured; these cells exhibit the key qualities of the two cell types: they will produce a single type of antibody and can be cultured indefinitely to manufacture that antibody on a large scale.

IV. **THE CELL-MEDIATED RESPONSE: A CLOSER LOOK** (*p. 863–865*)

The humoral immune response is that portion of the body's defenses that identifies and destroys free pathogens.

The *cell mediated immune response* is the defense mechanism that combats pathogens that have already entered cells.
☞ The key component of the cell-mediated immunity are T lymphocytes (T cells) which complete their maturation in the thymus and migrate to lymphoid organs such as the spleen and lymph nodes.

Why are T cells the key component of cell-mediated immunity? What is an MHC-antigen complex? What is the importance of helper T cells? What is a cytokine? What distinguishes interleukin-1 and interleukin-2?

T cells respond only to antigenic epitopes displayed on the surfaces of the body's own cells.
☞ Cannot be activated by free antigens in the body fluids.

Activation

Monoclonal antibodies

Hybridoma

Myelomas

576 The Body's Defenses

- T cell receptors (specific proteins embedded in the T cell plasma membrane) recognize the bound antigens displayed on body cells.
- The displayed antigen is coupled with class II MHC proteins to form a unique complex recognizable by specific T cells having a matching receptor.

The MHC-antigen complex displayed on an infected body cell activates two types of T cells [*helper T cells* (T_H) and *cytotoxic T cells* (T_C)] involved in the cell-mediated response.

- Remember, helper T cells also stimulate B cells to secrete antibodies against T-dependent antigens in a humoral response.

Helper T cells (T_H) are able to stimulate other lymphocytes by receiving and sending chemical signals called cytokines.

<u>Cytokine</u> = A molecule secreted by one cell that functions as a regulator of neighboring cells. (See Campbell, Figure 39.14)

- When a T_H cell binds to an antigen-presenting macrophage, the macrophage releases *interleukin-1* (a cytokine).
- The presence of interleukin-1 stimulates the T_H cell to release its own cytokine, *interleukin-2*.
- In a positive feedback mechanism, interleukin-2 stimulates T_H cells to grow and divide more rapidly; this results in the production of more T_H cells and an increased supply of interleukin-2.
- The humoral response against the antigen is enhanced because interleukin-2 and other cytokines secreted by T_H cells activate B cells.
- The increased levels of cytokines also increase the cell-mediated response by stimulating another class of T lymphocytes to differentiate into *cytotoxic T cells* (effector cells).

How are cytotoxic T cells formed? By what mechanism is a cytotoxic T cell able to recognize and destroy its targets? What is the function of a suppressor T cell?

Cytotoxic T cells (T_C) are the cells which actually destroy other cells.

- Host cells infected by viruses and other pathogens display antigens in complexes with *Class I MHC* molecules on their surfaces.
- T_C cells have specific receptors which recognize and bind to antigen-class I MHC markers. (Note that this differs from T_H cells which bind to antigen-class II MHC complexes.) (See Campbell, Figure 39.15)
- The T_C receptor can bind to any cell in the body displaying the antigen-class I MHC marker since class I MHC is present on all nucleated cells.
- When a T_C cell binds to a marker, it releases *perforin* which is a protein that forms a lesion in the infected cell's membrane.
- Cytoplasm escapes through the lesion and eventually results in cell lysis.
- Destruction of the host cell not only removes the site where pathogens can reproduce, but also exposes the pathogens to circulating antibodies form the humoral response.
- T_C cells continue to live after destroying the infected cell and may kill many others displaying the same antigen-class I MHC marker.

Cytotoxic T cells also function to destroy cancer cells which develop periodically in the body.

- Cancer cells possess distinctive markers not found on normal cells.

- T_C cells recognize these markers as nonself and attach and lyse the cancer cells.
- Cancers develop primarily in individuals with defective or declining immune systems.

A third type of T lymphocyte, *suppressor T cells* (T_S), has been found in the body. *Suppressor T cells*
- Probably function to suppress the immune system when an antigen is no longer present.
- Action is not well understood and some immunologists feel T_S cells are actually a form of T_H cells.

V. THE COMPLEMENT SYSTEM: A COMPONENT OF BOTH NON-SPECIFIC AND SPECIFIC DEFENSE (p. 865–867)

The 20 or so complement proteins circulate in the blood in inactive forms.
- Become activated in a series fashion and interact with both specific and nonspecific defense mechanisms.

What steps are involved in the classical pathway of complement system activation? How does the alternative pathway differ and why?

The *classical pathway* describes complement's activation in the specific defense mechanism. (See Campbell, Figure 39.17)   *Classical pathway*
- Initiated when antibodies bind to a specific pathogen in order to target the cell for destruction.
- A complement protein attaches to, and bridges the gap between, two adjacent antibody molecules.
- The antibody-complement association activates complement proteins to form, in a step-by-step sequence, a *membrane attack complex*.   *Membrane attack complex*
- The membrane attack complex lyses the pathogen's membrane producing a lesion.
- Lysis of the pathogenic cell then occurs.

The *alternative pathway* is how complement is activated in nonspecific defense mechanisms.   *Alternative pathway*
- Does not require cooperation with antibodies.
- Complement proteins are activated by substances found in many pathogens (yeasts, viruses, virus-infected cells, protozoans) to form a membrane attack complex.
- The complex lyses the pathogen without the aid of antibodies.
- Complement proteins also contribute to inflammation by binding to histamine-containing cells; this association triggers the release of histamine from those cells.
- Several complement proteins also attract phagocytic cells to infected sites.

How do complement and phagocytes cooperate to destroy pathogens?

Complement and phagocytes also work together in two ways to destroy pathogens.
- *Opsonization* is a cooperative mechanism in which complement proteins attach to a foreign ell and stimulate phagocytes to engulf the cell.   *Opsonization*
- In *immune adherence* complement proteins and antibodies coat a microbe which causes it to adhere to blood vessel walls and other surfaces; this makes the cell easy prey for circulating phagocytes.   *Immune adherence*

VI. SELF VERSUS NONSELF: SOME APPLICATIONS (p. 867–869)

The body's immune system distinguishes between self (the body's own cells) and nonself (foreign cells).

- Nonself includes not only the pathogens discussed earlier, but also cells from other individuals of the same species.

How many different blood types are possible in the human ABO blood groups? Which combinations would create a problem during a blood transfusion and why? Why is type O the universal donor and type AB the universal recipient?

A. Blood Groups (p. 868)

Individuals of blood group A have the A antigen and make anti-B antibodies.

Individuals of blood group B have the B antigen and make anti-A antibodies.

Individuals of blood group AB have the A and B antigen and make no antibodies.

Individuals of blood group O have neither the A nor B antigen and make anti-A and anti-B antibodies.

Blood group antibodies can cause blood of a different antigenic type to agglutinate, a life-threatening reaction.

- Type O individuals are universal donors since their blood has neither antigen.
- Type AB individuals are universal recipients since they produce neither antibody A or antibody B.

How does the immune response to Rh factor differ from the response to A and B blood antigens?

Another blood group antigen is *Rh factor*. Rh factor causes problems when a mother is Rh negative and her fetus is Rh positive (inherited from the father).

- When small amounts of fetal blood cross the placenta and come into contact with the mother's lymphocytes, the mother develops antibodies against the Rh factor.
- Usually only a problem in the second child since the response will be quick due to sensitization and formation of memory cells during the first baby's gestation; the mother's antibodies cross the placenta and destroy the red blood cells of the Rh positive fetus.
- Can be prevented by injection of anti-Rh antibodies which destroy Rh-positive red cells before the mother develops immunological memory.

B. Tissue Grafts and Organ Transplants (p. 868–869)

Why is it almost impossible for two individuals (other than identical twins) to have identical MHC markers? Why are rejections of tissue grafts and transplanted organs actually results of the proper functions of a healthy immune system?

MHC is a biochemical fingerprint unique to each individual.
- ☞ Remember there are at least 20 MHC genes and about 50 alleles for each gene.
- ☞ Complicates tissue grafts and organ transplants since foreign MHC molecules are antigens and cause cytotoxic T cells to mount a cell-mediated response.
- ☞ Cyclosporine and FK506 suppress cell-mediated immunity without crippling humoral immunity, thus increasing the chance of successful grafts and transplants.

Cyclosporine

Note the reaction of the immune system to transfusions, tissue grafts, and organ transplants are normal reactions of a healthy immune system, not disorders of the system.

VII. DISORDERS OF THE IMMUNE SYSTEM (p. 869–872)

A. Autoimmune Diseases (p. 869)

What is an autoimmune disease? What can cause such a disease? How common are such conditions?

Autoimmune Disease = The immune system reacts against self.
- ☞ Some cases involve immune reactions against components of the body's own cells which are released by the normal breakdown of skin and other tissues; especially nucleic acids in lupus erythematous.
- ☞ Rheumatoid arthritis is an autoimmune disease in which inflammation damages cartilage and bones in joints.
- ☞ Destruction of insulin-producing pancreas cells by an autoimmune reaction appears to cause insulin-dependent diabetes.
- ☞ Antibodies produced to repeated streptococcal infections may react with heart tissues and cause valve damage in some people.
- ☞ Other autoimmune diseases are Grave's disease and rheumatic fever.

Lupus erythematous

B. Allergy (p. 869)

What is an allergy? What immunoglobulin is normally involved in allergic reactions? How is degranulation involved in the reaction?

Allergies = Hypersensitivities of the body's defense system to environmental antigens called *allergens*.
- ☞ Some believe these reactions to be evolutionary remnants to infection by parasitic worms due to similarities in the responses.

Allergies, Allergens

IgE antibodies are commonly involved in allergic reactions; these antibodies recognize pollen as allergens. (See Campbell, Figure 39.18a)
- ☞ IgE antibodies attach by their stems to noncirculating mast cells found in connective tissues.
- ☞ When a pollen grain bridges the gap between two adjacent IgE monomers, the mast cell responds with a reaction called *degranulation*.

Degranulation

580 *The Body's Defenses*

☞ Degranulation involves the release of histamine and other inflammatory agents.
☞ Histamine causes dilation and increased permeability of small blood vessels which results in the common symptoms of an allergy.
☞ Antihistamines are drugs used to treat allergies since they interfere with the action of histamine.

What is anaphylactic shock? Why is this type of reaction serious?

Anaphylactic shock is a life-threatening reaction to injected or ingested antigens; the most serious type of acute allergic response.
☞ Occurs when mast cell degranulation causes a sudden dilation of peripheral blood vessels and a drastic drop in blood pressure.
☞ Death may occur in a few minutes.
☞ This hypersensitivity may be associated with foods (peanuts, fish) or insect venoms (wasp or bee stings).
☞ Epinephrine may be injected to counteract the allergic response.

C. **Immunodeficiency** (p. 869–870)

When is an individual considered to have an immunodeficiency? What types of immunodeficiency are recognized? What can contribute to the development of an immunodeficient condition?

Immunodeficiency refers to a condition where an individual is inherently deficient in either humoral or cell-mediated immune defenses.
☞ *Severe combined immunodeficiency* is a congenital disorder in which both the humoral and cell-mediated immune defenses fail to function.

Not all cases of immunodeficiency are inborn conditions.
☞ Some cancers, like Hodgkin's disease, depress the immune system and make the individual susceptible to infection.
☞ Some viral infections cause depression of the immune system.
☞ Physical and emotional stress may compromise the system; adrenal hormones secreted by stressed individuals affect the number of leukocytes and may suppress the system in other ways.

Some evidence suggests a direct link between the nervous system and the immune system.
☞ There is a network of nerve fibers which penetrates into lymphoid tissues including the thymus.
☞ Lymphocytes have also been found to possess surface receptors for chemical signals secreted by nerve cells.

D. **Acquired Immunodeficiency Syndrome (AIDS)** (p. 870–872)

What is the infective agent associated with acquired immunodeficiency syndrome (AIDS)? How does HIV infection progress in the body? What immuno-active cells become infected by HIV?

Acquired immunodeficiency syndrome is a severe immune system disorder caused by infection with the *human immunodeficiency virus* (*HIV*).

- ☞ Characterized by a reduction of T cells and the appearance of characteristic secondary infections.
- ☞ Mortality rate approaches 100%.
- ☞ HIV probably evolved from another virus in central Africa and may have gone unrecognized for many years.
- ☞ HIV infects T cells, including T_H cells, which carry the CD4 receptor on their surface.
- ☞ Macrophages and a few subclasses of B cells carrying the CD4 receptor can also be infected.
- ☞ Glycoproteins on the HIV envelope bind specifically to the CD4 receptor.
- ☞ After attaching, the HIV enters the cell and begins to replicate.
- ☞ Newly formed viruses bud continuously from the host cell, circulate, and infect other cells.
- ☞ Infected cells may be killed quickly by the virus or immune response; they may also live for an extended time.
- ☞ The HIV may also remain as a provirus in the infected cell genome for many years before becoming active.

How does HIV escape destruction by antibodies? Why is AIDS considered late-stage HIV infection and what characteristic symptoms are present?

HIV is not eliminated from the body by antibodies for several reasons:
- ☞ The latent provirus is invisible to the immune system.
- ☞ The virus undergoes rapid mutational changes in antigens which eventually overwhelms the immune system.
- ☞ The population of helper T-cells eventually declines to the point where cell-mediated immunity collapses.
- ☞ Secondary infections characteristic of HIV infection develop (Pneumocystis pneumonia and Kaposi's sarcoma).

AIDS is the late stage of HIV infection and is defined by a reduced T cell population and the appearance of secondary infections.
- ☞ Takes an average of about ten years to reach this stage of infection.
- ☞ During most of this time, only moderate symptoms are shown.
- ☞ Progression of infection is more rapid in infants who were infected in utero.
- ☞ Individuals exposed to HIV have circulating antibodies that can be detected; displaying these antibodies designates and individual as *HIV-positive*.

In what ways can HIV be transmitted between individuals? What treatments are used with AIDS patients? How effective are these treatments?

HIV is only transmitted through the transfer of body fluids, blood or semen, containing infected cells.
- ☞ Most commonly transmitted in the U.S. and Europe through unprotected sex between male homosexuals and unsterilized needles in intravenous drug users.

Kaposi's sarcoma

HIV-positive

- In Africa and Asia, transmission through unprotected heterosexual sex is rapidly increasing; especially in areas with a high incidence of other sexually transmitted diseases.
- Transmission to nursing infants through breast milk has been reported.
- Transmission through blood transfusions has also been reported, but the incidence has declined with implementation of screening procedures.

AIDS is currently considered an incurable disease.
- AZT, ddC, and ddI are antiviral drugs used to extend the lives of infected individuals, but they do not eliminate the virus.
- Other drugs are used to fight opportunistic infections common in AIDS patients.
- The best way to prevent additional infections is to educate people on how the disease is transmitted and how to protect themselves.

VIII. IMMUNITY IN INVERTEBRATES (p. 872)

What evidence is available that invertebrates have some form of immunity? What are coelomocytes? How well developed is self-nonself recognition in invertebrates? Do invertebrates show a memory response?

How invertebrates react against pathogens that enter their bodies is poorly understood. It is known that invertebrates have a well developed ability to distinguish self from nonself.
- Experiments have shown that if the cells from two sponges are mixed, the cells from each individual will aggregate in separate groups, excluding cells from the other individual.
- *Coelomocytes* which are amoeboid cells that destroy foreign materials have been found in many invertebrates.

A memory response has also been identified in earthworms.
- A body wall graft from one worm to another will survive for about eight months before rejection if the worms are from the same population.
- A graft involving worms from different populations is rejected in two weeks.
- A second graft from the same donor to the same recipient is rejected in less than one week due to coelomocyte activity.

40   CONTROLLING THE INTERNAL ENVIRONMENT

CHAPTER OUTLINE

Most animals can survive fluctuations in the external environment more extreme than any of their individual cells could tolerate.
- ☞ Accomplished by mechanisms of *homeostasis* which maintains their internal environment within ranges that the body cells can tolerate.

The majority of cells in most animals (all but sponges and cnidarians) are not exposed to the external environment, but are bathed by an internal fluid.
- ☞ Hemolymph is present in animals with open circulatory systems.
- ☞ Interstitial fluid, which interacts with the blood, is present in animals with closed circulatory systems.

I. OSMOREGULATION (p. 877–881)

Osmoregulation = Control of solute balance and the gain and loss of water.
- ☞ Even animals with specialized body coverings that retard water gain or loss have some unprotected structures exposed to the environment for gas exchange (lungs, gills).

Animals cells cannot survive a net gain (swell and burst) or loss (shrivel and die) of water.

A. Osmoconformers and Osmoregulators (p. 877)

Osmosis = Movement of water across a selectively permeable membrane.
- ☞ Occurs when two solutions separated by a membrane differ in *osmolarity* (total solute concentration).

If an organism with a body fluid osmolarity of 1.5% lives in water which has an osmolarity of 10%, is the organism isosmotic, hyperosmotic or hypoosmotic to its environment?

Isosmotic = Solutions are equal in osmolarity; no <u>net</u> osmosis occurs between the solutions.

Hyperosmotic = The solution having a greater concentration of solutes; <u>net</u> osmosis occurs into the solution.

Hypoosmotic = The solution having a lesser concentration of solutes (more water); net osmosis occurs out of the solution. If a selectively permeable membrane separates two solutions of differing osmolarities, water flows from the hypoosmotic solution to the hyperosmotic solution.

Margin notes: Osmoregulation; Osmosis; Osmolarity; Isosmotic; Hyperosmotic; Hypoosmotic

Osmoconformers

Osmoregulators

What is the distinction between an osmoregulator and an osmoconformer?

Osmoconformers = Animals that do not actively adjust their internal osmolarity.
☛ Many salt water animals; body fluids are isosmotic with surroundings.

Osmoregulators = Animals that regulate internal osmolarity by discharging excess water or taking in additional water.
☛ Many salt water animals, all freshwater animals, and terrestrial animals.

B. Problems of Osmoregulation in Different Environments
(p. 878–880)

Marine animals live in a saline environment consisting of about 96.5% water and 3.5% dissolved substances.
☛ The dissolved substances are collectively referred to as salts and the total amount of these dissolved substances in the water is salinity.

Most marine invertebrates are osmoconformers.
☛ While their body fluids are isosmotic to the environment, the composition of the body fluids usually differs due to internal regulation of specific ions.

What osmoregulatory adaptations are found in sharks which permit them to survive in the marine environment? In marine bony fishes?

Most cartilaginous fishes, including sharks, maintain internal salt concentrations lower than sea water by pumping salt out through rectal glands, yet their osmolarity is close to that of seawater due to retention of urea.
☛ Sharks also produce and retain trimethylamine oxide (TMAO), which protects their proteins from urea.
☛ Retention of these organic solutes (urea, TMAO) in the body fluids actually makes them slightly hyperosmotic.
☛ Do not drink, but balance osmotic uptake of water by copious urination.

Marine bony fishes are hypoosmotic and compensate water loss (by osmosis) by drinking large amounts of seawater and pumping excess salt out with their gill epithelium.

Many marine birds and reptiles have special salt glands that rid the body of excess salts.

What osmoregulatory adaptations are found in freshwater organisms which permit them to live in hypoosmotic fresh water?

Freshwater animals are hyperosmotic to their environment and constantly take in water by osmosis.
☛ Freshwater protozoa compensate with contractile vacuoles that pump out water.

Many freshwater animals, including fish, compensate by excreting large amounts of very dilute urine.
- Since salts are lost in this process, salt is replenished either by eating substances with a higher salt content or, in the case of some fish, by active uptake of sodium and chloride ions from the surrounding water by gill epithelium.
- Anadromous fishes such as salmon migrate between seawater and fresh water. While in the ocean, they osmoregulate like other marine fishes; when in fresh water, they alter their osmoregulation to that of freshwater fishes.

Euryhaline animals are those that can survive radical fluctuations in the osmolarity of surrounding waters.
- Some animals in this group are osmoregulators and minimize the osmotic shock of high saline conditions by drinking the salty water and then excreting the excess salts (e.g. brine shrimp).

Most animals cannot survive wide fluctuation in external osmolarity.
- These animals are referred to as *stenohaline* animals.

What is anhydrobiosis? What organisms exhibit this adaptation?

Anhydrobiosis is an adaptation found in a small number of aquatic invertebrates which permits them to survive when their habitat dries up.
- Best exemplified by the tardigrades. (See Campbell, Figure 40.4)
- Hydrated animals are about 1 mm long and are about 85% water.
- As the water around the animal disappears, water is lost from the tissues.
- The animal produces large amounts of trehalose (a disaccharide) which replaces the lost water associated with membranes and proteins.
- The presence of trehalose prevents cellular structures and molecules from being distorted by dehydration.
- The animals can lose up to 95% of their body water and enter a dormant state.
- Tardigrades can survive many years in this state and will rehydrate and become active when water returns.

What osmoregulatory problems face terrestrial animals? What adaptations are found in these organisms which allow them to maintain homeostasis?

Terrestrial animals live in a dehydrating environment and cannot survive desiccation.
- Humans die if 12% of their body water is lost.

Osmoregulatory mechanisms in terrestrial animals include protective outer layers, drinking and eating moist foods, behavioral adaptations and excretory organ adaptations.
- Arthropods have waxy cuticles, land snails possess shells, and vertebrates are covered by a multi-layer skin comprised of dead, keratinized cells.

- Drinking and eating moist foods replaces much of the water lost during gas exchange.
- Some desert animals are nocturnal; being active only at night reduces dehydration and some like the kangaroo rat produce large amounts of metabolic water.
- The excretory organs of terrestrial animals are adapted to conserve water while eliminating wastes.

C. Transport Epithelia and Osmoregulation (p. 880–881)

What is the importance of transport epithelia to the process of osmoregulation?

Most osmoregulators have different variations of *transport epithelia* to regulate the transport of salt. (See Campbell, Figure 40.6)
- Usually a single sheet of cells, joined by impermeable tight junctions, facing the external environment.
- The molecular composition of the epithelium's plasma membrane determines the specific osmoregulatory functions. (Remember that gill epithelium pumps salt out of marine fishes and pumps salts into freshwater fishes.)
- The transport epithelium in the nasal glands of marine birds are very efficient at eliminating the excess salts obtained from drinking seawater.
- May also function in excretion of nitrogenous wastes in some animals.

II. EXCRETORY SYSTEMS OF INVERTEBRATES (p. 881–883)

A. Protonephridia: The Flame-Cell System of Flatworms (p. 881–882)

How do flatworms osmoregulate?

Flatworms, which have neither circulatory systems nor coeloms, have a flame cell system that directly regulates interstitial fluid contents. (See Campbell, Figure 40.7)
- Consists of a branched system extending throughout the body.
- Interstitial fluid passes through a flame cell, propelled by a tuft of cilia (in the flame cell) along the branched system of tubules.
- In planaria, this fluid drains into excretory ducts that empty out of the body through numerous nephridiopores.
- Transport epithelium lining the tubules function in osmoregulation by absorbing salts before the fluid exits the body.
- Some parasitic flatworms are isosmotic to their hosts and this closed system is used mainly to excrete nitrogenous wastes.

The flame-cell system is one type of *protonephridim* (= a network of closed tubules lacking internal openings); other types can be found in other invertebrates and lancets.

Controlling the Internal Environment   587

   B.   **Metanephridia of Earthworms** (*p. 882−883*)

How do earthworms osmoregulate? What evolutionary advancement is found in metanephridia in comparison with the flame-cell system of flatworms?

   Each segment of most annelids, including earthworms, contain a pair of *metanephridia*, excretory tubules that have internal openings to collect body fluids. (See Campbell, Figure 40.8)   *Metanephridia*
   ☞ Fluid enters the nephrostome, passes through the metanephridium and empties into a storage bladder that empties outside the body   *Nephrostome*
      through the nephridiopore.
   ☞ The nephrostome collects coelomic fluid from the body segment just   *Nephridiopore*
      anterior.
   ☞ A network of capillaries envelops each metanephridium. These capillaries reabsorb essential salts pumped out of the collecting tubules by transport epithelium bordering the lumen.
   ☞ Excretion of hypoosmotic, dilute urine offsets the continual osmosis of water from damp soil across the skin.

   C.   **Malpighian Tubules of Insects** (*p. 882−883*)

How has the Malpighian tubule excretory system of insects contributed to their success in the terrestrial environment?

   <u>Malpighian tubules</u> = Excretory organs of insects and other terrestrial   *Malpighian tubules*
      arthropods that remove nitrogenous wastes from the hemolymph and
      function in osmoregulation. (See Campbell, Figure 40.9)
   ☞ Actually are outpocketings of the gut that open into the digestive tract at the midgut-hindgut juncture. The tubules dead-end at the tips away from the gut and are bathed in the coelomic fluid.
   ☞ Transport epithelium lining each tubule moves solutes (salts and nitrogenous wastes) from the body fluid into the tubule lumen.
   ☞ Accumulates nitrogenous wastes from the coelomic fluid and water follows by osmosis.
   ☞ The fluid in the tubule then passes through the hindgut to the rectum.
   ☞ Salts and water are reabsorbed across the epithelium of the rectum and dry nitrogenous wastes are excreted with feces.

III.   **THE VERTEBRATE KIDNEY** (*p. 883−893*)

Nephrons (vertebrate excretory tubules) are collected into *kidneys*, compact organs that remove nitrogenous wastes and function in osmoregulation by adjusting the   *Nephrons, Kidneys*
blood's salt concentration.
   ☞ Blood cycles through the kidney via an extensive capillary network.

   A.   **Anatomy of the Excretory System** (*p. 883−885*)

What function does the renal artery have in the excretory process of a vertebrate? The ureters? The urethra?

   Blood enters the kidney via the *renal artery* and exits via the *renal vein*.
   ☞ About 20% of the blood pumped by each heartbeat passes through   *Renal artery, Renal vein*
      the kidneys.

Urine, Ureter

Urinary bladder

Urine exits each kidney through the ureter and both ureters drain into a common urinary bladder.

Final excretion is through the urethra.
- ☞ Sphincter muscles near the junction of the urethra and bladder control micturition.

B. Structure of the Nephron (*p. 885*)

The two functional regions of the kidney are the outer *cortex* and inner *medulla*.

What is a nephron? What function does each part of the nephron perform in the excretory process?

Nephron

Nephron = Functional unit of the kidney, consisting of a renal tubule and its associated blood vessels.
- ☞ There are about one million nephrons in each human kidney.

Filtrate

Filtrate = Water, salts, urea and other small molecules that are separated from the blood passing through the capillaries and flow through the renal tubule.

The nephrons process 180 L of filtrate per day, and the transport epithelium, lining the renal tubule, processes this filtrate to form the approximately 1.5 L urine excreted daily.
- ☞ The rest of the filtrate is reabsorbed into the blood.

The blind end of the renal tubule that receives filtrate from the blood forms a cup-shaped *Bowman's capsule* which embraces a ball of capillaries, the *glomerulus*. (See Campbell, Figure 40.10)

Bowman's capsule

Glomerulus

- ☞ Filtrate then passes through the *proximal convoluted tubule*, the *loop of Henle* (a long hairpin turn with a descending limb and an ascending limb) and the *distal convoluted tubule*, which empties into a *collecting duct*.
- ☞ The collecting duct receives filtrate from many other nephrons.
- ☞ Filtrate, now called urine, passes from the collecting ducts into the renal pelvis. Urine then drains from the chamberlike pelvis into the ureter.

What distinguishes a cortical nephron from a juxtamedullary nephron? What significance is there to the fact that juxtamedullary nephrons are found only in birds and mammals?

The Bowman's capsules and the proximal and distal convoluted tubules are located in the cortex.
- ☞ The loops of Henle and collecting tubules extend into the medulla.

Cortical nephrons = Nephrons that have reduced loops of Henle and are confined to the cortex. 80% of the nephrons in humans are cortical nephrons.

Juxtamedullary nephrons = Nephrons that have long loops that extend into the medulla and are found only in mammals and birds. 20% of the nephrons are juxtamedullary nephrons.

Each nephron is closely associated with blood vessels:
- ☞ *Afferent arteriole* is a branch of the renal artery that divides to form the capillaries of the glomerulus.
- ☞ *Efferent arteriole* forms from the converging capillaries as they leave the capsule. This subdivides to form the *peritubular capillaries* which intermingle with the proximal and distal convoluted tubules.
- ☞ *Vasa recta* is the capillary system branching downward from the peritubular capillaries that serves the loop of Henle.

Materials are exchanged between capillaries and nephrons through interstitial fluid.

C. **General Physiology of the Nephron** (p. 885–887)

What are the differences among filtration, secretion, and reabsorption? Why are each of these processes important to excretion and osmoregulation?

Nephrons regulate blood composition by: filtration, secretion and reabsorption.

1. **Filtration**

Blood pressure forces fluid from the glomerulus across the Bowman's capsule epithelium into the lumen of the renal tubule.
- ☞ Porous capillaries and *podocytes* (specialized cells of the capsule) nonselectively filter out blood cells and large molecules; any molecule small enough to be forced through the capillary wall enters the renal tubule.

Proximal convoluted tubule
Loop of Henle, Distal convoluted tubule
Collecting duct

Cortex
Medulla
Cortical nephrons

Juxtamedullary nephrons

Afferent arteriole
Efferent arteriole
Peritubular capillaries

Vasa recta

Podocytes

☞ Now the filtrate contains a mixture of glucose, salts, vitamins, nitrogenous wastes, and small molecules in concentrations similar to that in blood plasma.

2. **Secretion**

Filtrate is joined by substances transported across the tubule epithelium from the surrounding interstitial fluid.
☞ Adds plasma solutes to the filtrate.
☞ Proximal and distal convoluted tubules are most common sites of secretion.
☞ Very selective. Involves both passive and active transport.
☞ For example, controlled secretion of H^+ ions helps maintain constant body fluid pH.

3. **Reabsorption**

Reabsorption is the selective transport of filtrate substances from the renal tubule back to the interstitial fluid.
☞ Reclaims small molecules essential to the body.
☞ Occurs in the convoluted tubules, the loop of Henle and the collecting duct.
☞ Nearly all sugar, vitamins, organic nutrients and, in mammals and birds, water are reabsorbed.
☞ Regulates salt concentrations.

The composition of the filtrate is modified by selective secretion and reabsorption.
☞ The concentration of beneficial substances is reduced as they are returned to the body.
☞ The concentration of wastes and nonuseful substances is increased and excreted from the body.

D. **Transport Properties of the Renal Tubule** (p. 887–889)

Reclamation of small molecules and water from the filtrate as it flows through the renal tubules and collecting duct converts the filtrate into urine.

How does the proximal convoluted tubule alter the volume and concentration of the filtrate?

1. The *proximal convoluted tubule* alters the volume and composition of filtrate by reabsorption and secretion.
   ☞ In this area ammonia, drugs and poisons processed in the liver are secreted to join the filtrate.
   ☞ Helps maintain a constant body fluid pH by controlled secretion of H^+.
   ☞ Nutrients such as glucose and amino acids are reabsorbed (active transport) from the filtrate and returned to the interstitial fluid.

Epithelial cells in this region have numerous microvilli facing the tubule lumen (a *brush-border*) that provide extensive surface area for reabsorption of potassium, nutrients, and NaCl into the interstitial fluid and from there into the peritubular capillaries.

Brush-border

☞ Na⁺ and Cl⁻ diffuse across the brush border into the epithelial cells; the membranes facing the interstitial fluid then actively pump Na⁺ out of the cells, which is balanced by passive transport of Cl⁻. Water follows passively by osmosis.

Epithelial cells facing the interstitial fluid (outside the tubule) have a small surface area to minimize leakage of salt and water back into the filtrate. (See Campbell, Figure 40.12)

What does water osmose out of the filtrate as it moves through the descending limb of the loop of Henle? What materials move into and out of the filtrate as it flows through the ascending limb of the loop of Henle?

2. In the *descending limb of the loop of Henle*, transport epithelium is freely permeable to water but not to salt and other small solutes.
    ☞ Filtrate moving down the tubule from the cortex to the medulla continues to lose water by osmosis since the interstitial fluid in this region increases in osmolarity; therefore, the NaCl concentration of the filtrate increases.

3. In the *ascending limb of the loop of Henle*, transport epithelium is very permeable to salt, but not to water.
    ☞ In a thin segment near the loop tip, NaCl diffuses out passively and contributes to the high osmolarity of interstitial fluids of the medulla.
    ☞ In the thick segment leading to the distal convoluted tubule, Cl⁻ is actively pumped out and Na⁺ flows passively.
    ☞ Here the filtrate becomes more dilute due to the removal of salts without lose of water.

What contributions do the distal convoluted tubule and the collecting ducts make toward the formation of urine?

4. The *distal convoluted tubule* regulates K⁺ and Na⁺ concentration of body fluids by regulating K⁺ secretion into the filtrate and Na⁺ reabsorption from the filtrate.
    ☞ This region also contributes to pH regulation by quantitative secretion of H⁺ and reabsorption of bicarbonate (an important body fluid buffer).

5. The *collecting duct* carries filtrate back towards the medulla and renal pelvis.
    ☞ The epithelium here is permeable to water but not to salt, so the filtrate loses water by osmosis to the hyperosmotic fluid outside the duct and urea is concentrated.
    ☞ The bottom portion of the duct is permeable to urea, some of which diffuses out. This contributes to the high osmolarity of the interstitial fluid of the kidney medulla, which enables the kidney to conserve water by excreting a hyperosmotic urine.

E. How the Mammalian Kidney Conserves Water: A Closer Look (p. 889–891)

How does the structure of the juxtamedullary nephron enhance water conservation in the kidney of a mammal?

The juxtamedullary nephron, with its urine-concentrating features, is a key adaptation to terrestrial life that enables mammals to excrete nitrogenous waste without squandering water.

- Urine is up to four times as concentrated as the blood.
- Filtrate passing from the Bowman's Capsule to the proximal tubule has the same osmolarity as blood (300 mosm/L). (See Campbell, Figure 40.13)
- A large amount of water and salt is reabsorbed as filtrate passes through the proximal convoluted tubule; the osmolarity remains about the same.
- Water moves out of the filtrate by osmosis as it flows from the cortex into the medulla through the descending loop of Henle; the osmolarity steadily increases due to loss of water until it peaks at the apex.
- As filtrate moves up the ascending loop of Henle back to the cortex, salt leaves the filtrate first by passive then by active transport; osmolarity declines to about 100 mosm/L at this point.
- Osmolarity changes very little as filtrate flows through the distal convoluted tubule.
- After entering the collecting duct, the filtrate passes back through the medulla; the filtrate loses water which increases osmolarity; some urea also leaks out.
- The passage of the filtrate back through the hyperosmotic medulla causes a gradual increase in osmolarity to about 1200 mosm/L; the remaining molecules are excreted in a minimal amount of water as urine which passes to the renal pelvis to the ureter.

NOTE: The lose of salt _from_ the filtrate passing through the ascending limb of the loop of Henle _to_ the interstitial fluid of the medulla contributes to the high osmolarity of the medulla. This, in turn, helps conserve water.

The *vasa recta* (capillary net) does not dissipate the crucial osmolarity gradient in the kidney.

- As the descending vessel conveys blood toward the inner medulla, water is lost from the blood and NaCl diffuses into the blood.
- These fluxes are reversed as blood flows back toward the cortex in the ascending vessel.

Urine, at its most concentrated, is isosmotic to the interstitial fluid of the inner medulla, but is hyperosmotic to body fluids elsewhere.

F. Regulation of the Kidneys (p. 891–893)

The kidney is a versatile osmoregulatory organ, excreting hyper- or hypoosmotic urine as necessary, subject to a combination of nervous and hormonal controls.

There is an elaborate system involving three mechanisms (antidiuretic hormone, the juxtaglomerular apparatus, and atrial natriuretic protein) that regulates the kidney's ability to change the osmolarity, salt concentration, volume, and pressure of the blood.

How does antidiuretic hormone affect the osmolarity of urine produced by the mammalian kidney?

Antidiuretic hormone (*ADH*), produced in the hypothalamus of the brain and stored and secreted from the pituitary gland, enhances fluid retention by increasing the permeability of the distal convoluted tubules and the collecting duct to water. (See Campbell, Figure 40.14a)

- ADH release is triggered when osmoreceptor cells in the hypothalamus detect increased blood osmolarity due to an excessive loss of water from the body.
- Increased water reabsorption reduces blood osmolarity and inhibits further secretion of ADH.
- Alcohol can inhibit ADH release, causing dehydration.

How does the juxtaglomerular apparatus help regulate blood pressure? Atrial natriuretic protein?

The *juxtaglomerular apparatus* (*JGA*) is a specialized tissue near the afferent arterioles leading to kidney glomeruli. It responds to a decrease in blood pressure or Na⁺ concentration by releasing the enzyme renin into the blood.

- Renin converts inactive *angiotensin* to active *angiotensin II* which functions as a hormone.
- Angiotensin II directly increases blood pressure by causing arteriole constriction; increased blood pressure increases the filtration rate.
- Angiotensin II acts indirectly by signaling adrenal glands to release *aldosterone*, which stimulates Na⁺ reabsorption across distal convoluted tubules (water follows by osmosis).
- The increased Na⁺ concentration in blood and increased blood volume and pressure suppresses further release of renin.

ADH and aldosterone cooperates in homeostasis. ADH alone would lower blood Na⁺ concentration by increasing water reabsorption, but aldosterone maintains balance by stimulating Na⁺ reabsorption.

Atrial natriuretic protein (*ANP*) is another hormone which opposes the renin-angiotensin-aldosterone system.

- Released by the heart's atrium wall in response to increased blood volume and pressure.
- Inhibits release of renin and reduces aldosterone release from the adrenal glands. This decreases Na⁺ reabsorption and lowers blood volume and pressure.

G. **Comparative Physiology of the Kidney** (*p. 893*)

How do the structures and physiological adaptations in the kidneys vary among animals which live in different environments?

Nephrons vary in structure and physiology to help different vertebrates in osmoregulation in various habitats.

- Desert mammals have very long loops of Henle to maintain steep osmotic gradients that conserve water by allowing urine to become very concentrated.
- Mammals living in aquatic environments (e.g. beavers) have nephrons with very short loops of Henle resulting in production of dilute urine.
- Birds have shorter loops of Henle and produce a more dilute urine than mammals.
- Reptiles have only cortical nephrons and produce isosmotic urine, but the epithelium of their cloaca conserves fluid by reabsorbing water from urine and feces. Also most excrete nitrogenous wastes as uric acid which conserves water.
- Freshwater fish (hyperosmotic to surroundings) nephrons use cilia to sweep the large volume of very dilute urine from the body. Salts are conserved by efficient ion reabsorption from filtrate.
- Amphibians excrete dilute urine and actively accumulate salt from the water via its skin. On land, body fluid is conserved by water reabsorption across urinary bladder epithelium.
- Bony seawater fishes (hypoosmotic to surroundings) excrete very little concentrated urine (many lack glomeruli and capsules) and kidneys function mainly to rid of divalent ions such as Ca^{2+}, Mg^{2+} and SO_4^{2-} taken in by drinking seawater. Monovalent ions like Na^+ and Cl^-, and the majority of nitrogenous waste (in the form of ammonium) is excreted mainly by the gills.

IV. NITROGENOUS WASTES (p. 893–894)

The metabolism of proteins and nucleic acids produces ammonia, a small and very toxic waste product. Some animals excrete the ammonia directly, while other first convert it to urea or uric acid which are less toxic.

What is the correlation between the type of nitrogenous waste produced (ammonia, urea, or uric acid) by an organism and the type of habitat in which it lives? The organism's mode of reproduction?

A. Ammonia (p. 894)

Ammonia is excreted directly by most aquatic animals.
- Easily permeates membranes since molecules are small and very water soluble.
- In soft-bodied invertebrates, ammonia just diffuses out.
- In freshwater fishes, it is excreted as ammonium ions across gill epithelium.
- Very toxic — excreted in very dilute solutions.

B. Urea (p. 894)

Urea is the nitrogenous waste excreted by mammals and most adult amphibians.
- Can be much more concentrated since it is much less toxic than ammonia; reduces water loss for terrestrial animals.
- Produced in liver by a metabolic cycle combining ammonia with CO_2. It is transported to kidneys via the circulatory system.

Controlling the Internal Environment 595

- Some urea is retained in the kidney where it contributes to osmoregulation by maintaining the osmolarity gradient in the medulla.
- Sharks also produce and retain urea in the blood as an osmoregulatory agent.
- Amphibians that undergo metamorphosis and move as adults to land switch from excreting ammonia to excreting urea.

C. Uric Acid (p. 894)

Uric acid is the primary form of nitrogenous waste excreted by land snails, insects, birds and some reptiles.

- Much less soluble in water than ammonia or urea; uric acid can be excreted as a precipitate after reabsorbing nearly all the water from the urine.
- Eliminated in a pastelike form through the cloaca (mixed with feces) in birds and reptiles.

The mode of reproduction is an important factor in determining whether uric acid or urea excretion evolved in a particular group.

- If an embryo released ammonia or urea within a shelled egg, the soluble waste would accumulate to toxic concentrations: uric acid precipitates out of solution to be stored as a solid.

The animals habitat, along with the phylogenetic position, influences the type of nitrogenous waste produced.

- Terrestrial reptiles excrete mostly uric acid; crocodiles excrete ammonia and uric acid; aquatic turtles excrete urea and ammonia.

Some animals can modify their nitrogenous wastes when the temperature or water availability changes.

V. REGULATION OF BODY TEMPERATURE (p. 894–903)

Metabolism and membrane properties are very sensitive to changes in an animal's internal temperature. Each animal lives in, and is adapted to, an optimal temperature range in which it can maintain a constant internal temperature when external temperatures fluctuate.

A. Heat Production and Transfer Between Organisms and Their Environment (p. 895)

What physical processes can be used by an organism to transfer heat between its body and the environment?

An organism exchanges heat with its environment by four physical processes:

1. *Conduction* is the direct transfer of thermal motion (heat) between molecules of the environment and a body surface.
    - Always conducted from a body of higher temperature to one of lower temperature.
    - Water is 50–100 times more effective than air in conducting heat.
    - For example, on a hot day, an animal in water cools more rapidly than one on land.

Margin notes: Uric acid; Conduction

596 Controlling the Internal Environment

Convection

Radiation

Evaporation

Ectotherms

Endotherms

Thermoregulation

Nonshivering thermogenesis

Brown fat

2. *Convection* is the mass flow of air or liquid past a body surface.
   - ☞ For example, breezes contribute to heat loss ("wind chill factor").

3. *Radiation* is the emission of electromagnetic waves produced by all objects warmer than absolute zero.
   - ☞ Can transfer heat between objects not in direct contact.
   - ☞ For example, an animal warmed by the sun.

4. *Evaporation* is the loss of heat from a liquid surface that is losing some molecules as gas.
   - ☞ Cooling increases greatly by production of sweat.
   - ☞ Can only occur if surrounding air is not saturated with water molecules.
   - ☞ Along with convection, is most variable cause of heat loss.

B. **Ectotherms and Endotherms** (*p. 895–896*)

What are the adaptive advantages of endothermy?

Ectotherms = Animals that warm their bodies mainly by absorbing heat from surroundings.
   - ☞ Generally are invertebrates, fishes, reptiles and amphibians.
   - ☞ May derive some body heat from metabolism.

Endotherms = Animals that derive most of their body heat from their own metabolism.
   - ☞ Usually maintain a consistent internal temperature even as the environmental temperature fluctuates.
   - ☞ May add some body heat by basking in the sun.
   - ☞ Generally are warmer than their surroundings. Requires active metabolism but, conversely, warm body temperature contributes to high levels of aerobic metabolism (cellular respiration) required for endurance of physical activity.

C. **Thermoregulation in Terrestrial Mammals** (*p. 896–897*)

1. Heat Production (*p. 896*)

How is body heat produced in terrestrial mammals? What two processes can increase the rate of heat production?

Heat production results from metabolism and insulating layers of fat and fur help retain body heat generated by this metabolism.

Rate of heat production can be increased in two ways:
   - ☞ Increased contraction of skeletal muscles (moving or shivering).
   - ☞ *Nonshivering thermogenesis* which is triggered by hormonal action.

Nonshivering thermogenesis = Hormonal triggering of heat production due to an increase in metabolic rate.
   - ☞ Although this occurs throughout body, some mammals have brown fat in the neck and between the shoulders that is specialized for rapid heat production.

- ☞ Brown fat cells store lipids and contain many mitochondria that make little ATP and release most energy from lipid oxidation as heat.

2. **Mechanisms of Thermoregulation** (*p. 896*)

Endothermy requires much more energy expenditure than ectothermy; land mammals maintain relatively constant body temperatures through four categories of physiological and behavioral adjustments:

What four physiological and behavioral adjustments can be used by terrestrial mammals to maintain relatively constant body temperatures?

1. Changing the rate of metabolic heat production through increased skeletal muscle activity and nonshivering thermogenesis can increase the amount of heat produced.
2. Adjusting the rate of heat exchange between an animal and its environment can increase or decrease body temperature.
   - ☞ In *vasodilation*, certain nerves to superficial (body surface) blood vessels decrease activity, relaxing blood vessel wall muscles and allowing blood flow to increase. More heat is transferred to the environment by conduction, convection and radiation.
   - ☞ *Vasoconstriction* reduces blood flow and heat loss by decreasing the diameter of blood vessels.
   - ☞ Most land mammals react to cold by raising their fur, which increases insulation.
3. Evaporative heat loss can cause substantial cooling in terrestrial mammals.
   - ☞ Evaporation by respiratory tract is increased by panting.
   - ☞ Evaporation across the skin can be increased by sweating, which is under nervous control.
   - ☞ Mammals lacking sweat glands may use saliva or urine to cool themselves by evaporation.
4. Behavioral responses such as relocation can also increase or decrease body heat loss.
   - ☞ Basking in the sun or moving to shady, damp areas helps regulate heat loss.

3. **The Thermostat** (*p. 897*)

How do the two thermoregulatory centers of the hypothalamus function to regulate the body's temperature?

The body's thermostat is a group of nerve cells concentrated in the hypothalamus of the brain.
- ☞ These cells are located in two thermoregulatory areas.
- ☞ The *heating center* controls vasoconstriction of superficial blood vessels, erection of fur, shivering and nonshivering thermogenesis.
- ☞ The *cooling center* of the hypothalamus controls vasodilation and sweating or panting.

Vasodilation

Vasoconstriction

Nerve cells that sense temperature are in the skin, hypothalamus and other parts of the nervous system.
- *Ruffini organs* (warm temperature receptors) increase nervous activity when temperatures increase and excite the cooling center and inhibit the heating center of the hypothalamus.
- *Bulbs of Krause* (cold receptors) increase activity when temperatures decrease and inhibit the cooling center and excite the heating center of the hypothalamus.

D. Thermoregulatory Adaptations in Other Animals (p. 897–902)

What thermoregulatory adaptations are found in animals other than terrestrial mammals?

Birds have no sweat glands but use panting to promote evaporative heat loss thus cooling the body.
- To reduce heat loss, birds have insulating feathers and a *countercurrent heat exchanger* to prevent loss of heat from feet and legs (arteries carrying warm blood to legs are in close contact with the veins returning blood to the trunk).
- The countercurrent arrangement allows the transfer of heat from the warm arterial blood to the cool venous blood.

Marine mammals live in water much cooler than body temperature.
- To conserve heat, many have an insulating layer of fat called blubber just under the skin (an anatomical adaptation rather than metabolic).
- In the tail and flippers, where there is no blubber, countercurrent exchange occurs between the arterial and venous blood.
- Can dissipate metabolic heat by dilating blood vessels that serve the skin to allow greater amount of blood to be cooled by the surrounding water.

Reptiles are generally ectotherms that warm themselves mainly by behavioral adaptations.
- Orient the body toward, and maximize the body surface exposed to, the heat source.
- Seek warm places or other favorable microclimates in the environment to regulate the body temperature.
- Can also increase thermogenesis (e.g. by shivering).

Amphibians produce little heat and most lose heat rapidly by evaporative cooling from their body surface, making thermoregulation difficult.
- Seek cooler or warmer microenvironments as necessary (behavioral adaptation).
- Some can vary the amount of mucus they secrete to regulate evaporative cooling.

Fish are generally ectotherms, but some are partial endotherms.
- These fish keep their swimming muscles warmer than their surface tissues by having some small blood vessels arranged into a countercurrent heat exchanger called the *rete mirabile* ("wonderful net").

Most invertebrates have little control over body temperature, but some do adjust temperature by behavioral or physiological mechanisms.
- ☞ The desert locust orients the body to maximize heat absorption from the sun.
- ☞ Some large flying insects (e.g. bees) can generate internal heat by contracting all flight muscles in synchrony (functionally analogous to shivering).
- ☞ The thorax of the winter moth has a countercurrent heat exchanger to warm flight and maintain the temperature of muscles even in cold weather.

Honeybees use social organization to regulate temperature.
- ☞ Increase movements and huddle to retain heat in cold weather. Maintain constant temperature by changing the density of huddling; heat is distributed by the movement of individuals from the core to the margins of the huddle.
- ☞ Cool hives by transporting water to it and fanning with wings to promote evaporation and convection.

E. Temperature Acclimation (p. 902)

What is acclimation? What changes can occur as an organism adjusts to a new temperature range?

Acclimation = Physiological adjustment to a new range of environmental temperature over a period of many days or weeks.
- ☞ Usually a multifaceted response to temperature changes such as those associated with seasons.
- ☞ Cells may increase production of enzymes to compensate for lower activity.
- ☞ Cells may produce enzyme variants having the same function but different temperature optima.
- ☞ Membranes may remain fluid by changing proportions of saturated and unsaturated lipids.

F. Torpor (p. 902–903)

What is torpor? What differences can be used to distinguish among hibernation, aestivation and diurnation?

Torpor = Alternative physiological state in which metabolism decreases and heart and respiratory systems slow down.

In *hibernation*, body temperature is lowered, allowing an animal to survive long periods of cold and diminished food supplies.
- ☞ May be triggered by seasonal changes in daylength.

Aestivation allows an animal to survive long periods of high temperature and diminished water.
- ☞ Characterized by a slow metabolism and inactivity.

Diurnation

Diurnation is physiologically similar to aestivation and hibernation, but lasts for much shorter times.

☞ Some animals with very high metabolic rates (e.g. hummingbirds) when active undergo a daily torpor not triggered by availability of food but by an internal biological clock.

VI. INTERACTION OF REGULATORY SYSTEMS (*p. 903*)

Regulation of body temperature may affect osmolarity, metabolic rate, blood pressure, tissue oxygenation and body weight. During times of physiological stress, one response may conflict with others, and physiological compromises result.

☞ Desert animals may have to endure hyperthermia during times when water is lacking due to water conservation by the body taking precedence over evaporative heat loss.

Normally, the various regulatory systems act in concert to maintain homeostasis; integrated feedback circuits involving nervous communication and hormones control the internal environment.

41

CHEMICAL SIGNALS IN ANIMALS

CHAPTER OUTLINE

The activities of the various specialized parts of an animal are coordinated by the two major systems of internal communication: the *nervous system* and the *endocrine system*.
- ☞ The nervous system is involved with high-speed messages.
- ☞ The endocrine system is slower and involves the production, release, and movement of chemical messages.

I. AN OVERVIEW OF CHEMICAL SIGNALS (p. 907–910)

What is a hormone? What are the primary differences which distinguish between a hormone and a pheromone? A hormone and a neurotransmitter?

All animals exhibit some coordination by chemical signals: pheromones are chemical signals used to communicate between different individuals, hormones convey information between organs of the body, and other messengers, such as neurotransmitters, act between cells on a localized scale.

A. Hormones (p. 908–909)

The endocrine system has 3 key components: hormones, endocrine glands, and target cells and their molecular receptors for hormones.

Target cells

Hormones = Specific molecules synthesized and secreted by a group of specialized cells, and released into the body fluids to travel to target cells where they elicit specific biological responses.

Hormones

What differences can be used to distinguish between endocrine glands and exocrine glands? Which type of gland is part of the hormonal control system of the body?

Endocrine glands = Ductless glands that secrete hormones into the bloodstream for distribution throughout the body.
- ☞ Composed of the specialized cells which secrete hormones.
- ☞ May comprise only a portion of organs that produce other materials (e.g. pancreas).

Endocrine glands

Exocrine glands = Glands that produce a variety of substances (e.g. sweat, mucus, digestive enzymes) and convey their products by means of ducts.
- ☞ Many organs perform both exocrine and endocrine functions.
- ☞ Exocrine glands are <u>not</u> a part of the endocrine system.

Exocrine glands

Endocrinology = The study of hormones.
- ☞ More than 50 hormones have been identified in humans.

Endocrinology

How can the three classes of hormones (steroids, amino acid derivatives, peptides) be distinguished from each other?

Hormones are grouped into 3 general classes according to similarities in chemical structure:

1. Steroid hormones.
   - Fat-soluble molecules fashioned from cholesterol; includes the sex hormones.
2. Hormones derived from amino acids.
   - Most frequently derived from tyrosine.
   - Are small and water soluble.
   - For example, epinephrine.
3. Peptide hormones.
   - Peptide hormones are chains of amino acids, or peptides.
   - May act as signal molecules in the nervous and endocrine systems.
   - The most diverse group of hormones.

Hormone receptors = Protein within or on the plasma membrane of the target cell that determines the specificity and action of hormones.
- The hormone and receptor are shape specific and must bind together to elicit a response to the hormone.

The actions of antagonistic hormones are important in maintaining homeostasis.
- Feedback mechanisms are usually involved in adjusting hormone actions.

B. **Pheromones** (*p. 909*)

Pheromones = Chemical signals that function <u>between</u> animals of the same species.
- Classified according to function (e.g. mate attractant, territorial marker, alarm substance).
- Are small, volatile, easily dispersed molecules that are active in minute amounts.

C. **Local Regulators** (*p. 909–910*)

What is a growth factor? A prostaglandin? What function does each have in the body?

Local regulators are chemical messengers that affect target cells close to their point of secretion.
- Includes *synaptic signaling* by neurotransmitters and *paracrine signaling* by histamine and interleukins.

1. **Growth Factors** (*p. 910*)

Growth factors = Proteins which must be present in the extracellular environment for certain cell types to grow and develop normally.
- Discovered during attempts to culture mammalian cells on artificial media, but also regulates development within the animal body.

- Binding of the growth factor to the receptor triggers the target cell's response.
- Some oncogenes code for proteins that mimic growth factor receptors that do not require binding of a growth factor to stimulate growth, thus leading to uncontrollable, abnormal growth.

2. **Prostaglandins** (*p. 910*)

Prostaglandins (PGs) = Modified fatty acids released into interstitial fluid to function as local regulators.
- Often derived from lipids of the plasma membrane.
- Very subtle differences in their molecular structure profoundly affect how these signals affect target cells (e.g. antagonistic actions of PGE and PGF).
- PGs secreted by the placenta help induce labor during childbirth.
- Other PGs help defend the body by inducing fever and inflammation.

II. MECHANISMS OF HORMONE ACTION (*p. 910–915*)

Hormones can act in very low concentration and can affect different target cells in an animal or different species of animals very differently.
- The specific change in a target organ cell which is triggered by a hormone results from one of two possible signal transduction pathways.
- Signal transduction pathways include a series of steps between hormone-receptor binding and target-cell response.

A. Steroid Hormones and Gene Expression (*p. 911–912*)

How do steroid hormones stimulate a target cell response? Why do different types of target cells give different responses to the same steroid hormone?

Steroids pass through the target cell membrane and bind to a receptor protein in the nucleus. (See Campbell, Figure 41.4a)
- One hypothesis states that then the hormone-receptor complex binds to specific acceptor proteins located at certain sites along the chromatin, activating expression of certain genes.
- mRNA molecules are produced, exit to the cytoplasm, and direct the synthesis of new proteins.
- Two cell types can respond differently to a single hormone since different acceptor proteins are associated with and activate transcription of different genes in the two kinds of cells.

B. Peptide Hormones and Second Messengers (*p. 912–915*)

Peptide hormones and most hormones derived from amino acids are unable to pass through the target cell plasma membrane, and have a different mechanism of action.
- They attach to their receptors which are on the target cell surface (average of 10,000 receptors/target cell) and influence activity within the cell through cytoplasmic intermediates called *second messengers*.

Oncogenes

Prostaglandins

Cytoplasmic intermediates

Second messengers

604 Chemical Signals in Animals

What is a second messenger? Why does the signal transduction pathway of peptide hormones differ from that of steroid hormones?

The most important secondary messengers are cyclic AMP and inositol triphosphate.

1. Cyclic AMP (p. 912–915)

Cyclic AMP was discovered to act as a second messenger by 1971 Nobel Laureate Earl W. Sutherland while studying epinephrine stimulation of glycogen hydrolysis within liver and muscle cells. (See Campbell, Figure 41.6)

- Binding of a hormone with its receptor activates a *G protein*, which binds GTP and hydrolyzes it to GDP to provide energy.
- This energy activates the G protein which binds to adenylate cyclase, an enzyme embedded in the plasma membrane, to convert ATP to cyclic AMP (cAMP).
- cAMP relays the signal from the membrane to the metabolic machinery of the cytoplasm.
- When the hormonal signal ceases, cAMP is converted to inactive AMP by the enzyme phosphodiesterase.
- A second type of G protein <u>lowers</u> the activity of adenylate cyclase when activated by an inhibitory hormone-receptor complex. This reduces cAMP concentration in the cytoplasm.

Many metabolic responses to hormones involve an *enzyme cascade*, where each step activates an enzyme that in turn activates the next enzyme in the series.

- cAMP may activate a *protein kinase* (an enzyme that catalyzes the transfer of a phosphate group from ATP to proteins, or phosphorylation), which in turn can activate other enzymes.
- The cascade thus amplifies response to the hormone since the number of activated products increases with each step.

Although a single basic mechanism is used to mediate responses of diverse cells to many different hormones, response is still specific. Specificity arises since:

- Particular cells can only interact with certain hormones — those for which they have receptors.
- cAMP-dependent protein kinases differ in structure and function from tissue to tissue.
- Post-protein kinase activation steps in the cascade differ between cell types; thus, individual cell types produce a unique set of proteins in response to cAMP-dependent protein kinase.

2. Inositol Triphosphate (p. 915)

How is calcium involved in the endocrine control mechanism of the body?

Many chemical messengers (neurotransmitters, growth factors, some hormones) stimulate responses in their target cells by increasing the calcium ion concentration of the cytoplasm. The second messenger in this pathway is *inositol triphosphate*.

Inositol triphosphate (IP$_3$) acts also as a second messenger as follows: (See Campbell, Figure 41.8)
- The hormone binds to its receptor, activating a G protein (different from those that function in the cAMP system) that in turn stimulates an enzyme in the plasma membrane called phospholipase C.

Phospholipase C

- Phospholipase C cleaves a membrane phospholipid into IP$_3$ and diacylglycerol; both may function as second messengers.

Diacylglycerol

- Diacylglycerol stimulates another membrane enzyme, protein kinase C, which triggers various target cell responses by phosphorylating specific proteins.
- IP$_3$ released form the plasma membrane stimulates movement of Ca^{2+} from the endoplasmic reticulum into the cytoplasm by interacting with the ER membrane.
- The Ca^{2+} regulates cellular protein activity either by itself or by first binding to calmodulin which binds to other proteins.

Calmodulin

III. INVERTEBRATE HORMONES (p. 915–917)

Do invertebrates produce hormones? Do they function independently or do they influence each other? What is an example of hormonal interaction in invertebrates?

Invertebrates possess a diversity of hormones which function in homeostasis, reproduction, development and behavior.

In many cases, the hormone may stimulate one activity while inhibiting another.
- In the sea slug, a peptide hormone secreted by specialized neurons stimulates egg laying while inhibiting feeding and locomotion.

All arthropods have extensive endocrine systems that regulate growth and reproduction, water balance, pigment movement in the integument and eyes, and metabolism.

Hormones may act together such as in the case of insect development and molting.
- *Ecdysone* is secreted by a pair of prothoracic glands; it triggers molting and favors development of adult characteristics and metamorphosis.

Ecdysone, Molting

- *Brain hormone* promotes development by stimulating production of ecdysone.

Brain hormone

- Brain hormone and ecdysone are balanced by *juvenile hormone* (JH), which actively promotes retention of larval characteristics; JH is secreted by the corpora allata.

Juvenile hormone

- When JH levels decrease, ecdysone-induced molting produces a pupa; in the pupa, adult anatomy replaces larval anatomy during metamorphosis.

IV. THE VERTEBRATE ENDOCRINE SYSTEM (p. 917–927)

Vertebrates possess a number of tissues and organs that secrete hormones. (See Campbell, Figure 41.10 and Table 41.1)
- Some hormones affect most tissues of the body (e.g. sex hormones).
- Some hormones affect only one or a few tissues.
- Some affect other endocrine glands; these are called *tropic hormones*.

Tropic hormones

What is the association of the hypothalamus and pituitary gland? Why are these two structures important to the control of the endocrine system?

A. The Hypothalamus and the Pituitary Gland (p. 917–921)

The *hypothalamus* is a region of the lower brain that receives information from peripheral nerves and the brain and initiates endocrine signals appropriate to the environmental conditions.

- ☞ Contains *neurosecretory cells* which are specialized neurons that receive impulses from other nerve cells and respond by releasing hormones into the blood instead of transmitting a nerve signal.
- ☞ Two sets of hormone-releasing neurosecretory cells are present: one set produces hormones that are stored in the posterior pituitary, the other produces *releasing hormones* that regulate the anterior pituitary.

What hormones are secreted by the different areas of the pituitary gland? Which of these hormones are steroids? Which are derived from amino acids? Which are polypeptides?

The *pituitary gland* is an appendage at the base of the hypothalamus consisting of two lobes: (See Campbell, Figure 41.11a)

1. The *neurohypophysis*, or posterior lobe, stores and secretes two hormones which are made by the hypothalamus: oxytocin and antidiuretic hormone.
   - ☞ Both have a similar structure consisting of nine amino acids, thus they are small peptides.
   - ☞ *Oxytocin* induces uterine muscle contraction and causes the mammary glands to eject milk during nursing.
   - ☞ *Antidiuretic hormone (ADH)* promotes reabsorption of water by the kidneys by increasing the permeability of collecting duct epithelial cells to water. More water can thus leave the collecting duct and enter nearby capillaries.

2. The *adenohypophysis*, or anterior lobe, produces many different protein and peptide hormones and is regulated by releasing factors from the hypothalamus.
   - ☞ *Growth hormone (GH)* is a protein hormone which affects a wide variety of tissues. It promotes growth of some tissues directly <u>and</u> promotes growth of others indirectly by stimulating the production of growth factors by other tissues. For example, *somatomedins* are hormones secreted by the liver which stimulate bone and cartilage growth; GH serves as a tropic hormone by stimulating the liver to secrete somatomedins.
   - ☞ *Prolactin (PRL)* is similar in structure to GH although their physiological roles are very different. PRL produces a diversity of effects in different vertebrates; it stimulates mammary gland development and milk synthesis in mammals; regulates fat metabolism and reproduction in birds; delays metamorphosis and may function as a larval growth hormone in amphibians; and regulates salt and water balance in freshwater fish.
   - ☞ *Follicle-stimulating hormone (FSH)* is a tropic hormone which affects the gonads (=gonadotropin) in males (necessary for spermatogenesis) and females (stimulates ovarian follicle growth).

- *Luteinizing hormone* (*LH*) is another gonadotropin tropic hormone which stimulates ovulation and corpus luteum formation in females and spermatogenesis in males.
- *Thyroid-stimulating hormone* is a tropic hormone which stimulates the thyroid gland to produce and secrete its own hormones.
- The remaining hormones from the anterior pituitary are formed by the cleaving of a single large protein, *pro-opiomelanocortin*, into short fragments. At least four of these fragments become active peptide hormones; *adrenocorticotropin* (*ACTH*) is a tropic hormone which stimulates the adrenal cortex to produce and secrete its steroid hormones; *melanocyte-stimulating hormone* (*MSH*) regulates the activity of pigment-containing skin cells in some vertebrates; and *endorphins* (or *enkephalins*) which inhibit pain perception.

Regulation of the anterior pituitary by the hypothalamus involves releasing hormones produced by neurosecretory cells in the hypothalamus.
- The releasing hormones are secreted into a capillary network in the median eminence, located above the stalk of the pituitary. Blood containing these hormones then travels through a second capillary network within the anterior pituitary, where they stimulate or inhibit release of specific hormones by pituitary cells.

B. The Thyroid Gland (p. 922)

How is the thyroid gland involved in vertebrate development and maturation? In regulating metabolism?

The *thyroid gland* consists of two lobes located on the ventral surface of the trachea in mammals, and on the two sides of the pharynx in other vertebrates.
- Produces T_3 (triiodothyronine) and T_4 (thyroxine) derived form the amino acid tyrosine, which differ in structure by only one iodine atom.
- T_3 is usually more reactive than T_4 in mammals.

The thyroid gland plays a major role in vertebrate development and maturation.
- Thyroid hormones control metamorphosis in amphibians.
- Normal function of bone-forming cells and the branching of nerve cells during embryonic brain development of animals also requires the presence of the thyroid hormones.

The thyroid gland is critical for regulating metabolism in mammals.
- *Hyperthyroidism* from excessive secretion of thyroid hormones causes high body temperature, sweating, weight loss, irritability and high blood pressure.
- *Hypothyroidism*, or low secretion of thyroid hormones, can cause cretinism in infants and weight gain, lethargy and cold-intolerance in adults.
- *Goiter* (enlarged thyroid) is caused by a dietary iodine deficiency and thyroid hormones cannot be synthesized.

608 Chemical Signals in Animals

Hormone secretion is regulated by the hypothalamus and pituitary through a negative feedback loop.

- ☞ The anterior pituitary produces TSH which, when bound to receptors in the thyroid, generates cAMP that triggers release of T_3 and T_4.
- ☞ TSH secretion is regulated by TRH produced in the hypothalamus.
- ☞ High levels of T_3 and T_4 inhibits TSH secretion.

Calcitonin

The thyroid gland in mammals also produces and secretes *calcitonin* which is a polypeptide hormone that lowers blood calcium levels.

C. The Parathyroid Glands (p. 922–923)

What role does the parathyroid glands have in maintaining homeostasis?

Four parathyroid glands are embedded in the surface of the thyroid and function in the homeostasis of calcium ions.

Parathyroid hormone

- ☞ Secrete *PTH* (*parathyroid hormone*) which raises blood Ca^{2+} levels (antagonistic to calcitonin) and needs vitamin D to function.
- ☞ PTH stimulates Ca^{2+} reabsorption in the kidney and induces osteoclasts to decompose bone and release Ca^{2+} into the blood.

PTH and calcitonin (which has the opposite affects) work in an antagonistic manner and their balance in the body maintains the proper blood calcium levels.

D. The Pancreas (p. 923–924)

Why is the pancreas considered both an endocrine gland and an exocrine gland? What hormones are produced by the pancreas and why are they considered antagonistic?

The pancreas is composed primarily of exocrine tissue which produces digestive enzymes; however, scattered among the exocrine tissue are clusters of endocrine cells called the *islets of Langerhans*.

Alpha cells
Glucagon, Beta cells, Insulin

- ☞ Each islet is composed of *alpha* (α) *cells*, which secrete the peptide hormone *glucagon*, and *beta* (β) *cells* which secrete *insulin*, a protein hormone.

Glucagon and insulin work together in an antagonistic manner to regulate the concentration of glucose in the blood. (See Campbell, Figure 41.14)

Glucose

- ☞ Blood glucose levels must remain near 90 mg/mL in humans for proper body function.
- ☞ Insulin lowers blood sugar concentration by stimulating glucose uptake from the blood in many tissues; it also slows glycogen breakdown in the liver and inhibits the conversion of amino acids and fatty acids to sugar.
- ☞ Glucagon increases blood sugar concentrations by stimulating the hydrolysis of glycogen and the conversion of amino acids and fatty acids to glucose in the liver.

Glucose homeostasis is critical due to its function as the major energy source for cellular respiration and a key source of carbon for the synthesis of other organic compounds.

How can Type I diabetes mellitus be distinguished from Type II diabetes mellitus? What effects do these conditions have on the body?

Serious conditions can result when glucose homeostasis is unbalanced. Diabetes mellitus is caused by a deficiency of insulin or a loss of response to insulin in target tissues. This condition occurs in two forms:

1. *Type I diabetes mellitus* is an autoimmune disorder that appears in juveniles and is insulin dependent.
   - ☞ Results from an immune system attack on the cells of the pancreas.
   - ☞ Usually occurs suddenly during childhood and destroys the ability of the pancreas to produce insulin.
   - ☞ Treated by insulin injections several times each day.

2. *Type II diabetes mellitus* occurs most frequently in adults over 40.
   - ☞ May be due to an insulin deficiency but more commonly results from reduced responsiveness in target cells because of changes in insulin receptors.
   - ☞ Is non-insulin-dependent and can be treated with exercise and dietary controls.

Both types of diabetes mellitus will result in high blood sugar concentrations if untreated.
- ☞ Kidneys excrete glucose, resulting in higher concentrations in the urine.
- ☞ More water is excreted due to the high concentration of glucose (results in the symptoms of copious urine production accompanied by thirst).
- ☞ Fat must serve as the major fuel source for cellular respiration since glucose does not enter the cells.
- ☞ In severe cases, acidic metabolites formed during fat metabolism may lower the blood pH to a critical level.

E. **The Adrenal Glands** (p. 925–926)

What are the adrenal glands and where are they located? What are the catecholamines and what function do they have in the body? What hormones are produced by the adrenal cortex and how do they function in the body?

Adrenal glands are located adjacent to kidneys. In mammals, each gland has an outer cortex and inner medulla which are composed of different cell types, have different functions, and are of different embryonic origin.

The adrenal medulla synthesizes *catecholamines* (*epinephrine* and *norepinephrine*) from the amino acid tyrosine.
- ☞ Epinephrine is released in times of stress when nerve cells excited by stressful stimuli release the neurotransmitter acetylcholine. Acetylcholine combines with chromaffin cell receptors, stimulating release of epinephrine.

Adrenal glands

Adrenal medulla, Epinephrine

Norepinephrine

Acetylcholine, Chromaffin cell

610 Chemical Signals in Animals

- Epinephrine release into the blood results in rapid and dramatic effects on several targets: glucose is mobilized in skeletal muscle and liver cells; fatty acid release from fat cells is stimulated (may serve as extra energy sources).
- Epinephrine works with norepinephrine to: increase the rate and stroke volume of the heartbeat; shunt blood away from the skin, gut, and kidneys to the heart, brain, and skeletal muscles by stimulating smooth muscle contraction in some blood vessels and relaxation in others.

The adrenal cortex synthesizes and secretes two main types of *corticosteroids* in humans: *glucocorticoids* (cortisol) and *mineralocorticoids* (aldosterone).
- Stressful stimuli cause the hypothalamus to secrete releasing factors that stimulate release of ACTH from the anterior pituitary, which stimulates release of glucocorticoids.
- Glucocorticoids promote glucose synthesis from noncarbohydrate substances such as proteins. The effect is slower, but of longer duration than that of epinephrine. Have immunosuppressive effects and are used to treat inflammation.
- Mineral corticoids affect salt and water balance. (Release regulated by hormones produced in the liver and kidneys).

F. The Gonads (*p. 926*)

Why are the gonads considered part of the endocrine system? What are androgens, estrogens and progestins? What effect does each have on development and reproductive cycles?

The testes of males and ovaries of females produce steroid hormones that affect growth and development as well as regulate reproductive cycles.

The gonads of both males and females produce androgens, estrogens, and progestins, although the proportions differ.

Androgens generally stimulate the development and maintenance of the male reproductive system.
- Produced in greater quantities in males than females.
- Primary androgen is testosterone.
- Androgens produced during early embryonic development determine whether the fetus will be male or female.
- High androgen concentrations at puberty stimulate development of male secondary sex characteristics.

Estrogens perform the same functions in females as androgens do in males.
- Estradiol is the primary estrogen produced.
- Maintain the female system and stimulate development of female secondary sex characteristics.

Progestins are primarily involved with preparing and maintaining the uterus for reproduction in mammals.
- Include progesterone.

The gonadotropins from the anterior pituitary (FSH and LH) control the synthesis of both androgens and estrogens.
- FSH and LH are in turn controlled by gonadotropin-releasing hormone (GnRH) from the hypothalamus.

G. **Other Endocrine Organs** (p. 926–927)

What link exists between the digestive tract and the endocrine system? The kidneys? The heart? What role do the pineal gland and the thymus gland have in the endocrine system?

Many organs with primarily nonendocrine functions also secrete hormones.
- The digestive tract secretes at least eight hormones (e.g. gastrin, secretin).
- Kidneys secrete erythropoietin which stimulates red blood cell production.
- The heart secretes atrial natriuretic hormone which regulates salt and water balance and blood pressure.

Pineal gland = A small mass near the center of the mammalian brain that secretes *melatonin*, which is a modified amino acid.
- Melatonin regulates functions related to light and seasons such as biological rhythms associated with reproduction.
- Melatonin is secreted only at night; larger amounts are thus secreted during the winter than the summer.

The *thymus* is located across the front of the neck and secretes several messengers including *thymosin* which stimulates T lymphocyte development and differentiation.
- Large in children, declines in adults.

V. **ENDOCRINE GLANDS AND THE NERVOUS SYSTEM** (p. 927–928)

How are the endocrine system and the nervous system related? What structural, chemical and functional relationships exist?

The endocrine system and nervous system interact and are often inseparable.
- They are <u>structurally</u> related: many endocrine glands are either made of nerve tissue (hypothalamus, posterior pituitary) or have evolved from the nervous system (adrenal medulla).
- They are <u>chemically</u> related: several vertebrate hormones are used as signals by both systems (e.g. epinephrine).
- They are <u>functionally</u> related: may coordinate biological processes with both nervous and hormonal components arranged in series, or one system can affect the output of the other.

Gonadotropins
Gonadotropin-releasing hormone

Gastrin, Secretin
Erythropoietin
Atrial natriuretic hormone

Pineal gland
Melatonin

Thymosin

42   ANIMAL REPRODUCTION

CHAPTER OUTLINE

An individual organism exists for only a length of time, its lifespan. For a species to remain viable, its members must reproduce. Only through reproduction can extinction be avoided.

I. MODES OF REPRODUCTION (p. 931–934)

Two principal modes of reproduction are found in animals: asexual reproduction and sexual reproduction.

A. Asexual Reproduction (p. 931–932)

Why is asexual reproduction said to produce clones of the reproducing individual? What are gemmules? How do the different forms of asexual reproduction differ? Under what conditions would asexual reproduction be advantageous?

Asexual reproduction — Asexual reproduction occurs when an individual produces offspring genetically identical to itself (a clone).

Clone

There are various types of asexual reproduction: budding, fragmentation, release of specialized cells and regeneration.

Budding occurs when a new individual grows out from the parental body.
- ☞ Offspring may detach from its parent or remain joined to form extensive colonies.

Fragmentation is the breaking of the body into pieces, each of which develops into a complete adult.
- ☞ In some annelids, the "new" anterior ends will develop with sense organs before the parent actually fragments.

Some invertebrates release specialized groups of cells that grow into new adults.
- ☞ Freshwater sponges produce *gemmules* which form from aggregates of several types of cells which are surrounded by a protective coat.

Gemmules

Regeneration following an injury may form two or more individuals from one original.
- ☞ When the arm of a sea star is removed, it will grow a new arm; asexual reproduction will have occurred if the free arm contains a portion of the central disc and it grows into a new individual.

Animal Reproduction

Advantages of asexual reproduction:
- Sessile animals that cannot seek mates or more active animals in conditions of low population density can reproduce.
- Allows production of many offspring in a short time.
- Most advantageous in stable, favorable environments because it perpetuates successful genotypes.

B. Sexual Reproduction (p. 932)

What distinguishes sexual reproduction from asexual reproduction? Under what conditions would sexual reproduction be advantageous?

In sexual reproduction, two individuals produce offspring having a combination of genes inherited from both parents.
- Usually, two haploid *gametes* fuse to form a diploid *zygote*.
- The female gamete (*ovum*) is usually larger and nonmotile.
- The male gamete (*spermatozoon*) is usually a smaller, flagellated cell (except in arthropods which produce nonmotile sperm).
- Increases genetic variability, thus advantageous in a fluctuating environment.

Sexual reproduction
Gametes, Zygote
Ovum, Spermatozoan

C. Reproductive Cycles and Patterns (p. 932–934)

How can parthenogenesis, hermaphroditism and sequential hermaphroditism be distinguished from one another? What advantages do animals with reproductive cycles have and to what are these cycles usually related?

Most animals have reproductive cycles, often related to changing seasons.
- Allows conservation of resources; so they reproduce when energy is above that needed for maintenance.
- Allows reproduction when environmental conditions favor offspring survival.
- Cycles are controlled by a combination of hormonal and seasonal cues.

Animals employ various patterns of reproduction and may use either sexual or asexual reproduction exclusively or alternate between the two.
- Asexual reproduction often occurs under favorable conditions and sexual reproduction during times of environmental stress in animals that employ both mechanisms.

Parthenogenesis = Development of an egg without fertilization, often producing haploid adults.
- Plays a role in the social organization of some insects; male honey bees (drones) are produced parthenogenetically, females develop from fertilized eggs.
- Some fish, amphibians and lizards reproduce via parthenogenesis; this requires a doubling of chromosomes after meiosis to create diploid zygotes.

Parthenogenesis

Hermaphroditism = Each individual has both functional male and female reproductive parts.
- Solves the problem of finding a mate of the opposite sex for some sessile, burrowing, and parasitic animals.

Hermaphroditism

614 *Animal Reproduction*

☞ Some self-fertilize, but most mate with another; each donates and receives sperm, thus potentially producing twice as many offspring from one mating.

Sequential hermaphroditism — Sequential hermaphroditism = An individual reverses its sex during its lifetime.

Protogynous, Protandrous — ☞ Some are *protogynous* (female first); other animals are *protandrous* (male first).

☞ Reversal often associated with age and size. In protogynous species such as the blue-head wrasse, the largest (=oldest) fish in the harem becomes a male and defends the harem. Some oysters are protandrous, the largest individuals become females and can produce more egg cells.

II. **MECHANISMS OF SEXUAL REPRODUCTION** (*p. 934–936*)

A. **Patterns of Fertilization and Development** (*p. 934–935*)

What types of organisms exhibit external fertilization? Internal fertilization? What advantages and disadvantages are associated with external and internal fertilization?

Two major mechanisms of fertilization have evolved in animals: external fertilization and internal fertilization. Each has specific environmental and behavioral requirements.

External fertilization — 1. In external fertilization, eggs are shed by a female and fertilized by a male in the environment.

☞ Occurs almost exclusively in moist habitats when development can occur without desiccation or heat stress.

☞ Environmental cues, pheromones and courtship behavior trigger release of mature gametes in close proximity which increases the probability of successful fertilization.

☞ In vertebrates exhibiting external fertilization, courtship behavior also permits mate selection.

Internal fertilization — 2. Internal fertilization occurs when sperm are deposited in or near the female reproductive tract and fertilization occurs within the female body.

☞ Usually requires more sophisticated reproductive systems and cooperative mating behaviors.

Once fertilization occurs, the zygote undergoes embryonic development. The degree to which these developing embryos are protected, varies with the mechanism of fertilization and the organisms involved.

1. Organisms using external fertilization do not have protective coverings for the embryos.

☞ Fish and amphibian eggs are covered with a gelatinous coat which permits the free exchange of gases and water.

☞ The water or moist habitat into which the gametes are released prevent desiccation and temperature stress.

☞ Very large numbers of zygotes are usually formed although only a small proportion survive to complete development.

2. Many types of protection are found in those animals which use internal fertilization.

- Eggs resistant to harsh environments are produced by many reptiles and all birds. The amniote egg of these animals has a protective shell of protein (reptiles) or calcium (birds) which makes them resistant to water loss and physical damage.
- Mammals (other than monotremes which lay reptilian-like eggs) do not produce a shelled egg, but possess reproductive tracts that permit embryonic development within the female parent. In marsupials, the embryo develops for a short time in the uterus, then moves to the mother's pouch to complete development. Placental mammals retain the embryo in the uterus until development is completed.
- Parental protection is also important to survival of the developing embryo. While some organisms using external fertilization show forms of parental care (nesting fish will guard the eggs against predators), parental care is most highly developed in animals using internal fertilization.
- Internal fertilization usually produces fewer zygotes than external, but survival through development is much greater due to the protection and care of the developing offspring.

B. **Diversity in Reproductive Systems** (*p. 935–936*)

In general, the complexity of the reproductive system is <u>not</u> related to the phylogenetic position of the animal.

1. **Invertebrate Reproductive Systems** (*p. 935–936*)

What is a spermatheca? How does the reproductive system of insects differ from that found in flatworms?

Diverse systems have evolved among invertebrates; they range from simple systems without distinct gonads to very complex systems.

Most polychaetes have separate sexes without distinct gonads. Gametes develop from undifferentiated cells lining the coelom and fill the coelom as they mature. In some species they are released through excretory openings; in others the parental body splits open to release the eggs to the environment as the parent dies.

Insects have separate sexes and complex reproductive structures. (See Campbell, Figure 42.6)
- In males, sperm develop in testes, pass to and are stored in seminal vesicles, and empty through the penis via an ejaculatory duct. Accessory glands may be present and add fluid to semen.
- In females, eggs pass from ovaries, through oviducts, to the vagina which opens to the outside. May have a *spermatheca*, a blind-ended sac in which sperm may be stored for a year or more.

Flatworms are hermaphrodites with very complex reproductive systems similar to that in insects. (See Campbell, Figure 42.7)
- The female system also includes yolk and shell glands and uterus.
- The male system includes a copulatory apparatus and copulation may occur in a variety of ways in different groups.

Gonads

Spermatheca

2. **Vertebrate Reproductive Systems** (*p. 936*)

What is a cloaca? A penis? How are these structures associated with the reproductive systems of nonmammalian vertebrates?

Vertebrate reproductive systems are all similar, but have some important variations:
- Most mammals have separate openings for digestive, excretory, and reproductive tracts while other vertebrates have only a common opening, the *cloaca*.
- Some mammals, birds and snakes have a uterus with only one branch, while most other vertebrates have a bicornate (2 branched) uterus.
- Nonmammalian vertebrates do not have well-developed *penes* and use other mechanisms to transfer spermatozoa.

III. **MAMMALIAN REPRODUCTION** (*p. 936-953*)

A. **The Anatomy of Reproduction in Humans** (*p. 937-940*)

The *human male reproductive system* includes the external *genitalia* and the internal reproductive organs.
- The external genitalia includes the *scrotum* and *penis*.
- The internal reproductive organs consist of the gonads (testes), accessory glands, and associated ducts.

What functions do the external genitalia (scrotum and penis) of the human male have in reproduction?

The external male genitalia (*scrotum* and *penis*) aid the reproductive process in different ways.
- Testes develop in the abdomen and descend into the scrotum just before birth. This is important since sperm cannot develop at normal body temperature.
- By having the testes hang outside the abdominal cavity, in most mammals in the scrotum, the temperature is 2°C lower and sperm production can occur.
- The penis serves as the male copulatory organ. The ejaculatory duct joins the urethra (from the excretory system) which opens at the tip of the penis.
- The movement of semen through the urethra during copulation results in the sperm being deposited directly in the female system.

What are the seminiferous tubules? What is the function of interstitial cells in the reproductive effort? The epididymis? The vas deferens? The urethra?

Internal male reproductive organs are the gonads and associated ducts: (See Campbell, Figure 42.8)

Testes = Tightly coiled tubes, or *seminiferous tubules* where sperm form, surrounded by connective tissue layers. *Interstitial cells* scattered between tubules produce androgens.

Epididymis = Contains coiled tubules into which sperm pass from the seminiferous tubules. Site of final sperm maturation and storage.

Vas deferens = Muscular duct running from epididymis to the *ejaculatory duct*. The ejaculatory duct forms by the joining of the two vas deferens ducts (one from each testis) and opens into the urethra.

Urethra = Tube that runs through the penis and drains both the excretory and reproductive system.

What contributions are made to the reproductive effort by secretions from the seminal vesicles, prostate gland and bulbourethral glands?

There are three sets of accessory glands associated with the male system. These glands add their secretions to the semen.

Seminal vesicles = The pair of glands below and behind the bladder; empty into the ejaculatory duct. Secrete fluid containing mucus, amino acids (cause semen to coagulate after deposited in female), fructose (provides energy for sperm) and prostaglandins (stimulate female uterine contractions to help move semen to the uterus). Seminal vesicle secretions make up 60% of the total semen volume.

Prostate gland = Gland that surrounds the upper portion of, and empties directly into the urethra. Secretes a thin alkaline fluid that balances the acidity of residual urine in the male system and the acidity of the vagina.

Bulbourethral glands = A pair of small glands that empty into the urethra at the base of the penis. Secrete a viscous fluid before sperm emission. Function is unclear.

The human penis is composed of 3 cylinders of spongy tissue that fill with blood during sexual arousal.
- Rodents, raccoons, walruses and some other mammals also possess a *baculum* (a bone that stiffens the penis).
- The head of the penis, the *glans penis*, is covered by *prepuce*, or foreskin.

Why is the structure of the human female reproductive system more complicated than that of males?

The *human female reproductive system* is more complicated than that of the male; it possesses structures not only for the production of female gametes, but also to house the embryo and fetus.
- Internal reproductive structures are the gonads (ovaries) and associated ducts and chambers which are involved with gamete movement and embryo development.
- External genitalia include the clitoris and the two sets of labia that surround the clitoris and vaginal opening.

What is a follicle? What functions do follicles have in the reproductive effort? What changes occur in the follicle following ovulation?

The female gonads, the *ovaries*, are located in the abdominal cavity and enclosed in a tough protective capsule.
- Contains many *follicles* (one egg cell surrounded by follicle cells, which nourish and protect the developing egg). All are formed at birth.
- Follicle cells also produce estrogens.

618 *Animal Reproduction*

Ovulation

Corpus luteum, Progesterone, Estrogen

Oviduct

During *ovulation*, the egg is expelled from the follicle. The remaining follicular tissue forms the *corpus luteum*, which secretes progesterone (pregnancy hormone) and additional estrogen. (See Campbell, Figure 42.9)
- If the egg is not fertilized, the corpus luteum degenerates.
- Egg is expelled into the abdominal cavity near the opening of the *oviduct*.
- Cilia lining the oviduct draw in the egg cell and convey it to the uterus.

What are the structural components of the human female reproductive system? What function does each component have in the reproductive effort?

Uterus

Endometrium

Uterus = Womb; thick muscular organ shaped like an inverted pear and approximately 7 cm long and 4–5 cm wide.
- The inner lining, the *endometrium*, is richly supplied with blood vessels.

The remaining female reproductive structures are:

Cervix

Vagina

Vestibule

Labia minora

Labia majora

Glans clittoris

Cervix = The narrow neck of the uterus.

Vagina = Thin walled yet muscular chamber that is the repository for sperm during copulation and forms the birth canal.

Vestibule = Chamberlike area formed by the two pairs of skin folds covering the vaginal orifice and urethral opening.

Labia minora = Thin erectile tissue bordering the vestibule.

Labia majora = Thick, fatty ridges protecting the labia minora.

Glans clittoris = Bulb of erectile tissue at the top of the vestibule; female equivalent of the glans penis.

Mammary glands

Alveoli

Mammary glands are important to mammalian reproduction, although not actually a part of the reproductive system.
- Consist of *alveoli*, small sacs of epithelial tissue that secrete milk.
- Alveoli drain into lactiferous ducts that open at the nipple.

Primordia

Androgens

The external genitalia of both sexes arise from common *primordia* (undifferentiated embryonic structures).
- Presence of a Y chromosome determines whether androgens are present; if so, male structures form.
- In the absence of androgens, female structures form.

B. Hormonal Control of Mammalian Reproduction (p. 940–945)

What hormones are involved in the hormonal control of male reproduction in mammals? What is the function of each?

In males, androgens are directly responsible for formation of primary sex characteristics (reproductive organs) and secondary sex characteristics (deepening of voice, hair growth pattern, muscle growth).
- Also potent determinants of sexual and aggressive behaviors.

GnRH, LH, FSH

- GnRH from the hypothalamus regulates release from the pituitary of LH (stimulates androgen production) and FSH (acts on seminiferous tubules to increase sperm production).

Animal Reproduction 619

Hormonal control in females is more complicated and reflects the cyclic nature of female reproduction. Female mammals display two different types of cycles.
1. Estrous cycles.
2. Menstrual cycles.

What differences can be used to distinguish between menstrual cycles and estrous cycles?

Estrous cycles occur in most non-primate mammals.
☞ Ovulation occurs after endometrium thickens and vascularizes, as in the menstrual cycle.
☞ If pregnancy does not occur, endometrium is reabsorbed.
☞ More pronounced behavioral changes; and seasonal and climatic changes effect the estrous cycle more than the menstrual cycle.
☞ *Estrus* is the period of sexual activity surrounding ovulation. The length and frequency varies widely with species.

Menstrual cycles occur in humans and many other primates, and consists of 3 phases:

<u>Menstrual flow phase</u> = Menstrual bleeding results from the shedding of the endometrium; persists for a few days.

<u>Proliferative phase</u> = Thickening of the endometrium; 1–2 weeks long.

<u>Secretory phase</u> = Endometrium thickens further, becomes more vascularized, and secretes a glycogen-rich fluid.

Paralleling the menstrual cycle is an *ovarian cycle*.
☞ Begins with the *follicular phase* in which several follicles on the ovaries begin to grow, but only one continues to enlarge and mature. This phase ends with ovulation.
☞ In the *luteal phase*, the follicular tissue remaining in the ovary after ovulation is transformed into the *corpus luteum*, which secretes female hormones.

How is the hypothalamus involved in the human female reproductive cycle? What hormones are active in the menstrual cycle and what functions do they carry out?

Five hormones work together (by positive and negative feedback mechanisms) to coordinate the menstrual and ovarian cycles. This coordination synchronizes follicle growth and ovulation with preparation of the uterine lining for embryo implantation. (See Campbell, Figure 42.13)
☞ During the follicular phase of the ovarian cycle, *GnRH* secreted by the hypothalamus stimulates the anterior pituitary to secrete small quantities of *FSH* and *LH*.
☞ FSH stimulates the immature follicles in the ovary to grow and these follicle cells secrete small amounts of *estrogen*.
☞ As the follicle continues to grow, the amount of estrogen secreted increases. The mature follicle secretes a large amount of estrogen.

<u>Estrous cycles</u>

<u>Estrus</u>

<u>Menstrual cycles</u>

<u>Menstrual flow phase</u>
<u>Proliferative phase</u>
<u>Secretory phase</u>

<u>Ovarian cycle</u>
<u>Follicular phase</u>

<u>Luteal phase</u>
<u>Corpus luteum</u>

☞ The high concentration of estrogen stimulates the hypothalamus to increase secretion of GnRH which results in a sudden increase in FSH and LH secretion. LH increases more than FSH because high estrogen levels increases the sensitivity of LH-releasing mechanisms (in the pituitary) to GnRH.
☞ The sudden surge in LH concentration stimulates final maturation of the follicle and ovulation (after about 24 hours). The high concentration of LH stimulates the ruptured follicular tissue to transform into the corpus luteum.
☞ The presence of LH during the luteal phase of the ovarian cycle stimulates the corpus luteum to continue to secrete estrogen, but to also secrete increasing amounts of *progesterone*.
☞ Increasing concentrations of estrogen and progesterone inhibits GnRH secretion by the hypothalamus resulting in a decrease in FSH and LH.
☞ As the LH concentration declines, the corpus luteum begins to atrophy; this results in a sudden drop in estrogen and progesterone concentrations.
☞ Decreasing levels of estrogen and progesterone removes the inhibition exerted on the hypothalamus, which begins to secrete small amounts of GnRH that stimulates the anterior pituitary to secrete low levels of FSH and LH.
☞ A new follicular phase begins at this point.

How are the menstrual cycle and ovarian cycle synchronized in female mammals?

Coordination of the menstrual cycle with the ovarian cycle depends primarily on the levels of estrogen and progesterone.
☞ Increasing amounts of estrogen secreted by growing follicles during the follicular phase stimulate the endometrium lining the uterus to thicken.
☞ After ovulation, the estrogen and progesterone secreted by the corpus luteum stimulates continued development and maintenance of the endometrium. Arteries supplying blood to the endometrium enlarge and the endometrial glands that supply the nutritional fluid to the early embryo grow and mature.
☞ Decreasing concentrations of estrogen and progesterone due to the disintegration of the corpus luteum reduce blood flow to the endometrium. The endometrium breaks down and passes out of the uterus as the menstrual flow.
☞ A new menstrual cycle begins with a new ovarian cycle.
☞ Estrogens are also responsible for development of the female secondary sex characteristics.

C. **Gamete Formation (Gametogenesis)** (*p. 945–947*)

What is spermatogenesis? Where does it occur and what processes are involved? How many functional products are produced from each cell which begins the process?

Spermatogenesis = Production of mature sperm cells. (See Campbell, Figure 42.14)
☞ A continuous process in adult males.
☞ Occurs in seminiferous tubules of the testes.

- ☞ Begins with the differentiation of primordial germ cells into spermatogonia in the embryonic testes. Both of these types of cells are diploid.
- ☞ The spermatogonia are located near the outer wall of the seminiferous tubules. They increase in numbers through repeated mitoses throughout development and early life.
- ☞ When the male matures, spermatogonia begin to differentiate into primary spermatocytes which are diploid.
- ☞ Primary spermatocytes will pass through several stages before giving rise to mature spermatozoa; each primary spermatocyte undergoes meiosis I to produce two haploid secondary spermatocytes; each secondary spermatocyte undergoes meiosis II to form two spermatids. Thus each primary spermatocyte forms four spermatids through meiotic division.
- ☞ Each spermatid becomes associated with a large Sertoli cell from which it receives nutrients.
- ☞ All four spermatids from each primary spermatocyte differentiate into mature spermatozoa.

In a spermatozoon, the thick head containing the haploid nucleus is tipped with the *acrosome*, which contains enzymes to aid in egg penetration. (See Campbell, Figure 42.15a)

Spermatozoon

Acrosome

- ☞ The neck contains many mitochondria to provide ATP for movement of the flagellum.

What is oogenesis? Where does it occur and what processes are involved? How many functional products are produced from each cell which begins the process?

Oogenesis = Development of mature ova. (See Campbell, Figure 42.16a)

Oogenesis

- ☞ Begins in the embryo when primordial germ cells undergo mitotic divisions to produce diploid oogonia.
- ☞ Each oogonium will develop into a primary oocyte by the birth of the female, resulting in all potential ova being present in the ovaries at birth.
- ☞ Between birth and puberty, primary oocytes enlarge and their surrounding follicles grow.
- ☞ After puberty, during each ovarian cycle, FSH stimulates a follicle to enlarge and the primary oocyte passes through meiosis I to produce a haploid secondary oocyte and the first polar body.
- ☞ LH triggers ovulation and the secondary oocyte is released from the follicle.
- ☞ If a sperm cell penetrates the secondary oocyte's membrane, meiosis II will occur and the second polar body will separate from the ovum.

How do spermatogenesis and oogenesis differ?

Spermatogenesis and oogenesis differ in three ways:

1. In spermatogenesis, all four products of meiosis I and II become mature spermatozoa; in oogenesis, the unequal cytokinesis which occurs during meiosis I and II results in most of the cytoplasm being in one daughter cell which will form the single ovum, the other cells (polar bodies will degenerate.

2. Spermatogenesis is a continuous process throughout the reproductive life of the male; all potential ova that can be produced via oogenesis will be present as primary oocytes in the ovaries at the time of the female's birth.
3. Spermatogenesis occurs as an uninterrupted sequence; in oogenesis, long "resting" periods occur between the formation of the initial steps and final production of the ovum.

D. Sexual Maturation (p. 947)

What changes occur at puberty in humans? How is the hypothalamus involved in these changes?

Puberty = The onset of sexual maturation.
- ☞ The hypothalamus secretes increasing amounts of GnRH, leading to higher FSH and LH levels. These hormones trigger maturation of the reproductive system and development of secondary sex characteristics.
- ☞ Usually occurs in humans between the ages of 8 and 14.

E. Human Sexual Physiology (p. 947)

Although humans display very variable sexual behavior, there is a common physiological sexual response cycle consisting of four phases:

What four phases occur in the sexual response cycle of humans? What physiological changes occur in each phase?

1. In the *excitement phase*, vasocongestion ("filling of blood") of the penis and clitoris occurs along with enlargement of testes, labia and breasts. Vaginal lubrication and myotonia (increased muscle tension) also occur.
2. The *plateau phase* continues these responses and the vagina forms a depression to receive sperm. The heart rate and breathing increases in response to stimulation of the autonomic nervous system.
3. *Orgasm* is characterized by rhythmic, involuntary contractions in the reproductive system of both sexes, and by emission and ejaculation of the male.
4. The *resolution phase* reverses the responses of earlier phases.

F. Conception, Pregnancy and Birth (p. 947-953)

What is conception? How are conception and zygotic cleavage associated with pregnancy? What is the function of the placenta during pregnancy?

Pregnancy (gestation) = Condition of carrying developing embryos in the uterus.
- ☞ Begins at *conception* (fertilization of an egg by a spermatozoon), ends at birth.
- ☞ Human pregnancy averages 266 days.
- ☞ Duration in other species correlates with body size and extent of development of young at birth.

After fertilization, the zygote travels via the oviduct to the uterus.
- *Cleavage* (cell division) begins in the oviduct 24 hours after fertilization.
- The embryo reaches the uterus in 3–5 days and develops into a hollow ball of cells called a *blastocyst*. This stage develops about one week after fertilization.
- The blastocyst implants into the endometrium.
- Embryonic tissues mingle with the endometrium to form the *placenta*, which functions in respiratory gas exchange, nutrient transfer, and waste removal for the embryo.

What hormones are involved in control of pregnancy in human females? Where are each of the hormones produced and how do their effects change as pregnancy progresses?

Pregnancy is divided into trimesters:
1. The first trimester is the main period of *organogenesis*, the development of organs. After eight weeks, the embryo becomes a *fetus* and possesses all organs of the adult in rudimentary form.
   - *Human chorionic gonadotropin* (HCG) is the embryonic hormone that maintains progesterone and estrogen secretion by the corpus luteum to prevent menstruation which would end the pregnancy.
   - High progesterone levels also stimulate formation of a protective mucous plug in the cervix, growth of the maternal part of the placenta, uterus enlargement, and cessation of ovulation and menstrual cycling.
2. In the second trimester, rapid growth occurs and the fetus is very active.
   - HCG declines, the corpus luteum degenerates, hormone levels stabilize and the placenta secretes its own progesterone to maintain the pregnancy.
3. In the third trimester, growth is rapid and fetal activity decreases.
   - The maternal abdominal organs become compressed and displaced.

What hormones are involved in parturition? What is the relationship between progesterone, prolactin and the mammary glands?

Birth, or *parturition*, occurs through strong, rhythmic contractions of the uterus.
- Regulated by uterine prostaglandins, oxytocin from the posterior pituitary and nervous reflexes.
- During the first stage of labor the cervix dilates; the second stage is birth of the baby; the final stage of labor is expulsion of the placenta.

After birth, decreasing levels of progesterone allow prolactin secretion which stimulates milk production after a delay of 2–3 days.

624 Animal Reproduction

Why is there no immune response mounted against the developing embryo by the mother's immune system?

Although the embryo has foreign antigens, the mother usually does not mount an immune response that rejects or harms the embryo.
- ☞ A protective trophoblast layer prevents the embryo itself from contacting maternal tissue.
- ☞ One hypothesis suggests that a special type of white blood cell (suppressor cell) in the uterus is stimulated by an initial immune response and acts to prevent other white blood cells from mounting further attack.

Contraception

Contraception = Prevention of pregnancy.

Why is chemical contraception preferred by many women over the rhythm method or intrauterine devices? Why are tubal ligation and vasectomy the most effective methods of contraception?

There are three major means of contraception: (See Campbell, Figure 42.21)

1. Preventing the egg and sperm from meeting in the female reproductive tract.
   - ☞ *Rhythm method* is refraining from intercourse when conception is most likely (failure rate = 10–20%).
   - ☞ Condoms, diaphragms, cervical caps, and contraceptive sponges are physical barriers fitting over the penis (condom) or into the vagina (failure rate < 10%). Most effective when used in conjunction with spermicidal foam or jelly.
   - ☞ Withdrawal is unreliable since sperm may be present in secretions preceding ejaculation.

Rhythm method

Intrauterine device

2. Preventing implantation of a blastocyst by using an *intrauterine device (IUD)*; a plastic device fitted into the uterine cavity. Probably works by irritating the endometrium but precise mechanism is unknown.
   - ☞ May have serious side effects.

Chemical contraception

3. Chemical contraception prevents the release of mature gametes from the gonads.
   - ☞ Birth control pills are combination of synthetic estrogen and progestin that stop release of GnRH, LH and FSH. By blocking LH release, progestin prevents ovulation. Estrogen inhibits FSH so no follicles develop (failure rates < 1%).

Tubal ligation, Vasectomy

Sterilization by *tubal ligation* (cutting the oviduct in women) or *vasectomy* (cutting the vas deferens of men) is nearly 100% effective, safe, but difficult to reverse.

G. **Reproductive Technology** (*p. 953*)

How have technological advancements improved the understanding of human reproductive problems? What advantages and disadvantages are associated with ultrasound, amniocentesis, chorionic villi sampling and in-vitro fertilization?

Scientific and technological advances have made it possible to determine if problems exist in the developing fetus.

Ultrasound = A noninvasive procedure in which a transducer emits high-frequency sound waves that are reflected back by tissues of varying densities to form an image of the fetus.

Amniocentesis = An invasive procedure in which a needle is inserted into the amnion and fluid is withdrawn. The fluid contains fetal cells that are cultured and analyzed for chromosomal defects and genetic disorders.
- 1% risk of spontaneous abortion following procedure.

Chorionic villi sampling = Another invasive procedure where a small portion of the chorion, the fetal part of the placenta, is removed, cultured and analyzed for genetic and metabolic disorders.
- 5%–20% risk of spontaneous abortion following procedure.
- Can be performed earlier in pregnancy and results are obtained in days rather than weeks.

Scientific discovery has also led to solving some fertility problems.

In vitro fertilization = Ova are surgically removed from hormonally prepared follicles, fertilized in petri dishes and the embryo placed in the uterus for implantation.
- Difficult and costly procedure.
- Success rate is 1 in 6 attempts.

43 ANIMAL DEVELOPMENT

CHAPTER OUTLINE

Animals develop throughout their lifetime. Development begins with the changes that form a complete animal from the zygote and continue as progressive changes in form and function.

I. AN OVERVIEW OF DEVELOPMENTAL PROCESSES (p. 956–957)

Two early views of how animals developed from an egg competed for supporters until modern techniques were developed. One view was termed *preformation* and the other *epigenesis*.

How does the concept of development by preformation differ from the concept of development by epigenesis?

Preformation suggested that the embryo contained all of its descendants as a series of successively smaller embryos within embryos.
- Based on this idea, by dissecting an egg an individual would be able to find smaller and smaller individuals as the dissection continued.
- Was a popular idea as recently as the eighteenth century.

Epigenesis proposed that the form of an embryo gradually emerged from a formless egg.
- Originally proposed by Aristotle.
- Gained support in the nineteenth century as improved microscopy permitted biologists to view embryos as they gradually developed.

What association exists between a zygote's genome and the egg cell's cytoplasm during the development of an animal?

Modern biology has found that an organism's development is mostly determined by the zygote's genome and the organization of the egg cell's cytoplasm.
- The heterogenous distribution of messenger RNA, proteins, and other components in the unfertilized egg greatly impacts the development of the embryo in most animals.
- Division of the zygote after fertilization partitions the heterogenous cytoplasm in such a way that nuclei of different embryonic cells are exposed to different cytoplasmic environments.
- These different cytoplasmic environments result in the expression of different genes in different cells.
- This leads to an emergence of inherited traits that is ordered in space and time by mechanisms controlling gene expression.

What developmental functions are associated with cell division? With differentiation? With morphogenesis? What effect would irregularities in these processes?

Embryonic development includes a succession of mitotic divisions which result in the single cell zygote becoming a complete organism consisting of trillions of cells. In order for an organism to develop instead of a ball of identical cells, other processes must be involved. The three component processes involved in embryonic development are: cell division, differentiation and morphogenesis.

1. *Cell division* results in an increase in the number of cells. All of the cells result from mitotic divisions beginning with the zygote.
2. *Differentiation* is the divergence in structure and function of cells as they become specialized during a multicellular organism's development. The changes that result in two cells with the same genes becoming two different types of cells depends upon which genes are expressed in the cell.
3. *Morphogenesis* is the development of body shape and organization during development. This involves the movement and arrangement of cells and tissues in the early embryo and produces the characteristic three-dimensional form of the organism.

Precise regulation of the three developmental processes is necessary for proper development. A breakdown in this regulation can result in developmental disorders.
- A cleft palate results from an improper morphogenesis.
- Retinoblastoma (fatal cancer of the retina) appears to be caused by a defect in cell division regulation.
- Some forms of testicular cancer are caused by improper regulation of differentiation.

II. FERTILIZATION (p. 957–959)

Fertilization forms a diploid zygote and triggers onset of embryonic development.

A. The Acrosomal Reaction (p. 957–958)

What is the acrosomal reaction and how does it ensure species specificity?

The acrosomal reaction (based on studies with sea urchins):
- Upon contacting the egg's jelly coat, the acrosomal vesicle in the head of the sperm releases hydrolytic enzymes that enable an *acrosomal process* to penetrate the jelly coat. The acrosomal process extends by polymerization of actin. (See Campbell, Figure 43.2)
- Bindin (a protein) molecules on the process surface attach to receptors on the egg's *vitelline layer* (just external to the plasma membrane).
- This provides species specificity for fertilization.
- Enzymes of the acrosomal process probably digest vitelline layer materials allowing the tip of the process to contact the egg plasma membrane.
- The sperm and egg's plasma membranes fuse, allowing the sperm nucleus to enter the egg and causing a depolarization of the plasma membrane that bars other sperm cells from also uniting with the egg (*fast block to polyspermy*).

B. The Cortical Reaction (p. 958)

What is the cortical reaction? How do the acrosomal and cortical reactions function sequentially to prevent an egg cell from being fertilized by more than one sperm cell?

The acrosomal reaction stimulates a subsequent cortical reaction: (See Campbell, Figure 43.3)
- Plasma membrane depolarization increases the cytoplasmic Ca^{2+} concentration which causes *cortical granules* in the viscous egg cortex to fuse with the plasma membrane. The cortical granules then release their contents by exocytosis into the space between the plasma membrane and vitelline layer.
- Enzymes and other materials released by the cortical granules cause the vitelline layer to elevate and harden. This forms a *fertilization membrane* that acts as a *slow block to polyspermy*.

C. Activation of the Egg (p. 958–959)

What changes occur in an egg once it is activated? What stimulates these changes to occur?

The sharp rise in cytoplasmic Ca^{2+} concentration also incites metabolic changes that *activates* the egg cell.
- Cellular respiration and protein synthesis rates increase.
- Cytoplasmic pH changes from 6.8 to 7.3 due to extrusion of H^+.
- Activation can be artificially induced by injection of Ca^{2+}.
- The sperm nucleus within the egg swells and merges with the egg nucleus to form the zygote (actual fertilization).
- DNA replication begins and the first division occurs in about 90 minutes.

III. EARLY STAGES OF EMBRYONIC DEVELOPMENT (p. 960–965)

The following information is based upon studies of the sea urchin and *Xenopus laevis*.

What is cleavage? Why is polarity of the embryo important during the process? What is the primary difference between a morula stage and a blastula stage?

A. Cleavage (p. 960–961)

Cleavage = Rapid cell divisions following fertilization that produces a ball of many smaller *blastomeres*. No growth occurs.
- Nourishment for embryos during early development is derived from yolk stored in the egg cell.
- Formation of smaller blastomeres through cleavage enhances oxygen uptake and exchange of other substances by increasing the surface-to-volume ratio of each cell; results in a reduction of the volume of cytoplasm controlled by each nucleus.
- Cleavage planes follow a species-specific pattern relative to the animal and vegetal poles of the zygote.

- The upper and lower hemispheres of the zygote are named for their respective poles: the animal hemisphere and vegetal hemisphere. The two hemispheres can be distinguished in *Xenopus* due to the heterogenous distribution of cytoplasmic contents: melanin granules embedded in the cortex gives the animal hemisphere a deep gray hue; yolk materials in the vegetal hemisphere result in a lighter yellow hue.
- Cleavage in deuterostomes is radial. The first two cleavages are vertical resulting in four cells that each extend from animal to vegetal pole. The third cleavage is horizontal and displaced toward the animal pole. This produces an eight-cell stage with four smaller animal pole cells aligned directly over four large vegetal pole cells. (See Campbell, Figure 43.7a)
- Some animals (most protostomes) exhibit spiral cleavage in which the upper tier of cell sits in the grooves between the cells of the lower tier.
- Cell division forms a solid ball of the cells, the *morula*, and a fluid-filled cavity (the *blastocoel*) forms in the center to create the *blastula*.

Animal hemisphere, Vegetal hemisphere

Radial cleavage

Spiral cleavage

Morula

Blastocoel

Blastula

B. **Gastrulation** (p. 961–963)

What is gastrulation? What changes occur in the developing embryo during the process? How does cell motility influence formation of the archenteron and the primary germ layers?

Gastrulation involves an extensive rearrangement of cells which transforms the blastula, a hollow ball of cells, into a multilayered embryo called the *gastrula*.

- A set of common cellular changes is involved in all animals: changes in cell motility; changes in cell shape; and changes in cellular adhesion to other cells and to molecules of the extracellular matrix.
- Specific details of processes may differ from one animal group to the next.

Gastrulation

Gastrula

Gastrulation in sea urchins begins at the vegetal pole. (See Campbell, Figure 43.8)
- The sea urchin blastula consists of a single cell layer.
- Vegetal pole cells form a flattened plate that buckles inward (= *invagination*).
- Cells near the plate detach and enter the blastocoel as migratory *mesenchyme* cells.
- The invaginated plate undergoes rearrangement to form a deep, narrow pouch, the *archenteron* or primitive gut. The archenteron opens to the surface through the *blastopore*.
- The archenteron extends about two-thirds of the way across the blastocoel at this point and some mesenchyme cells migrate to its tip.
- Some cells at the tip of the archenteron develop *filopodia* (thin cytoplasmic extensions) which extend toward and adhere to animal pole cells.
- Contraction of the filopodia pulls the archenteron completely across the blastocoel.

Invagination

Archenteron, Blastopore

630 Animal Development

☞ The embryo is now a gastrula with the three *primary germ layers* in place.

What are the three primary germ layers? What adult structures are derived from each?

Ectoderm, Mesoderm
Endoderm

The three primary germ layers are arranged from the outside to the inside of the embryo: *ectoderm* is the outermost layer; *mesoderm* is the middle layer (derived from mesenchyme cells); and *endoderm*, the innermost layer which lines the archenteron.

☞ The sea urchin embryo will develop into a larval stage. (See Campbell, Figure 43.9)
☞ The larval structures which develop from the primary germ layers are: the epidermis from the ectoderm; muscles, pigment cells, and skeletal spicules from the mesoderm; and the gut wall from the endoderm.
☞ The blastopore forms the anus and the anterior part of the archenteron fuses with the outer cell layer near the animal pole to form the mouth.

Gastrulation during <u>frog development</u> also results in an embryo with the three primary germ layers and an archenteron that opens through a blastopore. The mechanics of gastrulation are more complicated than in the sea urchin because of the large, yolk-laden cells in the vegetal hemisphere and the presence of more than one cell layer in the blastula wall. (See Campbell, Figure 43.10)

How does gastrulation in the frog differ from that in the sea urchin? What is the dorsal lip and why is it important to embryonic development? How does the process of involution influence embryonic development?

☞ A small crease forms on one side of the blastula due to the invagination of "bottle" cells which remain attached for a while to the surface by long necklike extensions (they take on the shape of a bottle).
☞ The tuck produced in the side of the embryo by this invagination of cells will become the *dorsal lip* (upper edge) of the blastopore.

Dorsal lip

☞ The position of the dorsal lip indicates the dorsal-ventral axis of the animal.
☞ *Involution* then occurs. This is a process in which cells on the surface of the embryo roll over the dorsal lip and move into the embryo's interior. These cells continue to migrate along the dorsal wall of the blastocoel.

Involution

☞ The surface area of the embryo is maintained by the spreading of cells in the multilayer blastopore wall near the animal pole. This is necessary since cells nearer the blastopore continue to involute and disappear from the surface.
☞ As gastrulation continues, more and more lateral cells involute with the dorsal cells resulting in a wider blastopore lip which eventually forms a complete circle.
☞ The circular blastopore lip surrounds a group of large, food-laden cells from the vegetal pole called the *yolk plug*.

Yolk plug

☞ At this point the blastocoel has been almost completely filled by cells and the archenteron has formed.

☞ The primary germ layers have also formed: the ectoderm is comprised of those cells remaining on the surface; the mesoderm is made up of those cells that involuted to the interior; and the endoderm are cells lining the archenteron.

C. **Organogenesis** (p. 963–965)

What is organogenesis? What association exists between the notochord, neural plate, neural tube, and neural crest? What are somites?

<u>Organogenesis</u> = An early period of rapid embryonic development in which the organs take form from the primary germ layers.
- ☞ The dorsal mesoderm above the archenteron forms the *notochord* in chordates.
- ☞ Ectoderm above the rudimentary notochord thickens to form a *neural plate* that sinks below the embryo's surface and rolls itself into a *neural tube*, which will become the brain and spinal cord. (See Campbell, Figure 43.11)
- ☞ The notochord elongates and stretches the embryo lengthwise; it functions as the core around which mesoderm cells that form the vertebrae gather.
- ☞ Blocks of mesodermal cells called *somites* form lateral to the notochord and give rise to the muscles associated with the axial skeleton.
- ☞ Ectoderm also gives rise to epidermis, epidermal glands, inner ear and eye lens.
- ☞ Mesoderm also gives rise to the notochord, coelom lining, muscles, skeleton, gonads, kidneys and most of the circulatory system.
- ☞ Endoderm forms the digestive tract linings, liver, pancreas and lungs.
- ☞ The *neural crest* forms from ectodermal cells which break loose from the border where the neural tube breaks off from the ectoderm; these cells migrate to other parts of the body and form pigment cells in the skin, some bones and muscles of the skull, teeth, adrenal medulla, and parts of the peripheral nervous system.

IV. **THE EMBRYOLOGY OF AMNIOTES** (p. 965–968)

What distinguishes an amniote vertebrate from a nonamniote vertebrate? What adaptive advantage does the presence of an amnion have for terrestrial vertebrates?

All vertebrate embryos require an aqueous environment for development.
- ☞ Fish and amphibians lay their eggs in water.
- ☞ Terrestrial animals live in dry environments and have evolved two solutions to this problem: the shelled egg in birds and reptiles and the uterus in placental mammals.
- ☞ The embryos of reptiles, birds, and mammals develop in a fluid-filled sac, the *amnion*.
- ☞ These three classes of animals are referred to as *amniotes* due to the presence of the amnion around the embryo.
- ☞ Important differences between cleavage and gastrulation occur in the development of birds and mammals, as well as differences between these amniotes and nonamniotes.

A. Avian Development (p. 965–966)

What are the functions of the yolk and egg white found in a bird egg? What is the blastodisc and how does it change during cleavage? What changes occur in the primitive streak during gastrulation?

The larger yellow "yolk" of a bird egg is actually the ovum containing a large food reserve properly called yolk.
- Surrounding this large cell is a protein-rich solution (the egg white) that provides additional nutrients during development.
- After fertilization, cleavage will be restricted to a small disc of yolk-free cytoplasm at the animal pole.
- Cell division partitions the yolk-free cytoplasm into a cap of cells called the *blastodisc* which rests on large undivided yolk mass.

Meroblastic cleavage = Incomplete division of a yolk-rich egg. Yolk is so abundant that cleavage is restricted to a small disc of cytoplasm at the animal pole.
- Blastodisc cells sort into an upper layer (epiblast) and lower layer (hypoblast) forming the blastocoel between them.

Gastrulation begins with the formation of a straight invagination, the *primitive streak*, in the upper cell layer along the antero-posterior axis. (See Campbell, Figure 43.15)
- Some cells in the epiblast migrate through the primitive streak and into the blastocoel to form mesoderm; other invade the hypoplast and contribute to endoderm.
- The ectoderm forms from cells remaining in the epiblast.
- Embryonic disc borders fold downward, coming together below the embryo forming a triple-layered tube attached to the yolk below by a stalk at midbody.

Subsequent organogenesis occurs as in the frog, except that primary germ layers also form four extraembryonic membranes: the *yolk sac*, *amnion*, *chorion* and *allantois*. (See Campbell, Figure 43.15)

B. Mammalian Development (p. 967–968)

What are the major differences between meroblastic cleavage and holoblastic cleavage? What is the embryonic disc which develops in mammals? How does the trophoblast layer contribute to development?

The mammalian egg stores little nutrients and the zygote displays *holoblastic cleavage*.

Holoblastic cleavage = Type of cleavage in which there is complete division of the egg).
- Cleavage planes appear randomly oriented and blastomeres are of equal size.

Cells form tight junctions arranged around a central cavity forming the blastocyst. (See Campbell, Figure 43.16)
- At one end is the *inner cell mass*, that subsequently develops into the embryo and some of its extraembryonic membranes.

- The outer *trophoblast* layer secretes enzymes that enable blastocyst implantation in the uterus and thickens to form the fetal portion of the placenta.
- During implantation, the inner cell mass forms a flat *embryonic disc* similar to that in birds; the embryonic disc will develop into the embryo proper.

What are the four extraembryonic membranes which develop in amniote vertebrates? How do the functions of each differ between birds and mammals?

Four extraembryonic membranes homologous to those in birds and reptiles form in mammals.
- The chorion forms from the trophoblast and surrounds the embryo and all extraembryonic membranes.
- The amnion forms above the embryonic disc and encloses the embryo in a fluid-filled cavity.
- The yolk sac membrane forms below the embryonic disc and forms blood cells that later migrate into the embryo.
- The allantois develops from an outpocketing of the rudimentary gut and is incorporated into the umbilical cord where it forms blood vessels that transport oxygen and nutrients from the placenta to the embryo and waste products from the embryo to the placenta.

Gastrulation occurs by inward movement of mesoderm and endoderm through a primitive streak.

Organogenesis begins with formation of the neural tube, notochord, and somites.
- Rudiments of all major organs have developed from the three primary germ layers by the end of the first trimester in humans. (See Campbell, Table 43.1)

V. MECHANISMS OF DEVELOPMENT (p. 968–979)

Early in the embryonic development of an animal, a sequence of changes takes place that establishes the basic body plan of that animal.
- Developmental biologists have investigated the general mechanisms of how these changes occur through a variety of experiments.

What are the two basic concepts of developmental mechanisms?

Two basic concepts about the mechanism of development have been formulated from the results of these experiments.

1. *The cytoplasmic organization of the unfertilized egg leads to regional differences in the early embryos of many animal species.*
   - Different blastomeres receive different substances (mRNA, proteins, etc.) during cleavage due to the partitioning of the heterogenous cytoplasm of the ovum.
   - Gene expression in, and the developmental fate of, cells in the early embryo are influenced by these local differences in cytoplasmic composition. (See Campbell, Figure 43.17)

Trophoblast

Embryonic disc

2. *Cell-cell interaction compound the influence of location on a cell's developmental fate.*
   - One cell of an embryo may transmit chemical signals to a neighboring cell which will induce cytoplasmic changes affecting gene expression in that cell.
   - Membrane interaction between cells that are in contact may also induce cytoplasmic changes affecting gene expression.

   A. **Polarity of the Embryo** (p. 968–969)

   In most animals except mammals, polarity of the embryo is established at the zygote stage or during early cleavage.

   What causes the polarity to develop in nonmammalian embryos? How do changes in the arrangement of the cytoplasm affect further development during cleavage?

   - Heterozygous distribution of cytoplasmic components of an egg usually polarize the egg along an animal-vegetal axis.
   - In amphibian eggs, *yolk platelets* are concentrated near the vegetal pole and cortical melanin granules near the animal pole.
   - At fertilization the cytoplasm in amphibian eggs becomes rearranged; the cortex of the egg rotates toward the point of sperm entry. (See Campbell, Figure 43.18)
   - This rotation is probably due to cytoskeleton reorganization by the centriole contributed by the sperm cell.
   - A *gray crescent* forms opposite of the sperm entry point due to contraction of the cortex exposing lighter-colored cytoplasm.
   - The gray crescent is bisectionally divided by the first cleavage, and the dorsal lip of the blastopore will later form where the gray crescent was located on the zygote.

   These changes establish the three axes of the embryo before cleavage begins in the zygote.
   - The animal-vegetal axis of the egg becomes the anterior-posterior axis of the embryo.
   - Location of the gray crescent determines both the dorsal-ventral axis and the left-right axis.
   - Experimental manipulation of the egg to prevent cortex rotation that forms the gray crescent, results in abnormal gastrulation with axial structures failing to develop.

   B. **Localized Cytoplasmic Determinants** (p. 970–971)

   How do cytoplasmic determinants influence the developmental fates of cells?

   Totipotent cells retain the potential to form all parts of the animal.
   - If cytoplasmic components are unevenly distributed in the egg, cleavage may localize substances in specific blastomeres. These substances function as *cytoplasmic determinants* that fix developmental fates of different embryonic regions very early.

Totipotency is lost at different times in different species due to uneven distribution of cytoplasmic determinants.
- ☞ The zygotes of many mollusks form a polar lobe that apparently contains cytoplasmic determinants. The polar lobe is restricted to one blastomere at the first cleavage division; if the blastomeres are separated, the cell with the polar lobe develops normally and the cell lacking the polar lobe shows abnormal development. (See Campbell, Figure 43.19a)
- ☞ Cytoplasmic determinants appear to be associated with the gray crescent in amphibian zygotes. A normal first cleavage bisects the gray crescent into the blastomeres; if the cells are separated, both develop normally. Experimental manipulation where the first cleavage misses the gray crescent produces blastomeres, that if separated, will only show normal development from the cell receiving the gray crescent. (See Campbell, Figure 43.19b)
- ☞ Cytoplasmic determinants appear to have a polar distribution in sea urchin eggs. At the eight-cell stage, splitting the embryo so that each four-cell group has two animal pole and two vegetal pole cells will result in development of two normal larvae. A horizontal splitting which separates animal pole cells from vegetal pole cells produces one mildly abnormal larva and one severely abnormal larva. (See Campbell, Figure 43.19c)

C. **Mosaic and Regulative Development** (*p. 971*)

What distinguishes mosaic development from regulative development?

In *mosaic development*, each cell is set to form specific parts of the embryo very early.
- ☞ Mollusk eggs develop polar lobes <u>before</u> the first cleavage, thus cells lose totipotency.

In *regulative development*, cells are totipotent longer.
- ☞ Distinction is relative since at some point all animal cells lose totipotency.
- ☞ Sea urchin eggs lose totipotency at the third cleavage.
- ☞ Mammalian embryo cells remain totipotent much longer than those of most other animals. Totipotency is not lost until cells become arranged into the trophoblast and blastocyst.
- ☞ Longer regulative development in mammals is probably related to a lack of egg polarity and randomness of early cleavage planes.

D. **Fate Maps** (*p. 971*)

What is a fate map and how can one be used?

A *fate map* is used to map the developmental fates of individual cells or cell groups of a zygote.
- ☞ First accomplished by W. Vogt in the 1920s, using vital dyes to color different regions of the amphibian blastula surface and subsequently sectioning the embryo to see where each color turned up.
- ☞ Modern techniques permit marking of individual blastomeres and the construction of detailed fate maps. (See Campbell, Figure 43.20)

Animal Development

E. Morphogenetic Movements (p. 971–974)

How do morphogenetic movements help shape an embryo? What are the relationships of cell extension, contraction and adhesion to this process?

Various morphogenetic movements help shape an embryo.
- Cell extension, contraction and adhesion are involved in these movements.
- Changes in shape usually involve reorganization of the cytoskeleton.
- Amoeboid movement based on extension and contraction is an important factor, especially during gastrulation.
- Morphogenetic movements are partially guided by the extracellular matrix, which contains adhesive substances and fibers that may direct migrating cells. (*Fibronectins* are extracellular glycoproteins that help cells adhere to their substratum as they migrate.)
- *Cell adhesion molecules* (*CAMs*) are substances on a cell surface that contribute to the selective association of certain cells with each other. CAMs vary either in amount or chemical identity from one type of cell to another.

F. Induction (p. 974)

What is induction? Why is this process important to embryonic development?

Induction = The ability of one cell group to influence the development of another.
- The rudimentary notochord induces the dorsal ectoderm of a gastrula to form the neural plate. Transplanting the chordamesoderm, which forms the notochord, to an abnormal sight in the embryo will cause the neural plate to develop in an abnormal location.
- The dorsal tip of the blastopore acts as a primary *organizer*, setting up the interaction between chordamesoderm and the overlying ectoderm.
- A growth factor called *activin* has been discovered that apparently serves as a chemical signal causing cells migrating inward over the dorsal lip to form chordamesoderm.
- Reciprocal inductions between different layers also occur. (See Campbell, Figure 43.23)

G. Differentiation (p. 974–976)

What is cellular differentiation? What are tissue-specific proteins and why are they important?

Cells become specialized during *cellular differentiation* which occurs as tissues and organs form in the embryo.
- First appears as alterations in cellular structure.
- Molecular changes first appear with production of *tissue-specific proteins*, proteins found only in a specific type of cell.

Cells continue to make certain proteins as they become specialized during differentiation.

Margin notes:
- *Fibronectins*
- *Cell adhesion molecules*
- *Induction*
- *Organizer*

- In vertebrates, developing lens cells produce large quantities of crystallins, proteins that will aggregate to form the transparent lens fibers.
- Immature lens cells are induced to begin crystallin production by immature retinal cells.
- Chemical signals from the retinal cells stimulate immature lens cells to activate their crystallin genes which transcribe specific mRNA molecules coding for crystallin.
- Those mRNA molecules accumulate in the immature lens cells' cytoplasm.
- Production of crystallin begins when lens cells mature and the cells flatten and elongate to accommodate the crystallin fibers.

Why do the cells of an organism exhibit different structures and functions? What is a determined cell? What changes occur as a cell becomes determined?

Genomic equivalence refers to the presence of all of an organism's genes in all of its cells.
- The cells of an organism differ in structure and function because they express different portions of a common genome.
- Transplantation experiments where nuclei from differentiated and undifferentiated cells were placed in anucleated eggs have shown that normal development occurs in eggs receive nuclei from undifferentiated cells while abnormal development occurs in 98% of eggs receiving nuclei from differentiated cells. (See Campbell, Figure 43.24)
- These results have led to two conclusions: 1) nuclei do change in some way as they prepare for differentiation; 2) the change is not always irreversible.

Determination is the progressive restriction of developmental potential.

A *determined* cell is committed to follow a particular course of development.   *Determined cell*
- A serial process; developmental options become narrower and narrower as development progresses.
- Involves control of the genome by the cell's cytoplasmic environment.
- Morphogenetic movements play a role by placing cells in differing chemical and physical embryonic neighborhoods.
- Experimentally distinguished by transplanting cells to another part of the embryo to check if their developmental fate can be altered.

H. Pattern Formation (p. 976–979)

Pattern formation = The emergence of a body form with specialized organs   *Pattern formation*
and tissues all in their characteristic places.

What are homeotic genes? How do they influence developmental patterns?

Homeotic genes are one class of regulatory genes that controls the developmental fate of a group of cells.
- Homeotic genes were first identified in fruit flies.

- All contained a specific sequence of 180 amino acids which was termed the *homeobox*.
- Further investigations have found virtually identical homeoboxes in almost all eukaryotic organisms examined.

Homeobox sequences are now believed to play a role in controlling developmental patterns.
- Homeobox sequences are found in a variety of protein-coding genes, most of which are involved in development.
- These sequences are coded into *homeodomains* which bind to specific DNA sequences; regulate development by coordinating transcription of groups of developmental genes.
- High degree of similarity among homeodomain sequences indicate a highly conservative evolutionary history.

How does the location of a cell influence its development? What is a zone of polarized activity? What relationship might morphogens have to pattern formation?

Positional information is received by genes that regulate development as signal's that indicate a cell's location relative to other cells in an embryonic structure.
- Vertebrate limbs all develop from rudiments called limb buds. (See Campbell, Figure 43.26)
- A specific pattern of tissues emerges as the limb develops.
- Every component has a precise location and orientation relative to three axes: the proximal-distal axis (from the limb base to tips of the digits); the anterior-posterior axis (front edge to rear edge of the limb); and the dorsal-ventral axis (top to bottom of the limb).
- For proper development to occur, embryonic cells in the limb bud must receive positional information indicating location along the three axes.

Experiments by Lewis Wolpert and others on development of chick limbs have provided information about positional information.
- A *zone of polarized activity* (ZPA) has been found to be the point of reference for the anterior-posterior axis.
- The ZPA is located where the posterior part of the bud is attached to the body.
- Cells nearest the ZPA give rise to posterior structures, while cells farthest from the ZPA give rise to anterior structures. Experiments involving transplanted cells supports this hypothesis. (See Campbell, Figure 43.27)
- A region of mesoderm located beneath an *apical epidermal ridge* at the wing bud tip is believed to assign embryonic tissue position along the proximal-distal axis. Transplantation experiments reversing the position of this tissue have produced limbs with reversed orientation.

Experiments by Wolpert and others have led to the conclusion that pattern formation requires cells to receive and interpret environmental cues that vary between locations.

- One possible cue could be the presence of *morphogens* in developing tissues.
- A morphogen is a chemical signal that varies in concentration along a gradient, allowing cells to resolve their position along the gradient.
- Gradients of different morphogens along the three orientation axes would provide cell positional information needed to determine its position in a three-dimensional organ.
- Christina Thaller and Gregor Eichele have identified a possible morphogen in animals, retinoic acid.

Other types of experiments suggest positional information functions in overall body form development, not just formation of individual parts.
- These experiments also indicate positional information may be received in installments, not all at one time.

What is a hierarchy of pattern formation? How could cytoplasmic cues influence pattern formation?

Many researchers believe a *hierarchy of pattern formation* exists. These scientists believe installments of positional information throughout development establishes gross form when the embryo is small and then refines the parts as development continues.
- *Drosophila* is being used to study this area of pattern formation.
- The heterogenous distribution of substances in unfertilized eggs determines overall polarity (anterior-posterior and dorsal-ventral axes) of a fruit fly.
- Egg-polarity genes in nurse cells that surround the egg cell transcribe mRNA molecules.
- These mRNA molecules are secreted into and become localized in the cytoplasm of the cell.
- It has been found that nurse cells at one end of the egg secrete mRNA that directs synthesis of the protein *bicoid* in that end.
- The presence of bicoid mRNA establishes that end as the future anterior of the animal.
- The bicoid mRNA is thought to be held in place by the cytoskeleton.
- After fertilization development begins; bicoid mRNA is translated early in development and a bicoid protein gradient provides positional information with the end of highest concentration becoming the head. (See Campbell, Figure 43.28)
- In addition to bicoid, there appears to be a "posterior" morphogen and a "dorsal-ventral" morphogen active in the first stage of pattern formation.
- These morphogens trigger regional differences in segmentation gene expression during the second stage of pattern formation resulting in a finer scale of formation.
- In the third stage of pattern formation, products from the segmentation genes determine which homeotic genes are expressed in each segment.

Morphogens

Retinoic acid

44 NERVOUS SYSTEMS

CHAPTER OUTLINE

The body of an animal possess two coordinating systems which often cooperate and interact to control physiology and behavior: the *endocrine system* and the *nervous system*.
- ☞ Differences in the structural organizations of these two systems result in important differences in their roles in the body.

What structural and functional differences are there between the endocrine and nervous systems with regards to coordinating the body's functions and responses?

The endocrine system relies on *chemical signals* to conduct its coordination functions. (See Campbell, Chapter 41)
- ☞ Consists of individual glands in various locations in the body that release hormones which travel to target cells by way of the bloodstream. Response to a hormone may take seconds or even minutes.
- ☞ Dispersal of hormones through the circulatory system exposes most of the body cells to the hormone although only target cells with specific receptors give a response.
- ☞ Structure of the system limits the type of stimuli to which the system usually responds.

The nervous system sends *electrical signals* to coordinate body functions and responses.
- ☞ Consists of a system of specialized cells (neurons) which branch throughout the body and conduct electrical signals directly to and from specific targets. Signals travel up to 100 m/sec, resulting in a response time measured in milliseconds.
- ☞ Due to branching and direct contact with targets, the nervous can control very fine movements by stimulating one or a few muscles.
- ☞ Structural complexity of the system permits the integration of a broad spectrum of information and stimulation of a very wide range of responses.

What are the two primary parts of the nervous system? What three functions do these components perform?

The nervous system has three overlapping functions:
1. *Sensory input* is the conduction of signals from sensory receptors to integration centers of the nervous system.

2. *Integration* is a process by which information from sensory receptors is interpreted and associated with appropriate responses of the body.
3. *Motor output* is the conduction of signals from the processing center to *effector cells* (muscle cells, gland cells) that actually carry out the body's response to stimuli.

Effector cells

These functions involve both parts of the nervous system:
1. *Central nervous system* (*CNS*): comprised of the brain and spinal cord; responsible for integration of sensory input and associating stimuli with appropriate motor output.

Central nervous system

2. *Peripheral nervous system*: consists of the network of nerves extending into different parts of the body that carry sensory input to the CNS and motor output away from the CNS.

Peripheral nervous system

I. CELLS OF THE NERVOUS SYSTEM (p. 983–986)

The nervous system includes two main types of cells: *neurons* which actually conduct messages; and *supporting cells* which provide structural reinforcement to the system as well as protecting, insulating and assisting the neuron.

A. Neurons (p. 983–985)

What is a neuron? What are the parts of the neuron and what function does each perform? What is a synapse? What relationship exists between neurotransmitters and synapses? What three major classes of neurons are found in the body and how are they distinguished?

Neurons = Cells specialized for transmitting chemical and electrical signals from one location in the body to another.
☞ Has a large cell body containing most of the cytoplasm, the nucleus, and other organelles. (See Campbell, Figure 44.3)

Neurons

Neurons have two types of fiberlike extensions (processes) that increase the distance over which the cells can conduct messages:
1. *Dendrites* convey signals to the cell body; are short, numerous and extensively branched to increase surface area where the cell is most likely to be stimulated.

Dendrites

2. Usually a single *axon* conducts impulses away from the cell body and is wrapped in concentric layers of *Schwann cells* which form an insulating *myelin sheath*.

Axon, Schwann cells

Myelin sheath

Axons extend from the *axon hillock* (where impulses are generated) to many branches called *telodendria*, which are tipped with *synaptic terminals* that release *neurotransmitters*.

Axon hillock

Telodendria, Synaptic terminals

Neurotransmitters = Chemicals that cross the *synapse* (gap between a synaptic terminal and dendrites of another neuron or an *effector* cell) to relay the impulse.

Neurotransmitters, Synapse

The cell bodies of most neurons are located in the CNS, although certain types of neurons have their cell bodies located in *ganglia* outside of the CNS.

642 Nervous Systems

There are three major classes of neurons: (See Campbell, Figure 44.5)

1. *Sensory neurons* convey information about the external and internal environments from sensory receptors to the central nervous system.
2. *Motor neurons* convey impulses from the CNS to effector cells.
3. *Interneurons* integrate sensory input and motor output; located within the CNS.

B. **Supporting Cells** (p. 985–986)

What types of supporting cells are associated with the nervous system? What function does each type of supporting cell perform and why are they important?

Supporting cells structurally reinforce, protect, insulate and generally assist neurons.
- Do not conduct impulses.
- Outnumber neurons 10- to 50-fold.

Glial cells = Supporting cells of the CNS; several types are present.
- *Astrocytes* encircle capillaries in the brain to help form the blood-brain barrier and also help control the ionic environment around neurons.
- *Oligodendrocytes* form myelin sheaths that insulate nerve processes.

In the peripheral nervous system, *Schwann cells* form the myelin sheath to provide electrical insulation and to increase the rate of nerve impulses.

Myelination of neurons in a developing nervous system occurs when Schwann cells or oligodendrocytes grow around an axon in such a manner that their plasma membranes form concentric layers. (See Campbell, Figure 44.6)
- Provides electrical insulation since membranes are mostly lipid which is a poor current conductor.
- Presence of the myelin sheath increases the speed of nerve impulse propagation.
- In multiple sclerosis, progressive loss of coordination occurs due to disruption of nerve impulse transmissions resulting from deterioration of myelin sheaths.

II. **TRANSMISSION OF ELECTRICAL SIGNALS ALONG A NEURON** (p. 986–993)

Signal transmission depends on electrical currents in the form of ionic fluxes across neuron plasma membranes.

A. **The Origin of Electrical Membrane Potential** (p. 986–989)

What is a membrane potential? A resting potential? An equilibrium potential? How is the selective permeability of the neuron's plasma membrane involved with producing and maintaining the resting potential? What ions and processes are involved?

All cells have a charge difference across their plasma membranes (thus are polarized); cytoplasm inside is negatively charged compared to the extracellular fluid.
- ☞ The difference in charge between the cytoplasm and extracellular fluids is called the *membrane potential* and is due to a differential distribution of ions.

Membrane potential

Resting potential = The membrane potential of a nontransmitting neuron.
- ☞ Is about −70 mV.

Resting potential

Membrane potential develops due to the differences in ionic composition of the intracellular and extracellular fluids and because of the selective permeability of plasma membranes.
- ☞ Intracellular and extracellular fluids contain a variety of ions (electrically charged substances) along with the other dissolved materials.
- ☞ In mammalian cells, the principal cation (positive charge) outside of the cell is sodium (Na^+) while a lower concentration of potassium (K^+) is also present; intracellular fluids show the reverse, K^+ is in much higher concentration than Na^+.
- ☞ Anions (negative charges) also show differential distributions: the primary anion in extracellular fluids is chloride (Cl^-); intracellular fluids have a low concentration of Cl^-, but the principal anion is a class of negatively charged substances (proteins, negative amino acids, etc.) that are impermeant and act as a single negative charge (A^-).

The membrane potential (or resting potential) of a neuron is retained by the cell which exerts energy to insure the concentration gradients of the ions are maintained. (See Campbell, Figure 44.7)
- ☞ K^+ diffuses out of the cell down its concentration gradient; this occurs because of the great difference in K^+ concentrations inside and outside of the cell and the fact that the membrane has a high permeability to potassium.
- ☞ K^+ diffusion out of the cell transfers positive charges from inside of the cell to outside.
- ☞ The cell's interior becomes progressively more negative as K^+ leaves because the molecules of the impermeant anion pool (A^-) are too large to cross the membrane.
- ☞ As the electrical gradient increases, the negatively charged interior attracts K^+ back into the cell.
- ☞ If K^+ was the only ion to cross the membrane, an *equilibrium potential* for potassium ions (= −85 mV) would be reached; this is the potential at which no movement of K^+ would occur since the electrical gradient attraction of K^+ would balance the K^+ loss due to the concentration gradient.
- ☞ However, K^+ is not the only ion to cross the membrane; although the membrane is less permeable to Na^+ than K^+, some Na^+ enters the cell due to its concentration gradient <u>and</u> the electrical gradient.
- ☞ The Na^+ trickle into the cell transfers positive charge to the inside resulting in a slightly more positive charge (−70 mV) than if the membrane were permeable only to K^+ (−85 mV).

Equilibrium potential

- If left unchecked, the Na⁺ trickle into the cell would cause a progressive increase in Na⁺ concentration and a decrease in K⁺ concentration (the electrical gradient would be weakened by the positive sodium ions and the potassium ions would diffuse down its concentration gradient).
- This shift in ionic gradients is prevented by sodium-potassium pumps (special plasma-membrane proteins) which use energy from ATP to pump sodium back out of the cell (against concentration and electrical gradients) while moving potassium into the cell, restoring its concentration gradient.

B. **The Action Potential** (p. 989–991)

What is an action potential? What is the relationship of gated ion channels to the action potential? How are hyperpolarization and depolarization related to an action potential?

All cells in the body exhibit the properties of membrane potential described above. However, only neurons can <u>change</u> their membrane potentials in response to stimuli.

The presence of *gated ion channels* in neurons permits these cells to change the plasma membrane's permeability and alter the membrane potential in response to stimuli received by the cell.
- Sensory neurons are stimulated by receptors that are triggered by environmental stimuli.
- Interneurons normally receive stimuli produced by activation of other neurons that make inputs to the cells.

The effect on the neuron depends on the type of gated ion channel the stimulus opens.
- *Hyperpolarization* of a neuron occurs when a stimulus opens a potassium channel and K⁺ effluxes from the cell; this increases the electrical gradient due to the interior of the cell becoming more negative.
- *Depolarization* of a neuron occurs when a stimulus opens a sodium channel and Na⁺ influxes into the cell; this reduces the electrical gradient and membrane potential because the inside of the cell becomes more positive.

How does the strength of a stimulus influence the membrane potential? What is a threshold? What effect does hyperpolarization have on the threshold potential?

Voltage changes caused by stimulation are called *graded potentials* because the magnitude of the change depends on the strength of the stimulus. (See Campbell, Figure 44.8)
- Each excitable cell has a *threshold* to which depolarizing stimuli are graded. This threshold potential is usually slightly more positive (−50 to −55 mV) than the resting potential.
- If depolarization reaches the threshold, the cell responds differently by triggering an *action potential*.

- ☞ Hyperpolarizing stimuli do not produce action potentials since they cause the potential to become more negative; actually reduces the probability an action potential will occur by making it more difficult for depolarizing stimuli to reach the threshold.

An *action potential* is the rapid change in the membrane potential of an excitable cell, caused by stimulus-triggered selective opening and closing of voltage-gated ion channels.
- ☞ Voltage-gated ion channels open and close in response to changes in membrane potential.
- ☞ Voltage-gated sodium channels have two gates: the activation gate opens rapidly at depolarization, the inactivation gate closes slowly at depolarization.
- ☞ The voltage-gated potassium channel has one gate that opens slowly in response to depolarization.

What are the four phases of an action potential? How do changes in the sodium and potassium gates influence each phase? What is a refractory period and why is it important in the functioning of a neuron?

An action potential has four phases. (See Campbell, Figure 44.9a)
- ☞ *Resting state*, no channels are open.
- ☞ *Large depolarizing phase* during which the membrane briefly reverses polarity (cell interior becomes positive to the exterior) due to opening of Na^+ activation gates allowing influx of Na^+ while potassium gates remain closed.
- ☞ The *steep repolaring phase* follows quickly and returns the membrane potential to its resting level; inactivation gates close the sodium channels and activation gates open potassium channels.
- ☞ The *undershoot phase* is a time when the membrane potential is temporarily more negative than the resting state; sodium channels remain closed but potassium channels remain open since the inactivation gates have not had time to respond to repolarization of the membrane.

A *refractory period* occurs during the undershoot phase; during this period, the neuron is insensitive to depolarizing stimuli. The refractory period limits the maximum rate at which action potentials can be stimulated in a neuron.

Refractory period

Action potentials are all-or-none events and their amplitudes are not affected by stimulus intensity. The nervous system distinguishes between strong and weak stimuli based on the frequency of action potentials generated.

All-or-none events

- ☞ Strong stimuli produce action potentials more rapidly than weak stimuli.
- ☞ Maximum frequency is limited by the refractory period of the neuron.

C. **Propagation of the Action Potential** (p. 991–992)

How is an action potential propagated along a neuron? In which direction does it travel?

A neuron is stimulated at its dendrites or cell body and the action potential travels along the axon to the other end of the neuron.

- Action potentials in the axon are usually generated at the axon hillock.
- Strong depolarization in one area results in depolarization above the threshold in neighboring areas.
- The action potential does not travel down the axon, but is regenerated at each position along the membrane.

The signal travels in a perpendicular direction along the axon to regeneration of the action potential. (See Campbell, Figure 44.10)
- Na^+ influx in the area of the action potential results in depolarization of the membrane just ahead of the impulse, surpassing the threshold.
- The voltage-sensitive channels in the new location will go through the same sequence previously described regenerating the action potential.
- Subsequent portions of the axons are depolarized in the same manner.
- The action potential moves in only one direction (down the axon) since each action potential is followed by a refractory period when sodium channel inactivation gates are closed and no action potential can be generated.

D. Speed of Propagation of the Action Potential (p. 992–993)

What factors influence the speed at which an action potential travels? How does saltatory conduction result in faster transmission of a nerve impulse?

Factors affecting the speed of action potential propagation:
1. The larger the diameter of the axon, the faster the rate of transmission since resistance to the flow of electrical current is inversely proportional to the cross-sectional area of the "wire" conducting the current.
2. Saltatory conduction.

Saltatory conduction = The action potential "jumps" from one *node of Ranvier* to the next, skipping the myelinated regions of membrane. (See Campbell, Figure 44.11)
- Nodes of Ranvier are gaps in the myelin sheath between successive glial cells.
- Voltage-sensitive ion channels are concentrated in node regions of the axon.
- Extracellular fluid only contacts the axon membranes at the nodes; restricts the area for ion exchange to these regions.
- Results in faster transmission of the nerve impulse.

III. THE SYNAPSE: TRANSMISSION BETWEEN CELLS (p. 993–999)

What is a synapse? What are the differences between electrical synapses and chemical synapses?

Synapse = Tiny space between neurons that control communication between neurons.

- ☞ Also found between sensory receptors and sensory neurons, and between motor neurons and muscle cells.
- ☞ *Presynaptic cell* is the transmitting cell; *postsynaptic cell* is the receiving cell.
- ☞ Two types of synapses exist: electrical and chemical.

A. Electrical Synapses (p. 993–994)

Electrical synapses allow action potentials to spread directly form pre- to postsynaptic cells via gap junctions (intercellular channels).
- ☞ Much less common than chemical synapses.
- ☞ Examples are the giant neuron processes of crustaceans.

B. Chemical Synapses (p. 994–995)

Where are synaptic vesicles located? How does calcium ion concentration influence the activity of synaptic vesicles? Why can impulses only travel in one direction across a chemical synapse?

At a *chemical synapse* a synaptic cleft separates the pre- and postsynaptic cells so they are not electrically coupled. (See Campbell, Figure 44.12)
- ☞ Within the cytoplasm of the synaptic terminal of a presynaptic cell are numerous *synaptic vesicles* containing thousands of neurotransmitter molecules.
- ☞ An action potential arriving at the synaptic terminal depolarizes the presynaptic membrane causing Ca^{2+} to rush through voltage-sensitive channels.
- ☞ The sudden rise in Ca^{2+} concentration stimulates synaptic vesicles to fuse with the presynaptic membrane and release neurotransmitter into the synaptic cleft by exocytosis.
- ☞ The neurotransmitter diffuses to the postsynaptic membrane where it binds to specific receptors, causing ion gates to open.
- ☞ Depending on the type of receptors and the ion gates they control, the neurotransmitter may either excite the membrane by depolarization or inhibit the postsynaptic cell by hyperpolarization.
- ☞ The neurotransmitter molecules are quickly degraded by enzymes and the components recycled to the presynaptic cell.

Chemical synapses allow transmission of nerve impulses in only one direction.
- ☞ Synaptic vesicles and their neurotransmitter molecules are found only in the synaptic terminals at the tip of axons.
- ☞ Receptors for neurotransmitters are located only on postsynaptic membranes.

C. Summation: Nervous Integration at the Cellular Level (p. 995–997)

One neuron may receive information from numerous adjacent neurons by way of its synapses. Some synapses are excitatory, others are inhibitory.

What is the difference between an excitatory postsynaptic potential and an inhibitory postsynaptic potential? What roles do sodium and potassium ions play in these potentials?

648 Nervous Systems

Excitatory postsynaptic potentials

Inhibitory postsynaptic potentials

Summation

Temporal summation

Spatial summation

Excitatory postsynaptic potentials (*EPSP*) occur when excitatory synapses release a neurotransmitter that opens gated channels allowing Na^+ to enter the cell and K^+ to leave (depolarization).

Inhibitory postsynaptic potentials (*IPSP*) occur when neurotransmitters released from inhibitory synapses bind to receptors which open ion gates that make the membrane more permeable to K^+ (which leaves the cell) and/or to Cl^- (which enters the cell) causing hyperpolarization.

EPSPs and IPSPs are graded potentials; they vary in magnitude with the number of neurotransmitter molecules binding to postsynaptic receptors.
- ☞ Change in voltage lasts only a few milliseconds since neurotransmitters are inactivated by enzymes soon after release.
- ☞ The electrical impact on the postsynaptic cell also decreases with distance from the synapse.

How can temporal summation and spatial summation affect an excitatory postsynaptic potential? An inhibitory postsynaptic potential?

A single EPSP is rarely strong enough to trigger an action potential, although an additive effect (= summation) from several terminals or repeated firing of terminals can change membrane potential.
- ☞ *Temporal summation* is when chemical transmissions from one or more synaptic terminals occur so close in time that each affects the membrane while it is partially depolarized and before it has returned to resting potential.
- ☞ *Spatial summation* is when several different synaptic terminals, usually from different presynaptic neurons, stimulate the postsynaptic cell at the same time and have an additive effect on membrane potential.
- ☞ EPSPs and IPSPs can summate, each countering the effects of the other.

At any instant, the axon hillock's membrane potential is an average of the summated depolarization due to all EPSPs and the summated hyperpolarization due to all IPSPs.
- ☞ An action potential is generated when the EPSP summation exceeds the IPSP summation to the point where the membrane potential of the axon hillock reaches threshold voltage.

D. Neurotransmitters and Receptors (*p. 997–999*)

Ten different molecules are known to be neurotransmitters and many others are suspected to function as such.

What criteria must be met for a molecule to be considered a neurotransmitter? What molecules are known to be neurotransmitters and where are they commonly found? What changes can occur in a postsynaptic cell due to the presence of a neurotransmitter molecules?

Criteria for recognition as neurotransmitter are:
- ☞ Must be present in and discharged from synaptic vesicles in the presynaptic cell when stimulated and affect the postsynaptic cell's membrane potential.

- Must cause an IPSP or EPSP.
- Must be rapidly removed from the synapse by an enzyme or uptake by a cell permitting the postsynaptic membrane to return to resting potential.

Types of neurotransmitters:

1. *Acetylcholine* may be excitatory or inhibitory depending on the receptor; functions in the vertebrate neuromuscular junction (between a motor neuron and muscle cell) and in the central nervous system.
    - The most common neurotransmitter in both vertebrates and invertebrates.

2. *Biogenic amines* are derived from amino acids.
    - Epinephrine, norepinephrine, dopamine are produced from tyrosine; serotonin is synthesized from tryptophan.
    - Commonly function in the central nervous system; imbalances in dopamine and serotonin are associated with mental illness. Norepinephrine also functions in the peripheral nervous system.

3. Amino acids *glycine*, *glutamate*, and *gamma aminobutyric acid* (*GABA*) function as neurotransmitters in the central nervous system.
    - GABA is the most abundant inhibitory transmitter in the brain.

4. *Neuropeptides* are short chains of amino acids.
    - Endorphins (or enkephalins) function as natural analgesics in the brain.

Neurotransmitters may affect the postsynaptic cell by altering the permeability of the postsynaptic membrane to specific ions (e.g. acetylcholine, amino acid transmitters) or by affecting postsynaptic cell metabolism (e.g. biogenic amines, neuropeptides) through production of cAMP.
- The action produced by a neurotransmitter depends on how the specific receptor on the postsynaptic membrane behaves when bound by the transmitter.

E. **Neural Circuits and Clusters** (*p. 999*)

How can you distinguish between convergent circuits, divergent circuits and reverberating circuits? What is a ganglion?

Neurons are arranged in groups referred to as circuits: (See Campbell, Figure 44.15)

Convergent circuits = Information from several presynaptic neurons come together at a single postsynaptic neuron. Permits integration of information from several sources.

Divergent circuits = Information from a single neuron spreads out to several postsynaptic neurons. Permits transmission of information from a single source to several parts of the brain.

Reverberating circuits = Circular circuits in which the signal returns to its source. Believed to play a role in memory storage.

Ganglia

Nucleus

Nerve cell bodies are arranged into functional *ganglia*.
- A *nucleus* is a ganglion within the brain.
- Allows coordination of activities by only part of the nervous system.

IV. INVERTEBRATE NERVOUS SYSTEMS (p. 999–1001)

There is great diversity in invertebrate nervous system organization. (See Campbell, Figure 44.16)

What is a nerve net? In what groups of organisms are nerve nets found? How do the nervous systems of flatworms and annelids differ from a nerve net? What is cephalization? How does the degree of cephalization correlate to the phylogeny, habitat and natural history of an organism?

Nerve net

The *Hydra*, a cnidarian, has a *nerve net*. This is a loosely organized system of nerves with no central control.
- Impulses are conducted in both directions causing movements of the entire body.
- Some Cnidaria, Ctenophora and echinoderms have modified nerve nets with rudimentary centralization.

Cephalization

Cephalization = Evolutionary trend for concentration of sensory and feeding organs on the anterior end of a moving animal; gave rise to the first brains.
- Different degrees of cephalization are found in the bilaterally symmetrical animals.

Flatworms have a simple "brain" containing many large interneurons that coordinate most nervous functions.
- Two or more nerve trunks travel posteriorly in a ladderlike system with transverse nerves connecting the main trunks.

Annelids and arthropods have a well-defined ventral nerve cord and a prominent brain.
- Often contain ganglia in each body segment to coordinate actions of that segment.

Cephalopods have the most sophisticated invertebrate nervous system containing a large brain and giant axons.

Nervous system complexity often correlates with phylogeny, habitat and natural history. For example, sessile animals such as clams show little or no cephalization.

V. THE VERTEBRATE NERVOUS SYSTEM (p. 1001–1011)

The complexity of the vertebrate nervous system results in it being divided into two components based on function for discussion: the *peripheral nervous system* and the *central nervous system*.

A. **The Peripheral Nervous System** (*p. 1002*)

What are the two components of the peripheral nervous system? What type of information is transmitted by each component?

The *peripheral nervous system* consists of: a *sensory (afferent) nervous system* which brings information from sensory receptors to the CNS; and the *motor (efferent) nervous system* which carries signals from the CNS to effector cells.

The peripheral nervous system of humans consists of 12 pairs of cranial nerves and 31 pairs of spinal nerves.
- ☞ Cranial nerves originate from the brain and innervate organs of the head and upper body; most contain both sensory and motor neurons, although some are sensory only (e.g. optic nerve).
- ☞ Spinal nerves innervate the entire body and contain both sensory and motor neurons.

What are the two basic functions of a nervous system? How is the sensory nervous system involved with these functions?

The two basic functions of a nervous system are: to control responses to external environment and to maintain homeostasis by coordinating internal organ functions.
- ☞ The sensory nervous system contributes to both functions by carrying stimuli from the external environment and monitoring the status of the internal environment.
- ☞ The motor nervous system has two separate divisions associated with these functions.

What two divisions comprise the motor nervous system? What is a reflex? Which division of the motor nervous system includes reflexes? Why are the two components of the autonomic nervous system considered antagonistic? How is the motor nervous system involved with the basic functions of a nervous system?

The motor nervous system has two divisions:

1. The *somatic nervous system*'s neurons carry signals to skeletal muscles in response to external stimuli. Includes *reflexes* (automatic responses to stimuli) and is often considered "voluntary" since it is subject to conscious control.

2. The *autonomic nervous system* controls primarily "involuntary," automatic, visceral functions of smooth and cardiac muscles and organs of the gastrointestinal, excretory, cardiovascular and endocrine systems.
   - ☞ Divided into a *parasympathetic* division that enhances activities that gain and conserve energy, and a usually antagonistic *sympathetic* division that increases energy expenditures.

Peripheral nervous system
Sensory nervous system
Motor nervous system

Somatic nervous system

Autonomic nervous system

Parasympathetic

Sympathetic

B. The Central Nervous System (CNS) (p. 1002–1008)

What components comprise the central nervous system? Why is the central nervous system considered a "bridge" between the sensory and motor functions of the peripheral nervous system? What are meninges and cerebrospinal fluid? What are their functions?

The CNS bridges the sensory and motor functions of the peripheral nervous system.
- ☞ Consists of the *spinal cord*, which is located inside the vertebral column and <u>receives</u> information from skin and muscles and <u>sends</u> out motor commands for movement; and the *brain*, which carries out complex integration for homeostasis, perception, movement, intellect and emotions.
- ☞ Covered with *meninges*, three protective layers of connective tissue.
- ☞ In the brain, white (*myelinated*) matter is in the inner and gray matter is in the outer regions. This orientation is reversed in the spinal cord.
- ☞ *Cerebrospinal fluid* fills the ventricles in the brain and the central canal of the spinal cord; it functions in circulation of hormones, nutrients and white blood cells and in absorption of shock.

1. The Spinal Cord (p. 1004–1005)

What types of responses are integrated by the spinal cord? What involvement does the spinal cord have with more complex functions?

The spinal cord integrates simple responses to certain stimuli (reflexes) and carries information to and from the brain.
- ☞ The patellar (knee-jerk) reflex is one of the simplest and involves only two neurons. A stretch receptor in the quadriceps muscle is stimulated stretching of the patellar tendon; a sensory neuron is activated and carries the information to the spinal cord where it synapses with a motor neuron; if an action potential is generated in the motor neuron, it travels back to the quadriceps which contracts and causes the forward knee jerk. (See Campbell, Figure 44.19)
- ☞ Larger-scale, more complex responses result when branches of a reflex pathway carry signals to other parts of the spinal cord or to the brain.

2. Evolution of the Vertebrate Brain (p. 1005–1006)

What evolutionary trend is evident in a comparison of the brains of different vertebrates? How has this trend changed the basic structure of the rhombencephalon, mesencephalon and prosencephalon?

The vertebrate brain has shown an evolutionary trend toward greater complexity which has resulted in more complex behavioral patterns.

All vertebrates posses a *rhombencephalon* (hindbrain), *mesencephalon* (midbrain) and *prosencephalon* (forebrain). (See Campbell, Figure 44.20)
- ☞ More complex brains have further subdivisions.

Trends in the evolution of the vertebrate brain are: (See Campbell, Figure 44.21)
- ☞ Relative brain size increases in certain evolutionary lineages.
- ☞ Increased compartmentalization of function with certain areas of the brain assuming specific responsibilities.
- ☞ Increasing complexity and sophistication of the forebrain; increased complexity of behaviors parallels an increase in growth of the cerebrum.

3. **The Human Brain** (p. 1006–1008)

What comprises the hindbrain of humans and what are the functions of each part? Of the human midbrain? Of the human forebrain?

The human brain weights about 1.35 kg and is one of the largest organs in the body.
- ☞ It develops at the anterior end of the spinal cord from three primary bulges that later differentiate into distinct structures with specific functions.

The human hindbrain consists of three parts: (See Campbell, Figure 44.22)
- ☞ The *medulla oblongata* and *pons* control visceral functions like breathing, heart and blood vessel activity, swallowing, vomiting and digestion; also coordinates large-scale body movements like walking.
- ☞ The *cerebellum* functions in balance and coordination of movement.

The human midbrain together with the hindbrain forms the *brainstem*.
- ☞ The *superior* and *inferior colliculi* are areas of the midbrain that function in the visual and auditory systems.
- ☞ The *reticular formation* regulates states of arousal.

The forebrain is divided into the lower *diencephalon*, containing the thalamus and hypothalamus, and the upper *telencephalon*, containing the cerebrum.
- ☞ The *thalamus* sorts incoming information from all the senses and sends it to appropriate higher brain centers for interpretation and integration.
- ☞ The *hypothalamus* regulates homeostasis by secreting posterior pituitary hormones and releasing hormones.
- ☞ The *cerebral cortex* is divided into 5 highly folded lobes. It is bilaterally symmetric, the two hemispheres connected by a thick fibrous band called the *corpus callosum*. It functions in vision, hearing, speech, reading, olfaction and has sensory and motor functions. (See Campbell, Figure 44.23)
- ☞ The *basal ganglia* are important centers for motor coordination and are located below the cerebral cortex.

C. **Integration and Higher Brain Functions** (p. 1008–1011)

What is the reticular formation? What is its structure and with what actions is it associated?

Margin notes:
- *Hindbrain*
- *Medulla oblongata, pons*
- *Cerebellum*
- *Midbrain, Brainstem*
- *Superior colliculi, Inferior colliculi*
- *Reticular formation*
- *Forebrain, Diencephalon*
- *Telencephalon*
- *Thalamus*
- *Hypothalamus*
- *Cerebral cortex*
- *Corpus callosum*
- *Basal ganglia*

Sleep, *Arousal*

Alpha waves

Beta waves, *Delta waves*

REM sleep

Sleep and arousal are controlled by several centers in the cerebrum and brainstem, the most important of which is the reticular formation.
- Different states of arousal produce different brainwaves: slow, synchronous alpha waves are produced in the relaxed closed-eye state; faster beta waves are produced when eyes are open; slow, highly synchronized delta waves are produced during sleep (although during REM sleep desynchronized waves result).
- The reticular formation is a group of 90 separate brain nuclei (brain ganglia) which extends from the medulla to the thalamus; almost all neuron processes which reach the cerebral cortex pass through the reticular formation.
- The reticular formation serves as a filter that selects what sensory information will reach the cortex.
- Specific centers also regulate sleep and wakefulness.

What is the cerebral cortex? How does it interact with the limbic system to produce emotional and behavioral responses? How can lateralization be described? What functions are associated with the left and right hemispheres of the cerebral cortex? What function does the corpus callosum perform with regards to lateralization?

Limbic system

Emotions depend on interactions between the cerebral cortex and the *limbic system* (a group of giant nuclei in the lower part of the forebrain).
- The limbic system appears to select emotional and behavioral responses to certain stimuli.

Lateralization

Lateralization (right brain/left brain) refers to the association areas of the cerebral cortex which are not bilaterally symmetric; each side of the brain controls different functions.
- The left hemisphere controls speech, language, and calculation.
- The right hemisphere controls artistic ability and spatial perception.
- The corpus callosum functions in the transfer of information between the left and right hemispheres.
- Severing the corpus callosum of a human will not alter perceptions but will dissociate sensory input from spoken response.

Where are Wernicke's area and Broca's area located in the brain? What functions are associated with these areas and what is their relationship?

Language and *speech* are controlled by two areas on the left hemisphere of the cerebral cortex.

Wernicke's area

Broca's area

- *Wernicke's area* stores information required for speech content (arrangement of learned words in grammatical order).
- *Broca's area* contains necessary information for speech production.
- Information for speech delivery is formulated in the Wernicke's area and verbalization is coordinated by the Broca's area which programs the motor cortex to move the tongue, lips, and speech muscles.

Aphasia

- Damage to either area results in some form of aphasia, the inability to speak coherently.

What is memory? What distinguishes short-term memory from long-term memory? What distinguishes fact memory from skill memory?

Memory is the ability to store and retrieve information related to previous experiences.
- Memory occurs in two stages: short-term memory and long-term memory.
- *Short-term memory* reflects immediate sensory perceptions of an object or idea and occurs before the image is stored.
- *Long-term memory* is stored information that can be recalled at a later time.
- Transfer of information from short-term to long-term memory is enhanced by rehearsal, favorable emotional state, and association of new information with previously learned and stored information.

Fact memory differs from skill memory.
- Fact memory involves conscious and specific retrieval of data from long-term memory.
- Skill memory usually involves motor activities learned by repetition which are recalled without consciously remembering specific details.

Fact memory involves a pathway in which sensory information is transmitted from the cerebral cortex to the *hippocampus* and *amygdala* which are two parts of the limbic system.
- The hippocampus and amygdala relay the impulses to other areas of the forebrain.
- The memory circuit is completed when the basal forebrain sends impulses back to the region of the cortex where the sensory perception first occurred.

What cellular changes are believed to be involved in memory and learning?

What cellular changes are actually involved in memory or learning is being investigated. A promising hypothesis is that structural changes in dendrites have a function in learning.
- This idea postulates that neural input causes a postsynaptic cell to take up calcium.
- The calcium influx then activates enzymes that alter the cytoskeleton and changes the shape of the dendrite.
- The shape change of the dendrite is such that future transmissions across the synapse are enhanced.

Studies have shown that simple learning increases the amount of calcium that enters the synaptic terminals of sensory neurons resulting in the release of more neurotransmitter at each action potential.

Short-term memory
Long-term memory

Hippocampus, Amygdala

45 SENSORY AND MOTOR MECHANISMS

CHAPTER OUTLINE

I. SENSORY RECEPTORS (p. 1015–1019)

An animal's interaction with its environment depends on the processing of **sensory** information and the generation of motor output.
- Sensory receptors receive information from the environment.
- Motor effectors carry out the movement in response to the sensory information.

A. Sensation and Perception (p. 1015–1016)

What is the difference between a sensation and a perception?

<u>Sensations</u> = Action potentials travelling along sensory neurons that are interpreted by the brain as perceptions.
- What one perceives depends on what part of the brain receives the impulse.

B. General Function of Sensory Receptors (p. 1016–1017)

What is a sensory receptor? What is the relationship between reception and transduction? What is amplification?

<u>Sensory receptors</u> = Structures that transmit information about changes in an animal's internal and external environment.
- Are usually modified neurons occurring singly or within groups in sensory organs.
- Specialized to respond to specific stimuli and convert stimuli energy into electrochemical energy of action potentials.
- All receptor cells have the same five functions: reception, transduction, amplification, transmission, and integration.

<u>Reception</u> = Ability of a cell to absorb stimuli energy.
- Each kind of receptor has a specific region which absorbs a particular type of energy.

<u>Transduction</u> = The conversion of stimulus energy into electrochemical activity of nerve impulses (action potentials).
- Stimulus energy may change membrane permeability of the receptor cell, open or close ion channel gates, or increase ion flow by stretching the receptor cell membrane.

Amplification of stimulus energy which is too weak to be carried into the nervous system often occurs.
- May take place in accessory structures or be a part of the transduction process.

What is receptor potential? How does summation of graded potentials result in integration of signals?

Initial responses to stimuli during transmission are graded changes in membrane potential (*receptor potential*); the stimulus modulates action potential *frequency*.

Receptor signals are *integrated* through summation of graded potentials.
- *Sensory adaptation* is a decrease in sensitivity during continued stimulation; a type of integration which results in selective information being sent to the CNS.
- The threshold for firing in receptor cells varies with conditions resulting in a change in receptor sensitivity.
- Sensory information integration occurs at all levels in the nervous system.

C. **Types of Receptors** (*p. 1017–1019*)

What distinguishes exteroreceptors from interoreceptors? What are the five types of receptors and to what type of energy does each respond? What is one example of each type of receptor?

Exteroreceptors monitor the external environment while *interoreceptors* monitor changes in the body's internal environment. All receptors can be grouped into 5 types depending on the type of energy to which they respond:

1. *Mechanoreceptors* are stimulated by their physical deformation caused by pressure, stretch, motion, sound, etc. (See Campbell, Figure 45.2)
   - Bending of the plasma membrane increases its permeability to Na^+ and K^+ resulting in a receptor potential.
   - In human skin, *Pacinian corpuscles* deep in the skin respond to strong pressure, while *Meissner's corpuscles* and *Merkel's discs*, closer to the surface, detect light touch.
   - Muscle spindles are stretch receptors (a type of interoreceptor) that monitor the length of skeletal muscles, as in the reflex arc.
   - *Hair cells* detect motion.

2. *Chemoreceptors* include general receptors that sense total solute concentration (e.g. osmoregulators of the mammalian brain), receptors that respond to individual molecules, and those that respond to categories of related chemicals (e.g. *gustatory* and *olfactory* receptors).

3. *Electromagnetic receptors* respond to electromagnetic radiation such as light (*photoreceptors*) and magnetic fields (*magnetoreceptors*).

4. *Thermoreceptors* respond to heat or cold and help regulate body temperature.
   - In human skin, *Ruffini's end organs* may be heat receptors, and *end-bulbs of Krause* may be cold receptors.
   - The interothermoreceptors in the hypothalamus function as the primary temperature control of the mammalian body.

5. *Nociceptors* = A class of naked dendrites that function as *pain receptors*.

Margin notes:
Transmission
Receptor potential
Integrated, Sensory adaptation
Exteroreceptors, Interoreceptors
Mechanoreceptors
Pacinian corpuscles
Meissner's corpuscles, Merkel's discs
Muscle spindles
Hair cells
Chemoreceptors
Gustatory receptors, Olfactory receptors
Electromagnetic receptors
Photoreceptors, Magnetoreceptors
Thermoreceptors
Ruffini's end organs
End-bulbs of Krause
Nociceptors
Pain receptors

658 *Sensory and Motor Mechanisms*

- Different groups respond to excess heat, pressure, or specific chemicals released from damaged or inflamed tissue.
- Prostaglandins increase pain by lowering receptor thresholds; aspirin reduces pain by inhibiting prostaglandin synthesis.

II. VISION (p. 1019–1026)

A. Light Receptors and Vision of Invertebrates (p. 1019–1020)

How does the structure differ among the eye cup of planarians, the compound eye of insects, and the single-lens eye of other invertebrates? What type of "vision" does each impart?

Eye cup

The *eye cup* of planarians is a simple light receptor that responds to light intensity and direction without forming an image. (See Campbell, Figure 45.5)

- An opening on one side of the cup permits light to enter; the opening to one cup faces left and slightly forward, the other cup opens right and slightly forward.
- Light enters the opening and stimulates photoreceptors that contain light-absorbing pigments.
- Planaria move away from light sources to avoid predators.
- The proper direction is determined by the brain, which compares the rate of nerve impulses coming from the two cups; the animal turns until the impulses from each cup are equal and minimal.

Two types of image-forming eyes evolved in invertebrates:

Compound eye, Ommatidia

1. A *compound eye* contains thousands of light detectors called *ommatidia* each with its own cornea and lens. (See Campbell, Figure 45.6)
   - Found in insects, crustaceans, and some polychaete worms.
   - Results in a mosaic image.
   - More acute at detecting movement partly due to rapid recovery of photoreceptors.
   - Superimposition eyes have ommatidia with lens that work like prisms and parabolic mirrors, focusing light entering several ommatidia onto photoreceptor (increases sensitivity to light).

Single-lens eye

2. In a *single-lens eye*, one lens focuses light onto the *retina*, a bilayer of photosensitive receptor cells.
   - Found in some jellyfish, polychaetes, spiders and many mollusks.

B. Vertebrate Vision (p. 1020–1026)

What are the component parts of the vertebrate eye? What is the function of each part?

1. Structure and Function of the Vertebrate Eye (p. 1021–1022)

The parts of the vertebrate eye are structurally and functionally diverse. (See Campbell, Figure 45.7)

Sclera, Choroid

- The vertebrate eye consists of a tough outer layer of connective tissue, the *sclera*, and a thin inner pigmented layer, the *choroid*.

- ☞ The *cornea* is located in front and is a transparent area of the sclera; it allows light to enter the eye and acts as a fixed lens.
- ☞ The anterior choroid forms the *iris*, which regulates the amount of light entering the *pupil*. The iris is pigmented and gives the eye color; the pupil is the hole in the center of the iris.
- ☞ The *retina* is the innermost layer of the eyeball; it contains photoreceptor cells which transmit signals from the optic disc, where the optic nerve attaches to the eye.
- ☞ The lens and ciliary body divide the eye into two chambers: a small chamber between the lens and cornea; and a large chamber within the eyeball.
- ☞ The *ciliary body* produces *aqueous humor* that fills the cavity between the lens and cornea.
- ☞ *Vitreous humor* fills the cavity behind the lens and comprises most of the eye's volume.
- ☞ Both aqueous humor and vitreous humor help to focus light onto the retina.

The *lens* is a transparent, protein disc that focuses an image onto the retina by changing shape (*accommodation*). (See Campbell, Figure 45.8)
- ☞ Is nearly spherical when focusing on near objects and is flatter when focusing at a distance.
- ☞ Controlled by the ciliary muscle.

2. **Signal Transduction in the Eye** (*p. 1022–1024*)

Cells in the retina transduce stimuli (caused by the lens focusing a light image onto the retina) into action potentials.

How do the rod cells and cone cells found in the vertebrate eye function? What other cells are involved in sending a nerve impulse from the retina, through the optic nerve, to the brain?

The photoreceptors of the eye are rod cells and cone cells. (See Campbell, Figure 45.9)
- ☞ Their relative numbers in the retina are partly correlated with whether an animal is diurnal or nocturnal.
- ☞ The light absorbing molecule is *retinal*, synthesized from vitamin A, which is bonded to a characteristic kind of protein, *opsin*.

Rod cells are sensitive to light but do not distinguish colors.
- ☞ Found in greatest density at lateral regions of the retina; completely absent from the *fovea* (visual field center).
- ☞ Contains *rhodopsin*, which is inactivated by light.

Cone cells are responsible for daytime color vision.
- ☞ Consists of 3 subclasses, red cones, green cones and blue cones, each with its own type of opsin associated with retinal for form visual pigments (*photopsins*).
- ☞ Most dense at the fovea.

Cornea

Iris

Pupil

Retina

Lens

Ciliary body, Aqueous humor

Vitreous humor

Accommodation

Retinal responds to light by changing shape; this triggers a chain of metabolic events that hyperpolarizes the photoreceptor cell membrane; thus, a decrease in chemical signal to the cells with which photoreceptors synapse serves as the message.
- ☞ Rod and cone cell axons synapse with neurons called *bipolar cells*, which in turn synapse with *ganglion cells*.
- ☞ *Horizontal cells* and *amacrine cells*, neurons in the retina, help integrate the information, after which ganglion cell axons convey action potentials along the optic nerve to the brain.

3. **Visual Integration** (*p. 1024–1026*)

What distinguishes a vertical pathway from a lateral pathway? How does lateral inhibition enhance visual integration?

Integration of visual information begins at the retina.

Rod and cone cell signals may follow vertical or lateral pathways.
- ☞ Vertical pathways involve information passing directly from receptor cells to bipolar cells to ganglion cells.
- ☞ Lateral pathways involve: horizontal cells carrying signals from one rod or cone to other receptor cells and several bipolar cells; amacrine cells spread the signals from one bipolar cell to several ganglion cells.

Lateral inhibition = Horizontal cells, stimulated by rod or cone cells, stimulate nearby receptors and inhibit more distant receptors and non-illuminated bipolar cells, thus sharpening image edges and enhancing contrast. (See Campbell, Figure 45.12)
- ☞ Occurs at all levels of visual processing.

What is the optic chiasma? Through what other parts of the brain do the impulses travel to their final destination?

Optic nerves from each eye meet at the *optic chiasma*. (See Campbell, Figure 45.13)
- ☞ Has nerve tracts arranged so that what is viewed in the left field of view is transmitted to the right side of the brain, and vice versa.
- ☞ Ganglion axons usually continue through the *lateral geniculate nuclei* of the thalamus, and these neurons continue back to the *primary visual cortex* in the occipital lobe of the cerebrum.
- ☞ Additional neurons carry information to more sophisticated visual processing centers in the cortex.

III. **HEARING AND BALANCE** (*p. 1027–1032*)

Hearing and balance are related in most animals and involve mechanoreceptors.

A. **The Mammalian Ear** (*p. 1027–1030*)

What parts comprise the outer ear of a mammal? The middle ear? The inner ear? What is the function of each part?

Sound waves are collected by the outer ear (the external *pinna* and the *auditory canal*) and are channeled to the *tympanic membrane* of the *middle ear*. (See Campbell, Figure 45.14)

The sound waves cause the tympanic membrane to vibrate at the same frequency; the tympanic membrane transmits the waves to three small bones — the *malleus*, *incus* and *stapes* — which amplify and transmit the mechanical movements of the membrane to the *oval window*, a membrane of the cochlea surface.
- ☞ Oval window vibrations produce pressure waves in the fluid (endolymph) in the coiled cochlea of the inner ear.

What occurs in the inner ear that results in an action potential traveling through the auditory nerve to the brain? What is the difference between the volume and pitch of a sound?

The pressure waves vibrate the *basilar membrane* (forms the floor of the cochlear duct) and the attached *organ of Corti*, which contains receptor *hair cells*. (See Campbell, Figure 45.15)
- ☞ The bending of the hair cells against the *tectorial membrane* depolarizes the hair cells and causes them to release neurotransmitter that triggers an action potential in a sensory neuron which then carries sensations to the brain through the auditory nerve.

Volume is determined by the amplitude of the sound wave; pitch is a function of sound wave frequency.
- ☞ The greater the amplitude of a sound, the more vigorous the vibrations; this results in more bending of the hair cells and more action potentials.
- ☞ Different sound frequencies affect different areas of the basilar membrane, thus some receptors send more action potentials than others.

How does the mammalian ear function to maintain body balance and equilibrium?

The sense of balance is centered in the inner ear; behind the oval window is a vestibule that contains two chambers, the *utricle* and the *saccule*; the utricle opens into three *semicircular canals*.
- ☞ Hair cells in the utricle and saccule respond to changes in head position with respect to gravity and movement in one direction.
- ☞ Hair cells are arranged in clusters with their hairs projecting into a gelatinous material containing numerous *otoliths* (small calcium carbonate particles).
- ☞ The otoliths are heavier than endolymph in the saccule and utricle; gravity pulls them down on the hairs of receptor cells, thus causing a constant series of action potentials indicating position of the head.
- ☞ Semicircular canals detect rotation of the head due to endolymph movement against the hair cells.

Notes:

Pinna, Auditory canal

Tympanic membrane, Middle ear

Malleus, Incus, Stapes

Oval window, Cochlea

Endolymph, Inner ear

Basilar membrane, Organ of Corti

Utricle, Saccule

Semicircular canals

Otoliths

662 Sensory and Motor Mechanisms

B. Hearing and Equilibrium in Other Vertebrates (p. 1030–1031)

How does the lateral line system of fishes and aquatic amphibians differ in structure and function from the mammalian ear? What is the Weberian apparatus?

Fishes and aquatic amphibians have a *lateral line system* running along both sides of the body. (See Campbell, Figure 45.18)
- Mechanoreceptors called *neuromasts* contain hair cell clusters whose hairs are embedded in a gelatinous cap, the *cupula*.
- Water enters the system through numerous pores on the animal's surface and flows along the tube past the neuromasts.
- Pressure of moving water bends the cupula causing an action potential in the hair cells.
- Provides information about the body's movement direction and velocity of water currents, and movements or vibrations caused by predators and prey.

The inner ear of a fish has no eardrum, does not open to the outside of the body and has no cochlea; a saccule, utricle, and semicircular canals are present.
- Sound waves are conducted through the skeleton of the head to the inner ear, this sets otoliths in motion, stimulating the hair cells.
- Some fish have a *Weberian apparatus*, a series of 3 bones which conducts vibrations from the swim bladder to the inner ear.
- Fish can hear higher frequencies due to their inner ears.

In terrestrial amphibians, reptiles and birds, sound is conducted from the tympanic membrane to the inner ear by a single bone, the stapes.

C. Sensory Organs for Hearing and Balance in Invertebrates (p. 1031–1032)

How do invertebrates detect sound and maintain their equilibrium?

The body hairs of many insects vibrate in response to sound waves of specific frequencies.
- Fine hairs on the antennae of male mosquitoes detect the hum produced by a female's wings.
- Vibrating body hairs of some caterpillars detect predatory wasps.

Many insects also have "ears," commonly located on their legs, consisting of a tympanic membrane stretched over an internal air chamber containing receptor cells that send nerve impulses to the brain. (See Campbell, Figure 45.19)

Most invertebrates have mechanoreceptors called *statocysts* that function in their sense of equilibrium. (See Campbell, Figure 45.20)
- Gravity causes statoliths (dense granules) to settle to the low point in a chamber, stimulating hair cells in that location.
- Statocysts are located along the bell fringe of many jellyfish and at the antennule bases in lobsters and crayfish.

IV. TASTE AND SMELL (p. 1032–1034)

Gustation (taste) and olfaction (smell) depend on chemoreceptors that detect specific chemicals in the environment.

How do insects detect taste? Where are the receptor cells for taste located in mammals? How do they function? Where are the human olfactory receptor cells located? How are the human senses of taste and smell integrated?

Insects have taste receptors within sensory hairs called *setae* on the feet and mouthparts; olfactory setae are usually located on antennae.
- Several chemoreceptor cells, each responding to a particular chemical, are located on each tasting hair; integrating impulses from the different receptors permits distinguishing many tastes.

In humans and other mammals, receptor cells for taste are organized into *taste buds* scattered in several areas of the tongue and mouth.
- Sweet, sour, salty and bitter are detected in distinct regions of the tongue.
- These tastes are associated with specific molecular shapes and charges that bind to separate receptor molecules.

In humans, olfactory receptor cells line the upper nasal cavity and send impulses along their axons directly to the olfactory bulb of the brain. (See Campbell, Figure 45.23)
- Receptive ends of the cells contain cilia that extend into the coating layer of mucus lining the nasal cavity.
- Specific receptors respond to certain odorous molecules by depolarizing.
- Respond to airborne chemicals that diffuse into the region.
- Taste and olfaction have different receptors but interact.

V. AN INTRODUCTION TO ANIMAL MOVEMENT (p. 1034–1035)

What are the advantages and disadvantages associated with movement by swimming? Walking or running? Flying?

Different modes of transportation (running, flying, swimming) have evolved along with adaptations of animals to overcome the difficulties associated with each type of locomotion.
- Swimming animals must overcome resistance; thus, many are fusiform in body shape.
- A walking or running land animal must support itself and move against gravity, and also overcome inertia with each step by accelerating a leg from a standing start; strong skeletal support and muscles evolved.
- Flying animals do not use a skeleton for support during motion, and must almost completely overcome gravity to become airborne; wings must provide enough lift to overcome gravity.

At the cellular level, all movements are based on the contractile systems of microfilaments and microtubules.

VI. SKELETONS AND THEIR ROLES IN MOVEMENT (p. 1035–1037)

What are the functions of a skeleton?

Skeletons function in support, protection and movement.
- Help maintain shape of aquatic animals.
- Hard skeletons protect soft body tissues.
- Skeletons provide a firm attachment against which muscles can work during movement.

A. Hydrostatic Skeletons (p. 1035–1036)

What is a hydrostatic skeleton? How does a hydrostatic skeleton function?

<u>Hydrostatic skeletons</u> consist of fluid held under pressure in a closed body compartment.
- Found in most cnidarians, flatworms, nematodes and annelids.
- Control form and movement by using muscles to change the shape of fluid-filled compartments.
- Provide no protection and could not support a large land animal.

B. Exoskeletons (p. 1036–1037)

What is an exoskeleton? How do the exoskeletons of mollusks and arthropods differ? What are the advantages and disadvantages of an exoskeleton?

<u>Exoskeleton</u> = Hard encasement deposited on the surface of an animal.
- Mollusks are enclosed in calcareous shells secreted by the mantle.
- Arthropods are enclosed in a *cuticle* secreted by the epidermis that contains *chitin* and also contains quinones that add extra strength in areas where protection is most important.
- Exoskeletons must be shed (molted) as the animals grow.

C. Endoskeletons (p. 1037)

What is an endoskeleton? What organisms possess some form of endoskeleton? What is the adaptive advantage of having different types of joints in different locations in the vertebrate skeleton?

<u>Endoskeleton</u> = Hard supporting elements buried within the soft tissues of an animal.
- Sponges possess hard spicules of inorganic material or softer protein fibers.
- Echinoderms have ossicles composed of magnesium carbonate and calcium carbonate forming hard plates beneath the skin.
- Chordates have cartilage and/or bone skeletons divided into several areas. (See Campbell, Figure 45.26)
- The vertebrate frame is divided into an axial skeleton (the skull, vertebral column, and rib cage) and an appendicular skeleton (limb bones, pectoral and pelvic girdles).
- Bones of the vertebrate act in support and as levers when their attached muscles contract.

Hydrostatic skeleton

Exoskeleton
Cuticle
Chitin

Endoskeleton

VII. MUSCLES (p. 1037–1045)

How does the skeleton combine with the antagonistic action of muscles to provide a mechanism for movement?

Animal movement is based on contraction of muscles working against some kind of skeleton.
- ☞ Muscles always contract actively; they can extend only passively.
- ☞ Ability to move a body part in opposite directions requires that muscles be attached to the skeleton in antagonistic pairs.

A. The Structure and Physiology of Vertebrate Skeletal Muscle (p. 1038–1043)

What is a skeletal muscle? What are the basic components of a skeletal muscle? What is a sarcomere? What produces the banded appearance seen in Figure 45.29?

Skeletal muscle = Bundle of long fibers running the length of the muscle; attached to and responsible for movement of bones.

Each fiber is a single cell with many nuclei; consists of bundles of smaller *myofibrils* arranged longitudinally. Two kinds of myofilaments are found in each myofibril.
- ☞ *Thin filaments* consist of two strands of actin and one strand of regulatory protein coiled together.
- ☞ *Thick filaments* are staggered arrays of myosin molecules.

Sarcomere = Unit of organization of skeletal muscle. (See Campbell, Figure 45.29)
- ☞ *Z lines* are the borders of the sarcomere; aligned in adjacent myofibrils.
- ☞ *I bands* are areas near the edge of the sarcomere containing only thin filaments.
- ☞ *A bands* are regions where thick and thin filaments overlap and correspond to the length of the thick filaments.
- ☞ *H zones* are areas in the center of the A bands containing only thick filaments.

1. Molecular Mechanism of Muscle Contraction (p. 1039–1040)

How does the sliding filament model explain muscle contraction?

Muscle contraction reduces the length of each sarcomere.

Sliding filament model = Mechanism of muscle contraction. (See Campbell, Figure 45.30)
- ☞ Thin filaments rachet across thick filaments to pull the Z lines together and shorten the sarcomere; the myofilaments themselves do not contract.
- ☞ Myosin molecules on thick filaments attach to actin on the thin filament to form a crossbridge. It then bends inward, pulling the thin filament toward the center of the sarcomere, breaks the crossbridge, and forms a new crossbridge further down.

666 Sensory and Motor Mechanisms

☞ Energy for crossbridge formation comes from the hydrolysis of ATP by the head region of myosin. (See Campbell, Figure 45.31)

2. **Excitation-Contraction Coupling** (*p. 1040–1042*)

Skeletal muscles contract when stimulated by motor neurons.

What processes are involved in the excitation-contraction coupling process of muscle contraction? What roles do calcium and phosphagens play?

In a muscle at rest, myosin-binding sites on the actin are blocked by the regulatory protein strand (*tropomyosin*) in the thin filament and by the *troponin complex* located at each binding site.

☞ An action potential in the motor neuron innervating the muscle cell causes the axon to release acetylcholine.

☞ Infoldings in the muscle cell plasma membrane (*transverse tubules*) carry the resulting graded depolarization deep into the muscle cell.

☞ The *sarcoplasmic reticulum* membrane becomes depolarized and releases its store of Ca^{2+}.

☞ The calcium ions bind to troponin, causing the thin filament to change shape and expose the myosin-binding sites; the muscle can then contract.

☞ The contraction is terminated as calcium is pumped out of the cytoplasm by the sarcoplasmic reticulum; as the calcium concentration falls, the tropomyosin-troponin complex again blocks the myosin-binding sites.

☞ *Phosphagens* (*creatine phosphate* in vertebrates) provide most of the necessary energy.

3. **Graded Contractions of Whole Muscles** (*p. 1042–1043*)

What is tetanus? How do summation of multiple motor unit activity and wave summation result in graded contractions of skeletal muscles?

Graded contractions of skeletal muscles are due to summation of multiple motor unit activity (*recruitment*) and wave summation.

☞ Motor neurons usually deliver their stimuli rapidly, resulting in smooth contraction typical of *tetanus* (sustained contraction) rather than the jerky actions of muscle twitches.

☞ A motor unit consists of a single motor neuron and all the muscle fibers it controls; all fibers in the motor unit contract as a group when the motor neuron fires.

☞ As more motor neurons are recruited by the brain, tension in the muscle progressively increases.

4. **Fast and Slow Muscle Fibers** (*p. 1043*)

How does cytoplasmic calcium concentration affect muscle contraction? What distinguishes slow muscle fibers from fast muscle fibers? Is there an adaptive advantage to possessing both slow and fast muscle fibers?

Duration of muscle contraction is controlled by how long the calcium concentration in the cytoplasm remains elevated.

Tropomyosin

Troponin complex

Transverse tubules

Sarcoplasmic reticulum

Phosphagens, Creatine phosphate

Summation

Motor unit, Wave summation

Tetanus

Muscle twitches

Slow fibers have longer-lasting twitches because they have less sarcoplasmic reticulum; thus, Ca^{2+} remains in the cytoplasm longer.
- ☞ Have many mitochondria, a rich blood supply, and the oxygen-storing protein *myoglobin*.
- ☞ Used to maintain posture since they can sustain long contractions.

Fast fibers have short duration twitches and are used in fast muscles for rapid, powerful contractions.
- ☞ Some are able to sustain long periods of repeated contractions without fatiguing.

B. Other Types of Muscles (p. 1043–1045)

What structural and functional properties distinguish skeletal muscle from cardiac muscle? From smooth muscle?

Vertebrate *cardiac muscle* is found only in the heart.
- ☞ Is striated.
- ☞ Muscle cells are branched, and the junction between cells contain *intercalated discs* that electrically couple all heart muscle cells, allowing coordinated action.
- ☞ Cells can also generate their own action potentials.
- ☞ Rhythmic depolarizations due to pacemaker channels in the plasma membrane trigger action potentials which last up to 20 times longer than those for skeletal muscle and have long refractory periods.

Smooth muscles lack striations and contain less myosin; the myosin is not associated with specific actin strands.
- ☞ Generates less tension but can contract over a greater range of lengths.
- ☞ Does not have a transverse tubule system or a well developed sarcoplasmic reticulum; calcium ions must enter the cytoplasm through the plasma membrane during an action potential.
- ☞ Contractions are relatively slow but there is greater range of control.
- ☞ Found mainly in the walls of blood vessels and digestive tract organs.

How do the muscles of invertebrates differ from those of vertebrates?

Invertebrates have muscles similar to the skeletal and smooth muscles of vertebrates, but with some interesting adaptations.
- ☞ Arthropod skeletal muscles are very similar to vertebrate skeletal muscle.
- ☞ Insect wings actually beat faster than action potentials arrive from the CNS since the flight muscles are capable of independent, rhythmic contractions.
- ☞ The thick filaments of muscles in clams which hold the shell closed contain *paramyosin*; this unique protein allows the muscles to stay in a fixed state of contraction for up to a month.

Myoglobin

Cardiac muscle

Intercalated discs

Smooth muscles

46   ECOLOGY: DISTRIBUTION AND ADAPTATIONS OF ORGANISMS

CHAPTER OUTLINE

Ecology

Abiotic, Biotic

Ecology = The scientific study of the interactions of organisms and their environments.
- ☞ Abiotic and biotic factors are included in an organism's environment.
- ☞ Organisms and their environments affect one another.

I. THE SCOPE AND DEVELOPMENT OF ECOLOGY (p. 1053–1054)

    A.   The Questions of Ecology (p. 1053)

Ecology is the study of the distribution and abundance of organisms.

What are the four levels of ecological inquiry? Is there an hierarchical association among these levels?

There are four levels of inquiry in ecology:

Organismal ecology

1. *Organismal ecology* studies how individual organisms tolerate environmental stresses that determine where they can live; includes study of behavioral physiological, and morphological ways individuals meet environmental challenges.

Population ecology

2. *Population ecology* studies groups of individuals of the same species in a particular geographic area.
 - ☞ Questions of population ecology concern factors that affect population size and composition.

Community ecology

3. *Community ecology* studies all organisms that inhabit a particular area.
 - ☞ Questions concern predation, competition and other interactions that affect community structure and organization.

Ecosystem

4. *Ecosystem ecology* studies all abiotic factors as well as communities in an area.
 - ☞ Questions concern energy flow and chemical cycling.

    B.   Ecology as an Experimental Science (p. 1053–1054)

Why are mathematical models important tools in the study of ecology?

Ecology has a long history as a descriptive science but is young as an experimental science.
- ☞ Very difficult to conduct experiments and control variables.
- ☞ Some test hypotheses in lab experiments and by manipulating communities in field experiments.
- ☞ Some devise mathematical models which include important variables and hypothetical relationships; usually studied with the aid of a computer.

C. **Ecology and Evolution** (*p. 1054*)

What is the relationship between ecology and evolution?

Short term interactions of organisms with their environments could have long term effects through natural selection. Ecological time translates into effects over evolutionary time.
- For example, predator-prey interactions may affect gene pools where individuals with protective coloration would become more prevalent.
- Continental drift helps explain the distribution of many animal groups.

II. **TERRESTRIAL BIOMES** (*p. 1054–1062*)

What is the biosphere? A biome?

The *biosphere* is that portion of Earth inhabited by life and represents the sum of all communities and ecosystems.   *Biosphere*
- A thin layer consisting of seas, lakes, rivers, streams, the land to a soil depth of a few meters, and the atmosphere to an altitude of a few kilometers.

Biomes are the major types of communities that are typical of broad geographic areas.   *Biomes*

Terrestrial biomes are often named for the predominant vegetation but each is also characterized by animals adapted to that particular environment. (See Campbell, Figure 46.3)
- Species composition may vary from one location to another within a biome.
- Biomes grade into each other without sharp boundaries.
- May be patchy, with several communities represented in one biome.
- Prevailing climate, particularly temperature and rainfall, is most important factor in determining what kind of biome develops.

A. **Tropical Forest** (*p. 1056–1057*)

What environmental conditions result in the formation of a tropical thorn forest? A tropical deciduous forest? A tropical rain forest?

Tropical forest is found near the equator (within 23.5° latitude) where temperature varies little from approximately 23°C and the length of daylight varies from 12 hours by less than one hour.   *Tropical forest*

Rainfall is variable and the amount determines what vegetation is found in an area.
- In lowlands with prolonged dry seasons and scarce rain, *tropical thorn* forests occur; these are a mixture of thorny threes and shrubs, and succulents.
- In regions with distinct wet and dry seasons, *tropical deciduous* forests occur. Trees releaf following heavy rains and drop leaves during the dry season.

Near the equator where rainfall is abundant (> 250 cm/year) and the dry season lasts less than a few months is *tropical rain forest*. (See Campbell, Figure 46.5)
- Harbors more plants and animal species than any other community.
- Widespread individuals depend on animals for pollination and dispersal of fruits and seeds.
- Competition for light is strong selective force in plant communities.
- Soils are typically poor due to rapid nutrient recycling.
- Animals are typically tree-dwellers; numerous exothermic animals are found due to warm temperatures.

Destruction of tropical rain forests, occurring rapidly due to human intervention, may cause large-scale changes in world climate as well as destruction of many species.

B.   Savanna (p. 1057–1058)

Where are the major savanna biomes located? What seasonal changes, soil conditions and other factors combine to produce a savanna?

Savanna is grassland with scattered individual trees. (See Campbell, Figure 46.6)
- Covers areas of central South America, central and southern Africa and parts of Australia.
- Generally 3 distinct seasons: cool and dry, hot and dry, warm and wet, in that order.
- Soils generally porous with a thin humus layer; water drainage is rapid.
- Usually have a simple physical structure but are often rich in species numbers.
- Frequent fires inhibit invasion by trees to maintain the grasses (wind pollinated) and forbs (often insect pollinated).
- Large herbivores (zebras, giraffes) and burrowing animals are commonly most active in the rainy season and many are nocturnal.
- During the dry season, many small animals are dormant or survive on seeds and dead plant parts.

The term *savanna* also applies to areas where forest and grassland biomes integrate: the climatic conditions and community features are intermediate between grassland and forest.

C.   Desert (p. 1058–1059)

What conditions characterize a desert? What adaptations are found in desert plants and animals which permit them to survive in this environment?

Desert is characterized by low and unpredictable precipitation (< 30 cm/year), not by temperature: both cold and hot deserts exist.
- Hot deserts occur in S.W. United States, W. South America, North Africa, the Middle East, and Central Australia.
- Cold deserts occur in E. Argentina, central Asia and west of the Rocky Mountains.

The cycles of growth and reproduction are keyed to rainfall.
- ☞ Scattered shrubs, cacti and other succulents that store water are common. Many bloom abundantly after a rainfall.

Reptiles and seed eaters such as ants, birds and rodents are common. Many live in burrows, are nocturnal or have other adaptations for conserving water.
- ☞ Light coloration reflects sunlight; some mice derive all of their water from the metabolic breakdown of the seeds they eat and never drink.

D. Chaparral (*p. 1059*)

How does a chaparral differ from a desert and a savanna? What major abiotic factor helps maintain the chaparral? What special adaptations are found in chaparral plants?

Chaparral (scrubland) are regions of dense, spiny shrubs with tough evergreen leaves found along coasts where cool ocean currents circulate offshore to make mild, rainy winters and long, hot, dry summers (between 30° and 40° latitude).
- ☞ Occur in Mediterranean and coastlines of California, Chile, S.W. Africa, and S.W. Australia. (See Campbell, Figure 46.8)

Maintained by periodic fires.
- ☞ Many shrubs have root systems and seeds adapted for fire; root crowns may be fire resistant and resprout quickly, others have seeds that only germinate after a fire.
- ☞ Other plants are colonial (asexual reproduction).

Browsers such as deer, fruit-eating birds, rodents, snakes and lizards are common.

E. Temperate Grasslands (*p. 1059*)

How do temperate grasslands differ from tropical savannas?

Temperate grasslands are similar to tropical savanna but occur in regions with relatively cold winters.
- ☞ The veldts of southern Africa, the pampas of Uruguay and Argentina, the steppes of the former Soviet Union and the plains and prairies of the U.S. are examples.
- ☞ Occasional fires and seasonal drought prevent encroachment of trees upon the grassland.
- ☞ Soil tends to be thick and rich; annual rainfall amounts influence the height of vegetation.
- ☞ Large grazing mammals and large carnivores are common. (See Campbell, Figure 46.9)
- ☞ Burrowing rodents and other small mammals are present in dense populations.

Chaparral

Temperate grasslands

F. **Temperate Deciduous Forest** (p. 1060–1061)

What conditions result in the development of a temperate deciduous forest?

Temperate forests grow throughout midlatitude regions with sufficient moisture to support growth of large, broad-leaved deciduous trees. (See Campbell, Figure 46.10)
- ☞ Occur in Eastern U.S., Middle Europe and E. Asia.
- ☞ Temperatures range from very cold in winter to very hot in summer (−30°C to 30°C), with a 5 to 6 month growing season.
- ☞ Precipitation is fairly high and evenly distributed throughout the year.
- ☞ Soil is rich in nutrients; decomposition rates are slow and a thick layer of leaf litter usually accumulates.

Several layers of vegetation including herbs, shrubs and one or two strata of trees (species composition varies widely) are present.
- ☞ Variety and abundance of food and habitat supports a rich diversity of animal life.

G. **Taiga** (p. 1061)

What are the characteristics of a taiga biome?

Taiga (coniferous or boreal forest) is characterized by harsh winters and short, wet (and occasionally warm) summers.
- ☞ Occurs in North America, Europe, Asia and at high elevations in more temperate latitudes.
- ☞ Considerable precipitation in the form of snow that insulates the soil to reduce the permafrost; also provides a protective layer for small mammals.
- ☞ Conifers like spruce, pine, fir and hemlock and deciduous oak, birch, willow, alder and aspen are common; usually one or very few species are present in dense stands.
- ☞ Animals include: seed eaters such as squirrels, birds, and insects; browsers such as elk, moose, deer, beaver, and porcupine; predators such as grizzly bears, wolves and wolverines.
- ☞ Soil is thin, acidic and forms slowly due to low temperatures and slow decomposition of waxy needles.

H. **Tundra** (p. 1061–1062)

What distinguishes an arctic tundra from an alpine tundra?

Tundra is at the northern-most limits of plant growth and at high altitudes plant forms are limited to low, shrubby or matlike vegetation. (See Campbell, Figure 46.12)

There are two types of tundra:

1. *Arctic tundra* encircles the North Pole and is very cold with little light for long periods.
   - ☞ Brief warm summers are marked by nearly 24 hours of daylight; plant growth and reproduction occur rapidly during the summer.
   - ☞ Some areas are characterized by permafrost (permanently frozen ground) which contributes to the absence of taller plant forms.

- Soil is continuously saturated, further restricting plant forms.
- Dwarf perennial shrubs, sedges, grasses, mosses and lichens are common.

2. *Alpine tundra* occurs at high elevations in all latitudes.
   - Near the equator, daylight varies little from 12 hours throughout the year and vegetation exhibits slow, steady rates of photosynthesis year round.

Many animals are migratory and exhibit adaptations to the cold. Large animals are herbivores like musk oxen and caribou. Smaller animals are lemmings, white fox and snowy owl.
- Insects are prevalent during the summer but spend the winter as immature stages.

III. FRESHWATER BIOMES (p. 1062–1065)

Freshwater biomes have several characteristics that distinguish them from marine biomes.
- Salt concentration is less than 1%.
- Freshwater biomes are closely linked to the terrestrial biomes that surround them or through which they flow.
- Overall characteristics are influenced by the pattern and speed of water flow, and the local climate.

A. Ponds and Lakes (p. 1063–1064)

These standing bodies of water vary greatly in size with *ponds* being smaller than *lakes*.

How does light penetration and water temperature produce a vertical stratification in ponds and lakes? How does this stratification effect the distribution of plants and animals in the system?

Ponds and lakes usually exhibit a significant vertical stratification in light penetration and water temperature.
- Light is rapidly absorbed by the water and microorganisms in the water resulting in a rapid decrease in light intensity as depth increases.
- This divides the pond or lake into two layers: the *photic zone* is the upper layer where light is sufficient for photosynthesis; the lower *aphotic zone* receive little light and no photosynthesis occurs.
- Temperature stratification also occurs in deeper ponds and lakes during summer in temperate zones. Sunlight heats the upper layers of water as far as it penetrates; the deeper waters remain cold.
- As depth increases, the separation point of the warmer upper water form the lower colder water is noticeable as the *thermocline*; the thermocline is a narrow vertical zone between the warmer and colder waters where a rapid temperature change occurs.

The distribution of plants and animals within a pond or lake also shows a stratification based on water depth and distance from the shore.

Alpine tundra

Photic zone

Aphotic zone

Thermocline

674 Ecology: Distribution and Adaptations of Organisms

Littoral zone
- The *littoral zone* is shallow, well-lighted, warm water close to shore. Characterized by the presence of rooted and floating vegetation, a diverse attached algae community, and a very diverse animal fauna including suspension feeders (clams), herbivorous grazers (snails), and herbivorous and carnivorous insects, crustaceans, fishes, and amphibians. Some reptiles, water fowl, and mammals also frequent this zone.

Limnetic zone
- The *limnetic zone* is the open, well-lighted waters away from shore. Occupants of this zone include phytoplankton (algae and cyanobacteria) which are photosynthetic, zooplankton (rotifers and small crustaceans) that grazes on phytoplankton, and small fish which feed on the zooplankton. Occasional visitors to this zone are large fish, turtles, snakes, and piscivorous birds.

Profundal zone
- The *profundal zone* is the deep, aphotic zone lying beneath the limnetic zone. This is an area of decomposition where *detritus* (dead organic matter that drifts in from above) is broken down. Water temperature is usually cold and oxygen is low due to cellular respiration of decomposers; mineral nutrients are usually plentiful due to decomposition of detritus.

Waters of the profundal zone usually do not mix with surface waters due to density differences related to temperature.
- Mixing of these layers usually occurs twice each year in temperate lakes and ponds; this results in oxygen entering the profundal zone and nutrients being cycled into the limnetic zone.

What characteristics can be used to distinguish between an oligotrophic lake and a eutrophic lake? What conditions can result in an oligotrophic lake becoming eutrophic?

Lakes are often classified as *oligotrophic* or *eutrophic*, depending on the amount of organic matter produced.

Oligotrophic lakes
- *Oligotrophic lakes* are deep, nutrient-poor lakes in which the phytoplankton is not very productive. The water is usually clear; the profundal zone has a high oxygen concentration since little detritus is produced in the limnetic zone to be decomposed.

Eutrophic lakes
- *Eutrophic lakes* are shallow, nutrient-rich lakes with very productive phytoplankton. The waters are usually murky due to large phytoplankton populations and the large amounts of detritus being decomposed may result in oxygen depletion in the profundal zone during the summer.

Oligotrophic lakes may develop into eutrophic lakes over time.
- Runoff from surrounding terrestrial habitats brings in mineral nutrients and sediments.
- Human activities increase the nutrient content of runoff due to lawn and agricultural fertilizers; municipal wastes dumped into lakes dramatically enriches the nitrogen and phosphorus concentrations which increases phytoplankton and plant growth.
- Algal blooms and increased plant growth results in more detritus and can lead to oxygen depletion due to increased decomposition.

B. **Streams and Rivers** (p. 5064–1065)

Streams and rivers are bodies of water that move continuously in one direction.

How does the structure of a river or stream change from its headwaters to its mouth? What factors influence the nutrient and oxygen content, turbidity and rate of flow of a river or stream?

The structure of these bodies of water changes from their point of origin (headwaters) to where they empty into a larger body of water (mouth).
- ☞ At the headwaters, the water is cold and clear, carries little sediment, and has few mineral nutrients. The channel is narrow with a rocky substrate and the water flows swiftly.
- ☞ Near the mouth, water moves slowly and is more turbid due to sediment entering from other streams and erosion; the nutrient content is also higher; the channel is usually wider with a silty substrate that has resulted from deposition of silt.

The nutrient and oxygen content, turbidity, and rate of flow in rivers and streams are influenced by many factors.
- ☞ Rough, shallow bottoms produce rapid turbulent flow known as *riffles*; smooth, deep bottoms result in a slower, smooth flow in areas called *pools*.
- ☞ Nutrient content of the water is higher in streams and rivers flowing through densely vegetated regions (leaves and other vegetation entering the water add organic matter) and where erosion takes place (increases inorganic nutrient content).
- ☞ Oxygen content of the water is affected by the flow rate; turbulent flow constantly oxygenates the water while slow waters contain relatively little oxygen.
- ☞ Turbidity reflects the amount of material suspended in the water; streams and rivers flowing through areas of high erosion will have more suspended materials than those surrounded by hard substrates. Large amounts of suspended organic matter also increases turbidity.

How do the biological communities of a river or stream differ from the headwaters to the mouth? What conditions contribute to these changes?

The biological communities found in rivers and streams differ as you move from headwaters to mouth; they also differ from those found in ponds and lakes.
- ☞ Due to the current, large plankton communities are not found in rivers and streams; photosynthesis which supports the food chains is carried out by attached algae and rooted plants.
- ☞ Organic material washed into the system also provides an important food source, especially where dense vegetation along the shore blocks out sunlight or high turbidity prevents light penetration.
- ☞ In upstream areas where water is cool, clear, and has a high oxygen content, many insects are found which require these conditions; fish such as trout are also found in these areas.

676 *Ecology: Distribution and Adaptations of Organisms*

- Near the mouth where water is murky and warmer with a lower oxygen content, fish such as carp and catfish are prominent; a different group of insects is also found in this area.
- Various adaptations are found in organisms which permit them to inhabit their specific areas of rivers and streams.

IV. MARINE BIOMES (p. 1066–1069)

In what ways do the Earth's oceans contribute to the conditions of the terrestrial biomes?

Ocean cover nearly 75% of the Earth's surface and contribute greatly to conditions on the other 25%.
- Most of the planet's rainfall is provided by evaporation of sea water.
- The world's climate and wind patterns are affected by ocean temperatures.
- Marine algae produce a large portion of the Earth's oxygen and consume large amounts of carbon dioxide.
- Salinity varies over time and area, but averages 3%.

How are the different marine communities classified? What zones are recognized and how are they distinguished from each other?

Marine communities are classified based on the depth at which they occur, their distance from shore, and light penetration. (See Campbell, Figure 46.16)
- A photic zone is present and extends to the depth at which light penetration supports photosynthesis; occupied by phytoplankton, zooplankton, and many fish species.
- The aphotic zone is below the level of effective light penetration and represents a majority of the ocean's volume.
- The *intertidal zone* is the shallow zone where the terrestrial habitat meets the ocean's water.
- The *neritic zone* extends from the intertidal zone, across the shallow regions, to the edge of the continental shelf.
- *Oceanic zones* extend over deep water from one continental shelf to another.
- *Pelagic zones* refer to open waters of any depth.
- *Benthic zones* refer to the seafloor.

A. Estuaries (p. 1066–1067)

What conditions are found in an estuary? What organisms are usually found in an estuary and why are they usually so diverse?

An *estuary* is the area where a freshwater stream or river merges with the ocean.
- Often bordered by saltmarshes or intertidal mudflats.
- Salinity varies within the estuary from nearly fresh water to ocean water; varies daily in areas due to rise and fall of tides.
- Estuaries are very productive due to nutrients brought in by rivers.

Intertidal zone

Neritic zone

Oceanic zones

Pelagic zones, Benthic zones

Estuary

Estuaries have a diverse flora and fauna due to their productivity.
- Saltmarsh grasses, algae, and phytoplankton are the major producers.
- Many species of annelids, oysters, crabs, and fish are also present.
- Many marine invertebrates and fish breed in estuaries.
- A large number of water fowl and other semiaquatic vertebrates use estuaries as feeding areas.

B. The Intertidal Zones (p. 1067)

What is an intertidal zone? What causes the zonation evident in a rocky intertidal zone? What special adaptations are found in intertidal zone organisms?

Intertidal zones, where land and sea meet, are alternately submerged and exposed by the daily tide cycles. (See Campbell, Figure 46.18)
- Organisms in this zone are exposed to greater variations in availability of seawater and temperature.
- These organisms are also subjected to the mechanical forces of wave action.

Rocky intertidal zones are vertically stratified and inhabited by organisms that possess structural adaptations that allow them to remain attached in this harsh environment.
- The uppermost zone is submerged only by the highest tides and is occupied by relatively few species of algae, grazing mollusks, and suspension-feeding barnacles; these organisms have various adaptations to prevent dehydration.
- The middle zone is exposed at low tide and submerged at high tide; many species of algae, sponges, sea anemone, barnacles, mussels, and other invertebrates are found in this area. The diversity is greater here due to the longer time spans this area is submerged.
- Tide pools are often found in the middle zone. These are depressions which are covered during high tide and remain as pools during low tide; tidepool organisms face dramatic salinity increases as water evaporates at low tide.
- The low intertidal zone is exposed only during the lowest tides and shows the greatest diversity of invertebrates, fishes, and seaweeds.

Sandy intertidal zones and *mudflats* do not show a clear stratification.
- Wave action continually shifts sand or mud particles and few algae or plants are present.
- Many suspension-feeding worms and clams bury themselves in the sand or mud and feed when the tide submerges the area.
- Predatory crustaceans also burrow into the substrate and feed when the tide is in.
- Predatory or scavenging crabs and shorebirds often feed on burrowing organisms in these areas.

C. Coral Reefs (p. 1067–1068)

What conditions favor the formation of coral reefs? How does the structure of coral reefs result in such a diverse community?

Intertidal zones

Coral reefs

Coral reefs are found in warm tropic waters of the neritic zone where sunlight penetrates to the ocean floor.
- ☞ Sunlight penetration permits photosynthesis and a constant supply of nutrients is provided by currents and waves.

Coral reefs are dominated by the structure of the coral. This diverse group of cnidarians secretes a hard, calcium carbonate external skeleton that provides a substrate on which other corals and algae grow.
- ☞ The cnidarian coral animals feed on microscopic organisms and organic debris even though they are dependent on the photosynthetic products of their symbiotic dinoflagellates.
- ☞ Coral reefs are very diverse and productive; a large variety of microorganisms, invertebrates, and fish are found among the corals and algae.
- ☞ Many herbivorous snails, sea urchins, and fish are present along with predators such as the octopus, sea stars, and many carnivorous fish.

Although many coral reefs are very large, they are delicate and can be severely damaged or destroyed by pollution, human induced damage, or introduced predators.

D. **The Oceanic Pelagic Biome** (*p. 1068*)

What conditions are found in the oceanic pelagic biome? What comprises the plankton community of this biome? What special adaptations are found in these organisms? What organisms comprise the nekton?

Oceanic pelagic biome

The *oceanic pelagic biome* consists of the open waters far from shore.
- ☞ Constantly mixed by circulating ocean currents.
- ☞ Nutrient content is usually very low except in those areas where periodic upwelling carry nutrients from the bottom to the surface.
- ☞ Waters are generally cold, although temperatures vary with latitude and depth.

Plankton is found in this zone.
- ☞ Photosynthetic phytoplankton grow and reproduce in the photic region (top 100m) of this biome.
- ☞ Zooplankton (which graze on the phytoplankton) includes protozoans, copepods, krill, jellyfish, and larvae of many invertebrates and fishes.
- ☞ Most species stay afloat in this zone through the aid of morphological structures like bubble-trapping spines, lipid droplets, gelatinous capsules, and air bladders.

Nekton

Nekton (free-swimming animals) are also found in the pelagic biome.
- ☞ Large squid, fishes, sea turtles, and marine mammals feed in this area.
- ☞ Many fish are adapted to and live in the aphotic region of the pelagic zone; some have large eyes allowing them to see in dim light, while others have luminescent organs used to attract mates and prey.
- ☞ Many bird species also feed in this region.

Ecology: Distribution and Adaptations of Organisms 679

E. **Benthos** (*p. 1068–1069*)

What is benthos? How do conditions differ between the neritic benthic zone and the abyssal zone? How do these conditions affect the diversity of organisms present?

Benthos refers to those organisms which inhabit the benthic zone, the ocean bottom below both the neritic and pelagic zones. ——— Benthos

☞ The benthic zone receive many nutrients in the form of detritus which settles to this area from the waters above.
☞ Light and temperature decline rapidly as you move from shallow, near-shore benthic areas to the ocean's depths.
☞ The benthic zone substrate may be sand or very fine sediment composed of silt and shells of dead microscopic organisms.

Neritic benthic communities are diverse and productive.

☞ Many bacteria, fungi, seaweeds, filamentous algae, and sponges are found in this area.
☞ Sea anemones, worms, clams, various echinoderms and fishes are also usually present.
☞ Many burrowing forms are present.
☞ The composition of the community varies with distance from shore, water depth, and bottom composition.

Deep benthic communities living in the abyssal zone are exposed to very different conditions than those under the neritic. ——— Abyssal zone

☞ Water temperature is continuously cold (3°C), water pressure is extremely high, there is very little (if any) light, and low nutrient concentrations are typical.
☞ Oxygen is usually present and a fairly diverse community of invertebrates and fishes are found.
☞ The deep-sea hydrothermal vent communities are found along midocean ridges in this region.
☞ These vent communities include chemoautotrophic bacteria as the primary producers in place of photosynthesizing organisms. These bacteria obtain energy by oxidizing H_2S which forms by a reaction of the hot vent water with dissolved sulfate.
☞ The bacteria are consumed by a variety of giant polychaete worms, arthropods, echinoderms, and fishes. (See Campbell, Figure 46.20)

V. **THE ENVIRONMENTAL DIVERSITY OF THE BIOSPHERE**
(*p. 1069–1075*)

Abiotic factors such as temperature, precipitation, and light influence the distribution of organisms.

☞ The patchiness of the global biosphere illustrates how the different physical environmentals produce a mosaic of habitats.

A. **Important Abiotic Factors** (*p. 1070*)

How do the major abiotic factors affect the distribution of organisms?

Some of the important abiotic factors that affect distribution of species include:

1. *Temperature* affects biological processes and the ability of most organisms to regulate their body temperature. Temperature affects metabolism: few organisms have active metabolisms at temperatures close to 0°C and temperatures above 45°C denature most essential enzymes.
   - The actual body temperature of ectotherms is affected by heat exchange with the environment; most animals maintain a body temperature only a few degrees above or below ambient temperature.
   - Even endotherms function best within the temperature range to which they are adapted.

2. *Water* is essential and adaptations for water balance and conservation help determine a species' habitat range.
   - Marine and freshwater animals face the problems of regulating intracellular osmolarity; terrestrial animals face the problem of desiccation.

3. *Sunlight* provides the energy that drives nearly all ecosystems although only photosynthetic organisms use it directly as an energy source.
   - In aquatic environments, water selectively reflects and absorbs certain wavelengths; therefore, most photosynthesis occurs near the water surface.
   - The physiology, development, and behavior of many animals and plants are often sensitive to photoperiod.

4. *Rocks and soil.* The physical structure, pH, and mineral composition of soil limit distribution of plants and hence animals that feed on those plants.
   - The composition of the substrate in a stream or river greatly influences the water chemistry, which in turn influences the plants and animals.
   - The type of substrate influences what animals can attach or burrow in intertidal zones.

5. *Wind* amplifies the effects of temperature by increasing heat loss by evaporation and convection; wind also increases the evaporation rate of animals and transpiration rate of plants, resulting in more rapid water loss.
   - Mechanical pressure of wind can effect plant morphology (for example, inhibiting growth of limbs on windward side of trees).

6. *Periodic disturbances* such as fire, hurricanes, typhoons, and volcanic eruptions can devastate biological communities, after which the area is recolonized by organisms or repopulated by survivors.
   - May go through a succession of changes.
   - Those disturbances that are infrequent (volcanic eruptions) do not illicit adaptations. Adaptations do evolve to periodically recurring disturbances such as fires.

B. Climate and the Distribution of Biomes (*p. 1070–1071*)

What are the primary components of climate? How can a climatograph be used in studying the distribution of biomes?

Climate is the prevailing weather conditions at a locality.

Climate

☞ The major components of climate are temperature, water, light, and wind.

A *climatograph* plots temperature and rainfall and shows the impact of climate on the distribution of biomes.
  ☞ Plots temperature and rainfall in a region in terms of annual means.
  ☞ Must be careful to distinguish a correlation between climate variables and biomes from causation.
  ☞ Overlapping regions on climatographs show factors other than mean temperature and rainfall, such as soil composition, which play roles in biome patterns. (See Campbell, Figure 46.22)

C. Global Climate Patterns (*p. 1071–1072*)

Solar energy input and the Earth's movement in space determine the planet's global climate patterns.
  ☞ About 50% of the solar energy that reaches the atmosphere's upper layers is absorbed before it reaches the surface.
  ☞ Ultraviolet and certain other wavelengths are more readily absorbed by oxygen and ozone than by other molecules.
  ☞ Some of the solar energy that reaches the Earth's surface is reflected back into the atmosphere; large amounts are absorbed by land and water.
  ☞ The atmosphere, land, and water are heated when they absorb solar energy; this heating establishes the temperature variations, air movement cycles, and evaporation of water responsible for the latitudinal variations in climate.

What is the association of the Earth's shape and tilt on its axis with the latitudinal variation in sunlight intensity?

Latitudinal variation in the intensity of sunlight results from the Earth's spherical shape; seasonal variation in solar radiation in the Northern and Southern Hemispheres are due to the Earth's tilt of 23.5° relative to its plane of orbit. (See Campbell, Figures 46.23 and 46.24)
  ☞ Only the *tropics* (23.5°N to 23.5°S) receive sunlight from directly overhead year round; the tilt causes solar radiation to change daily as the Earth rotates around the sun.
  ☞ The tropics receive the greatest annual input of solar radiation and show the least seasonal variation; only small variations in daylength and temperature occur.
  ☞ Seasonal variation in light and temperature increases steadily toward the poles; polar regions have long, cold winters with periods of continual darkness and short summers with periods of continual light.

What affect does the intense solar radiation near the equator have on precipitation and wind patterns? How do these patterns influence the distribution of terrestrial biomes?

A global circulation of air which creates precipitation and winds results from the intense solar radiation near the equator. (See Campbell, Figure 46.25)

Climatograph

- Evaporation of surface water due to high tropical temperatures causes warm, wet air masses to rise near the equator; this rising air creates an area of light, shifting winds (doldrums) along the equator.
- As these warm air masses rise, they expand and cool; cool air can hold less water vapor so the rising air masses drop large amounts of rain in the tropics.
- The cool, dry air masses flow toward the poles at high altitudes; they continue to cool as they move farther from the equator.
- Air mass density increases as they become cooler; they being to descend toward the surface as cool, dry air masses at about 30° latitude.
- The air masses absorb water as they descend, thus creating arid climates around 30°N and 30°S.
- Some of the descending air masses flows toward the poles at low altitudes, the rest flows toward the equator.
- The air masses flowing toward the poles is warmed and rises again at about 60°N and 60°S; as these masses of air begin to cool with increased altitude, the water vapor is lost as precipitation in this region.
- As air from this second cell reaches the higher altitude and cools, it flows toward the poles where it descends and flows back toward the equator.

The Earth's predictable wind patterns are established by air flowing in the circulation cells. The rotation of the Earth deflects the winds from a vertical path since land near the equator is moving faster than that at the poles.
- Tropical and subtropical tradewinds blow from east to west.
- In temperate zones, the predominating winds flow from west to east.

D. Local and Seasonal Effects on Climate (p. 1072–1075)

How can large bodies of water and topographical variation influence the climatic conditions of terrestrial biomes? What conditions cause the development of a rainshadow?

Although global climate patterns explain the geographic distribution of major biomes, local variations due to bodies of water and topographical features create a regional patchiness in climatic conditions.
- The local variation in climate and soil have a major influence on less widely distributed communities and individual species.

Proximity of large bodies of water affect local climates.
- Ocean currents are generated by the Earth's rotation and may heat or cool (depending on whether they are tropical or polar currents) air masses passing over them toward land; evaporation is also greater over the ocean than over land.
- Coastal areas are moister than inland areas at the same latitude; the California current flows from North to South along the west coast and helps form the cool, moist climate in this area.
- During warm summer days, air over land heats faster than that over the ocean or large inland lakes; this warmer air rises, drawing a cool breeze from the water across the land.

Topographical variations, such as mountains, also exert an influence on solar radiation, local temperature and rainfall.
- ☞ In the Northern Hemisphere, south-facing slopes receive more sunlight and are therefore warmer and drier than north-facing slopes; the vegetation differs with south-facing slopes being covered by shrubby, drought resistant plants while the north-facing slopes have forests.
- ☞ Air temperature declines 6° C for each 1000m increase in elevation; this parallels the decline in temperature associated with increasing latitude. For this reason, mountain communities are similar to those at lower elevations farther from the equator. (See Campbell, Figure 46.27)
- ☞ Deserts commonly occur on leeward sides of mountains; this results from the movement of air masses over the mountain which produces a rainshadow.
- ☞ When warm, moist air moves over a mountain, its altitude is increased and it cools; cooling causes the air to release its moisture as rain on the windward side. The cooler, drier air then flows down the leeward side of the mountain; the cool dry air is warmed and absorbs moisture, producing the rainshadow.

How do local conditions vary with seasonal changes? What is an upwelling? What is a microclimate and why do conditions vary in these areas?

The Earth's orbit around the sun causes seasonal changes in local conditions.
- ☞ The changing angle of the sun causes slight shifts in the wet and dry air masses on either side of the equator; these shifts result in the wet and dry seasons at 20° latitude where tropical deciduous forests grow.
- ☞ Seasonal changes in wind patterns produce variations in ocean currents, sometimes causing *upwellings* that bring nutrient rich water from the deep ocean layer to the surface.
- ☞ Seasonal temperature changes also cause the temperature profiles that develop during the summer in temperate zone ponds and lakes; these profiles reverse in the autumn and spring resulting in biannual mixing that brings nutrient-rich water from the bottom to the top. (See Campbell, Figure 46.28)

Climate also varies on a smaller scale, the *microclimate*. Microclimate refers to small areas within a habitat that may have very different conditions than the overall area (e.g. under a rock, a forest floor).
- ☞ Clear areas in a forest generally show greater temperature extremes than the shaded forest floor due to greater solar radiation and wind currents.
- ☞ Low lying areas are usually moister than high ground and support different forms of vegetation.
- ☞ The area under a large stone or log is protected from extremes of temperature and moisture; a large number of small organisms usually live in such sheltered areas.

Microclimate

VI. RESPONSES OF ORGANISMS TO ENVIRONMENTAL VARIATION
(p. 1075–1080)

Organisms live in an integrated environment which includes various combinations of abiotic factors. The success of an organism reflects the overall tolerance to the complete set of variables it encounters.

- ☞ Some polar organisms tolerate air temperatures of −70°C, while desert organisms survive at 45°C or higher.
- ☞ Aquatic organisms are found in waters with near zero salinity, while some tolerate salinities several times that of seawater.
- ☞ No organism can survive the full range of environmental conditions present on Earth; each population or species is distributed, at least in part, based on its tolerance for a specific subset of environmental conditions.
- ☞ Various structures and physiological mechanisms have evolved as adaptations to environmental constraints.

Organisms survive and reproduce in areas where the environmental conditions to which they are adapted are found.

- ☞ The ability to tolerate one factor may be dependant on another factor; for example, many aquatic organisms can tolerate reduced oxygen at low temperatures, but not at high temperatures which cause higher metabolic rates.

A. Homeostasis and the Principle of Allocation (p. 1076–1077)

Organisms faced with a fluctuation in an environmental variable may maintain the homeostasis of their bodies through behavioral and physiological mechanisms (*regulators*) or by allowing their internal conditions to vary with external conditions (*conformers*).

- ☞ Some species are conformers under certain conditions and become regulators under others. (See Campbell, Figure 46.29)

What is the Principle of Allocation? Why are regulators more affected by this principle than conformers?

Energy expenditure by the animal is necessary if behavioral or physiological mechanisms are used to maintain homeostasis.

- ☞ For organisms to survive and reproduce, the energy "cost" of regulation cannot exceed the benefits of homeostasis.
- ☞ Since few organisms are perfect regulators or perfect conformers, a majority of organisms represent a group of evolved strategies which permits them to live in their specific environment.

The *Principle of Allocation* is an important concept for assessing the responses of organisms to a complex environment.

- ☞ This principle holds that each organism has a limited amount of energy that can be allocated for obtaining nutrients, escaping predators, coping with environmental fluctuations, growth, and reproduction.
- ☞ Energy expended for one function reduces the amount of energy available for other functions.
- ☞ If an organism expends a large amount of energy to maintain homeostasis, less is available for growth, reproduction, and other functions.

The distribution of organisms and the homeostatic mechanisms they possess establish different priorities for energy allocation.
- ☞ Conformers living in a stable environment may have more energy available for growth and reproduction; however, their geographic distribution is restricted due to intolerance to environmental change.
- ☞ Regulators that allocate a large amount of energy to survive environmental changes have less available for other functions so they grow and reproduce less efficiently; however, they can survive and reproduce over a wider range because they can cope with changing environmental conditions.

B. **Environmental Grain** (p. 1077–1078)

How can you distinguish between a coarse-grained environment and a fine-grained environment? How does the environmental grain vary in a temporal and spatial manner? What adaptations to temporal and spatial variations in environmental grain are found in organisms?

The concept of *environmental grain* is used by ecologists to define the use of spatial variation in the environment by organisms of different sizes.
- ☞ A *coarse-grained* environment is one in which environmental patches are so large in relation to the size and activity of the organism that an individual organism can choose among patches.
- ☞ A *fine-grained* environment is one in which environmental patches are small relative to the size and activities of an organism, and the organism may not behave as though patches exist.
- ☞ An environment may be course-grained to one organism and fine-grained to another: a field of wild flowers is coarse-grained for small herbivorous insects since some plants may be preferential reproductive sites or food sources; the same field would be fine-grained to a large herbivorous mammal that feeds indiscriminately on all the plants.

Temporal variation in an environment may also be coarse-grained or fine-grained depending on the periodicity of the variation relative to the organism's lifespan.
- ☞ Daily environmental variations are usually fine-grained for most organisms.
- ☞ Seasonal environmental variations are usually coarse-grained to all organisms since they represent long term shifts in climate.

Organisms exhibit a variety of adaptations as responses to spatial and temporal variations in the environments. Some of these adaptations have short activation times while others take much longer.
- ☞ Behavioral adaptations are instantaneous and easily reversed.
- ☞ Physiological adaptations are activated and reversed on a time scale from seconds to weeks, depending on the specific response.
- ☞ Morphological adaptations may occur during an individual's lifespan.
- ☞ Adaptive genetic changes in populations take generations to appear.

Environmental grain

C. Behavioral Responses (p. 1078)

What types of behavioral responses do animals have to unfavorable environmental changes? In what manner does each aid the organism's survival?

Behavioral responses to unfavorable environmental changes include an animal's movement to a new, more favorable location and modification of the immediate environment by cooperative social behavior.
- Fish descend into deeper, cooler waters of a lake during summer when the upper layer becomes too warm.
- Desert animals return to burrows or move into the shade during the day when heat is most intense.
- Migratory birds migrate to warmer climates to overwinter.
- Honeybees seal the hive during cold periods to conserve heat and cool the hive on hot days by the collective beating of their wings.

D. Physiological Responses (p. 1078–1079)

How do physiological responses to environmental change differ from behavioral responses? What is the basis of physiological adaptations? What is acclimation?

Physiological responses to environmental change are slower than behavioral responses.
- An example of a slow change: when moved to an area of less oxygen, a person responds in several days to a few weeks with an increased number of red blood cells.
- An example of a faster change would be when blood vessels in the skin constrict within seconds to reduce loss of body heat.

Physiological adaptation is centered around regulation and homeostasis.
- Both regulators and conformers function most efficiently under certain environmental conditions which are optimal for the organism.
- Efficiency declines both above and below optimal values. (See Campbell, Figure 46.30)

Physiological responses to changing environments can shift tolerance limits of organisms (*acclimation*).
- Acclimation is a gradual process and is related to the range of environmental conditions experienced under natural conditions.

E. Morphological Responses (p. 1079)

How do morphological responses to environmental change differ from behavioral and physiological responses? What are two examples of morphological responses?

Organisms can react to environmental change with developmental or growth responses that alter body form or internal anatomy.
- Slower than behavioral or physiological responses.
- Plants are more morphologically plastic than animals.

F. **Adaptation over Evolutionary Time** (*p. 1079–1080*)

What is the relationship between adaptation over evolutionary time and the responses to environmental changes? How can adaptation affect distribution of organisms?

It is important to remember that the behavioral, physiological, and morphological responses to environmental change have evolved over time to their current levels through natural selection.
- What appear to be short-term adjustments are actually evolutionary adaptations to maintain homeostasis.
- Natural selection also places constraints on the distribution of populations by adapting them to localized environments.
- Organisms adapted to one type of environment may not survive if dispersed to a foreign environment or may become extinct if the local environment changes to beyond their tolerance limits.
- The existence of a species in a particular location depends on the species reaching that location and being able to survive and reproduce after getting there.

47   POPULATION ECOLOGY

CHAPTER OUTLINE

What is a population? With what characteristics of a population is population ecology concerned?

<u>Population</u> = Individuals of one species simultaneously occupying the same general area, utilizing the same resources, and influenced by similar environmental factors.

Population ecology is concerned with fluctuations in population size and the factors that regulate populations.

I.   DENSITY AND DISPERSION (p. 1084–1086)

How can you distinguish between population density and population dispersion? How can population density be measured?

<u>Population density</u> = The number of individuals per unit area or volume.

<u>Population dispersion</u> = The pattern of spacing for individuals within the boundaries of the population.

A.   Measuring Density (p. 1084)

It is usually impractical or impossible to count all individuals in a population, so ecologists use a variety of sampling techniques to estimate densities and total population size.

- May count all individuals in a sample plot, or *quadrant*. Estimates become more accurate as sample plots increase in size or number.
- May estimate by indirect indicators such as number of nests or burrows, or by droppings or tracks.
- May use the *mark-recapture method*.

In the mark-recapture method, animals are trapped within boundaries, marked and after time retrapped.

- The number of individuals in a population (N) is figured by the formula:

$$N = \frac{(\text{number marked}) \times (\text{total catch the second time})}{\text{number of recaptures}}$$

- Assumes marked individuals have same probability of being trapped as unmarked individuals. This assumption is not always valid.

B. Patterns of Dispersion (p. 1085–1086)

How would you define a population's range? What are the three patterns of spacing found in populations? What conditions would result in each of these patterns?

Range = Geographic limits within which a population lives.
☞ Local densities may vary substantially because not all areas of a range provide equally suitable habitat.

Individuals exhibit a continuum of three general patterns of spacing in relation to other individuals:

1. *Clumping* is when individuals are aggregated in patches.
   ☞ May result from the environment being heterogenous, with resources concentrated in patches.
   ☞ May be associated with mating or other social behavior.

2. *Uniform* is when the spacing of individuals is even.
   ☞ May result form antagonistic interactions of individuals of the population.
   ☞ For example, competition for some resource or social interactions that set up individual territories for feeding, breeding or nesting.

3. *Random* is when individual spacing varies in an unpredictable way.
   ☞ Occurs in the absence of strong attractions or repulsions among individuals.
   ☞ Not very common in nature.

II. DEMOGRAPHY (p. 1087–1089)

Demography = The study of the vital statistics affecting population size.
☞ Reflects relative rates of processes that add individuals to a population (birth immigration) versus processes that eliminate individuals (death, migration).
☞ Birth and death rates vary among population subgroups depending on age and sex.

A. Age Structure and Sex Ratio (p. 1087)

How can age structure, generation time and sex ratio in a population affect the population's growth?

Age structure = Relative numbers of individuals of each age in a population.
☞ Generation overlap when average life span is greater than the time it takes to mature and reproduce.
☞ Death rate often greatest for the very young and very old.
☞ Birth rate greatest for those of intermediate age.

In general, a population with more older, nonreproductive individuals will grow more slowly, than a population with a larger percentage of young, reproductive age individuals.

Generation time

Sex ratio

A shorter *generation time* (span of time between an individual's birth and that of their offspring) results in faster population growth, assuming birth rate is greater than death rate and all other factors being equal.

The *sex ratio* of a population is the proportion of individuals of each sex. This ratio affects population growth since the number of females is usually directly related to the expected number of births; the number of males may be less significant since one male may mate with several females.
- ☞ In strictly monogamous species, the number of males may be more significant in effecting the birth rate.

B. Life Tables and Survivorship Curves (p. 1087–1089)

What is a life table? A cohort? What characteristics of a population would result in a Type I survivorship curve? A Type II curve? A Type III curve?

Life tables

Cohort

Life tables describe how mortality varies with age over a time period corresponding to maximum life span. (See Campbell, Table 47.2)
- ☞ Constructed by following the fate of a group, or *cohort*, of new organisms until all are dead or by using the age at death of a sample of individuals.

Survivorship curves

Survivorship curves plot the numbers in a cohort still alive at each age.

Organisms may follow a continuum of three general types of survivorship curves: (See Campbell, Figure 47.6)

1. Type I is flat during early and middle life and drops suddenly as death rate increases among the elderly.
   - ☞ Associated with species such as humans and other large mammals that produce few offspring that are well cared for.
2. Type II is intermediate with mortality more constant over life spans.
   - ☞ Seen in *Hydra* and the gray squirrels.
3. Type III shows very high death rates for the young followed by lower death rates.
   - ☞ Associated with organisms, such as oysters, that produce very many offspring but provide little or no care.

Some invertebrates show a "stair-stepped" curve with brief periods of high mortality during molts, followed by periods of lower mortality when the exoskeleton is hard.

III. THE EVOLUTION OF LIFE HISTORIES (p. 1089–1092)

Life histories

Birth, reproduction and death, fundamental components of life histories, affect population growth over ecological time.

A life history pattern is the result of natural selection operating over evolutionary time.

What three life history characteristics affect the number of offspring produced by an individual? What effect could each have?

Three major life history characteristics affect the number of offspring produced by an individual, which in turn affects a population's intrinsic rate of increase:

1. Clutch size is the number of offspring produced at each reproductive episode.
    - Generally, large clutch size means smaller eggs and offspring; typical of organisms with a Type III survivorship curve.
    - Small clutch size usually produces larger offspring; found in organisms with Type I and Type II survivorship curves.

2. Number of reproductive episodes per lifetime.
    - Some organisms reproduce only one time during their lifespan; these organisms invest their entire energy budget into the production of a large number of offspring; reproducing organisms do not survive to reproduce again (e.g. Pacific salmon, annual plants).
    - Many plants and animals reproduce several times during their lifespan; partition their energy into maintenance, growth, and reproduction; continue to live and have subsequent reproductive episodes.

3. Age at first reproduction.
    - Reproduction at a younger-than-average age may reduce a female's reproductive potential by reducing the amount of energy available for growth and maintenance.
    - Female's that delay reproduction tend to be larger due to energy used for growth and maintenance; older (larger) females produce larger clutches and appear to maximize their reproductive output by delaying.

What distinguishes the two basic life history patterns? How could natural selection have shaped these two patterns?

Two concepts have been proposed with regards to how evolutionary forces have produced the basic life history patterns. (See Campbell, Table 47.3)

1. An *opportunistic life history* is based on the production of a large number of offspring during a single reproductive episode.
    - Usually exhibited by small species that mature rapidly.
    - Low numbers of offspring usually survive and population sizes fluctuate dramatically.
    - In variable environments many individuals reproduce when conditions are good; many die without reproducing when conditions are bad.
    - Natural selection has emphasized the production of large numbers of offspring in these organisms rather than individual survival.
    - Opportunistic species take advantage of environmental opportunities by dispersing to open or disturbed habitats where rapid maturation and reproduction permits them to establish a large population.
    - Desert annuals and garden weeds are opportunistic species.

2. An *equilibrial life history* is based on repeated reproductive episodes that produce smaller numbers of well-endowed offspring likely to survive to adulthood.
    - Usually larger species that mature slowly.
    - High survival rate of offspring results in more stable population sizes that vary around an equilibrium point.

☞ Well developed mechanisms to maintain homeostasis reduces the influence of environmental variation.
☞ Natural selection has emphasized survival of well-endowed offspring rather than production of large numbers.
☞ Slow maturation and parental care are typical of these species.
☞ Large terrestrial vertebrates are equilibrial species.

Why do most species exhibit life histories somewhere along the opportunistic-equilibrial continuum? What effect can environmental conditions and interspecific interactions have on life history patterns?

Designations of opportunistic or equilibrial life histories represent the two extremes in life history patterns. Research studies have indicated that most species vary along an opportunistic-equilibrial continuum due to factors which influence the populations.

☞ Different populations of the same species living in different environments show variations: dandelions in frequently mowed areas are smaller and produce more seeds than those in undisturbed areas.
☞ A stressful environment may be one where conditions remain severe for long periods with little fluctuation (extreme cold or dim light); species in these environments mature slowly and produce few offspring infrequently, characteristics of equilibrial species that have homeostatic mechanisms to survive fluctuating environments.
☞ Predation, and other interspecific interactions, may also affect life history patterns. *Daphnia* is a freshwater crustacean which develops a round body form with a large brood chamber in the spring when predators are scarce; when predation is high during the summer, *Daphnia* develops long spines and an enlarged "helmet" which reduces the chance of being eaten but compresses the size of the brood chamber so only 50% of the number of eggs can be carried.

IV. MODELS OF POPULATION GROWTH (p. 1092-1097)

Indefinite increases in population size do not occur. A population may increase rapidly from a low level under favorable environmental conditions, but this increase in numbers will eventually approach the level where resources cannot support continued increases. The combination of limited resources and other factors will stop the growth of the population.

☞ Mathematical models based on accurate assumptions provide predictions about how populations respond to changes in factors which affect population growth.

A. Exponential Population Growth (p. 1092-1093)

What is exponential population growth? What is the relationship between population growth and the birth and death rates?

A population consisting of a few individuals which lives in an environment with no limiting factors (no restrictions on available energy, growth, or reproduction) will increase over time in proportion to the birth and death rates:

Change in population size during time interval = Births during time interval − Deaths during time interval

In a mathematical form, this equation becomes:

$$\frac{\Delta N}{\Delta t} = B - D$$

where N = population size, t = time, B = absolute number of births during the time interval, and D = absolute number of deaths during the time interval.

Since populations differ in size, the basic model must be altered in order to apply it to other populations. This alteration involves a conversion from absolute birth and death rates to average numbers of births and deaths per individual during the specified time interval.

- ☞ If b = the average birth rate, its value would equal 0.034 in a population of 1000 individuals where 34 births occurred (34/1000 = 0.034) in a year.
- ☞ If d = the average death rate, its value would equal 0.016 in a population of 1000 individuals where 16 deaths occurred (16/1000 = 0.016) in a year.

Including average (or per capita) birth and death rates would alter the previous equation so that expected numbers of births and deaths which would change the population could be predicted for a population of any size:

$$\frac{\Delta N}{\Delta t} = bN - dN$$

- ☞ The change in size of our 1000 individual population would be:

    $\Delta N/\Delta t = (0.034)(1000) - (0.016)(1000)$

    $\Delta N/\Delta t = 34 - 16$

    $\Delta N/\Delta t = 18$

- ☞ If another population of the same species (same b and d values) contained 1500 individuals, then:

    $\Delta N/\Delta t = (0.034)(1500) - (0.016)(1500)$

    $\Delta N/\Delta t = 51 - 24$

    $\Delta N/\Delta t = 27$

What is the relationship between the differences in per capita birth and death rates (r) and population growth?

Population ecologists study overall changes in population sizes and use r to represent the difference between per capita birth rates and per capita death rates:

$$r = b - d \quad \text{thus} \quad \frac{\Delta N}{\Delta t} = rN$$

- ☞ *Zero population growth* (ZPG) occurs where per capita birth and death rates are equal, thus $r = 0$. (Note that births and deaths still occur, but are equal in number.)
- ☞ A population is increasing in size if $r > 0$ (more births than deaths); it is decreasing in size if $r < 0$ (fewer births than deaths).

Zero population growth

Many population ecologists use a slightly different equation based on differential calculus to express population growth in terms of instantaneous growth rates:

$$\frac{dN}{dt} = rN$$

☞ This expresses the population growth over very short time intervals.

What is an intrinsic rate of increase? What are the characteristics of an r-selected species? What shape is their growth curve and why?

A population living under ideal conditions will increase at the fastest rate possible; nutrients are abundant and only the physiological capacity of the individuals limits reproduction.

☞ The maximum population growth rate is called the *intrinsic rate of increase* and is symbolized by r_{max}.

Exponential population growth is the population increase under ideal conditions due to intrinsic rate of increase. It is expressed as:

$$\frac{dN}{dt} = r_{max}N$$

☞ The size of the population increases rapidly due to ideal conditions of unlimited resources.
☞ Produces a J-shaped growth curve. (See Campbell, Figure 47.9a)
☞ Although the intrinsic rate of increase is constant, more new individuals accumulate when the population is large than when it is small; this is due to the fact that N gets larger.

Opportunistic species often exhibit periods of exponential population growth.
☞ Sometimes referred to as *r-selected* species or populations because their growth rates are close to r_{max}.
☞ Usually have short generation times and high reproductive potentials.
☞ Generation time and r_{max} are inversely related. (See Campbell, Figure 47.10)

B. Logistic Population Growth (p. 1094–1097)

What is the carrying capacity of a habitat? How can carrying capacity affect the intrinsic rate of increase of a population? What factors help determine the carrying capacity?

As a population increases in size, the higher density may influence the ability of individuals to obtain resources necessary for maintenance, growth and reproduction.

☞ Populations inhabiting environments which contain a finite amount of available resources reach a size where further increases in number reduces the share of resources available to each individual.
☞ The *carrying capacity* of a habitat is the maximum stable population size that the particular environment can support over a relatively long time period.
☞ Carrying capacity is an environmental property that varies over space and time with the abundance of limiting resources.

Crowding and resource limitation can greatly effect the population growth rate.
- ☞ Insufficient resources may reduce the number of births occurring in a population (lower *b*).
- ☞ If enough energy cannot be obtained for maintenance, the death rate increases (higher *d*).
- ☞ A decrease in *b* and/or an increase in *d* results in a lower overall population growth rate (smaller *r*).

What is logistic population growth? What is the relationship among carrying capacity (K), population size (N) and population growth (r)?

Logistic population growth assumes the rate of population growth (*r*) slows as the population size (N) approaches the carry capacity (K) of the environment.
- ☞ A mathematical model for logistic population growth incorporates the effect of population density on *r*, allowing it to vary from r_{max} when resources are plentiful to zero when the carrying capacity is reached.

The equation for logistic population growth is:

$$\frac{dN}{dt} = r_{max}N \left(\frac{K - N}{K} \right)$$

- ☞ K = carrying capacity; the maximum sustainable population.
- ☞ K−N = the number of new individuals the environment can accommodate.
- ☞ (K−N)/K = percentage of K available for population growth.
- ☞ Multiplying r_{max} by (K−N)/K reduces the value of *r* as N increases. (See Campbell, Table 47.6)

The implications of the logistic growth equation at varying population sizes for a growing population are:
- ☞ When N is low, (K−N)/K is large and *r* is only slightly changed from r_{max}.
- ☞ When N is large and resources are limiting, (K−N)/K is small; this reduces *r* substantially from r_{max}.
- ☞ When N=K, (K−N)/N is 0 and *r*=0; this means the number of births is equal to the number of deaths and zero population growth occurs.

Why does the logistic model of population growth produce a sigmoid growth curve? What characterizes a K-selected species?

The logistic model of population growth produces a sigmoid (S-shaped) growth curve. (See Campbell, Figure 47.9b)
- ☞ Intermediate population sizes add new individuals most rapidly since the breeding population is of a substantial size and the habitat contains plentiful amounts of resources and available space.
- ☞ As N approaches K, the growth rate slows due to limitations in available resources.

Logistic population growth

K-selected

Equilibrial species may experience population growth similar to that predicted by the logistic model.
- ☞ Long generation times and small clutch sizes limit reproductive potential.
- ☞ Populations usually do not fluctuate drastically.
- ☞ Sometimes referred to as *K-selected* since their populations tend to stabilize around the carrying capacity.

Carrying capacities can be determined by many factors:
- ☞ Energy limitations (food resources) is the most common determinant of K.
- ☞ Other factors include: availability of specialized nesting sites required by some birds; roosting sites as for some bats; shelters and refuges from potential predators; accumulation of toxic metabolic wastes.

What permits the logistic growth model to be applied to studies of intraspecific competition?

Intraspecific competition

The prime implication of the logistic growth model is that increasing population density reduces resource availability and resource limitations ultimately limits population growth. This model can thus be applied to *intraspecific competition*: the reliance of two or more individuals of the same species on the same limited resource.
- ☞ Competition becomes more intense as population size increases, and r is reduced in proportion to the intensity of competition.
- ☞ *Territoriality* (the defense of a well-bounded physical space) is a behavioral mechanism to reduce intraspecific competition since each individual protects resources only within their own territory; competition still occurs when individuals compete for space to establish their territories.

Territoriality

Laboratory populations of paramecia and *Daphnia* show population growth rates which fit the predicted S-shaped curve fairly well. (See Campbell, Figures 47.13a and 47.13b)
- ☞ These represent relatively unnatural situations of idealized conditions without predators and other species.
- ☞ Studies of wild organisms which have been introduced into new habitats and populations which are rebounding after near elimination by disease or hunting, provide general support for logistic population growth. (See Campbell, Figure 47.13c)

Why do the assumptions of the logistic model not apply to all populations? What is the Allee effect?

Some assumptions of the logistic model that do not hold true for all populations:
1. Even at low levels, each individual added may have the same negative effect on population growth rate.
   - ☞ The *Allee effect* points out that individuals may benefit by population increase.
   - ☞ For example, a plant standing alone would suffer from excessive wind, but be protected in a clump of individuals from wind and/or predation.

Allee effect

- Solitary animals like rhinoceri have a greater chance of locating a mate during breeding season if populations are higher.
- When a population is at low levels, there is a greater possibility that chance events will eliminate all individuals or inbreeding will lead to reduction in fitness.

2. Assumes that populations approach carrying capacity smoothly.
    - Often see a lag time before the negative effects of an increasing population are realized, causing the population to overshoot carrying capacity. Eventually, deaths exceed births and population size drops below carrying capacity; therefore, many populations appear to oscillate about some general carrying capacity.

3. Populations do not necessarily remain at, or even reach, levels where population density is an important factor; so, in these cases, the idea of carrying capacity does not really apply.
    - Seen in short-lived, quickly reproducing insects and opportunistic species.

Although the logistic model fits few if any real populations, it incorporates ideas that apply to many.

IV. THE REGULATION OF POPULATIONS (p. 1097–1101)

Populations are regulated by density-dependent factors and density independent factors, either separately or in combination. The relative importance of these factors varies between opportunistic and equilibrial species and their specific circumstances.

A. Density-Dependent Factors (p. 1097–1099)

What criteria can be used to classify a condition or interaction as a density-dependent factor with regards to population growth? What are some density-dependent factors and how do they effect population growth?

In restricting population growth, a *density-dependent factor* affects a greater percentage of individuals in a population as the number of individuals increases; it will also affect each individual more strongly.

- Population growth declines because death rate increases, birth rate decreases or both.
- Resource limitation is one such factor.
- A reduction in available food often limits reproductive output as each individual produces fewer eggs or seeds.
- Health and survivorship also decrease as crowding results in smaller, less robust individuals.
- Many predators concentrate on a particular prey when its population density is high, taking a greater percentage than usual.

Intrinsic factors may also play a role in regulating population size.
- Population growth rate decreases may occur even when food and shelter are abundant since high densities may cause physiological changes that delay sexual maturation or otherwise inhibit reproduction.

Density-dependent factor

Density-independent factors

B. Density-Independent Factors (p. 1099–1100)

What criteria can be used to classify a condition or interaction as a density-independent factor with regards to population growth? What are some density-independent factors and how do they affect population growth?

Density-independent factors affect the same percentage of individuals regardless of population size.

- ☞ Weather, climate and natural disasters such as freezes, seasonal changes, hurricanes and fires are examples; the severity and time of occurrence being the determining factor on what number of organisms is affected.
- ☞ In some natural populations, these effects routinely control population size before density-dependent factors become important.

C. Interaction of Regulating Factors (p. 1100)

How can density-dependent and density-independent factors work together to control a population's growth?

Many populations remain fairly stable in size, close to carrying capacity determined by density-dependent factors, yet display short term fluctuations due to density-independent factors.

- ☞ Long term population size may remain the same, masking short term effects of some factor such as an extremely cold winter.

The relative importance of density-dependent and density-independent regulation, which usually act together to regulate a population, may vary seasonally.

- ☞ A severe winter may greatly reduce a population due to cold temperatures (density-independent) and intraspecific competition for limited food (density-dependent).
- ☞ This reduction in population size may benefit the surviving adults by reducing competition for food in the following spring.

D. Population Cycles (p. 1100–1101)

What conditions and interactions may have resulted in the evolution of population cycles? What advantages and disadvantages are associated with these cycles?

Some bird, mammals and insect populations fluctuate with regularity (for example, lemmings show a 4-year cycle, snowshoe hares show a ten-year cycle).

Several hypotheses to explain population cycles are:

1. Crowding affects the organisms endocrine systems: stress from high density may alter hormone balance and reduce fertility, increase aggressiveness, and induce mass emigrations.

2. Caused by a time lag in the response to density-dependent factors, which cause the population to overshoot and undershoot the carrying capacity.
   - ☞ High density of snowshoe hares causes a deterioration of food quality.
   - ☞ May involve predation as an accessory factor.

3. Periodical cicadas have population cycles of 13 or 17 years.
   - May be an adaptation to avoid predator-prey interactions.
   - Cicada populations are controlled in some local regions by a fungus whose spores can survive in soil for the years between cicada outbreaks.

VI. HUMAN POPULATION GROWTH (p. 1101–1104)

What effect has the development of agriculture and the Industrial Revolution had on human population growth?

Human population has been growing exponentially for centuries. (See Campbell, Figure 47.18)
- Advent of agriculture 10,000 years ago increased birth rates and decreased death rates.
- Since the Industrial Revolution, exponential growth resulted mainly from a drop in deaths, especially infant mortality. The growth was due to improved nutrition, better medical care and sanitation.

How many humans the Earth can support cannot be calculated.
- Agriculture and industrial technology have increased K at least twice.

Worldwide population growth is a mosaic with some countries having near zero population growth while others have relatively high growth rates.

How has variations in age structure among countries produced a mosaic of population growth rates around the globe? Why is human reproduction considered unique among the animals?

Age structure within each country appears to be a major factor in the variation of population growth rates. (See Campbell, Figure 47.20)
- A relatively uniform age distribution, where individuals of reproductive age or younger are not disproportionately represented, lends itself to a stable population size.
- A population which contains a large proportion of reproductive age or younger individuals will face a sudden increase in rate of population growth in the future.

A unique feature of human reproduction is the ability to be consciously controlled by voluntary contraception or government sanctions.
- Social change, such as women delaying reproduction in favor of employment and advanced education, also affects birth rates.

How has the Earth's carrying capacity for humans changed?

Technology increased the Earth's carrying capacity for humans.
- What that carrying capacity is and how we approach it are questions generating concern and debate.

The human population will eventually stop growing due to social changes, individual choice, government intervention, or increased mortality due to environmental limitations.
- Hopefully, if the human population reaches K, it will approach it smoothly and level off.
- If the population fluctuates around K, periods of increase will be followed by times of mass death due to famine and disease.

48

COMMUNITY ECOLOGY

CHAPTER OUTLINE

A *community* consists of all the organisms inhabiting a particular area; an assemblage of populations of different species living close enough together for potential interaction.

I. TWO GENERAL VIEWS OF COMMUNITIES (p. 1106–1107)

What is a community? How do the views of H.A. Gleason and F.E. Clements with respect to communities differ?

Among the pioneers of community ecology, there were two divergent views on why certain combinations of species are found together as members of a community.

1. H.A. Gleason depicted a community as a chance assemblage of species found in an area because they have similar environmental requirements; the individualistic hypothesis.
   - Predicts plant species will have an independent distribution along an environmental gradient — no discrete boundaries between communities.
   - In most cases, the composition of plant communities does change independently along some environmental gradient.
2. F.E. Clements saw each community as an assemblage of closely linked species having mandatory biotic interactions that cause the community to function as an integrated unit — a "superorganism"; the interactive hypothesis.
   - Predicts species should be clustered, with discrete boundaries between communities. Since the presence or absence of one species is governed by the presence or absence of other species with which it interacts.
   - Not generally seen in plants.

Animals in a community are often linked more rigidly to other organisms.
   - Some animals feed primarily on certain food items so their distribution is linked to distribution of their food source.

Distribution of almost all organisms is affected by both abiotic gradients and interactions with other species.

II. POPULATION INTERACTIONS (p. 1108–1118)

While adaptations to abiotic factors largely determine the geographic distributions of many species, all organisms are influenced by biotic interactions with other individuals.

- Intraspecific competition for resources was discussed in Chapter 42.
- *Interspecific interactions* are those that occur between populations of different species living in a community; take a variety of forms.

A. Coevolution (p. 1108–1109)

What is coevolution? How could this type of interspecific interaction result in an evolutionary change?

Coevolution = A change in one species acts as a new selective force on another species, and counteradaptation of the second species, in turn, affects selection of individuals in the first species.
- Studied most extensively in predator-prey relationships and in symbiosis.
- Often, what appears to be coevolution, may actually result from interactions with many species (not just the obvious two) where it is difficult to determine importance of various selective forces.

The association between passionflower vines and the butterfly *Heliconius* is believed to be an example of coevolution.
- The vines produce toxic chemicals to reduce damage to young shoots and leaves by herbivorous insects.
- Butterfly larvae of *Heliconius* can tolerate these chemicals due to digestive enzymes which break down the toxic chemicals (a counteradaptation).
- Females of some *Heliconius* species avoid laying eggs (which are bright yellow) on leaves where other yellow egg clusters have been laid; reduces intraspecific competition on individual leaves.
- Some species of passionflowers develop large, yellow nectaries which resemble *Heliconius* eggs; an adaptation that may divert egg-laying butterflies to other plants.
- These nectaries, as well as smaller ones, also attract ants and wasps which prey on butterfly eggs and larvae.

B. Predation (p. 1109–1110)

Predation = A community interaction where one species, the predator, eats another, the prey. Includes both animal-animal interactions and animal-plant interactions.

What common adaptations have evolved in predators? What advantages are associated with these adaptations?

Adaptations of predators to this interaction are obvious and familiar. Predators have:
- Acute senses that are used to locate and identify prey items (e.g. heat-sensing pits of rattlesnakes, and chemical sensors of herbivorous insects).
- Structures such as claws, teeth, fangs, stingers, and poisons which function to catch, subdue, or chew the prey item.
- Speed and agility to pursue prey or camouflage which permits them to ambush prey.

Coevolution

Predation, Predator
Prey

C. Plant Defenses Against Herbivores (p. 1110)

What are some common defensive adaptations found in plants which reduces predation by herbivores? How do these protect the plant from herbivores?

Since plants cannot escape herbivores, they have evolved several ways to protect themselves from predation.

Plants have mechanical and chemical defenses against herbivores.
- Thorns or microscopic hooks, spines, or crystals may be present to discourage herbivores.
- May produce *secondary compounds* that make vegetation distasteful or harmful (formed through metabolic diversion of major metabolic pathways).
- May produce analogues of insect hormones that cause abnormal development in insects that prey on them.

Plant defenses may act as selective factors leading to counteradaptations in herbivores that can then nullify those defenses and consume the plant.

D. Animal Defenses Against Predators (p. 1110–1112)

What are some common defensive adaptations found in animals? How does cryptic coloration differ from aposematic coloration? How do they help prey avoid predators?

Animal defenses against predators include: hiding, escaping and physical or chemical defense. Other forms of defense involving adaptive coloration have evolved repeatedly among animals.
- *Cryptic coloration* (coloration making prey difficult to spot against its surroundings) is common and only requires the animal to remain still to avoid detection.
- Deceptive marking such as large, fake eyes can startle predators, allowing prey to escape, or cause predators to strike a nonvital area.
- Some mechanical and chemical defenses actively discourage predators (e.g. porcupine quills and skunk spray).
- Some insects acquire chemical defense passively by accumulating toxins from plants they eat; these make them distasteful to predators.
- *Aposematic coloration* (bright coloration that acts as a warning of effective physical or chemical defense) appears to be adaptive since predators quickly learn to avoid prey with this coloration.

1. Mimicry (p. 1111–1112)

Mimicry is a phenomenon in which a *mimic* bears a superficial resemblance to another species, the *model*.
- Occurs in both predatory and prey species.
- Defensive mimicry in prey usually involves aposematic models.

How does Batesian mimicry differ from Mullerian mimicry? How do predators use mimicry to obtain prey?

- Batesian mimicry = A palatable species mimics an unpalatable model.
  - ☞ Mimic must be much less abundant than the model to be effective since predators must learn the coloration indicates a bad or harmful food item.

- Mullerian mimicry = Two or more unpalatable species resemble each other.
  - ☞ Gain additional advantage since predators learn more quickly to avoid prey with this coloration.

- Predators also use mimicry to lure prey.
  - ☞ The tongue of snapping turtles resembles a wriggling worm which attracts small fish into capture range.

E. **Interspecific Competition** (p. 1112–1116)

What is interspecific competition? The Competitive Exclusion Principle? An ecological niche? What is the relationship between a fundamental niche and a realized niche?

- Interspecific competition = When two or more species rely on similar limiting resources. They may have negative effects on one another.
  - ☞ May take the form of *interference competition* (actual fighting) or *exploitative competition* (consumption or use).
  - ☞ As the population density of one species increases, it may limit the density of the competing species as well as its own.

- Competitive exclusion principle = Concept that two species competing for the same limiting resources cannot coexist in the same community. One will use resources more efficiently, thus reproducing more rapidly and eliminating the inferior competitor.
  - ☞ Derived independently by A.J. Lotka and V. Voterra who modified the logistic model of population growth to incorporate interspecific competition.
  - ☞ Experiments by G.F. Gause confirmed competitive exclusion.

- Ecological niche = Sum total of an organism's use of biotic and abiotic resources in its environment; how it "fits into" an ecosystem.
  - ☞ May apply to species, populations or even individuals.
  - ☞ A *fundamental niche* is the resources an organism or population is theoretically capable of using under ideal circumstances.
  - ☞ Biological constraints such as competition restrict organisms to their *realized niche*: the resources an organism or population actually uses.
  - ☞ Two species cannot coexist in a community if their niches are identical.

How can interspecific competition and competitive exclusion result in an evolutionary change such as character displacement? How can resource partitioning affect community diversity?

Robert MacArthur

Resource partitioning

Character displacement

Symbiosis

Parasitism

Evidence demonstrating the importance of competition to community structure was provided by Robert MacArthur, who described *resource partitioning*.

Resource partitioning = The division of environmental resources by coexisting species populations such that the niche of each species differs by one or more significant factors from the niches of all coexisting species populations.
- ☞ Sympatric species may eat slightly different foods or utilize other resources in different ways.

Character displacement = The divergence of overlapping characteristics in two species living in the same environment as a result of resource partitioning.
- ☞ Another line of evidence indicating the importance of competition when comparing sympatric and allopatric populations of the similar species.
- ☞ Two closely related species of Galapagos finches (*Geospiza fuliginosa* and *G. fortis*) have beaks of similar size when the populations are allopatric; however, on the island where they are sympatric, a significant difference in beak depth has occurred.

F. Symbiotic Interactions (*p. 1116–1118*)

What is symbiosis? How can you distinguish between the different forms of this interspecific interaction? How has natural selection and coevolution influenced symbiotic relationships?

Symbiosis is a form of interspecific interaction in which a *host* species and a *symbiont* maintain a close association.
- ☞ Parasitic, commensal, and mutualistic interactions are also important determinants of community structure.

Parasitism is the form of symbiosis where one organism, the parasite, harms the host.
- ☞ The parasite derives its nourishment from the host by absorbing organic nutrients from the body fluids.
- ☞ Parasites are usually smaller than their hosts.
- ☞ *Endoparasites* live within the host tissues or body cavities.
- ☞ *Ectoparasites* attach to or briefly feed on the external surface of the host.

Natural selection favors parasites that are best able to locate and feed on a host.
- ☞ Infection by a parasite (locating a host) may be passive as when an endoparasite's egg is swallowed by a host.
- ☞ Active location of a host may involve thermal or chemical cues that help the parasite identify the host.

Defensive capabilities in potential hosts have also been selected for by natural selection.
- ☞ Secondary plant products toxic to herbivores may also be toxic to parasites.

- ☞ The vertebrate immune system provides defense against endoparasites.
- ☞ Many parasites are adapted to specific hosts, often a single species.
- ☞ Coevolution generally results in stable relationships in which the host is not killed (host death would also kill the parasite).

Commensalism is the form of symbiosis in which one partner benefits without significantly affecting the other.
- ☞ Evolutionary changes involve only the beneficiary.

Mutualism may involve two-way evolutionary adaptations because change in one species is likely to affect the other since both benefit from the relationship.
- ☞ Mutualism may be derived from interactions originally involving predation or parasitism.
- ☞ Examples include photosynthesis by algae in the tissues of corals and the specific interactions of pollinators and flowering plants.

III. COMMUNITY STRUCTURE: PATTERNS AND PROCESSES
(*p. 1118–1121*)

A. Community Characteristics (*p. 1118–1119*)

What four factors influence the structure of a community and how?

The characteristics exhibited by communities result from the species composition of the community and biological interactions among the populations.

1. The form of vegetation in a community influences which animals are present.
   - ☞ Large grazing mammals are found in grasslands and savannas; tree-nesting birds in forests.
   - ☞ The more structurally complex the vegetation, the greater the diversity of microhabitats.
2. *Trophic structure* (the various feeding relationships) is determined by predator-prey, host-parasite, and plant-herbivore interactions.
   - ☞ Determines energy flow and cycling of nutrients through consumers and decomposers.
3. Species *richness* (number of species in a community), species *relative abundance* (a measure of the community proportion of different species), and *diversity indexes* (mathematical formulas combining data on species richness and relative abundance) are used to provide a measure of species diversity for a community.
   - ☞ No agreement on which of several indexes provides the most appropriate balance of the two measures since the relative abundance of species is not necessarily a measure of its importance in overall community structure.
4. *Stability* refers to the community's ability to resist change and recover to its original composition after some disturbance.
   - ☞ Depends on community type and nature of disturbance.

Commensalism

Mutualism

Trophic structure

Species richness, Relative abundance

Diversity

Stability

B. Determinants of Community Characteristics (p. 1119–1121)

Population interactions influence the characteristics of particular communities.

1. The Role of Competition (p. 1119)

How can interspecific competition affect the characteristics of a community?

The importance of competition in developing community characteristics is related to whether population sizes are approaching carrying capacity and if resources are limiting.
- Many studies have documented interspecific competition, but it did not always result in competitive exclusion; competing species sometimes coexist at reduced densities.
- Competition is recognized as an important factor in regulating relative abundance in many species, and possibly species diversity in communities, but no one knows just how important it is.

2. The Role of Predation (p. 1120)

What role does predation play in shaping the structure of a community?

Predation plays a complex role in helping shape community structure.
- Predators rarely drive prey species to extinction or they would themselves disappear.
- Defensive abilities of prey prevent them from becoming extinct.
- Prey switching also protects prey species as predators change to different prey as the population of one is depleted due to predation.
- A *keystone predator* may actually exert a regulating effect on other species and maintain a higher community species diversity by reducing the densities of strong competitors; this would prevent competitive exclusion of poorer competitors.

3. The Role of Environmental Patchiness (p. 1120)

How does environmental patchiness affect the species diversity of a community? Why is it difficult to determine what factor is most important in structuring a community?

Environmental patchiness can increase a community's species diversity by facilitating resource partitioning among potential competitors.
- Results from different species being better adapted to different local conditions.

4. Evaluating Causative Factors (p. 1121)

Evaluating what factors are important in developing the characteristics of a community is difficult since predation, competition, symbiosis, and abiotic factors that cause patchiness all have significant impact.
- Communities are structured by multiple interactions of organisms with the biotic and abiotic factors in their environment.

Keystone predator

Community Ecology

☞ Which types of interactions are most important varies from one type of community to another, and even among different components of the same community.

IV. SUCCESSION (p. 1121–1124)

What is ecological succession? How does primary succession differ from secondary succession? How would you define a sere? A climax community?

Ecological succession = A transition in species composition over ecological time.
- ☞ Called *primary succession* if it begins in areas essentially barren of life due to lack of formed soil (e.g. volcanic formations).
- ☞ Called *secondary succession* if an existing community has been cleared by some disturbance that leaves the soil intact.
- ☞ In some cases, the community passes through a series of predictable transitional stages (*seres*) to reach a relatively stable, final stage called the *climax community*.

A. Causes of Succession (p. 1121–1123)

What factors influence the course of succession? How would you distinguish between autogenic and allogenic factors? How can biotic interactions such as inhibition and facilitation influence succession?

A variety of interrelated factors determines the course of succession.
- ☞ Such as tolerance for general abiotic conditions of the area, which can differ on a fairly local level, as on opposite sides of a mountain.

Early species are typically good colonizers with high fecundity and excellent dispersal mechanisms that may not compete well in established communities.

Changes in community structure during succession may be *autogenic*, induced by the organisms themselves.
- ☞ Organisms also affect the abiotic environment by modifying local conditions.
- ☞ Direct biotic interactions include *inhibition* of some species by others through exploitative or interference competitions (or both).
- ☞ Colonizers may inhibit some species while facilitating colonization by others.
- ☞ In *facilitation* the group of organisms representing one successional stage "paves the way" for species typical of the next stage (e.g. older leaves lower soil pH as they decompose, and lower soil pH facilitates the movement of spruce and hemlock into the area).
- ☞ Sometimes facilitation for other species makes the environment unsuitable for the very species responsible for the changes.

At climax stage, environmental conditions are such that one set of species can continue to maintain themselves.
- ☞ Communities may never truly reach a climax, but just appear to because further change is slow.
- ☞ Many communities are also routinely disturbed by *allogenic* (outside) factors during the course of succession.

Key terms (margin): Ecological succession; Primary succession; Secondary succession; Climax community; Autogenic; Inhibition; Facilitation; Allogenic factors

Community Ecology

B. Human Disturbance (p. 1123)

How do human disturbances affect community succession?

Human disturbance affects community succession.
- Logging and clearing for farmland have reduced and disconnected forests.
- Agricultural development has disrupted grasslands.
- After a community disturbance, the early stages of succession, characterized by weedy and shrubby vegetation, may persist for many years.
- Centuries of overgrazing and agricultural disturbance have contributed to current famine in parts of Africa by turning grasslands into barren areas.

C. Equilibrium and Species Diversity (p. 1124)

What is the relationship between species diversity and the climax community within the traditional view of ecological succession? How do K-selected and r-selected species fit into this view? How does the traditional view differ from the other view which includes the polyclimax?

The traditional concept of ecological succession predicts that species diversity reaches an equilibrium in the climax community.
- New colonists would be balanced by localized extinctions.
- Species diversity increases during succession.
- K-selected species replace r-selected species as population density increases and vegetation modifies the site.
- Emphasizes the importance of population interactions in shaping community structure.
- Interactions become more extrusive and varied, increasing diversity.
- The climax is reached when the web of interactions is so intricate that the community is saturated; no new species can enter unless a localized extinction occurs.

An opposing view sees communities as in continual nonequilibrium.
- The identity and numbers of species change in all successional stages, even climax.
- Emphasizes the importance of unpredictable factors (dispersal, disturbance) in shaping community structure.
- Disturbances affect species diversity and may prevent the community from reaching equilibrium.
- A mature community is a *polyclimax*, an unpredictable mosaic of patches at different successional stages.
- Local environmental heterogeneity contributes to the polyclimax structure as different species inhabit different patches.

What is the intermediate disturbance hypothesis? How does it relate the frequency and intensity of disturbances to species diversity?

Intermediate disturbance hypothesis states that species diversity is greatest where disturbances are moderate in frequency and severity because organisms typical of several successional stages will be present.

Equilibrium

Polyclimax

Intermediate disturbance hypothesis

☞ Supported by studies of clearings formed by trees falling in tropical rain forests; in these disturbed areas, immigrations and extinctions occur rapidly and species from different successional stages coexist.

D. Diversity and Community Stability (*p. 1124*)

No simple relationship exists between diversity and stability.
☞ Researchers do not agree on the definition of stability: some mean resistance to change, others mean resiliency.
☞ Is high diversity a cause of stability or does it result from a certain amount of instability?

V. BIOGEOGRAPHICAL ASPECTS OF DIVERSITY (*p. 1124–1128*)

What is biogeography? With what aspects of the community is it associated?

Biogeography = The study of the past and present distribution of individual species, entire floras, and entire faunas.
☞ Concerned with species diversity and community composition.

Terrestrial life can be divided into biogeographical realms having boundaries associated with continental drift patterns. (See Campbell, Figure 48.19)
☞ Thus species distribution reflects past history as well as present interactions of organisms with their environment.

A. Limits of Species Ranges (*p. 1125*)

What three factors usually limit a species to a particular range? What evidence is available to support the importance of these factors?

Limitations of a species to a particular range is due to:
1. Species may never have dispersed beyond its current range.
2. Those that spread beyond the observed range failed to survive.
3. The species has retracted from a once larger range to its current boundaries.

The first two limitations are distinguished by transplant experiments in which specimens of a species are moved outside of this range.
☞ If the specimens survive, they probably never dispersed there; if they die, it indicates an intolerance to abiotic conditions or an inability to compete limits their range.
☞ Most species fail to survive outside their range, but there are notable exceptions.

The third limitation is well supported by data from paleontology and historical biogeography.
☞ The fossil record includes evidence that close relatives of some living animals with limited ranges once had much wider ranges.
☞ For example, fossils of close relatives of elephants and camels have been found in North America.

Biogeography

B. Global Clines in Diversity (p. 1125–1126)

There are major clines in species diversity over large areas of Earth.
- Generally, diversity increases in moving from far northern and southern latitudes toward the equator.
- Diversity of terrestrial organisms at a given latitude decreases with altitude.
- In the ocean benthic community, diversity increases from shallow coastal waters toward deep sea.

What hypotheses have been proposed as possible explanations for the high diversity of tropical communities and the existence of latitudinal gradients in species diversity? Why do these hypotheses not apply to marine benthic communities?

Many hypothesis have been proposed as to why tropical communities have such a high diversity and why latitudinal gradients exist.

1. Tropical communities are very old and rarely experience major natural disturbance. This fosters rapid and frequent speciation and their age has allowed complex population interactions to evolve and develop to a greater extent than in temperate zones.
2. Tropical regions usually experience intermediate levels of disturbance and have greater environmental patchiness. This permits a greater diversity of plant species which forms a resource base for diverse animal communities.
3. Stability and predictability in tropical climates may permit organisms to specialize on a narrow range of resources. Small niches would reduce competition and permit a finer level of resource partitioning, both of which increases species diversity.
4. Increased solar radiation in the tropics increases the photosynthetic activity of plants. This would provide a larger resource base for other animals.
5. The structural complexity of tropical forests creates a great variety of microhabitats that other plants and animals can partition.
6. Diversity is self-propagating because complex predator-prey and symbiotic interactions in a diverse community prevent any populations from becoming dominant.

While most of these hypotheses will apply to some terrestrial organisms and communities, complex combinations of these factors probably operate under most circumstances.
- None of these hypotheses apply to the increased diversity of marine benthic communities with increased depth; long term stability in these habitats may explain the many relationships among these organisms.

C. Island Biogeography (p. 1126–1128)

Why are islands important to the study of factors affecting species diversity? What relationship between island size and distance from the mainland was developed by R. MacArthur and E.O. Wilson? What evidence supports their prediction?

Islands provide opportunities to study factors affecting species diversity of communities due to their isolation and limited size.
- Islands can be oceanic islands or habitat islands on land (lakes or mountain peaks separated by lowlands).

A general theory of island biogeography, developed by R. MacArthur and E.O. Wilson, predicts that species number on an island will reach an equilibrium that is directly proportional to island size and inversely proportional to island distance from mainland.

E.O. Wilson

- Number of species to inhabit islands is determined by immigration and extinction rates, themselves affected by island size and distance from the mainland.
- The smaller the island and the more distant from mainland, the lower the immigration rate due to difficulty in finding the island and the distance colonizers must travel.
- Smaller islands also have higher extinction rates since they have fewer resources and less diverse habitat to be partitioned; these factors increase the probability of competitive exclusion.
- Immigration and extinction rates also are affected by the number of species already present. As species numbers increase, immigration rate of additional new species decreases since new arrivals are less likely to represent a new species. Extinction rate increases due to increased chance of competitive exclusion.
- Eventually, an equilibrium will be reached where immigration rates match extinction rates. The number of species present at the equilibrium point being correlated to island size and distance from the mainland. (See Campbell, Figure 48.21)
- Theory supported by experimental evidence obtained through the research of E.O. Wilson and D. Simberloff who studied small mangrove islands off the tip of South Florida.

D. Simberloff

VI. CONSERVATION BIOLOGY: LESSONS FROM COMMUNITY ECOLOGY AND BIOGEOGRAPHY (*p. 1128–1129*)

What was the purpose of the Biodiversity Treaty presented at the United Nations Earth Summit in 1992?

Efforts to conserve as much of the Earth's biological diversity as possible led to the Biodiversity Treaty which was presented at the 1992 United Nations Earth Summit in Rio de Janeiro.
- The treaty was designed to protect a wide variety of species around the world.
- Conservation efforts developed out of: esthetic appreciation for other life forms; recognition that yet undiscovered species may provide useful products; and the understanding that natural environments are necessary to maintain proper functioning of the biosphere.
- Conservation efforts are complicated by religious and cultural opposition to human population control and the socio-economic problems facing humans in every region.
- Of 172 nations attending the conference, only the United States failed to sign the Biodiversity Treaty.

What contributions can studies of community ecology and island biogeography make to the conservation of biodiversity?

Conservation efforts currently focus on creating natural preserves which are protected from development and pollution and are designed to preserve particular species and communities. Information from community ecology and biogeography can be applied to the preserves in order to maximize the conservation of biodiversity.

1. Preserves can be viewed as islands surrounded by disturbed habitat.
    - ☞ Studies of island biogeography provided advance knowledge about what species a preserve with a particular set of characteristics can successfully protect.
    - ☞ For example, we know the number of species that can be protected in a preserve is directly related to the size of the preserve. Species in large preserves are less likely to encounter competitive exclusion.

2. Detailed knowledge of species requirements is essential to preserve design.
    - ☞ Minimum sizes of preserves are known for the successful preservation of some species, particularly large predators requiring large amounts of space.
    - ☞ To preserve a population of 500 leopards in Sri Lanka, an area of 5000 km^2 of undisturbed habitat would be necessary. A population of 500 tigers would require about twice as much habitat.
    - ☞ This situation is complicated since many birds and mammals would use only the central areas of a suitable habitat patch. This means the peripheral areas would not contribute to their conservation although other species might use these areas.

3. Would one large preserve be more beneficial to conservation of species than several small preserves representing the same area?
    - ☞ One large preserve would be preferable if the smaller preserves all contained the same representation of species.
    - ☞ Several small preserves would retain a larger total number of species than one large preserve if the community being preserved included environmental heterogeneity resulting in different species occupying different areas.
    - ☞ Several small preserves would also be less susceptible to disease epidemics which might not be able to cross unsuitable habitat.
    - ☞ Spatial relationships between small preserves must also be considered. Closely spaced preserves or those connected by land corridors might encourage the spread of protected species among the preserves or permit recolonization by a protected species in case of local extinction in one preserve.

In your opinion, why is it important to preserve the Earth's biodiversity?

49 ECOSYSTEMS

CHAPTER OUTLINE

What is an ecosystem? What two processes can only be studied at the ecosystem level?

<u>Ecosystem</u> = All organisms living in a given area along with the abiotic factors with which they interact; a community and its physical environment.
- ☞ Boundaries are usually not discrete.
- ☞ The most inclusive level of biological organization.
- ☞ Involves two processes which cannot be described at lower levels: energy flow and chemical cycling.

I. TROPHIC LEVELS AND FOOD WEBS (p. 1132–1133)

How would you define the trophic structure of an ecosystem? What are the five trophic levels possible in an ecosystem? What criteria can be used to distinguish among these five trophic levels?

Each ecosystem has a *trophic structure* which represents the different feeding relationships that determine the route of energy flow and the pattern of chemical cycling.
- ☞ Ecologists divide the species in a community or ecosystem into different *trophic levels* based on their main source of nutrition.

The five trophic levels are:

<u>Primary producers</u> = Autotrophs (usually photosynthetic) that support all other trophic levels either directly or indirectly by synthesizing sugars and other organic molecules using light energy.
- ☞ For example, terrestrial plants, aquatic photosynthetic protists, and cyanobacteria.
- ☞ An exception is communities of organisms living around hot water, deep sea vents where producers are chemosynthetic bacteria that oxidize H_2S (driven by geothermal energy).

<u>Primary consumers</u> = Herbivores that consume primary producers.
- ☞ Examples include terrestrial insects, snails, grazing mammals, seed-eating birds, aquatic zooplankton, and some fish.

<u>Secondary consumers</u> = Carnivores that eat herbivores.
- ☞ Includes terrestrial spiders, frogs, insects-eating birds, lions, many fish, and sea-stars.

<u>Tertiary consumers</u> = Carnivores that eat other carnivores.

<u>Detritivores</u> = Consumers that derive energy from organic wastes and dead organisms.

Ecosystem

Trophic structure

Trophic levels

Primary producers

Primary consumers

Secondary consumers

Tertiary consumers

Detritivores

☞ For example, bacteria and fungi.
☞ Also included are scavengers such as cockroaches and bald eagles.
☞ Often form a major link between primary producers and higher-level consumers.
☞ Important components of the recycling process.

What is a food chain? How does it differ from a food web?

Food chain Food chain = Transfer of food from trophic level to trophic level. (See Campbell, Figure 49.2)
☞ Rarely are unbranched since several different primary consumers may feed on the same plant species and a primary consumer may eat several species of plants.
Food webs ☞ Feeding relationships are usually woven into elaborate *food webs* within an ecosystem. (See Campbell, Figure 49.3)

II. **ENERGY FLOW** (p. 1133–1140)

What determines an ecosystem's energy budget? In what forms does energy flow through an ecosystem?

Energy for growth, maintenance, and reproduction is required by all organisms; some species also require energy for locomotion.
☞ Light energy is used by primary producers to synthesize organic molecules (photosynthesis) which are latter broken down to produce ATP (cellular respiration).
☞ Energy is obtained by consumers in the form of organic molecules that were produced at the previous trophic level. Thus, energy flows to higher trophic levels through food webs.
☞ Since only primary producers can directly utilize solar energy, an ecosystem's entire energy budget is determined by the photosynthetic activity of the system.

A. **The Global Energy Budget** (p. 1134)

What factors influence the amount of solar radiation available to an ecosystem?

The amount of solar radiation which strikes the Earth's surface shows dramatic regional variation which limits the photosynthetic output of ecosystems in different places.
☞ Earth receives an estimated 10^{22} joules (J) of solar radiation each day.
☞ Most of the solar radiation is reflected, absorbed, or scattered by the atmosphere, clouds, and dust particles in the air; this amount varies over different regions.
☞ The intensity of solar radiation also varies with latitude resulting in the tropics receiving the most input.
☞ Only a fraction of the solar radiation which reaches the biosphere strikes plants and algae (much hits bare ground or is absorbed or reflected by water) and these primary producers can only use some wavelengths for photosynthesis.

☞ Only about 1% of the visible light reaching primary producers is converted to chemical energy by photosynthesis; the photosynthetic efficiency also varies with the type of plant, light levels, and other factors.
☞ Even with all the variations mentioned above, the primary production of Earth collectively creates about 170 billion tons of organic material each year.

B. **Primary Productivity** (p. 1134–1138)

What is primary productivity? How is gross primary productivity allocated by the producers of an ecosystem? What is a standing crop biomass?

Primary productivity = The rate at which light energy is converted to chemical energy by autotrophs of an ecosystem.
☞ The total is known as *gross primary productivity* (GPP) which may be determined by measuring the total oxygen produced by photosynthesis.

Net primary productivity (NPP) = GPP − Rs (energy used by producers for respiration)
☞ NPP accounts for the organic mass of plants (growth) and represents storage of chemical energy available to consumers.
☞ The NPP:GPP ratio is generally smaller for larger producers with elaborate nonphotosynthetic structures (such as trees) which support large metabolically active stem and root systems.
☞ Can be expressed as biomass (expressed as dry weight since water contains no usable energy) added to an ecosystem per unit area per unit time ($g/m^2/yr$) or as energy per unit time ($J/m^2/yr$).

What factors influence the contribution of each ecosystem to the Earth's total productivity? Is there a difference between the most productive systems and those that contribute most to global productivity?

Primary productivity should not be confused with *standing crop biomass*.
☞ Primary productivity is the rate at which new biomass is synthesized by vegetation.
☞ Standing crop biomass is the total biomass of plants present at a given time.

Primary productivity varies among ecosystems and their sizes affect each ecosystem's contribution to the Earth's total productivity. (See Campbell, Figure 49.5)
☞ Tropical rainforests are very productive and contribute a large proportion to the planet's overall productivity since they cover a large portion of the Earth's surface.
☞ Estuaries and coral reefs are also very productive but make only a small contribution to planetary productivity since they do not cover an extensive area.
☞ The open ocean has a relatively low productivity but makes the largest contribution to overall productivity of any ecosystem due to its very large size.

Primary productivity

Gross primary productivity

Net primary productivity

Biomass

Standing crop biomass

716 Ecosystems

What factors can limit the productivity of an ecosystem? How does the importance of each factor vary among the different types of ecosystems?

Factors important in limiting productivity depend on the type of ecosystem and temporal changes such as seasons.

1. Generally, precipitation, temperature, and light intensity are factors limiting productivity in terrestrial ecosystems.
    - ☞ Productivity in terrestrial systems increases as latitudes approach the equator because availability of water, heat, and light increases in the tropics.

2. Productivity in terrestrial ecosystems may also be limited by availability of inorganic nutrients.
    - ☞ Plants require a variety of nutrients, some in large quantities and some in small quantities. Exact proportions vary among plants.
    - ☞ Primary production sometimes removes nutrients from the system faster than they can be replenished.
    - ☞ If a nutrient is removed in such quantities that sufficient amounts are no longer available, it becomes the *limiting nutrient*.
    - ☞ The limiting nutrient is one present in insufficient quantities to support further primary production which will slow down or cease.
    - ☞ Nitrogen and phosphorus are usually limiting nutrients since they are needed in large quantities but are often present in small or moderate amounts in natural environments.
    - ☞ Carbon dioxide availability sometimes limits productivity.

3. An aquatic ecosystem's productivity is usually determined by light intensity, water temperature, and availability of inorganic nutrients.
    - ☞ Light intensity and temperature affect primary productivity of phytoplankton in the open oceans; productivity is highest near the surface and decreases with depth.
    - ☞ Inorganic nutrients are limiting at the surface of open ocean waters with nitrogen and phosphorus in especially short supply; this is a primary reason for the relatively low productivity of open oceans.
    - ☞ Marine phytoplankton is most productive where upwellings bring nutrient rich waters to the surface; these areas (usually in polar seas) are more productive than tropical seas.
    - ☞ Freshwater ecosystem productivity also varies from the surface to the depths in relation to light intensity; water temperature is important and seasonal fluctuations in productivity occur in temperate zones; availability of inorganic nutrients is sometimes limiting, but biannual turnovers bring nutrients to the surface waters.

C. Energy Transfers and Ecological Pyramids (*p. 1138–1140*)

What is secondary productivity? Why does productivity decline at each successive trophic level?

Secondary productivity = The rate at which consumers incorporate organic material (the food they eat) into new biomass, which can be equated to chemical energy.

Productivity declines with each transfer of energy to the next trophic level. Not all energy stored in biomass can be converted to productivity at the next trophic level due to loss of organized energy dissipated as heat.

☞ Herbivores only consume a small fraction of available plant material.

☞ Typically, 2/3 of organic material absorbed by herbivores is used for cellular respiration which degrades compounds into inorganic waste products and heat. The other 1/3 adds biomass to the trophic level.

☞ Carnivores are more efficient at converting food into biomass but more is used for cellular respiration, further decreasing energy available to the next trophic level.

How would you define ecological efficiency? In what ways can energy loss be represented? How can you distinguish among pyramids of productivity, biomass pyramids and pyramids of numbers? What does each represent?

Ecological efficiency = Ratio of net productivity at one trophic level compared to net productivity at the level below.

☞ Can vary greatly depending on the organisms involved, but is roughly 10%.

☞ This means that 90% of the energy available at one trophic level never transfers to the next.

Loss of energy in a food chain can be represented diagrammatically:

1. A *pyramid of productivity* has trophic levels stacked in blocks proportional in size to the energy acquired from the level below.
    ☞ Usually bottom heavy since only 10% of energy is transferred. (See Campbell, Figure 49.6)

2. A *biomass pyramid* has each tier symbolizing the total dry weight of all organisms in an ecosystem's levels at any given time. (See Campbell, Figure 49.7)
    ☞ Biomass represents chemical energy stored in the organic matter of a trophic level.
    ☞ Most narrow sharply from producers at the base to top-level carnivores at the top.
    ☞ Some aquatic systems are inverted since producers can have high turnover rates. They grow rapidly but are consumed rapidly, leaving little standing biomass.
    ☞ Biomass of top-level carnivores is usually small compared to the total biomass of producers and lower-level consumers.

Food chains are limited to 3–5 links due to the multiplicative loss of energy.

☞ Limits the biomass of top-level carnivores that can be supported.

☞ Only about 1/1000 of the chemical energy fixed by photosynthesis flows through a food web to a tertiary consumer.

☞ Only 3–5 trophic levels can be supported since biomass at the apex is insufficient to support another level.

718 Ecosystems

- The fact that top-level consumer biomass is concentrated in a small number of individuals can be reflected in a *pyramid of numbers* where each tier represents the number of organisms at each trophic level. (See Campbell, Figure 49.8)
- Predators (top-level consumers) are highly susceptible to extinction when their ecosystem is disturbed due to their small population and wide spacing within the habitat.

III. CHEMICAL CYCLING (p. 1140–1145)

Despite an inexhaustible influx of energy in the form of sunlight, continuation of life depends on recycling of essential chemical elements.
- These elements are continually cycled between the environment and living organisms as nutrients are absorbed and wastes released.

What are biogeochemical cycles? What are the four main reservoirs of elements and what processes transfer elements between these reservoirs?

<u>Biogeochemical cycles</u> = Nutrient circuits involving both biotic and abiotic components of ecosystems.
- Water is an important vehicle for transfer of chemicals and itself moves in a global hydrologic cycle. (See Campbell, Figure 49.9)
- Elements such as carbon, oxygen, sulfur, and nitrogen have gaseous forms, thus, their cycles are global in character and the atmosphere serves as a reservoir.
- Elements less mobile in the environment like phosphorus, potassium, calcium and trace elements generally cycle on a more localized scale over the short term. The soil serves as the main reservoir for these elements.

A general scheme of nutrient cycling includes the four main reservoirs of elements and the processes that transfer elements between reservoirs. (See Campbell, Figure 49.10)

1. Reservoirs are defined by two characteristics: whether they contain organic or inorganic materials; and whether or not the materials are directly available for use by organisms.
   - *Available organic reservoir.* This reservoir contains the living organisms and the nutrients are readily available when organisms feed on one another.
   - *Unavailable organic reservoir.* This reservoir is comprised of coal, oil, and peat which formed from organisms that died and were buried millions of years ago; these nutrients are unavailable since they cannot be directly assimilated.
   - *Available inorganic reservoir.* Included in this reservoir are all elements, ions, and molecules present in the soil or air and those dissolved in water. Organisms can directly assimilate these nutrients from the soil, air, or water.
   - *Unavailable inorganic nutrients.* Elements in this reservoir are tied up in limestone and minerals of other rocks. These nutrients cannot be assimilated until released by weathering or erosion.

<u>Biogeochemical cycles</u>

2. Various processes are involved in the transfer of nutrients between the four reservoirs which form the basis for biogeochemical cycling. The general schemes were determined by adding small amounts of radioactive tracers to systems in order to follow the movement of elements.
   - ☞ Weathering and erosion are the primary processes which move nutrients from the unavailable inorganic reservoir to the available inorganic reservoir.
   - ☞ Erosion is also important, along with the burning of fossil fuels, in moving nutrients from the unavailable organic reservoir to the available inorganic reservoir.
   - ☞ Nutrients are transferred from the available organic reservoir to the unavailable organic reservoir only by the covering of detritus by sediments and its eventual fossilization to oil, coal, or peat.
   - ☞ Sedimentary rock formation is the process which moves nutrients from the available inorganic reservoir to the unavailable inorganic reservoir.
   - ☞ Nutrients enter the available organic reservoir from the available inorganic reservoir through photosynthesis and assimilation by living organisms.
   - ☞ Nutrients are transferred from the available organic reservoir to the available inorganic reservoir by respiration, decomposition, excretion, and leaching.

   A. **The Carbon Cycle** (p. 1142–1143)

What occurs during the carbon cycle? Why can it be considered the reciprocal processes of photosynthesis and cellular respiration? How is carbon diverted from the cycle?

During the carbon cycle, autotrophs acquire CO_2 from the atmosphere by diffusion through leaf stomata, incorporating it into their biomass. Some of this becomes a carbon source for consumers and respiration returns CO_2 to the atmosphere.
   - ☞ Photosynthesis and cellular respiration form a link between the atmosphere and terrestrial environments. (See Campbell, Figure 49.11)
   - ☞ Carbon cycles fast. Plants have a high demand for CO_2, yet CO_2 is present in the atmosphere at a low concentration (0.03%).
   - ☞ Carbon loss by photosynthesis is balanced by carbon release during respiration.

Some carbon is diverted from cycling for longer periods of time, as when it accumulates in wood or other durable organic material.
   - ☞ Decomposition eventually recycles this carbon to the atmosphere.
   - ☞ Can be diverted for millions of years, such as in the formation of coal and petroleum.

The amount of atmospheric CO_2 decreases in the Northern Hemisphere in summer due to increased photosynthetic activity.
   - ☞ Amounts increase in the winter when respiration exceeds photosynthesis.

Atmospheric CO_2 is increased by combustion of fossil fuels by humans, disturbing the balance.
- ☞ The ocean may act as a buffer to absorb excess CO_2.

Why is the carbon cycle more complex in aquatic systems than in terrestrial systems?

In aquatic environments photosynthesis and respiration are also important but carbon cycling is more complex due to interaction of CO_2 with water and limestone.
- ☞ Dissolved CO_2 reacts with water to form carbonic acid, which reacts with limestone to form bicarbonates and carbonate ions.
- ☞ As CO_2 is used in photosynthesis, bicarbonates convert back to CO_2 thus bicarbonates serve as a CO_2 reservoir and some aquatic autotrophs can use dissolved bicarbonates directly as a carbon source.
- ☞ The ocean contains about 50 times the amount of carbon (in various inorganic forms) as is available in the atmosphere.

B. The Nitrogen Cycle (p. 1143–1144)

What four processes are involved in the nitrogen cycle? Why is nitrogen fixation important to all organisms? What is the primary nitrogen source for plants? In what form is nitrogen passed through the consumer levels?

The nitrogen cycle involves four primary processes: fixation, nitrification, denitrification, and ammonification. (See Campbell, Figure 49.12)
- ☞ Although the Earth's atmosphere is almost 80% N_2, it is unavailable to plants since they cannot assimilate this form.

Nitrogen fixation is the reduction of atmospheric nitrogen to ammonia (NH_3) which can be used to synthesize nitrogenous organic compounds such as amino acids.
- ☞ Only certain prokaryotes can fix nitrogen: in terrestrial ecosystems some nonsymbiotic soil bacteria and some symbiotic (*Rhizobium*) bacteria fix nitrogen; cyanobacteria fix nitrogen in aquatic ecosystems.
- ☞ Nitrogen fixing prokaryotes are fulfilling their own metabolic needs, but other organisms benefit since excess ammonia is released into the soil or water.
- ☞ Small amounts of nitrogen are also fixed by lightening in the atmosphere.
- ☞ Industrial fixation in the form of fertilizer makes significant contributions to the nitrogen pool in agricultural regions.

Nitrification is a metabolic process by which certain aerobic soil bacteria use ammonia as an energy source by first oxidizing ammonia (NH_3) to nitrite (NO_2^-) and then to nitrate (NO_3^-).
- ☞ While plants can use NH_3 directly, the nitrifying bacteria use most of the available NH_3 as an energy source.
- ☞ Plants assimilate the NO_3^- released from these bacteria and convert it to organic forms like amino acids and proteins.
- ☞ Animals can only assimilate organic nitrogen which they obtain by eating plants and other animals.

Denitrification is a process that returns nitrogen to the atmosphere by converting NO_3^- to N_2.
- ☞ Some soil bacteria obtain the oxygen necessary for their metabolism from NO_3^- rather than O_2.

Ammonification is the decomposition of organic nitrogen back into ammonia.
- ☞ Carried out by many decomposer anaerobic and aerobic bacteria and fungi.
- ☞ Process is especially important since it recycles large amounts of nitrogen to the soil.

Why are prokaryotes considered a vital link in the nitrogen cycle?

Some important aspects of the Nitrogen Cycle to note:
- ☞ Prokaryotes serve as vital links at several points in the cycle.
- ☞ Most of the nitrogen cycling involves nitrogenous compounds in the soil and water.
- ☞ While atmosphere nitrogen is plentiful, nitrogen fixation contributes only a small fraction of the nitrogen assimilated by plants; however, many species of plants depend on symbiotic, nitrogen-fixing bacteria in their root nodules as a source of nitrogen in a form that can be assimilated.
- ☞ Denitrification returns only a small amount of N_2 to the atmosphere.
- ☞ Most assimilated nitrogen comes from nitrate which is efficiently recycled from organic forms by ammonification and nitrification.

C. The Phosphorus Cycle (*p. 1144–1145*)

Why is the cycling of phosphorus less complex than other cycles? How is phosphorus recycled on a local basis in most ecosystems?

The phosphorus cycle is relatively simple since it does not have a gaseous form and occurs in only one important inorganic form, phosphate.

Phosphorus cycles locally as follows: (See Campbell, Figure 49.13)
- ☞ Weathering of rock adds phosphate to the soil.
- ☞ Plants absorb the soil phosphate and incorporate it into molecules.
- ☞ Phosphorus is transferred to consumers in organic form.
- ☞ Phosphorus is added back to the soil by excretion and decomposition of detritus.

Phosphorus cycling is localized since humus and soil particles bind phosphate.
- ☞ Some leaching does occur and phosphate is lost to the oceans through the water table.
- ☞ Weathering of rocks keeps pace so terrestrial systems are not depleted.

In a sedimentary cycle, phosphorus is removed by leaching into the water table and inputed by weathering of rock over geological time.

Phosphate that reaches the oceans accumulates in sediments and becomes incorporated into rocks which may eventually be exposed to weathering.

Denitrification

Ammonification

722 Ecosystems

D. Variations in Nutrient-Cycling Time (p. 1145)

What factors influence the amount of time necessary for nutrients to be cycled? How do these times vary among the different ecosystems? Why does the soil in tropical forests contain lower nutrient levels than those of temperate forests?

Factors influencing nutrient-cycling times in ecosystems are: rate of decomposition, size of standing crop, local soil chemistry, and frequency of fires.

- ☞ In temperate forests, a larger store of nutrients (5–20% of all organic material) remains in detritus, yet to be decomposed, and nutrients remain in soil longer before being assimilated.
- ☞ In tropical forests, only a small store of nutrients (1 to 2% of organic material) is found in detritus due to rapid decomposition and the large biomass of the forest removes nutrients from the soil quickly.

IV. HUMAN INTRUSIONS IN ECOSYSTEM DYNAMICS (p. 1145–1153)

The growth of human populations, human activities, and our technological capabilities have disrupted the trophic structure, energy flow, and chemical cycling of ecosystems in most areas of the world.

- ☞ Some effects are local, some regional, and a few are global.

A. Habitat Destruction and the Biodiversity Crisis (p. 1146–1147)

How has destruction of natural habitats impacted on biodiversity?

The destruction of natural systems due to human encroachment has resulted in only a small proportion of natural, undisturbed habitat remaining in existence.

- ☞ 75% of the Earth's original forests have been cleared or severely disrupted.
- ☞ Only 15% of the original primary forest and just 1% of original tallgrass prairie remain in the United States.
- ☞ Tropical rain forests are being cut at a rate of 500,000 Km^2 per year and will be eliminated in a few decades.

One result of the destruction of natural habitat will be the extinction of many species as their ecosystems disappear. This biodiversity crisis has many aspects which must be considered in order to protect endangered species.

- ☞ Not only is localized protection necessary, but many migratory species are facing habitat destruction in both their northern breeding grounds and their tropical wintering grounds.

B. Effects of Deforestation on Chemical Cycling: The Hubbard Brook Forest Study (p. 1147–1148)

How does deforestation disrupt water and nutrient cycles? What evidence is available to prove this point?

Deforestation not only eliminates trees and associated animals from a region, it also disrupts water and nutrient cycles.

Scientists led by Herbert Bormann and Gene Likens have conducted a 30 year study of nutrient cycling in a forest ecosystem under natural conditions and after severe human intrusion. The study site is the Hubbard Brook Experiment Forest in New Hampshire.
- Determined mineral budget by measuring influx and outflow of several key nutrients.
- Collected rainwater to measure amounts of water and dissolved minerals added to the ecosystem.
- Monitored water and mineral loss by using concrete weirs across the creek at the bottom of each valley.
- Found that mineral influx and outflow were nearly balanced and were small compared to minerals being recycled within the forest ecosystem.
- Found that 60% water added by rainfall exits through streams and 40% is lost by plant transpiration and evaporation from soil.

After logging an experimental area, comparisons were made.
- Water runoff increased by 30–40% (no trees left to absorb and transpire water).
- Net losses of minerals were huge — nitrate loss increased sixty fold (water nitrate levels made the water unsafe for drinking).

This is an example of long-term ecological research (LTER) which is designed to tract the dynamics of natural ecosystems and the effects of human intrusion.

C. **Agricultural Effects on Nutrient Cycling** (*p. 1148–1149*)

What effects can agricultural practices have on nutrient cycling?

Human activity often removes nutrients from one part of the biosphere and adds them to another, depleting key nutrients in one area and disrupting the natural equilibrium of both areas.

Farming exhausts the natural store of nutrients as crop biomass is removed from an area, this greatly reduces the amount of nutrients recycled. Supplements must then be added in the form of fertilizer.
- Nutrients in crops soon appear in human and livestock wastes, and then turn up in lakes and streams through sewage discharge and field run-off.
- Once in aquatic systems, these nutrients may stimulate excessive algal growth.
- Researchers are attempting to safely reclaim sewage nutrients.

D. **Accelerated Eutrophication of Lakes** (*p. 1149*)

What processes are involved in the natural eutrophication of lakes? How can "cultural eutrophication" interfere with the natural process?

Lakes undergo a naturally occurring change in chemical composition and character.
- Young oligotrophic lakes have low primary productivity because nutrient levels will not support large phytoplankton populations.
- The lake will gradually mature (become more *eutrophic*) due to runoff from surrounding land which contains nutrients.

Eutrophic

- These additional nutrients are assimilated by phytoplankton and productivity increases; once in the lake the nutrients are continually recycled through the lake's food webs.
- Lakes reach an equilibrium where nutrient input is balanced by losses due to outflow and sedimentation.

Sewage, factory wastes, livestock runoff, and fertilizer leaching increases inorganic nutrient levels in waters and results in *cultural eutrophication*.
- Results in explosive growth of photosynthetic organisms.
- As the plants die, metabolism of decomposers consume all the oxygen in the water and species die out.

E. Poisons in Food Chains (*p. 1149–1151*)

How are toxic compounds passed through food chains? What is biological magnification? Why do toxic compounds usually have the greatest effect on top-level carnivores?

A variety of toxic chemicals, including unnatural synthetics, have been and are dumped into ecosystems.
- Many cannot be degraded by microbes and persist for years or decades.
- Some are harmless when released but are converted to toxic poisons by reactions with other substances or metabolism of microbes.

Organisms acquire toxic substances along with nutrients or water, some of which accumulate in their tissues.

Biological magnification = Toxins become more concentrated with each link in a food chain. Results from biomass at each trophic level being produced from a much larger biomass ingested from the level below.
- Top level carnivores are usually most severely affected by toxic compounds released into the environment.

The pesticide DDT is a well known example of biological magnification. (See Campbell, Figure 49.18)
- Used to control mosquitoes and agricultural pests.
- DDT persists in the environment and is transported by water to areas away from the point of application.
- It is soluble in lipids and collects in fatty tissues of animals.
- The concentration is magnified at each trophic level and reached such high concentrations (10×10^6 increase) in top-level carnivorous birds that calcium deposition in eggshells was disrupted.
- Reproductive rates declined dramatically since the weight of nesting birds broke the weakened shells.

The release of radioisotopes by nuclear accidents and the unsafe storage of nuclear wastes present a serious environmental threat.
- These contaminants can last for many years due to long half-lives and are subject to biological magnification.

F. Intrusions in the Atmosphere (p. 1151–1153)

Human activities have resulted in two very pressing problems with regard to the atmosphere: rising levels of carbon dioxide and degradation of atmospheric ozone.

In what ways can increased atmospheric concentrations of carbon dioxide affect the Earth? What is the greenhouse effect?

Carbon dioxide emissions have caused atmospheric CO_2 concentrations to increase 11% in 30 years. This increase is due to combustion of fossil fuels and burning of wood removed by deforestation.

<u>Effects of Increased CO_2 Levels</u>:

1. Increased productivity by vegetation
   - ☞ C_3 plants are more limited than C_4 plants by CO_2, so spread of C_3 species into habitats previously favoring C_4 species may have important natural and agricultural implications.
2. Temperature increases with increased CO_2 concentration since CO_2, though transparent to visible light, absorbs infrared radiation and slows its escape from Earth.
   - ☞ Called the *greenhouse effect*.
   - ☞ Some predict warming near the poles resulting in melting of polar ice and flooding of current coastal areas.
   - ☞ A warming trend will probably alter geographical distribution of precipitation which could have major agricultural implications.

Greenhouse effect

What is the ozone layer? Why is it important? How do chlorofluorocarbons affect the ozone layer? What consequences may result from the depletion of the ozone layer?

Depletion of atmospheric *ozone* weakens a protective layer in the stratosphere which absorbs ultraviolet radiation.
- ☞ Much of the ultraviolet radiation which strikes the Earth is absorbed by an ozone layer 17 to 20 miles above the surface.
- ☞ Destruction of the ozone layer is largely due to accumulation of chlorofluorocarbons used as aerosol propellants and in refrigeration.
- ☞ Breakdown products of chlorofluorocarbons include chlorine which rises to the stratosphere where it reacts with ozone and reduces it to atmospheric oxygen.

Ozone depletion could have serious consequences.
- ☞ Increases are expected in lethal and nonlethal forms of skin cancer and cataracts among humans.
- ☞ Unpredictable effects on crops and natural communities (especially phytoplankton) are expected.

What problems can result from the introduction of exotic species? What impact can they have on endemic species? What ecological problems could result from the release or escape of genetically engineered organisms?

The *introduction of exotic species* can cause a variety of problems. Although most transplanted species fail to survive, there are notable exceptions.

- Gypsy moths were imported into the United States from Europe for cross-breeding experiments with silk-producing moths. Some of these moths escaped in 1869 and have been spreading southward ever since, defoliating large areas of forest along the way.
- Africanized honeybees were imported by Brazilian honey producers in 1956. These hybrid bees (a cross between African and European varieties) are very aggressive and have occasionally swarmed to attach humans and livestock. In the past 30 years, the range of Africanized bees has spread across South America, through Central America and Mexico, and into the southern United States.

The release of genetically engineered organisms into the environment poses some potential ecological problems.
- Genetically engineered plants might escape the plots and become weeds or pests in other areas.
- These plants might interbreed with wild plants and pass on their "special" traits (such as herbicide resistance) to offspring that could not be controlled by standard means.
- Genetically engineered insect resistance might result in a coevolutionary change in insect pests.

V. BIOTIC EFFECTS AT THE BIOSPHERE LEVEL (p. 1153–1154)

What is the Gaia hypothesis? Is there evidence to support this view? In what ways might human interference alter the biosphere?

The Gaia hypothesis, first enunciated by James Lovelock, states that life plays a major role in regulating the overall environment and maintaining it within certain suitable limits.
- For example, transpiration by tropical rain forests produces clouds that affect global weather patterns.

Chemical cycling has generated O_2 and maintained it within a range of 15–25% of the atmosphere for the past 200 million years.
- Release of methane by bacteria and termites controls O_2 levels since methane reacts with O_2 to form CO_2 and H_2O.
- Organisms may also moderate fluctuations in atmospheric CO_2 concentrations to some extent.

A more moderate view of this hypothesis views the regulation as a natural consequence of checks and balances inherent in the dynamics of ecosystems.

Studies of global change including interactions between climate and ecological processes, biological diversity and its role in ecological processes, and the ways in which productivity can be sustained are critical to our understanding of natural systems.

50

BEHAVIOR

CHAPTER OUTLINE

Behavior is influenced by innate and learned factors.
- ☞ Baby sounds (human and bird) are coded by innate neural templates, but young animals must learn and practice the sounds which make up adult repertoires.
- ☞ Motor and memory elements are essential to behavior.
- ☞ Animal behavior is one of the oldest studied areas in biology. Pre-agricultural societies needed to understand animal behavior in order to survive.

<u>Innate</u> = Inborn or present at birth.

<u>Behavior</u> = what an animal does and *how* it does it.

I. THE EVOLUTIONARY LOGIC OF BEHAVIORAL ECOLOGY (p. 1159)

How can innate behavior be distinguished from learned behavior? What is optimal behavior?

Fitness is a central concept in animal behavior. Since natural selection works on genetic variation caused by mutation and recombination, organisms should have features that maximize fitness over time.
- ☞ Animals are expected to engage in optimal behaviors.

<u>Optimal behavior</u> = A behavior that maximizes individual fitness.

What evidence supports the hypothesis that optimal behavior is genetically influenced and subject to natural selection?

Optimal behavior is a valid concept if behavior is genetically influenced and subject to natural selection.
- ☞ Genes influence behavior. (See Campbell, Figure 50.2)

<u>Experiment</u>: Two species of lovebirds were interbred. Female fischer's lovebirds cut long strips of nesting material which are carried individually to the nest. Female peach-faced lovebirds cut short strips and carries several at a time by tucking them into her back feathers.

Results: Hybrid females cut intermediate length strips and tried, but failed, to transport them by tucking into back feathers. They learned to carry strips in their beaks, but never gave up all tucking behavior.

Conclusions: Phenotypic differences in the behavior of the two species are based on different genotypes. Innate behavior can be modified by experience.

Innate

Behavior

Optimal behavior

Learned behaviors are typically based upon gene created neural systems that are receptive to learning.

Behavioral ecologists assume animals use optimal behavior to increase fitness.
- ☞ Because behavioral ecology is based on the assumption that behavior increases fitness, it leads logically into hypothesis testing and the scientific method for problems in behavior research.

<u>Behavioral ecology</u> = A field of study that assumes animals increase fitness through optimal behavior.

II. PROXIMATE VERSUS ULTIMATE CAUSATION (p. 1160–1162)

Because behavior is assumed to increase fitness, questions about ultimate causation, or the reasons <u>why</u> behaviors exist are evolutionary questions.

How do ultimate causation of behavior and proximate causation of behavior differ? What examples of ultimate and proximate causes can you describe?

<u>Ultimate causation</u> = The evolutionary reason for the existence of a behavior.

The immediate mechanism or <u>how</u> a behavior is expressed is the proximate cause. Proximate causes may be internal processes or environmental stimuli.

<u>Proximate causation</u> = The immediate cause and/or mechanism underlying a behavior.
- ☞ Proximate causation limits the range of behaviors upon which natural selection can act.
- ☞ Proximate mechanisms produce behaviors that ultimately evolved because they increased fitness.

<u>Example</u>: Behavior — Bluegill sunfish breed in spring and early summer.
  Proximate cause: Breeding is triggered by the effect of increased day length on a fish's pineal gland.
  Ultimate cause: Breeding is most successful when water temperatures and food supplies are optimal.

<u>Example</u>: Behavior — Human "sweet tooth".
  Proximate cause: Sweet taste buds are a proximate mechanism that increases the chances of eating high-energy foods.
  Ultimate cause: Sweet, high-energy, foods were rare prior to mechanized agriculture. Increased fitness associated with consuming these foods is the ultimate reason for the natural selection of a "sweet tooth".

III. INNATE COMPONENTS OF BEHAVIOR (p. 1162–1165)

Ethology pre-dates behavioral ecology. Relying on descriptive studies, ethologists discovered many behaviors were innate.

What is ethology? Who are Konrad Lorenz, Niko Tinbergen and Karl von Frisch?

Ethology = Descriptive science based on studies of animals in the natural environment.

Innate behaviors may seem purposeful, but animals with innate behaviors are unaware of the significance of their actions. (See Campbell, Figure 50.5)
- ☞ Some cuckoos are brood parasites. Females lay eggs in the nests of other species. Within hours of hatching, the otherwise helpless cuckoo pushes its host's eggs and/or chicks out of the nest.

Konrad Lorenz, Niko Tinbergen and Karl von Frisch shared the 1973 Nobel Prize for physiology or medicine for their work in ethology.

    A.   **Fixed-Action Patterns** (*p. 1162–1165*)

The most fundamental concept in classical ethology is the fixed-action pattern (FAP).

What is a fixed action pattern? What stimulates a fixed action pattern? Is a fixed action pattern innate or learned behavior?

    Fixed-action pattern = A highly stereotyped, innate, behavior.
- ☞ Once a fixed-action pattern is triggered, the behavior continues until completion even in the presence of other stimuli or if the behavior is inappropriate.
- ☞ A fixed-action pattern is triggered by an external sign stimulus or releaser. (See Campbell, Figures 50.6 and 50.7)
- ☞ Fixed-action patterns are adaptive responses to <u>natural</u> stimuli. Strange responses can be initiated by presenting unnatural situations to animals with fixed-action patterns.

    Sign stimulus = An external sensory stimulus which triggers a fixed-action pattern.

    Example: Niko Tinbergen noticed male three-spined stickleback fish responded aggressively to red trucks passing by their tank.
        Fixed-Action Pattern: Male sticklebacks attack other males that enter their territories.
        Sign stimulus: The red belly of the invading male. Sticklebacks attacked nonfish-like models with red on the ventral surface.

    Example: Parent/young feeding behavior in birds. (See Campbell, Figure 50.8)
        Fixed-Action Pattern: The begging behavior of newly hatched chicks (raised heads, open mouths, loud cheeps).
        Sign stimulus: Parent landing at the nest.

    Example: Greylag goose egg retrieval behavior:
        Fixed-Action Pattern: Rolls the egg back to the nest using side-to-side head motions.
        Sign stimulus: The appearance of an object near the nest.

- If the goose loses the egg during the retrieval process, it stops the head motion, but continues the "pulling" motion of retrieval. It must sit down before it notices the egg at which time another retrieval FAP is initiated.
- If an inappropriate object (toy dog, doorknob) is placed near the nest, the goose retrieves it but may not keep or incubate it.

Example: Protective behavior in hen turkeys:

Fixed-Action Pattern: Mothering behavior.

Sign stimulus: Cheeping sound of chicks.
- A deaf turkey cannot hear the sign stimulus releasing mothering behavior and kills her chicks.

Example: The human infant grasping response is a fixed-action pattern released by a tactile stimulus. Human babies smile when they hear certain sounds, or see a figure consisting of two dark spots on a circle (rudimentary representation of a face).

Example: Female digger wasps place a paralyzed cricket in their nests which serves as food for the young wasp after it hatches.

Fixed-Action Pattern: She places the cricket 2.5 cm from the nest.

Sign stimulus: Nest site.

Fixed-Action Pattern: She enters the nest and inspects it.

Sign stimulus: Presence of cricket 2.5 cm from nest.

Fixed-Action Pattern: She exits the nest and retrieves the cricket.

Sign stimulus: Presence of cricket 2.5 cm from nest.
- If the cricket is moved during the nest inspection stage, the wasp will retrieve the cricket and repeat the FAP from the first step. She cannot get past the inspect the nest step if the cricket is not where she left it when she tries to retrieve it.

B. The Nature of Sign Stimuli (*p. 1165*)

What can serve as a sign stimulus? What is a supernormal stimulus?

Sign stimuli are usually simple characteristics (e.g. ultrasonic bat sounds trigger avoidance behavior in moths).

Sign stimuli may be specific choices from an array of possibilities.

Example: Herring gull chick feeding behavior. The adult lowers its head and moves its beak. The chick pecks the red spot on the beak, causing the adult to regurgitate.

Fixed-action pattern: Pecking the red spot on the beak.

Sign stimulus: Red spot swung horizontally at the end of a long, vertical object.

Natural selection favors cues associated with the relevant behavior or object. Some randomness is probable in fixing upon one of many possible sign stimuli for a FAP.

An animal's sensitivity to general stimuli and its sign stimuli are correlated.
- ☞ For example, frogs' retinal cells are sensitive to movement. Movement is the sign stimulus that releases the tongue-shooting FAP in frogs.
- ☞ A frog starves if surrounded by motionless flies.

A supernormal stimulus may elicit stronger responses than natural stimuli.
- ☞ For example, when given a choice between an egg and a volleyball, a greylag goose ignores the egg and tries to retrieve the volleyball.

<u>Supernormal stimulus</u> = Artificial stimulus that elicits a stronger response than occurs naturally.

IV. LEARNING AND BEHAVIOR (p. 1165–1170)

A. Learning (p. 1165)

<u>Learning</u> = is the modification of behavior by experience.

B. Nature Versus Nurture (p. 1165–1166)

While European ethologists were discovering innate behaviors in nature, American psychologists were finding learning abilities in lab animals. Thus began the debate over nature versus nurture.

What is learning? Which is more important to an animal's behavior, instinct or learning? How can maturation of an animal improve behaviors?

<u>Nature versus nurture</u> = The debate over whether instinct or learning is of primary importance in animal behavior.
- ☞ Modern biologists feel that behavior is a result of genetic and environmental factors.

Some behavioral aspects seem to be due primarily to either instinct or learning.
- ☞ Language differences among people are learned behavioral variations on an innate ability to vocalize.
- ☞ Songbirds have regional "dialects".
- ☞ Most *Drosophila* (fruit flies) have an activity cycle of 24 hours. There are individuals with 19 hour cycles. The difference is due to genetic variation.

C. Learning Versus Maturation (p. 1166)

Individuals may improve behaviors over time. This is often attributed to learning but, in some cases, may be due to developmental changes in neuro-muscular systems as animals mature.

<u>Maturation</u> = Development of neuro-muscular systems that allows behavioral improvement.

D. Habituation (p. 1166)

What is habituation? What adaptive advantage could habituation produce?

732 Behavior

Habituation

Animals stop responding to stimuli that do not provide appropriate feedback.
☞ Gray squirrels respond to the alarm calls of other squirrels. They stop responding if the calls are not followed by an attack ("cry-wolf" effect).

Habituation = Learning to ignore irrelevant stimuli or stimuli that do not provide proper feedback.

E. Imprinting (p. 1166–1168)

What is imprinting? What is the importance of the imprinting stimulus? The critical period?

Imprinting is a form of learning closely associated with innate behavior.

Imprinting

Konrad Lorenz conducted an experiment with greylag geese. (See Campbell, Figure 50.9)

Experiment: A clutch of goose eggs was divided between the mother and an incubator.
  Results: Goslings reared by the mother behaved normally and mated with other geese. The incubator goslings spent their first hours of life with Lorenz and preferred humans for the rest of their lives. They even tried to mate with humans.
  Conclusions: Greylags have no innate sense of "mother" or "gooseness". They identify with and respond to the first object with certain characteristics they encounter. The ability or tendency to respond is innate.
    ☞ The object to which the response is directed is the imprinting stimulus. For Lorenz's geese, the imprinting stimulus was movement of an object away from the young.

Imprinting stimulus

Imprinting stimulus = An object in the environment to which the response is directed.
☞ For example, salmon return to the stream they were hatched in to spawn. The imprinting stimulus is the unique chemical composition (odor) of the hatching stream.
☞ Imprinting is distinguished from other types of learning by its irreversibility and the presence of a critical period during which imprinting can occur.

Critical period

Critical period = A limited time during which imprinting can occur.
☞ Imprinting is usually thought to involve very young animals and short critical periods.
☞ Imprinting may occur at different ages with critical periods of varying durations.
☞ Adults require a critical period to "know" their young. Prior to imprinting on their young, adult herring gulls defend strange chicks. After the imprinting period, they kill and eat strange chicks.
☞ Sexual imprinting (or species identity) occurs later than parental imprinting and has a longer critical period. Male finches cross-fostered by two different finch species imprinted on the wrong species when they developed a sexual identity.

- While irreversibility and critical period characterize imprinting, they are not always fixed. The cross-fostered finches eventually mated with females of their own species.

F. **Song Development in Birds: A Study of Imprinting** (*p. 1168–1169*)

What is the relationship between innate behavior and imprinting with regards to song development in birds?

Song development is highly responsive to selective pressures.

In male white-crowned sparrows:
- Birds raised in soundproof chambers developed abnormal songs.
- Birds exposed to normal song recordings at 10–50 days old developed normal songs.
- Birds deafened after they were exposed to the recordings, but before they began to sing, developed songs more abnormal than birds reared in isolation.
- White-crowned sparrows learn to sing by hearing adults and matching those songs while listening to themselves.
- 10–50 days was the critical period for hearing adult song. Birds who heard a recording at 50+ days never sang properly.
- Birds exposed to other species' recordings did not learn those songs. There was an innate, genetic predisposition, for their own song.
- The innate predisposition was overcome by social interaction. Sparrows 50+ days old, when in social contact with another species, learned that species' song.
- The critical period may be flexible. It was longer for a live stimulus than for a recording.

Song development may be more or less fixed in other species:
- Song sparrows reared in isolation develop normal songs although males have larger repertoires if they hear other birds.
- Mockingbirds have repertoires of 150+ songs. Genetic programming places few or no restrictions on their song development.

G. **Classical Conditioning** (*p. 1169*)

What is associative learning? How would you distinguish between classical conditioning and operant conditioning?

Associative learning = The learning process where animals associate one stimulus with another.

Classical conditioning is a type of associative learning where a specific, or conditioned, response is elicited by a specific stimulus.
- A Russian physiologist, Ivan Pavlov induced dogs to salivate when they heard a bell. The dogs learned to associate the bell with powdered meat which was the normal stimulus for the salivary response.

Associative learning

Ivan Pavlov

Classical conditioning

B.F. Skinner

Operant conditioning

Observational learning

Play

Classical conditioning = A process in which an animal learns to respond to an external stimulus which does not normally elicit that response.

H. Operant Conditioning (p. 1169)

Operant conditioning, or trial-and-error learning, is another type of associative learning.

- B.F. Skinner put lab animals in a box with a variety of levers. Test animals learned to choose only those levers which yielded food.

Operant conditioning = A process where an animal learns to associate one of its behaviors with a reward or punishment and then tends to repeat or avoid that behavior.

- English tits learned to open milk bottles left on doorsteps and drink the cream.
- One or more of the birds discovered its probing behavior was rewarded when directed at the bottles.

Operant conditioning is the basis for most animal training and is common in nature.

- Genes influence the outcome of operant conditioning.

I. Observational Learning (p. 1169)

How does observational learning differ from associative learning? In what way does observation learning permit the establishment of new behaviors?

Observational learning = The ability of animals to learn by observing the actions of others.

- Observational learning allows new behaviors to become established and passed on to succeeding generations.
- The speed with which the tit milk drinking behavior spread throughout England indicates observational learning on the part of the birds.
- Song development in many birds involves observational learning as individuals listen to and eventually adopt the songs of older birds.

J. Play (p. 1169–1170)

What are the advantages and disadvantages to play behavior? What is the Play hypothesis? The Exercise hypothesis?

Play has no apparent goal but uses movements closely associated with goal-directed behaviors.

- Young predators playfully stalk and attack each other using motions similar to those used to capture and kill prey. (See Campbell, Figure 50.11)
- Play occurs in the absence of distracting external stimuli.

Play is potentially dangerous or costly.

- Young vervet monkeys are at higher risk of being caught and eaten by baboons when they are at play.
- In a study of young goats, 1/3 sustained play injuries that resulted in limps.

What is the selective advantage of play?
- Practice hypothesis — play is a type of learning that allows the perfection of survival behaviors. However, play movements rarely improve after the first few practices.
- Exercise hypothesis — play keeps the cardiovascular and muscular systems in condition and is common in young animals because they do not exert themselves in other ways while they are under parental care.

K. Insight (*p. 1170*)

What is insight learning? How does it differ from conditioning?

Insight learning is also called reasoning.

Insight learning = The ability of animals to perform appropriate behaviors on the first attempt in situations with which they have no prior experience.
- Insight learning is best developed in primates. (See Campbell, Figure 50.12)
- A chimpanzee placed in an area where a banana is hung too high to reach, but where boxes are scattered about, will "size" up the situation, stack the boxes, climb up and retrieve the banana.

L. The Ultimate and Proximate Bases of Learning (*p. 1170*)

What are the ultimate bases of learning?

Ultimate bases of learning
- Behaviors are relatively fixed patterns controlled by simple stimuli.
- Fixed pattern responses can be modified in their details by simple kinds of learning.
- The capacity to learn is a product of evolution and adaptation that enhances fitness.
- Learning is subject to natural selection in that learning is programmed to occur at specific times in the maturation process and is based on a limited set of stimuli.

Proximate bases of learning are poorly known. Some studies link simple learning to internal biochemical or physiological changes.

V. BEHAVIORAL RHYTHMS (*p. 1170–1171*)

Animals repeat behaviors at regular intervals (daily, seasonally).

Animals carry out behaviors when their particular ecological niches can be exploited most profitably from a survival and fitness standpoint. This can cause behavioral rhythms.

What types of behavioral rhythms are found in animals? Which rhythms are regulated by exogenous factors and which by endogenous factors? How do exogenous and endogenous factors influence behavioral rhythms?

The proximate bases for rhythms are not obvious.

Insight learning

736 Behavior

Circadian rhythms

Exogenous, Endogenous

Biological clock

☞ Circadian (daily) rhythms are regulated by environmental cues.
☞ Rhythmic behavior is based on exogenous (external) and endogenous (internal) factors.
☞ Circadian rhythms are strongly influenced by endogenous factors, or the biological clock. Internal rhythms do not match environmental time so exogenous cues are used to keep the cycles in sync.
☞ Light is the most common exogenous cue.

Experiment: Flying squirrels are normally active at night. Flying squirrels were placed in constant darkness and their activity monitored. (See Campbell, Figure 50.13)

Results: Rhythmic activity continued. Since normal cycles deviate from the 24 hour clock, the test squirrels eventually were out of phase with the environment.

Conclusions: Biological clocks are intrinsic, but the timing of rhythmic behavior must be adjusted to environmental cues to remain in sync.

Even when endogenous factors are present, the internal, proximate, timing mechanism is unknown. A biochemical, possibly molecular, clock has been hypothesized.

It is unclear if endogenous clocks are important in longer cycle rhythmic behaviors. Breeding and hibernation are partially based on physiological and hormonal changes linked to exogenous factors such as day length.

VI. MOVEMENT (*p. 1171–1174*)

Studies have concentrated on the proximate mechanisms animals use to detect and respond to external cues that guide movements.

The ultimate bases for oriented movements are related to animal adaptations to different environments and the development of behaviors that bring them to those environments at the proper time of the year.

What cues are used by animals to direct their movements? How can you distinguish between a kinesis, a taxis and a migration?

Animals use a variety of cues to guide them in their movements.

Kinesis involves a change in activity rate in response to a stimulus.
☞ Sowbugs are more active in dry areas and less active in humid regions. This behavior tends to keep them in moist areas.

Kinesis

Kinesis = A randomly directed change in activity rate in response to an environmental stimulus.

Taxis is a directed movement toward or away from a stimulus.
☞ Housefly larvae are negatively phototactic after feeding. Presumably this makes them less visible to predators.
☞ Trout are positively rheotactic. Swimming against the current keeps them from being swept downstream.

Taxis

Taxis = A semi-automatic, oriented, movement toward or away from a stimulus.

Migration is the most commonly known type of oriented animal movement.
- Migrants generally make an annual round trip between two regions (e.g. birds, whales, some butterflies, some pelagic fish). (See Campbell, Figures 50.14 and 50.15)

<u>Migration</u> = The seasonal movement of animals over relatively long distances.

Migrating animals use three mechanisms: *piloting*, *orientation* and *navigation*.

What criteria can be used to distinguish among piloting, orientation and navigation?

<u>Piloting</u> = Movement of animals from one landmark to another.
- Is used over short distances.
- It is not useful at night or over the ocean.

<u>Orientation</u> = Movement of animals along a compass line.
- Animals that use orientation can detect compass directions and travel in a straight-line to a destination.

<u>Navigation</u> = The ability of animals who can orient along compass lines to determine their location in relation to their destination.
- Migrant starlings captured in the Netherlands were released in Switzerland. Juvenile birds oriented in a straight-line to Spain. Adults navigated a new route to their wintering grounds in northern Europe.
- Many birds use celestial points for orientation and navigation. These animals need an internal clock to compensate for the movement of the sun and stars. The indigo bunting avoids the need for an internal clock by fixing on the north star.
- Some birds, bees and bacteria orient to the Earth's magnetic field. The mechanisms are poorly known, but magnetite, an iron-containing ore, has been found in animals that orient to the magnetic field.

VII. FORAGING BEHAVIOR (p. 1174–1175)

What is a search image? What distinguishes a generalist foraging strategy from a specialist foraging stratgy? Is there an adaptive advantage to having a search image?

Generalists feed on many items. They are not efficient collectors of any single food, but take advantage of multiple options when foods are scarce.
- Generalists concentrate on abundant prey.
- Generalists develop a search image for a favored item. If the item becomes scarce, a new search image is developed. Search images let generalists combine efficient short-term specialization with generalist flexibility.

<u>Search image</u> = The ability of a generalist feeder to learn the key visual characters of a prey item.

Specialists feed on specific items and usually have highly specific morphological and behavioral adaptations. They are extremely efficient foragers.

Food habits are fundamental to an animal's niche and may be shaped by interspecific competition and evolutionary factors.

How do optimal foraging strategies maximize the caloric return? What characteristics of the prey are important to an optimal foraging strategy?

Natural selection should favor foraging strategies that maximize gains and minimize costs in terms of calories gained and expended. Other criteria such as nutrient gain may be equally important.

- ☞ Foraging costs include: energy needed to locate, catch and eat food; the risk of being caught by a predator while feeding; time taken from other activities such as courtship and breeding.
- ☞ Behavioral ecologists analyze tradeoffs to predict optimal foraging strategies.
- ☞ Tradeoffs include density and size of prey versus foraging distances or prey catchability.
- ☞ For example, small mouth bass eat minnows and crayfish. Since no preference is shown, each may be optimal under different conditions. Minnows have more useable energy per unit weight than crayfish, but are harder to catch. Crayfish are easier to catch, but more difficult to subdue.

Animals modify behavior to keep the ratio of energy gain to loss high.
- ☞ This ability is probably innate, although learning may be involved.

Experiment: Bluegill sunfish eat small crustaceans. (See Campbell, Figure 50.16) Optimal foraging theory predicts the proportion of small to large prey will vary with the density of the prey population. At low densities, sunfish should not be selective but at high densities, they should concentrate on larger prey.

- *Results*: Sunfish were more selective at higher prey densities, but not to the extent predicted. Young fish were less efficient than adults. Younger fish may be less able to judge size and distance due to incompletely developed neural systems.
- *Conclusion*: Maturation and learning may result in increased foraging efficiency in adults.

VIII. SOCIAL INTERACTIONS (p. 1176–1178)

Social behavior is any interaction between two or more animals and may involve conflict or cooperation. Cooperating individuals usually maximize individual benefits even if it costs other participants (See Campbell, Figure 50.17).

A. Agonistic Behavior (p. 1176–1177)

What is agonistic behavior? How would you describe a ritual? Is there an evolutionary advantage to ritual behavior?

Agonistic behavior = A contest of threat displays which continues until a participant submits and yields access to a resource (e.g. mate, food, territory).
- ☞ Ritual behaviors are prevalent so it is rare that participants are seriously injured. (See Campbell, Figure 50.18)

<u>Ritual</u> = Symbolic behavior that minimizes the possibility of serious injury to the antagonists.
- Natural selection favors ending a contest as soon as a winner is established because further conflict could injure the victor as well as the vanquished.
- Canines show agonistic behavior by trying to look larger. They bare teeth; erect ears, tail and fur; stand upright; and make eye contact. The loser submits by sleeking its fur, tucking its tail and looking away.

B. Dominance Hierarchies (p. 1177)

What is a dominance hierarchy? What are the advantages to the individuals in a hierarchy?

The top ranked member of a social group controls the behavior of the members of the group. The second ranked animal controls everyone except the top individual and so on down the line to the lowest ranked animal.

<u>Dominance hierarchy</u> = A linear social organization within a group.
- Top ranked animals are assured access to resources. Low ranked animals do not waste energy or risk harm in combat.
- Wolf packs typically have a female dominance hierarchy. The top female relies on food availability to control mating in the pack.

C. Territoriality (p. 1177–1178)

What is a territory? How can dominance hierarchies and territories help stabilize population densities?

Territories are defended areas typically used for feeding, mating, rearing young etc.

<u>Territory</u> = An area defended from conspecifics.
- Territory size varies with the species, territory function, and the amount of resources available. (See Campbell, Figure 50.19)
- Some species defend territories during the breeding season and form social groups at other times of the year.
- Territories are not home ranges. Home ranges are areas which animals inhabit but do not defend. Territories and home ranges may overlap.

Territories are usually successfully defended by their owners.
- Owners usually win because a territory is more valuable to the owner since he is familiar with it.
- Ownership is continually proclaimed — a primary function of bird song, red squirrel chattering and the bellowing of sea lions. Others use scent marks or patrols to announce their presence. (See Campbell, Figure 50.20)
- Defense is usually directed at conspecifics who are most likely to compete directly for the same resources.

Dominance hierarchies and territoriality tend to stabilize population densities by assuring enough individuals reproduce to result in relatively stable populations from year to year.

IX. MATING BEHAVIOR (p. 1178–1180)

The correlation between mating behavior and reproductive fitness is vital to behavioral ecology.

A. Courtship (p. 1178–1180)

What is a courtship ritual? What are the advantages of courtship behavior?

Species often have a complex courtship ritual unique to that species.
- ☞ Courtships are a series of fixed-action patterns alternately triggered by the participants.
- ☞ Ritual courtships probably evolved from behaviors that once had a direct meaning.

For example, male balloon flies spin oval silk balloons which they carry while flying in a swarm. The swarm is approached by females seeking mates. A female accepts a male's balloon when they fly off to copulate.
- ☞ In a related species, the male brings a dead insect for the female to eat while they mate.
- ☞ In another species, the insect is presented inside a silk balloon, possibly because silk helps subdue the insect or makes it look larger.
- ☞ Balloon flies eat nectar. The ritual has evolved into bringing something that was once associated with food.
- ☞ It is as if, over evolutionary time, a suitor wooed a lady with diamonds, then with a box containing diamonds and finally with an empty box.

Courtship assures each partner that the potential mate is not a threat, is the proper species, the proper sex and in the correct physiological condition.
- ☞ Courtship may allow one or both sexes to choose a mate from a number of candidates.

What is parental investment? Does it play a role in mate selection? On what bases are mates selected?

Females are usually more discriminating than males because they normally have a greater parental investment.
- ☞ Eggs are usually larger and more costly to produce than sperm.
- ☞ Gametes of placental mammals are closer in size, but females invest considerable time and energy carrying young before birth.

Parental investment = The time and resources an individual expends to produce an offspring.

Competition among individuals of the same sex (usually males) may determine which individuals of that sex will mate.
- ☞ Most males mate with as many females as possible. They compete with other males for mates and may try to impress females.
- ☞ Males perform more intense courtship displays than females.
- ☞ Secondary sex characteristics may be highly developed in males (e.g. deer antlers, bird colors).

Behavior 741

There are two ultimate bases for mate selection.

1. If the other sex gives parental care, it is best to choose the most competent mate.
   - ☞ For example, male common terns bring fish to potential mates as part of the courtship ritual. This behavior may be a proximate indicator of his ability to feed the chicks.
   - ☞ Some females prefer males with the most extreme and energetic courtship displays or secondary sex characteristics. These characteristics may be proximate indicators of the male's health.

2. Genetic quality is important when males provide no parental care and sperm are their only contribution to offspring.
   - ☞ Lek species have a communal area where males display. Females visit the lek and choose a mate. The proximate basis for her choice is a preference for males that court the most vigorously and have the most extreme secondary sex characteristics.

It may be difficult to determine if differential mating success among males is due to male-male competition, female choice or both.
   - ☞ Three-spined stickleback courtship is based on stereotyped releasers and FAPs. (See Campbell, Figure 50.21) Despite this, the female can back out of the courtship anytime.
   - ☞ Female choice is probably ultimately based on the quality of the male's parental care, because only male sticklebacks give parental care.

B. Mating Systems (p. 1180)

What are the three mating systems? How do they differ? How can polygyny be distinguished from polyandry?

<u>Promiscuous</u> = A mating system with no strong pair-bonds or lasting relationships.

<u>Monogamous</u> = A mating system where one male mates with one female.

<u>Polygamous</u> = A mating system where an individual of one sex mates with several of the other.
   - ☞ *Polygyny* is a mating sub-system where one male mates with multiple females.
   - ☞ *Polyandry* is a mating sub-system where one female mates with multiple males.

How have the needs of the young and the certainty of paternity influenced the evolution of mating systems?

The needs of the young are an important ultimate factor in the evolution of mating systems.
   - ☞ Most birds are monogamous. Young birds often require significant parental care. A male may ultimately increase his reproductive fitness by helping a single mate rear a brood than by seeking additional mates.

Promiscuous

Monogamous

Polygamous

Polygyny

Polyandry

☞ Polygyny is common in birds where the young are able to care for themselves soon after hatching. Males can maximize their fitness by seeking additional mates.

Another factor influencing mating systems and parental care is the certainty of paternity. (See Campbell, Methods Box, page 1181)
☞ Young born or eggs laid by a female definitely contain the female's genes, but even in monogamous species, the young could have been fathered by a male other than the female's normal mate.
☞ The certainty of paternity is relatively low in species with internal fertilization because mating and birth (or egg laying) are separated over time. Exclusive male parental care is rare in birds or mammals.
☞ Certainty of paternity is higher when egg laying and mating occur together, as in external fertilization. Parental care, when present, in fishes and amphibians is as likely to be by males as by females.
☞ When parental care is given by males, the mating system may be polygynous with multiple females laying eggs in a nest tended by a male.

X. COMMUNICATION (p. 1180–1183)

What is communication? In what ways can animals communicate? Does an animal's lifestyle influence the way in which it communicates?

Communication = The intentional transmission of information between individuals.

Behavioral ecologists assume communication has occurred when an act by a "sender" produces a change in the behavior of another individual, the "receiver".

Ethologists assumed communication evolved to maximize the quantity and accuracy of information. Behavioral ecologists argue that communication evolved to maximize the fitness of communicators.

Animals lie. Mimicry often is adaptive to the sender and maladaptive for the receiver.
☞ Male and female *Photinus* fireflies communicate by a characteristic pattern of flashes. Females of the predatory firefly genus *Photurus* mimic the female *Photinus* flash pattern, attracting male *Photinus* fireflies which they kill and eat.
☞ In some mammals, a new dominant male kills young born too soon to be his offspring. Without dependent young, females ovulate sooner, allowing the new male to father their young. Hanuman langur females in the early stages of pregnancy solicit copulations from new dominant males. When they give birth shortly before young fathered by these males would appear, they may deceive the male into treating their young as his own.

An evolutionary consideration is the mode used to transmit information. Animals use visual, auditory, chemical, tactile and electrical signals.

The mode used to transmit information is related to an animal's lifestyle.
☞ Most mammals are nocturnal and use olfactory and auditory signals.

- ☞ Animals that communicate by odors emit chemical signals called pheromones. Pheromones are important releasers for specific courtship behaviors and are the cues ant scouts release that guide other ants to food.
- ☞ Birds are mostly diurnal and use visual and auditory signals. Diurnal humans also use visual and auditory signals. If we could detect the chemical signals of mammals, then mammal sniffing might be as popular as bird watching.

Pheromones

A complex communication system is found in honeybees. (See Campbell, Figure 50.22)

- ☞ To maximize foraging efficiency, workers communicate the location of food sources which change as flowers bloom and new patches are found.
- ☞ Karl von Frisch studied honeybee communication. He found individual bees communicated to other bees when they returned to the hive.
- ☞ Returning bees "dance" to indicate the location of food.
- ☞ If the source is < 50m, the bee does the "round dance", moving rapidly sideways in tight circles and regurgitating nectar. Workers leave the hive and forage nearby.
- ☞ If the food is farther away, the bee does a "waggle dance", a half-circle swing in one direction, followed by a straight run and then a half-circle swing in the other direction. This dance indicates location in two ways:
  1. The angle of the run in relation to the vertical surface of the hive is the same as the horizontal angle of the food in relation to the sun.
  2. Distance to the food is indicated by variations in the speed at which a bee wags its abdomen during the straight run.

XI. SELFISH VERSUS ALTRUISTIC BEHAVIOR (*p. 1183–1185*)

Behavior that maximizes individual reproductive success will be favored by selection, regardless of how much damage such behavior does to another individual, local population, or species.

Animals occasionally exhibit apparently unselfish or altruistic behavior.

What is altruistic behavior? How could it have evolved?

<u>Altruistic behavior</u> = A behavior that reduces an individual's personal welfare but benefits others.

Altruistic behavior

- ☞ When parents sacrifice their well-being to produce and aid offspring, they increase their fitness because it maximizes their genetic representation in the population.
- ☞ Like parents and offspring, siblings share 1/2 their genes, so selection might favor helping one's parents produce more siblings, or even helping siblings directly.
- ☞ Selection might result in animals increasing their genetic representation in the next generation by "altruistically" helping close relatives.
- ☞ Inclusive fitness includes assistance to relatives that maximizes individual fitness.
- ☞ Coefficient of relatedness is the proportion of genes that are identical in two individuals because of common ancestry. (See Campbell, Figure 50.23) The higher the coefficient of relatedness, the more likely an individual is to aid a relative.

Behavior 743

What is the relationship between inclusive fitness and kin selection? What is reciprocal altruism?

Inclusive fitness = The reproductive fitness of an individual as measured by its offspring and assistance to the reproductive efforts of close relatives.

Kin selection = The mechanism of increasing inclusive fitness.
- ☞ The contribution of kin selection to inclusive fitness varies among species. It may be rare or nonexistent in species that are not social or disperse widely.
- ☞ In predicting if an individual will aid relatives, behavioral ecologists have derived a formula that combines coefficients of relatedness, costs to the altruist and benefits to the recipient.

If kin selection explains altruism, then examples of unselfish behavior should involve close relatives.
- ☞ Belding's groundsquirrels give alarm calls when danger appears. These calls alert other squirrels but increase the risk to the alarm givers. Females remain near their birth sites and are usually related to other members of the group. Only females give alarm calls. (See Campbell, Figure 50.24)
- ☞ Worker bees are sterile. They labor on behalf of a single fertile queen. Workers sting intruders, a behavior that defends the hive but results in the death of the worker. The queen is the mother of all the bees in the hive.
- ☞ Nesting red-cockaded woodpeckers are aided by 2-4 nonbreeders that assist in all aspects including incubation and feeding young. Nest helpers are older offspring of the breeding pair or siblings of one parent that have been unable to establish a breeding territory. Helpers may eventually inherit the territory.

Altruistic behavior toward non-relatives sometimes occurs. This behavior is adaptive if there is a reasonable chance of the aid being returned in the future.
- ☞ Reciprocal altruism only occurs in stable social groups where individuals have many opportunities to exchange aid.

XII. ANIMAL COGNITION (p. 1185-1186)

What are cognition, behaviorism and cognitive ethology? Do any relationships exist among the three?

Cognition = Knowing, including awareness and judgement.

It is difficult to determine if animals are aware of themselves and their surroundings. The method used, behaviorism, does not test for cognitive function.

Behaviorism = A mechanistic approach which describes behavior in terms of stimulus and response.

Cognitive ethologists think cognitive ability arises through natural selection and forms a phylogenetic continuum stretching into evolutionary history.

Cognitive ethology = A view that sees conscious thinking as an inherent part of animal behavior.

XIII. HUMAN SOCIOBIOLOGY (*p. 1186–1187*)

What is the underlying thesis of sociobiology?

Sociobiology

The book, *Sociobiology* by E.O. Wilson (1975) presented the thesis that social behavior has an evolutionary basis.
- ☞ Behavioral characteristics are expressions of genes favored by natural selection.
- ☞ Rekindled the nature-versus-nurture controversy.

An example of the debate involves cultural taboos on incest.
- ☞ Incest avoidance is adaptive because inbreeding may increase the frequency of genetic disorders.
- ☞ Many species avoid incest.
- ☞ Most human cultures have taboos forbidding incest.

Is there an innate aversion to incest or is this an acquired behavior?
- ☞ The argument in favor of the "nurture or learned behavior" position is: cultural taboos are unnecessary if the behavior is innate, therefore incest avoidance is a learned behavior and the social stigma attached to incest is based on experience.
- ☞ The argument in favor of the "nature or genetic behavior" position is: the occurrence of incest taboos in many cultures is evidence for an innate component and taboos are simply proximate mechanisms that reinforce a behavior that ultimately evolved because of its effect on fitness.